Forensic Anthropology

Forensic Anthropology

An Introductory Lab Manual

CHRISTOPHER M. STOJANOWSKI
AND ANDREW C. SEIDEL

University of Florida Press
Gainesville

Published in the United States of America

28 27 26 25 24 23 6 5 4 3 2 1

Library of Congress Cataloging-in-Publication Data
Names: Stojanowski, Christopher M. (Christopher Michael), 1973– author. | Seidel, Andrew C., author.
Title: Forensic anthropology : an introductory lab manual / Christopher M. Stojanowski and Andrew C. Seidel.
Description: 1. | Gainesville : University of Florida Press, 2023. | Includes bibliographical references.
Identifiers: LCCN 2022045785 (print) | LCCN 2022045786 (ebook) | ISBN 9781683403562 (paperback) | ISBN 9781683403685 (pdf)
Subjects: LCSH: Forensic anthropology—Handbooks, manuals, etc. | Forensic sciences—Handbooks, manuals, etc. | Human body—Identification.
Classification: LCC GN69.8 .S76 2023 (print) | LCC GN69.8 (ebook) | DDC 614/.17—dc23/eng/20220926
LC record available at https://lccn.loc.gov/2022045785
LC ebook record available at https://lccn.loc.gov/2022045786

University of Florida Press
2046 NE Waldo Road
Suite 2100
Gainesville, FL 32609
http://upress.ufl.edu

UF PRESS

UNIVERSITY
OF FLORIDA

A Note to Readers

1) Forensic anthropology is a field that inherently deals with death; therefore, the content of this book may be disturbing for those who have recently lost a loved one or themselves have been a victim of a violent crime. In addition, due to the subject matter covered in this book it contains images, including those of the skeletal remains of deceased individuals, that some readers might find disturbing.

2) To the best of the authors' knowledge, this book contains no images of Native American individuals.

3) The authors recognize that the estimation of ancestry and its interpretation as social race is a contested topic within the field of forensic anthropology. At the time of this writing, ancestry estimation remains a required component of the certifying examination of the American Board of Forensic Anthropology. For this reason, we have included a chapter on ancestry estimation and contextualized it within a broader discussion emphasizing the difference between geographic patterns of human biological variation and socially constructed racial categories. This discussion, as well as an overview of the contemporary critiques of ancestry estimation, are provided at a level of discourse consistent with an introductory general science course. The inclusion of this chapter is meant to provide a balanced discussion of the topic; it is not intended as an endorsement of any particular perspective.

Contents

Figures

Tables

Preface

This lab manual was born from a series of online labs that were created over the course of three years to serve a large and diverse student body at Arizona State University. A majority of the enrollments for this class come from ASU online, a national and international initiative designed to expand the opportunities for higher education, and the societal and personal benefits that come with a college degree, to all communities by minimizing barriers to educational attainment. By standard measures, the approach we adopted was a success. Enrollments grew and evaluations were overall strongly positive. This history is important for understanding the rationale and choices we made here. By using those earlier versions of online labs as a basis, we hope this manual represents an important next step in the development of forensic anthropology as a key component of anthropology curricula nationwide. The field is tremendously popular with students, but also offers a rare opportunity to pair public fascination with key learning outcomes such as science literacy and basic numeracy. One obstacle, however, is the lack of sufficient teaching resources, which we hope this manual helps address, and we note that the design of the book means it can be tailored depending on each course's specific needs and goals.

Our primary goal in writing this book was to align forensic anthropology lab experiences with basic science competency as part of the general education requirements at most universities. A focus on basic science is important for two reasons. The first is the growing challenge of anti-science sentiment throughout the world, which as we prepare this book for submission in the fall of 2021 is often in the news and the subject of concerted political and social discourse. This is unlikely to change in the near future. The second is that the vast majority of students who complete an undergraduate forensic anthropology course will never be asked to assign sex based on a pubic bone. But they may very well have to interpret charts and tables of data, understand different data scales, and be asked to serve on a jury where rules of evidence and critical thinking skills have real world repercussions. It is important that we as educators acknowledge what is really important about our teaching. Nuances of anatomy, while fascinating, are not what most students will remember in a year's time. And this is certainly the case for large lower division courses. Therefore, as best as possible, we have tried to maintain an emphasis on science literacy throughout the assessments while acknowledging that there is plenty of room for expansion into more hands-on experiences.

Another important consideration relates to the scale of the class and the varying experiences students bring with them to the classroom. Some students come with some anatomy background while others are humanities or social sciences majors with no prior knowledge of the human skeleton. In our classes, very few are anthropology majors and even fewer have a strong science background. Our students are highly diverse and include working parents, active members of the military, 18-year-old college freshmen, mid-career people looking to finish their degree that they had to abandon some decades prior, and international students where cultural norms may differ from those here. For the vast majority of these students, this will be their first experience with the entire field of anthropology. Therefore, extra effort was made to provide very clear instructions on the lab exercises and to include more discussion of basic principles throughout the text.

Furthermore, many of the subtleties of forensic anthropology rely on hands-on knowledge of and experience with real human remains, which is not possible for an online class, not possible for a class with hundreds of students, and not possible at universities that lack resources to buy anatomical specimens. In addition, there are ethical concerns with using human remains in our teaching, especially those of unclear provenance. While we strongly feel that real bone osteology can never fully be replaced and certainly must center advanced and graduate level training in the field, we are unsure of the value added for a survey course for non-majors. Therefore, we approached this book with the goal of providing the necessary content to teach forensic anthropology at the lower division undergraduate level regardless of the budgetary situation at any specific college or university. This also helps serve the access mission of the book.

Therefore, this book serves a dual purpose in that it is part lab manual and part textbook. It gets to some points rather quickly, however, because it is technical in focus and may leave out key topics that appear in a fully developed text. We recommend using supplemental readings for these topics. On the other hand, the book also goes into greater explanation of some topics that might be glossed over in other manuals where it is assumed a higher baseline knowledge of basic science and human anatomy. We have found students do often need more explicit, step-wise instructions, which of course can be supplemented with in-class lectures and additional readings. We have found, however, that students appreciate our straightforward approach.

Another important consideration in adopting a text–lab manual hybrid is minimizing costs to students. Doing so serves a social justice purpose by reducing economic barriers to education. There is a national movement away from expensive textbooks, assigning multiple texts, and burdening students with more debt. We don't think this trend will reverse and students will increasingly seek courses that minimize their expenses while satisfying their course requirements and (hopefully) piquing their interest. We have had no complaints about using this single resource in our class, especially when paired with lectures and supplemental readings, case studies, online videos, and other freely available resources. That said, this book can certainly be paired with more extensive texts on the subject and we recommend that of Christensen et al. (2019) for quality, contemporary relevance, and maximum alignment in terms of level, theme, and tone. However, we see this book as a first step for students, and those interested in pursuing forensic anthropology should minimally plan on taking coursework in osteology, anatomy, and, where available, advanced courses in forensic anthropology.

The text itself follows a standard progression of topics—basic science, the biological profile, and trauma are all covered in that order. As a reflection of our emphasis on science literacy and the intended audience of this book, we intentionally omitted chapters on forensic archaeology and scene processing as well as the maintenance of chain of custody. While these are undoubtedly important considerations within the field of forensic anthropology, we consider them to be specialist topics and to offer limited opportunities for student engagement in many of the classroom contexts discussed above.

Each chapter follows a similar structure with explanatory text and images, in-text questions, and end-of-chapter exercises. The in-text questions are meant to test comprehension and mastery of a topic *as a student is working through the material.* Most of these questions were designed to assess basic comprehension of a concept and not to test whether the student has a complete mastery of a specific technique. For example, it is important for a student taking this class to understand the general progression of morphological change in the pubic symphysis for the purposes of age estimation. However, in an online class using static 2D images, it may be too much to ask that student to identify the specific phase an unknown specimen would be assigned to. Similar concerns are certainly the case even for in-person instruction where basic concepts such as os coxa orientation can be a challenge. The instructor is free to count these questions for credit, for participation, or not at all.

The end-of-chapter exercises weave together topics that were discussed in each chapter as well as integrate concepts across chapters. They are intended to be more difficult and to provide students with opportunities to apply the concepts that they have learned. Many of these exercises are framed in such a way as to provide illustrations of how the topics covered in this book are operationalized within the context of forensic anthropology. As such, many exercises also provide instructors with points of departure for more in-depth and nuanced discussions of the topics covered in this book. How each exercise is evaluated should be determined by the instructor for their particular course and instructors are free to augment or extend the exercises provided here using those resources that are available to them. The intended purpose of each exercise as well as a detailed explanation and, where applicable, suggestions for expanding upon it are provided in a separate resource for instructors hosted online by the University of Florida Digital Collections.

Acknowledgments

The authors would like to thank the numerous individuals and institutions that supported our trips to their collections to supply images for this volume. In particular, we thank Carmen Moseley at the Maxwell Museum of Anthropology at the University of New Mexico and Daniel Wescott, director of the Forensic Anthropology Center at Texas State University. We also wish to thank Laura Fulginiti for generously sharing her expertise with us as well as providing images for use within this work. We are grateful to the numerous individuals who produced freely available images on Wikimedia without whom much of this work would not be possible. Lastly, we wish to recognize the contributions of those individuals whose stories form not only the backbone of this work but the foundation of the discipline.

1

Forensic Anthropology as a Science

An Introduction to Data Types and Basic Statistics

Learning Goals

By the end of this chapter, the student will be able to:

- Describe the key principles of a scientific investigation.
- Describe the levels of measurement or data scales used in forensic anthropology.
- Describe the difference between metric and non-metric data.
- Describe the difference between continuous and discrete data.
- Define nominal, ordinal, ratio, and interval scale data types.
- Associate different measurement scales with basic quantification procedures appropriate for that scale of measurement.

Introduction

In this book, the student will learn techniques for determining basic identifying information about a decedent based on observations made from their skeleton and dentition. This is called the **biological profile**, which minimally consists of demographic information such as age, sex, ancestry, and stature that can be used to inform searches through missing persons databases and aid in the overall death investigation. Forensic anthropologists perform this service for the law enforcement community through the application of scientific principles. That is, forensic anthropology *is* a science. It is a subfield of biological anthropology that focuses on the application of principles from evolutionary biology (among other fields) to aid in legal death investigations. As such, it is important to understand that forensic anthropology is a field that is constantly improving, a feature of all scientific disciplines, and this improvement can only come from continued research. Those interested in reading the latest peer-reviewed research in forensic anthropology should access key academic journals such as the *American Journal of Biological Anthropology, Forensic Science International,* the *Journal of Forensic Sciences,* and the *Journal of Forensic and Legal Medicine.*

Before exploring how forensic anthropologists aid medicolegal death investigations, we first need to understand what constitutes a scientific analysis. Science is a way of knowing based on empirical hypothesis testing, experimentation, and repeatability. A scientist uses existing knowledge to offer a hypothesis about the natural world. A **hypothesis**, according to Merriam-Webster, is "a tentative assumption made in order to draw out and test its logical empirical consequences." In other words, scientists generate an explanation for a phenomenon and then systematically generate data that can determine whether their hypothesis is likely to be true or likely to be false. **Repeatability** is key to a scientific investigation. A scientist can never (or very rarely) *prove* anything because there are always more data that can be generated that is relevant to the hypothesis. Instead, scientists accept or reject hypotheses based on the data they have collected. In some cases, a natural phenomenon gains a high level of support from repeated hypothesis testing. It then becomes a scientific theory. A **scientific theory**, according to Merriam-Webster, is "a principle that has been formed as an attempt to explain things that have already been substantiated by data . . . [whose] likelihood of truth is much higher than that of a hypothesis." Evolutionary explanations for the diversity of life on earth have gained such support from decades of experiments that the principles guiding research in biology reach the level of theory (i.e., evolutionary theory).

Research Design in Forensic Anthropology

Let's consider a basic research design that is commonly used in forensic anthropology to examine how science works in practice. Forensic anthropologists are principally concerned with generating data that speak to the identity of the decedent. This is called the biological profile, as defined above. Because forensic methods are scientific, they are constantly being improved. New methods are being identified and existing methods are subject to continued testing on samples from different time periods or different parts of the world or using different measurement methods. For the sake of simplicity, we will use the example of sex and outline a basic research design. How does a researcher use the scientific method to develop new and better methods of sex estimation from the skeleton?

We start with the hypothesis that a particular skeletal observation is *sexually dimorphic*, that is, the observation or measurement differs between males and females in a systematic way. Our null hypothesis (H_0) would be that the measurement does *not* differ between males and females, or in other words the male average and the female average are equal to each other. We hope that by collecting and analyzing new data that we can *falsify* this null hypothesis in favor of the alternative hypothesis (H_a) that the average male and female measurement are *not* equal. If we collect data that show the two averages are not equal, we will *reject* the null hypothesis (of equal averages) and *accept* the alternative hypothesis (of unequal averages) until more data are collected that suggest otherwise. We can never fully prove that males and females have different averages for this measurement because that would require knowing the measurement values for all males and all females currently alive and that have ever lived! A tall order. This is an important point. Science does not prove things to be true. Science rejects null hypotheses in favor of alternative hypotheses until more data accumulate suggesting a high likelihood that a hypothesis is true, which then becomes a theory. Forensic anthropological research rarely reaches the level of general acceptability to produce a theory.

Now let's consider a basic research design. A forensic anthropologist hypothesizes that a new measurement is sexually dimorphic and potentially useful for sex estimation in forensic case work. How do they go about designing an experiment to test this hypothesis?

First, the researcher needs to ensure the new measurement is easy to measure. This is usually accomplished by defining clear landmarks on the bone or ensuring that the bone is oriented the same way each time it is measured. A new measurement that is difficult to consistently measure results in high levels of error (see chapter 2) and is not useful for forensic case work.

Second, the researcher needs to collect data to test the null hypothesis. This requires access to a documented collection of skeletons in which the demographic characteristics are known based on medical or other antemortem records. If one knows who is male and who is female, the researcher can simply record data on their newly proposed measurement method to generate ranges of variation for a new measurement. This comparative collection forms a *training sample* in the analysis to "train" statistical analyses to "know" what the range of variation is for males and for females. Statistical analyses are then used to estimate how well the measurement identifies the sex of an unknown individual. The process described above is so fundamental to basic research in forensic anthropology that it appears in countless research articles in the field.

Levels of Measurement and Data Scales

When most hear the word "data" they think of measurements, and in forensic anthropology this isn't a bad assumption. However, there are other types of data that appear in forensic research, and it is important to be able to identify the **data type** one is using as it determines how to proceed with an analysis. At the most basic level, data can be classified as either: (1) **continuous** (metric) or (2) **discontinuous** (non-metric, discrete, categorical).

Continuous data represent positions along continuous number lines, such as measurements in inches or centimeters. Continuous data are **quantitative** and can theoretically vary from 0 to positive infinity with infinite levels of accuracy possible (2 cm, 2.5 cm, 2.5434748484 cm, etc.). The number of decimal places you observe is limited only by your measuring equipment; one can always divide a continuous measure into increasingly smaller units of measures (from meters, to centimeters, to millimeters, etc.).

Discontinuous data do not vary continuously, but rather occur in discrete categories (hence *categorical* data).

For example, *count data* are considered discontinuous and discrete data because the number of possible values is finite and cannot be subdivided. The number of fingers or toes you have is a discrete variable, as is the number of cervical, thoracic, and lumbar vertebrae present in your spine. You cannot have 5.345 fingers or 12.2 thoracic vertebrae. Counts occur as whole integers, and as such are quantitative variables, much like the simple measurements discussed above.

Discrete data can also be qualitative, that is, non-numerical. For example, male/female, old/young, short/tall, and alive/dead are qualitative data. You cannot assign a specific measurement value to these categories and you generally cannot be in both categories at the same time (hence these are discrete data). Because there are only two choices, the variables are considered to be *binary data* or *dichotomous data*. However, qualitative data can also be *non-dichotomous data*. For example, class grades (A, B, C, D, E) are qualitative but non-dichotomous.

Note that continuous scale data can easily be converted to discontinuous scale data, but not vice versa. For example, using measurements of a bone to estimate the sex of an individual (as discussed above) is based on continuous scale data, but these measurements could be converted to a discontinuous scale (small, medium, large) using arbitrary measurement cutoffs. A tremendous amount of information would be lost in doing so; however, it is possible. If you only recorded the data as a discontinuous scale variable, you would not be able to extract specific lengths (continuous scale data) of the bones. *If time allows, it is always better to collect continuous scale quantitative data for this reason.*

Also note that just because numbers are used to describe a data scale does *not* mean the data type is continuous. For example, you can arbitrarily assign male = 1, female = 2, but this does not make the variable "sex" continuous. You cannot have a value of 1.342344 for sex. Just like you cannot be in a state of "sort of alive, sort of dead" (zombies notwithstanding).

The difference between continuous and discontinuous scales of data is one way to categorize the types of variables used in any scientific endeavor. However, these distinctions alone do not account for the full range of variable types one might encounter in a forensic analysis.

The Stevens' Data Types

In 1946, psychologist Stanley Stevens devised a four-level categorization of data types that further differentiated variable scales beyond a simple continuous/discontinuous distinction. The approach he defined to categorize data continues to be popular today and is useful for thinking about how basic research is conducted. Four levels of measurement are recognized: **nominal, ordinal, interval,** and **ratio**.

Nominal Data

Nominal data are *discrete* and *discontinuous* data that are classified according to some qualitative measure that *cannot be placed in a specific order;* that is, there is not a way in which membership in one category has more or less of a specific quality that the variable is capturing or representing. Nominal data types are simply *names or labels* to describe variables and observations. Another term for this is *categorical data.* In forensic anthropology, sex is a nominal variable when considered in a dichotomized form (male/female OR female/male—it doesn't matter which order you list them in, which is exactly the point of nominal data). Colors (red, blue, green), parts of speech (verb, noun, adjective), religion (Christian, Jewish, Muslim), language (English, Spanish, German), political preference (Republican, Democrat, Independent), and nationality (U.S., Canadian, Mexican) are all nominal variables because they cannot be placed in a specific order or ranking.

A variable can be nominal even when numerically characterized. For example, a student ID number is a nominal variable even though it has the appearance of being continuous. Please note that these categorizations of variable types are not absolute, but rather can change with the specific question you are asking of the data. For example, color is something that tends to be treated as nominal (red, blue, green) but does exist in a continuous state.

In forensic anthropology, bones may simply be categorized as "burned" or "unburned," but the degree of burning could also be characterized by noting fine gradations of color that relate to the temperature of the fire used to cremate the remains. In **Figure 1.1** one could simply categorize the bone as "burned," or one could note the degree of burning by comparing it to charts that relate color to the temperature of the fire.

Ordinal Data

Ordinal measurements are *discrete* data that *can be ranked* (i.e., put in order). Ordinal data can be counted and ordered, but not measured. The distance between categories is also not fixed and may vary from one point on the scale to another.

Consider drink sizes at a coffee shop. These come in sizes that are ordinal in scale:

Small, medium, large;
Medium, large, extra-large; or
Large, extra-large, grande

Note that it doesn't matter what the labels are ("regular" sounds better than "small" due to marketing), but you can place these labels in a specific order that has meaning. You know you

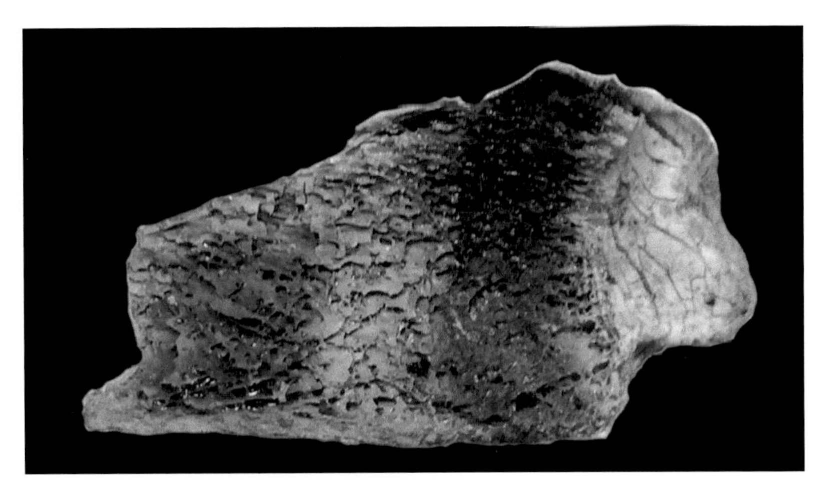

Figure 1.1. A fragment of burned human bone.

are getting more if you order a large vs. a small coffee, but are you getting the same amount of extra coffee if you go from a small to a medium vs. a medium to a large? For this, you would need to know the drink size in ounces, and most places will provide this information. An ounce is a quantitative measure and tells you what you need to know to make a decision when buying a coffee. Ordinal labels do not.

But what about popcorn at a movie theater? Again, the sizes used are ordinal scale (kiddie, small, medium, large), but the moviegoer has no idea *how much more popcorn they get* in a large vs. a medium, or in a medium vs. a small popcorn. This is what is meant when we say that the distance between categories is not fixed in ordinal-scale data.

Another example of ordinal-scale data that one encounters in everyday life is a *Likert scale,* where for example, you are asked to rate your satisfaction with a product (or college class!) (**Table 1.1**).

Note how one could ask the rater to provide a numerical value here instead (rate from 1 to 5 with 5 being the highest) and doing so would create the temptation to treat the data as continuous, but they are not. It is also unclear if the level of satisfaction increases to the same degree as it moves from Neutral to Satisfied to Very Satisfied. In fact, if the Neutral value is considered an arbitrary 0 point, then Likert scales can be mistaken for interval scale data (see below).

Grades are another good example of the use of an ordinal-scale data measurement when using a continuous scale is possible. In assigning grades some schools do use the percentile earned (0–100 percent), which provides a continuous scale variable and a better sense of student performance, but most use an ordinal-scale ranking (A, B, C, D, E). The order, of course, matters, but each grade represents a range of possible percentages.

Consider the grades in **Table 1.2**. Using an ordinal scale, the fact that Student 1 and 2 were actually very close in course performance is hidden (the difference between an A and a B was only 3 percent), while Student 2 and 3 also differ in only one grade scale yet had very different course performances (19 percent). Note that pass/fail would be considered a nominal variable.

Previously in this chapter sex was described as a nominal variable in forensics research. However, estimating the sex of an unknown individual often produces results that have some

Table 1.1. Example of a Likert scale of ordinal data

Very Dissatisfied	Dissatisfied	Neutral	Satisfied	Very Satisfied

Table 1.2. Relationship between a continuous and ordinal scale variable

	Student 1	Student 2	Student 3
Percentile	92%	89%	70%
Grade	A	B	C

Table 1.3. Ordinal scale ranking of sex assessments

Female	Probable Female	Unknown	Probable Male	Male

degree of uncertainty associated with the estimate that reflects the degree of confidence in one's sex estimation. Therefore, it is also common to see sex estimates that use an ordinal scale such as in **Table 1.3**. This scale is ordered from most to least "female" in appearance, but one cannot be sure if "femaleness" is consistently being measured along this scale.

Interval Data

Interval data are continuous scale data that are ordered and quantified with *exact differences between the values.* In other words, the *interval* between values is known (hence the name). However, there is no absolute zero for interval data that is non-arbitrary. In other words, zero does not mean the absence of the feature being measured, but instead represents an arbitrary value.

Temperature measured in degrees Celsius or Fahrenheit is an interval scale variable that can be quantified. The difference between 50, 51, and 52 degrees is the same; however, the zero point is arbitrary because 0°Celsius and 0°Fahrenheit are different temperatures and neither indicates the absence of temperature. In fact, both scales can have negative temperatures, which does not indicate a lack of temperature. You cannot multiply two temperatures and have a result that makes sense. Furthermore, 80 degrees is not twice as hot as 40 degrees. If your variable has similar properties, it is likely interval scale.

Note that temperature measured in Kelvin scale is not interval data because 0 degrees Kelvin is absolute 0; there are no negative values and 0 degrees Kelvin does indicate the absence of temperature. This shows that it is possible to measure the same phenomenon (temperature) using variables of different data types (**Figure 1.2**).

Time measured as dates from an arbitrary point (BC or AD) are also interval scale variables. A date of AD 0 does not mean an absence of time, and AD 200 is not twice as old as

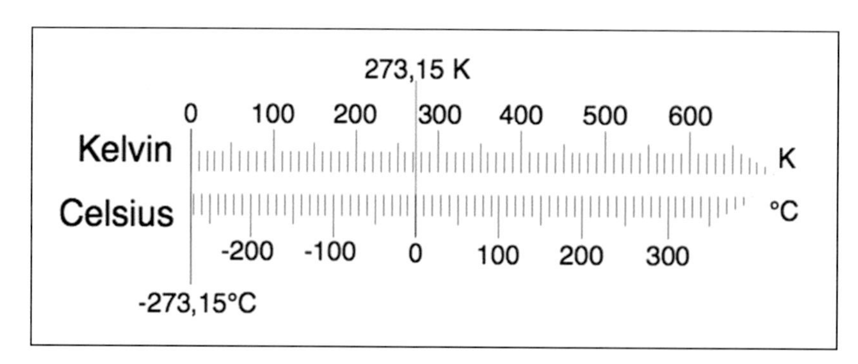

Figure 1.2. The Kelvin scale of temperature represents ratio scale data.

AD 100. Multiplying dates makes no intuitive sense. These are all features of interval scale data. However, time measured since the Big Bang does have a real 0 value. Therefore, the age of the universe is not considered interval scale data.

Ratio Data

Like interval data, ratio scale data are ordered, quantified, and have exact known differences between values. Unlike interval scale data, however, ratio scale data *do* have an absolute zero that is meaningful and not arbitrarily determined. Height, weight, head size, and most other variables that can be measured or counted are ratio scale data. A value of zero for height means the absence of height. An individual that is six feet tall is twice as tall as someone who is three feet tall. Ratio scale data are the most commonly encountered variable type in forensic analyses; however, certain ordinal-scale variables are also common. Ratio scale variables have all of the properties of nominal (labeling), ordinal (ranking), and interval (proportionate intervals) data, but the possibility of an absolute 0 value means many more statistical techniques are available for research.

The relationships between the four levels of measurement are summarized in **Figure 1.3**.

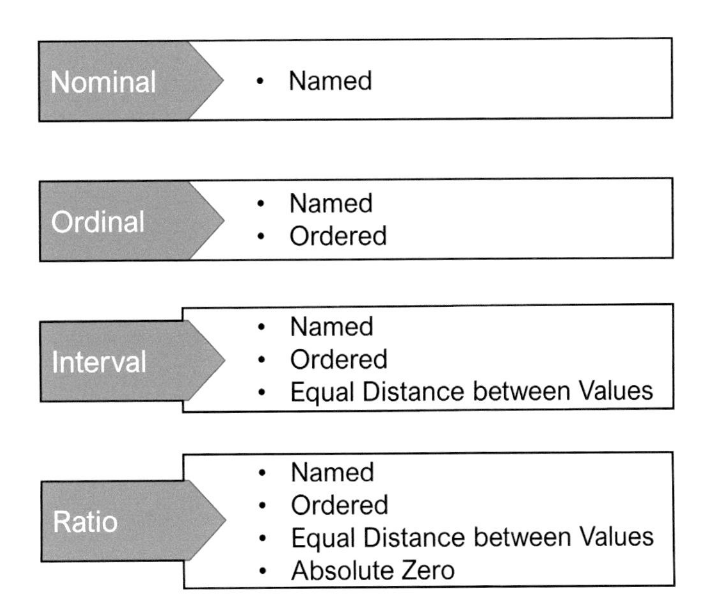

Figure 1.3. Summary of four scales of data.

LEARNING CHECK

Refer to **Figure 1.4** in answering the following questions:

Q1. How can the boxes be observed as nominal data?
A) By color: green, orange, blue
B) By order of size: small, medium, large
C) By the lengths of one of the sides: 1 cm, 2 cm, 3 cm

Q2. How can the boxes be observed as ordinal data?
A) By color: green, orange, blue
B) By order of size: small, medium, large
C) By the lengths of one side of the squares: 1 cm, 2 cm, 3 cm

Q3. How can the boxes be observed as ratio scale data?
A) By color: green, orange, blue
B) By order of size: small, medium, large
C) By the lengths of one side of the squares: 1 cm, 2 cm, 3 cm

Q4. Which of these is an example of *interval* scale data? Remember that interval scale data are continuous; however, a zero value does NOT indicate the absence of that condition.
Please read each choice carefully and select the one that meets this definition.
A) Temperature as measured along the Kelvin scale.
B) Age measured since the Big Bang.
C) Measurement of elevation based on sea level.
D) Count of the number of fingers.

Figure 1.4. Three squares of different colors and sizes.

Forensic Anthropology and Levels of Measurement

Questions 1–4 introduce the basic types of data scales used in any of the sciences. The use of colored boxes is a good way to show that the same objects can be "analyzed" in completely different ways depending on what your research question is: color, relative sizes, or absolute sizes. While it is easy to imagine why color may be the variable of interest (for example, if you were testing for color recognition/blindness), you might be asking, why not just measure the squares if size is of interest. That is, why would anyone use the ordinal ranking scale?

The answer is often pragmatic: How much time (and money) does one have to record the data? As noted above, you can always create an ordinal-scale ranking from your ratio scale raw measurements, but you cannot do the reverse. If you only record the box sizes as small, medium, and large, you cannot subsequently convert these to actual measurements. The important point here is that science, including forensic science, occurs within a specific social and political context. Resources are always limited! And this is one of the many things that forensic science TV shows get wrong—science is not quick, and budgets are always limited.

LEARNING CHECK

Q5. Consider the differences between the male and female pelvis as shown in **Figure 1.5**. For a series of pelves (plural of pelvis) one would assign each to a category of "male" or "female." How would the data (male, female) be categorized?
A) Nominal
B) Ordinal
C) Interval
D) Ratio

Q6. What types of data are being generated in **Figure 1.6**? Note the caliper readout is in millimeters.
A) Nominal
B) Ordinal
C) Interval
D) Ratio

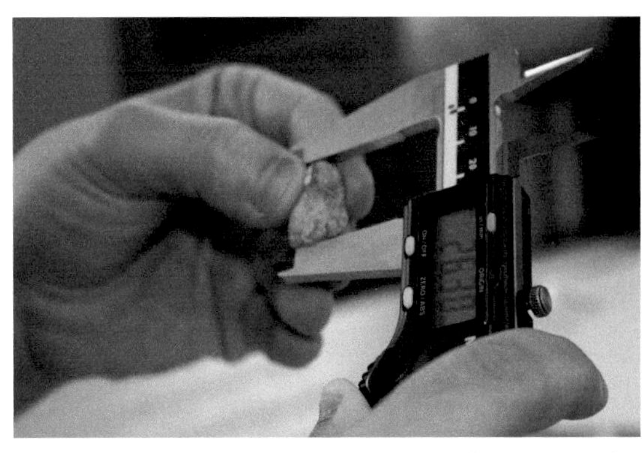

Figure 1.6. The author (CMS) measuring a human carpal bone with sliding calipers.

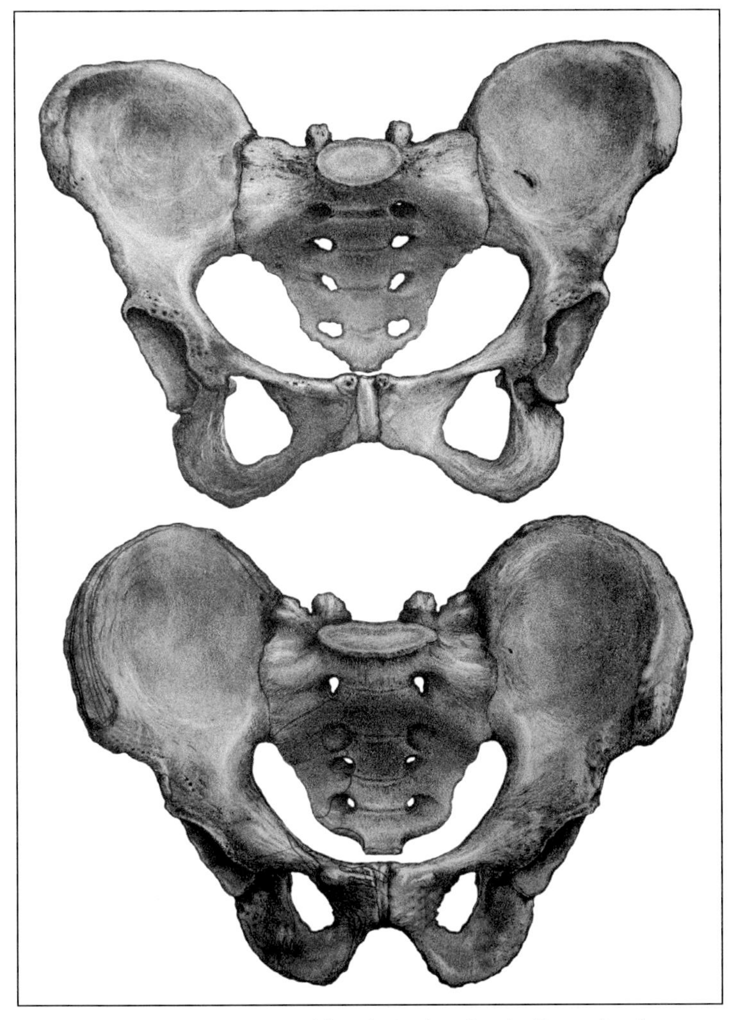

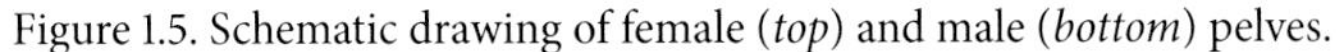

Figure 1.5. Schematic drawing of female (*top*) and male (*bottom*) pelves.

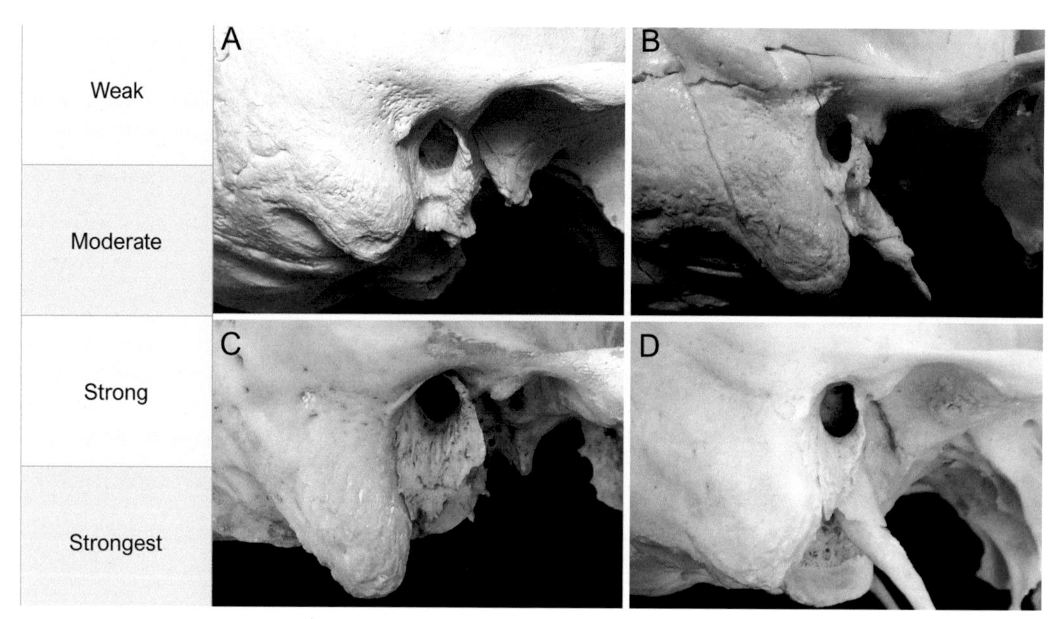

Figure 1.7. Four mastoid processes of varying size and thickness.

Q7. The mastoid process is one of the features used to assess the sex of an individual. This large mass of bone is located behind your ear and can be felt with your fingers by pressing behind your earlobe. Consider **Figure 1.7**. The table on the left side of the figure is how one records information on mastoid size.
What type of data are being depicted in this table?
A) Nominal
B) Ordinal
C) Interval
D) Ratio

Q8. Time-since-death is a critical concept that will be discussed later in this book. What kind of data are represented by the times 1:24 p.m., 3:15 a.m., 3:30 p.m., and 4:00 p.m.?
A) Nominal
B) Ordinal
C) Interval
D) Ratio

Q9. Based on the previous definitions, which of the following data series represents continuous scale data? (Select all applicable answers; there is more than one correct response.)
A) 12 ft, 9 ft, 7 ft, 3 ft
B) Red, green, yellow, purple
C) Male, female, female, male, female
D) 12°C, 7°C, 32°C, 18°C

Q10. Based on the previous definitions of variable types, which of the following data series represents discrete data?
A) Red, green, yellow, purple
B) 8 cm, 11 cm, 2 cm, 5 cm
C) 20 years old, 50 years old, 100 years old
D) 1 in, 4 in, 10 in, 16 in

Basic Research Statistics

Individual observations are combined to form a *sample*. It is assumed that if the sample is relatively large, it accurately represents the total *population*. In the example above, a bone length was used to assess whether males on average had larger bone lengths than females. If true, the bone length can then be used to assess the sex of an unknown individual found at a crime scene. This forms the null hypothesis: average male bone length equals average female bone length. If the collected data suggest otherwise, then this hypothesis is rejected, and one can conclude that the bone length does differ between the sexes.

To test this hypothesis, data are collected from a sample of males and females. These data comprise the male sample and the female sample. If a reasonable number of individuals are measured, then the sample should be an accurate reflection of the total population of males and females in the United States, which is what we really want to know. The sample will always be much smaller than the population, but research has shown that most samples accurately represent the population even with comparatively small sample sizes. For example, a sample of 50 males and 50 females is often sufficient to determine whether they differ statistically.

One of the first things to do after the data are collected is to describe the characteristics of the sample. In particular, there are two things to record (**Figure 1.8**):

1) Where the center of the sample is located along the measurement scale (vertical dashed lines in **Figure 1.8**).
2) How spread out (or variable) the observations are (horizontal lines with arrows in **Figure 1.8**).

The first of these characteristics is called the **measure of central tendency**. The most commonly used measure of central tendency is the *mean* or *average* value, but the specific measure used depends on the level or scale of measurement. In addition to the mean, there is the *median* and the *mode* (**Figure 1.9**). Some data types are characterized using a frequency rather than a mean.

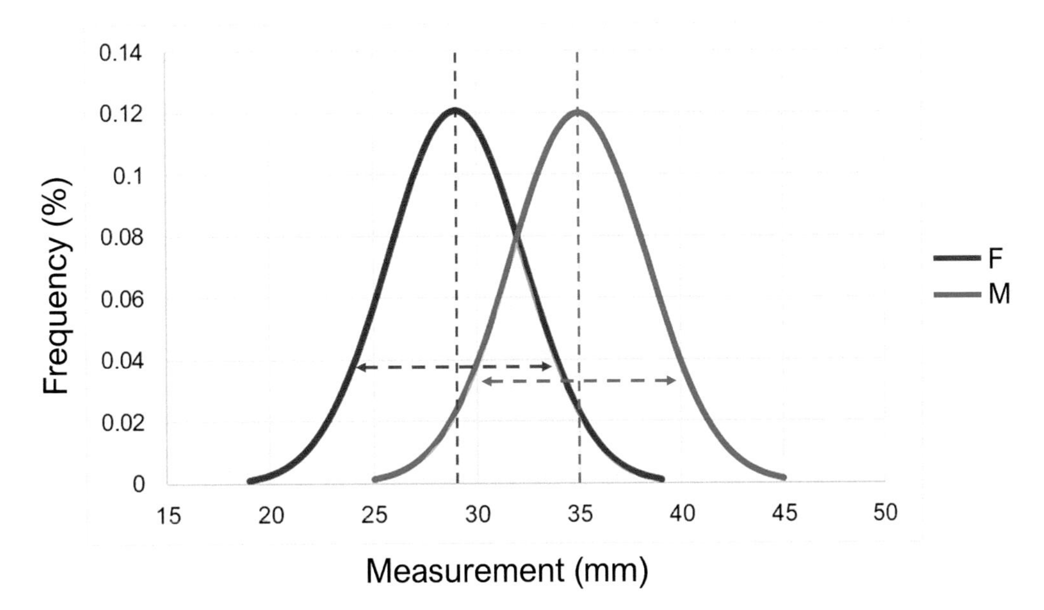

Figure 1.8. A bimodal statistical distribution of data.

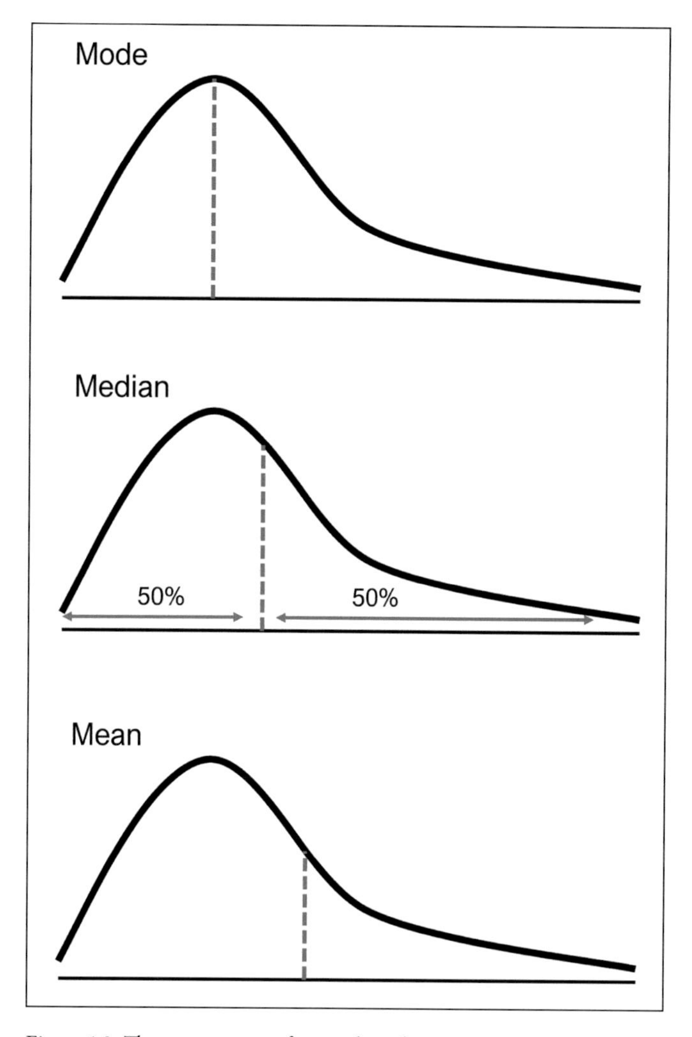

Figure 1.9. Three measures of central tendency.

Mean, Median, and Mode

1. Mode: The value that is most common in the data set. *To calculate, simply determine the frequency of each observation. The mode is the observation category with the highest frequency in the sample.* When plotted in a distribution (**Figure 1.9**), the mode is represented by the peak of the distribution.
2. Median: The middle value in an ordered list of numbers. Half the observations are greater than the median and half the observations are less than the median. *To calculate, order the observations from smallest to largest, and find the middle value (for 13 data points the median is the 7th largest—or smallest—value).*
3. Mean: the average of a set of numbers. *To calculate, sum all data values and divide this sum by the total number of values, or the sample size.* The mean is simply the average value.

Each measure of central tendency has strengths and weaknesses, which are not a concern here. It is important to note, however, that in **Figure 1.9,** the three statistics are quite different even though the data set has remained the same. The mode is lower than the median, which is lower than the mean. Although the hypothetical data in this image are unknown, consider which of the statistical values (mean, median, or mode) best represents that data set. Which estimate is most typical of the underlying data? We would argue the mode best

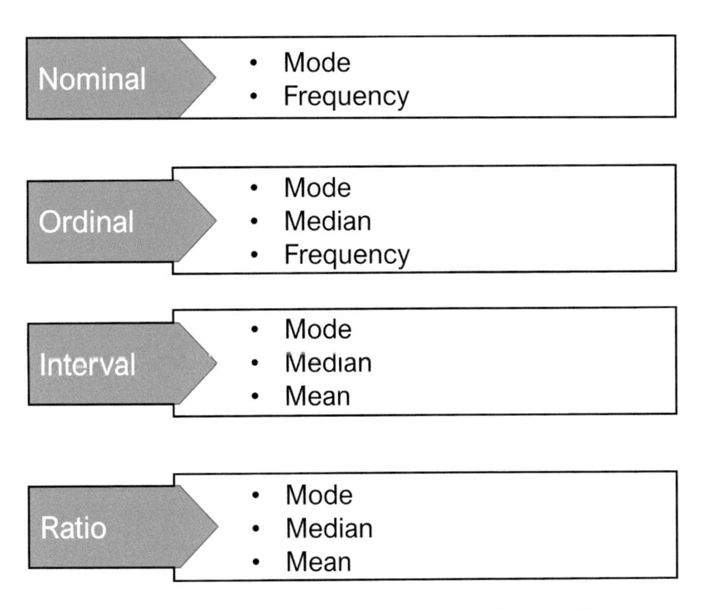

Figure 1.10. Relationship between variable scale and applicable measure of central tendency.

represents this data set because the mean and median have higher values than the peak due to the presence of high value *outliers* in the data. Note how the right side of the distribution is more drawn out than the left side.

Different scales of data (nominal, ordinal, interval, ratio) relate to the different measures of central tendency (mode, median, mean). That is, different measures of central tendency are appropriate to use to describe samples based on different data types. This information is summarized in **Figure 1.10**.

Nominal data, which are simply labels, can only be characterized in terms of the *mode* (the most common value) or the *frequency* of each categorical level in the data set. In the sex assessment example, the mode could be MALE if males are more common, *or* the frequency of each category could be characterized (males = 51 percent, females = 49 percent). The difference between the *mode* and *frequency*, in terms of how well each characterizes a data set, becomes more apparent as more categories are added. For example, the mode for world languages is Mandarin Chinese, but this does not tell you if this is a majority of people, how many total languages are spoken, or how many people speak each language.

Ordinal data can be characterized using the *mode* or the *median*, but not the *mean*. With letter grades, the median might give you a better sense of the typical grade in a class than the mode; however, this depends on how the grades are distributed among the possible scores.

For *interval* and *ratio* scale data, you generally want to use the *mean*. A *frequency*, while technically valid, has no real value with interval and ratio scale data. In some cases, the *median* might better describe the central tendency of a data set. For example, mean annual income can be highly biased by a few individuals with exceedingly high incomes, thus pulling the mean higher. Yet, most people earn much less. Here, the median is better reflective of the typical income in the data set.

LEARNING CHECK

Q11. The mode is relatively easy to calculate. Assuming a standard grading scale (A=90–100, B=80–89, C=70–79, D=60–69, E<60), what is the mode of the grades in the course as depicted in **Figure 1.11**?

A) A
B) B
C) C
D) D
E) E

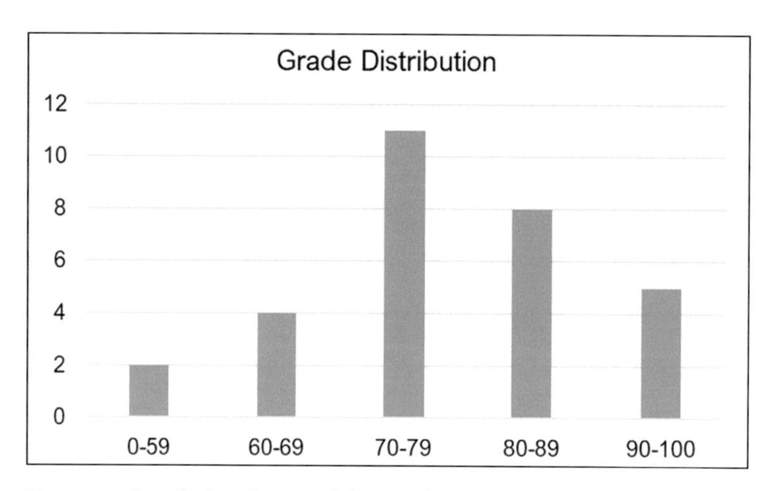

Figure 1.11. Sample distribution of class grades.

Calculating Simple Statistics

The median and average can easily be calculated in a program like Microsoft Excel.

Use the following data set and enter it into Excel: 1, 2, 5, 4, 3, 1, 1.

It is easy to see that the mode = 1. But what is the median and mean?

In an empty cell, type "=average()" and with the cursor inside the brackets select the cells that contain your data set. Press the enter key and the cell will now show the mean value for your data set. You can do the same for the median with "=median()."

LEARNING CHECK

Q12. Given the following data (**data** = 1, 2, 5, 4, 3, 1, 1), what is the *median* of the data set?

A) 1
B) 2
C) 3
D) 4

Q13. Given the following data (**data** = 1, 2, 5, 4, 3, 1, 1), what is the *mean* of the data set?

A) 2.43
B) 3.43
C) 4.33
D) 1.34

End-of-Chapter Summary

This chapter introduced basic concepts relevant for measuring and recording data that are applicable across the sciences. In particular, this chapter discussed the different scales of data or levels of measurement, defining a primary distinction between continuous and discontinuous scales, and further categorizing data types as nominal, ordinal, interval, or ratio scale data. Basic statistical concepts are introduced, such as the frequency of a trait or the mean, median, or mode of a data set. These statistics help describe what is called the *central tendency* of a data set that forms the basis of testing research hypotheses. Forensic anthropology has professionalized further in recent years and scientific rigor is now of primary importance in generating new methods and testing existing methods of forensic identification.

End-of-Chapter Exercises

Exercise 1

Materials Required: **Figure 1.12**

Scenario: A researcher is collecting data on the size of a cranial feature called the anterior nasal spine (indicated by the arrows in **Figure 1.12**). The researcher has decided to record data based on a graded scale of expression, with "1" indicating a slight expression and "3" indicating a marked expression of this feature.

Question 1: Given how the researcher is recording the data, are they continuous or discontinuous?

Question 2: Which of the Stevens' data types does this manner of recording represent?

Question 3: Can these data be recorded such that it represents other Stevens' data types? If so, how?

Question 4: Recording these data as which of the Stevens' types would capture the *most* information?

Exercise 2

Materials Required: None.

Scenario: Based on their own observations, a colleague thinks that, on average, individuals who are 65 years of age or older have shorter vertebral columns than those who are in their 30s. They have asked you to help them conduct some research to test their hypothesis.

Question 1: For this experiment, what would be your null hypothesis?

Question 2: For this experiment, what would be your alternative hypothesis?

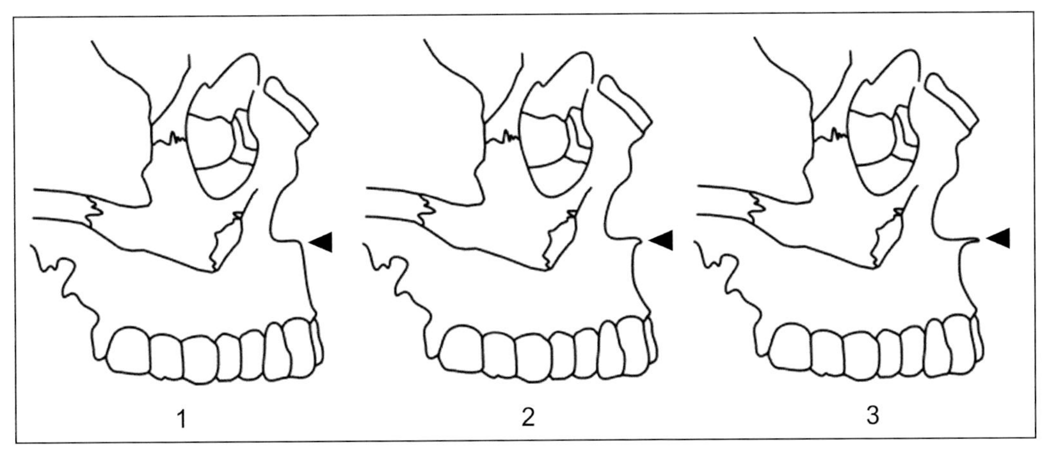

Figure 1.12. Three lateral views of the human cranium showing the anterior nasal spine (black arrow).

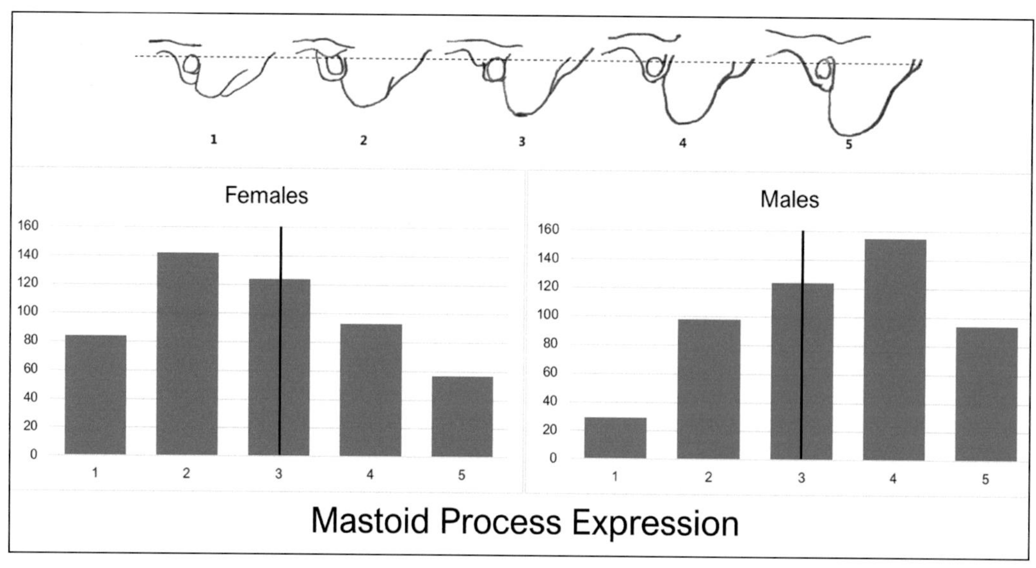

Figure 1.13. Variation in mastoid process scoring and sex-specific distributions of scores.

Question 3: Collecting your data as which of Stevens' types would be the most informative?

Question 4: Given your colleague's idea, which measure of central tendency is most appropriate to test their hypothesis?

Exercise 3

Materials Required: **Figure 1.13**

Scenario: You have collected data regarding the size of the mastoid process for 500 females and 500 males using the scoring system shown at the top of **Figure 1.13.** To see if this trait differs in size between males and females (i.e., is sexually dimorphic), you have plotted the sex-specific distributions of the scores that you recorded. The vertical black lines represent the median for each dataset.

Question 1: The data that you have recorded represent which of the Stevens' data types?

Question 2: What is the median score for females? What is the median score for males? Does the median score for mastoid expression appear to be sexually dimorphic?

Question 3: What is the modal score for females? What is the modal score for males? Does the modal score for mastoid expression appear to be sexually dimorphic?

Question 4: Given how you have recorded your data, can you calculate an average mastoid size for each sex? Why or why not?

References

Hefner JT. 2007. *The Statistical Determination of Ancestry Using Nonmetric Traits.* PhD Dissertation, University of Florida.

2

Rules of Evidence

Accuracy, Precision, and Error

Learning Goals

By the end of this chapter, the student will be able to:

Describe the importance of the *Daubert* ruling for guiding a scientific research strategy in forensic anthropology.
Contrast accuracy and precision as basic concepts in science.
Identify ways to increase accuracy and precision by reducing observer error.
Differentiate random error and systematic error (bias).
Define and identify the different types of observer error.
Provide examples of inter- and intra-observer error.
Define metadata and describe how it is used in forensic anthropological research.

Introduction

Although this book does not focus on the legal aspects of criminal investigations, it is important for forensic anthropologists to be aware of the issues that may arise if called to testify in court. Forensic anthropology is a relatively young field, born out of biological anthropology's articulation with the medicolegal community during the mid-twentieth century. Initially, the field applied the methods of the biological profile developed in skeletal biology to crime scene investigations. The question, however, remains whether these methods are truly appropriate for use in legal settings.

Imagine finding yourself in court as a defendant. You are innocent of the crime for which you are accused. The prosecution calls a famous witness, a superstar, who provides evidence that your specific bite marks were found imprinted on a piece of torn duct tape at the crime scene. The expert witness testifies that you used your teeth to bite and tear the duct tape found around the hands of the decedent. Because the expert witness is so well regarded, the jury believes her. You lose the case and are sentenced to life in prison for a crime you did not commit. As you sit in prison, you have plenty of time to consider the following:

1. What makes someone an expert? Who makes that determination?
2. What is the scientific basis behind the analysis of tooth impressions on torn duct tape?
3. How often is someone wrongfully matched to a piece of torn duct tape based on their teeth?

These are all fair questions. They are questions of basic science. They are questions that reflect a moving target in our legal system about what information is admitted as evidence and how scientifically valid is the testimony based on this evidence? This chapter addresses these questions by covering two topics.

The first half of the chapter briefly traces the history of **rules of evidence** in the U.S. judicial system. The outcome of this is a Supreme Court decision that currently stands as guidance for how forensic evidence can be submitted (and challenged) to a court. This ruling, called the ***Daubert* ruling**, dates to 1993 and initiated a flurry of scientific testing of methods throughout the forensic sciences. Some methods were discarded as pseudoscience and others were greatly improved. This process of improvement and refinement continues today and anchors much of the primary research in forensic anthropology (Christensen, 2004; Christensen and Crowder, 2009).

The second half of this chapter focuses on how forensic anthropology, as a field, has responded to the *Daubert* ruling by improving the scientific rigor of research. Key to this improvement is understanding the basic rules of scientific testing, the importance of repeatability of one's results, and the importance of being able to estimate the error rate for any forensic technique. Therefore, the second half of this chapter introduces the concepts of **accuracy** and **precision**, **random error** and **systematic error**, and **inter**- and **intra-observer error**.

Rules of Evidence

What makes a piece of forensic evidence admissible in court? Standards of admissibility vary from state to state, and depending on where you live, there may be a different standard compared with the rules of evidence applied in federal courts. That is, state and federal courts *may* use different standards of admissibility. The rules are ever changing as new court decisions are handed down.

Consider the following scenario. An expert witness testifies that you can determine whether a given set of remains belongs to a male or female based on the taste of the bones. This expert testifies that the decedent was male, and the prosecutor is alleging the defendant on trial killed a female. The expert says the prosecution has the case all wrong based on this "taste test" to determine sex. Should this piece of forensic evidence be admissible? Who decides? The judge can either declare the evidence inadmissible entirely *or* the judge could

Table 2.1. Summary of major court cases determining rules for admissibility of evidence, after Grivas and Komar (2008)

Case	Relevance
Frye v. United States (1923)	Evidence must be *generally accepted* in a scientific community.
Federal Rules of Evidence, Rule 702 (1975)	Guidelines for admitting testimony but did not address the *general acceptance* criterion.
Daubert v. Merrell Dow Pharmaceuticals, Inc. (1993)	Affirmed that FRE 702 supersedes *Frye,* judges are gatekeepers, provided guidelines for evaluating evidence.
United States v. Starzecpyzal (1995)	Non-scientific (technical) evidence can be admitted under looser standards.
General Electric v. Joiner (1997)	Affirmed the *Daubert* standards use in not allowing testimony.
Kumho Tire Company, Ltd. v. Carmichael (1999)	Closes the "non-scientific" loophole of technical testimony, *Daubert* should apply to all, but be flexible.

allow the jury to hear the testimony but direct them to give it little weight because the science is shaky or untested. Because it is almost certain that the "taste test" is not valid, one would hope the judge would not let the jury hear the testimony because it could bias their decision-making. But what if that decision is wrong? What if the "taste test" turns out to be valid? How can new techniques like this ever get admitted into court if the scientific community considers them "junk science?" This is important to keep in mind as the scientific community has been wrong before.

This long preamble contextualizes the difficult decision-making process that judges and juries must wade through when hearing expert witness testimony. The primary literature discussing changes in the rules of evidence over time is dense and lengthy. Specific milestone cases are summarized in **Table 2.1**; however, the nuances of each case are noteworthy. For those interested in further reading please see the references below. The most significant case that has directly affected the practice of forensic anthropological research is the *Daubert* ruling.

The *Daubert* Ruling

In the nineteenth century, the reputation alone of an expert witness could determine the value of their testimony, even if the methods and statements were completely unjustified from a scientific basis.

The ***Frye* ruling** established what has become known as the *general acceptance criterion,* which stipulates that forensic evidence can be admitted if the technique used is generally accepted within the scientific community. While a step in the right direction, it wasn't clear how one defined a scientific community or measured general acceptance. Furthermore, the general acceptance criterion prevented cutting-edge methods from being used in criminal cases because, by definition, a community has not had time to evaluate the method and generally accept it.

The Federal Rules of Evidence (FRE) provided other criteria for considering whether expert testimony was admissible based on whether the evidence was relevant to the case and reliably based on scientific principles. However, the issues with the general acceptance criterion were *not* addressed under the Federal Rules of Evidence criteria, which many considered a problem.

The *Daubert* ruling solved this by asserting that FRE does supersede *Frye.* The importance of the general acceptance criterion was reduced, thus allowing more cutting-edge techniques to be admissible. *Daubert* also directed that judges should be the *gatekeepers* of what is admissible and non-admissible, and further provided *flexible* guidelines for judges to use to determine admissibility. These guidelines have come to be known as the *Daubert Criteria* **(Table 2.2).** It is important to note that not all criteria must be met for a piece of expert testimony to be admissible. They are simply guidelines.

Table 2.2. *Daubert* criteria for admissibility of expert witness testimony

1	The technique has been or could be tested
2	The technique has been through the peer review process and published
3	The technique has a known error rate, or at least an error rate that can be determined
4	The technique is standardized and able to be implemented reliably
5	The technique is generally accepted within the relevant scientific community

Subsequent cases (**Table 2.1**) affirmed or clarified certain aspects of the *Daubert Criteria.* As with most legal entanglements, the application of these standards is subject to continued revision, debate, and nuance. For example, although Criterion 2 states that a method must have a known or knowable error rate, this does not mean the error rate must be low. Expert testimony can be admitted even if the error rate is high and the jury must decide how they view that evidence. The weight of evidence should not be based on the reputation of the person testifying. Finally, it is important to note that the above cases apply to court decisions at the federal level. Many states have adopted the *Daubert* criteria for use in the state courts, but not all of them have (**Table 2.3**).

The *Daubert* ruling prompted consideration of whether the methods used within forensic anthropology would continue to be admissible in court under the *Daubert* criteria. As a

Table 2.3. Evidence admissibility standards by state, after Lesciotto (2015)

Daubert standard of admissibility	*Frye* standard of admissibility	Other standards
Alabama	Arizona	Utah
Alaska	California	Virginia
Arkansas	District of Columbia	Wisconsin
Colorado	Florida	
Connecticut	Illinois	
Delaware	Kansas	
Georgia	Maryland	
Hawaii	Minnesota	
Idaho	Missouri	
Indiana	New Jersey	
Iowa	New York	
Kentucky	North Dakota	
Louisiana	Pennsylvania	
Maine	Washington	
Massachusetts		
Michigan		
Mississippi		
Montana		
Nebraska		
Nevada		
New Hampshire		
New Mexico		
North Carolina		
Ohio		
Oklahoma		
Oregon		
Rhode Island		
South Carolina		
South Dakota		
Tennessee		
Texas		
Vermont		
West Virginia		
Wyoming		

Note: this list is subject to change as states reconsider their admissibility standards.

result, modern forensic anthropologists are now more concerned with technical rigor, which is a noteworthy development in the field. Older methods are being refined and revised so that error rates can be estimated. And there is greater emphasis on empirical testing and quantitative methods as replacements for simple pattern recognition approaches that depend on expert opinion, experience, and impression. Accuracy, precision, data collection best practices, and acknowledging and mitigating error in our analyses has become a top priority. We turn to these topics next.

Accuracy and Precision

Although often used interchangeably, there is a difference between accuracy and precision. **Accuracy** is the degree to which a measurement conforms to the correct (true) value of what is being measured. **Precision** is the ability of a measurement to be consistently reproduced. Accuracy is independent of precision; meaning you can be accurate but not precise and vice versa. A precise but inaccurate measurement means you are recording measurements consistently (precisely) but doing so incorrectly (inaccurately). An accurate but imprecise measurement means you are recording measurements correctly (accurately) but you cannot do so consistently (imprecisely).

Note that the true value of a measurement is never really known by the researcher because a measurement can be stated using an infinite number of decimal places. Therefore, all research (and measurement) has a practical level of accuracy and precision. This means, to some extent, that assessing whether accuracy and precision levels are acceptable is relative. In a forensics case, measurement to two decimal places is likely acceptable because it will not impact the nature of your decision about a case. If one is building an interstellar spaceship, however, then the measurements must be confirmed with a high degree of accuracy and precision because small errors become big problems very quickly and lives are at stake. Therefore, accuracy and precision levels in an analysis are both arbitrarily determined as acceptable or not depending on the scope of your research question.

LEARNING CHECK

Consider the four dartboards shown in **Figure 2.1**. *Each response will only be used once for the four questions;* consider all four questions before answering.

Q1. Consider the placement of the darts with respect to the green bullseye (the true value) in **Figure 2.1A**. Are the darts accurate and precise? Before answering please make sure to view the images for questions 1–4.
A) The darts are accurate but not precise.
B) The darts are precise but not accurate.
C) The darts are both accurate and precise.
D) The darts are neither accurate nor precise.

Q2. Consider the placement of the darts with respect to the bullseye (the true value) in **Figure 2.1B**. Are the darts accurate and precise? Before answering please make sure to view the images for questions 1–4. Choose the best answer possible.
A) The darts are accurate but not precise.
B) The darts are precise but not accurate.
C) The darts are both accurate and precise.

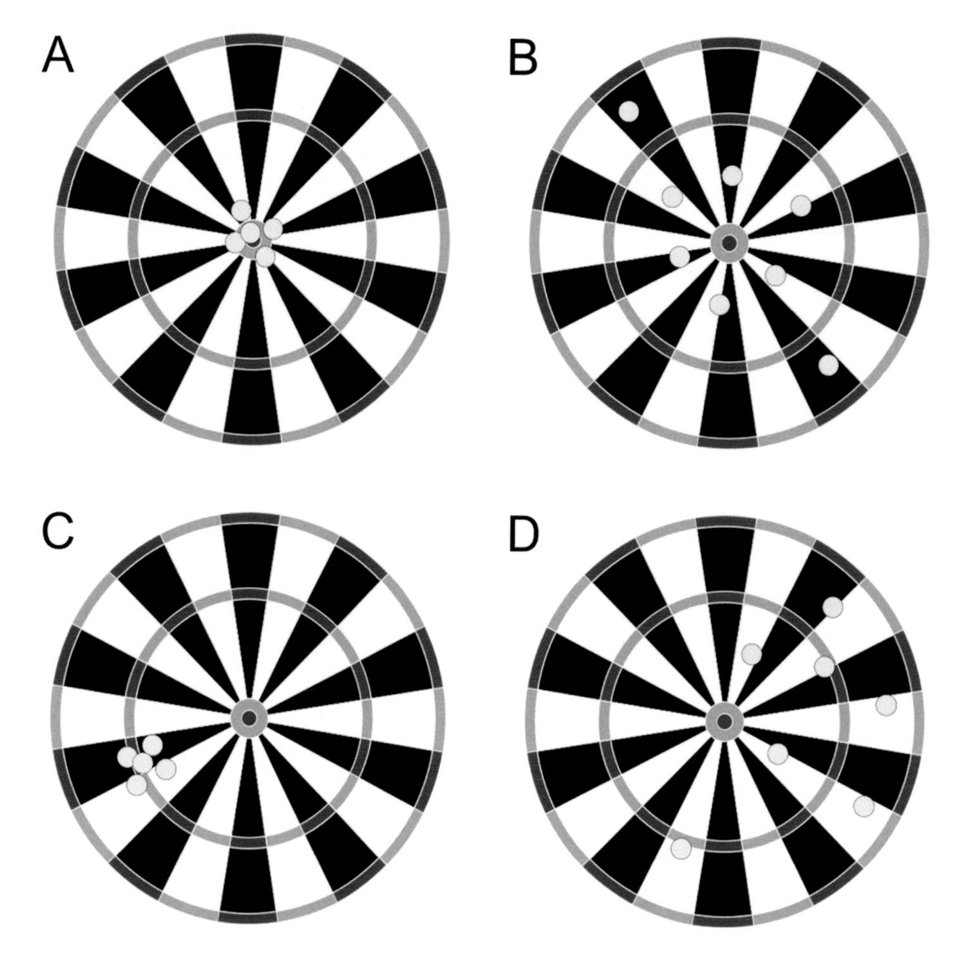

Figure 2.1. Four dart boards with yellow dots indicating throw attempts.

Q3. Consider the placement of the darts with respect to the bullseye (the true value) in **Figure 2.1C**. Are the darts accurate and precise? Before answering please make sure to view the images for questions 1–4.
A) The darts are accurate but not precise.
B) The darts are precise but not accurate.
C) The darts are both accurate and precise.
D) The darts are neither accurate nor precise.

Q4. Consider the placement of the darts with respect to the bullseye (the true value) in **Figure 2.1D**. Are the darts accurate and precise? Before answering please make sure to view the images for questions 1–4.
A) The darts are accurate but not precise.
B) The darts are precise but not accurate.
C) The darts are both accurate and precise.
D) The darts are neither accurate nor precise.

Q5. Darts and dartboards are not used in forensic anthropology. Rather, measurements are taken with precision equipment to generate research that helps solve crimes. Consider accuracy and precision again within a more statistically grounded approach.

Figure 2.2 shows a dartboard with a bullseye (the target or true value) and the locations where the darts were thrown (yellow dots). The bottom image shows a distribution where the horizontal axis represents the measurement value on the dartboard and the vertical axis represents how common those values are. The vertical line labeled "Reference Value" is directly beneath the bullseye. This is the *true* value that is trying to be accurately and precisely measured. The horizontal line A is the difference between the true value and the *measured* value and line B represents how spread out the darts are.

In this diagram, lines A and B represent what?

A) A = Accuracy, B = Precision

B) A = Precision, B = Accuracy

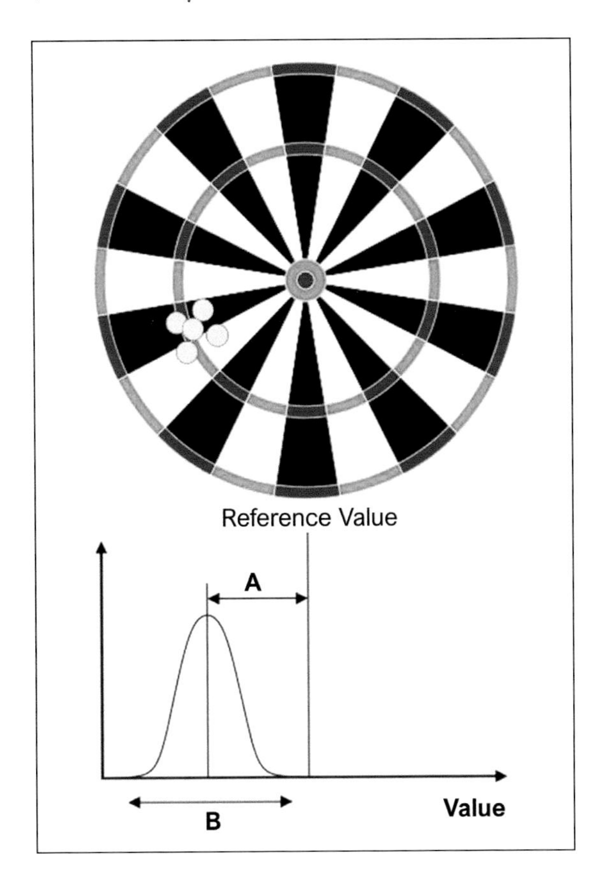

Figure 2.2. Statistical representation of accuracy and precision.

Q6. Which diagram in **Figure 2.3** represents a more precise measurement set?

A) A

B) B

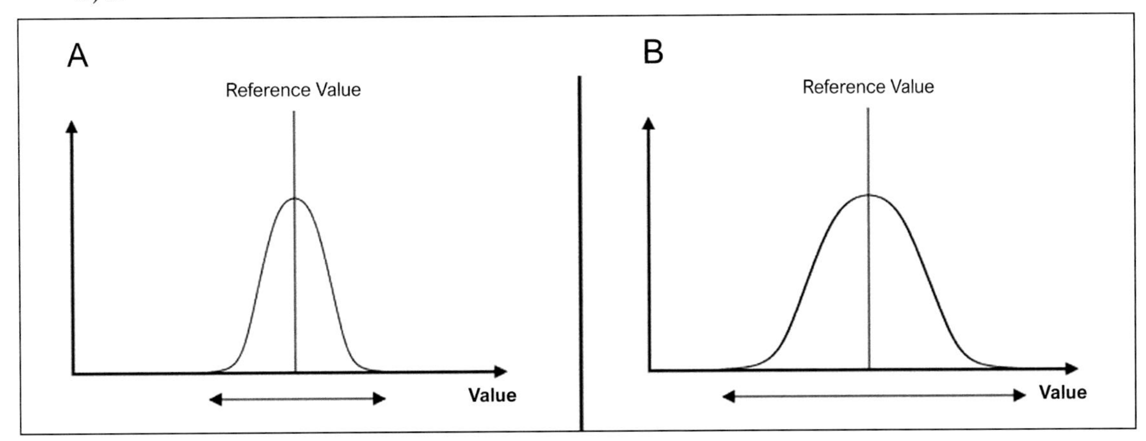

Figure 2.3. Two data distributions showing variation around a desired value.

Random Error

In any scientific study, measurements should be both accurate and precise. To increase accuracy and precision, measurement error must be minimized. *Random error* is error introduced through practices or sources that are unpredictable or perhaps even unknown. It is fair to say that most measurements are, to some degree, affected by random error. Random error decreases precision and causes minor statistical fluctuations around the true value you are trying to measure. Consider the calipers in **Figure 2.4**. It is not easy to read the lines associated with each number. If the researcher is tired at the end of a long day, they may read the calipers incorrectly for some specimens included in the study design. This introduces error that is random and difficult to predict or correct. An important feature of random error is that it can either *increase* or *decrease* a measured value with respect to the true value. This is different from systematic error (see below). One way to increase precision is to reduce observer error, which has two forms: intra-observer error and inter-observer error.

Intra-observer Error

Intra-observer error refers to the precision of data recorded by the *same* observer. Imagine measuring the same specimen three times on three different days. Due to variation in how you hold the calipers in your hand or how tightly you press the caliper tips onto the specimen during the measurement process you could (and likely will) record three different measurement values. This is intra-observer error.

Several factors increase intra-observer error in forensic anthropology:

—Humidity at time of observation: bones shrink and swell (ever so slightly) depending on humidity levels.
—Lighting conditions: this affects how well you see measurement landmarks.
—Temperature: this also causes slight variation in bone size.
—Observer's mood: data collection can be tedious, and one gets sloppy when tired.
—Tool failure/calibration: cheap tools are not worth the risk and may not be calibrated correctly. Note in **Figure 2.5** the readout says 0.02 mm when the tips are fully closed. These calipers need calibration.

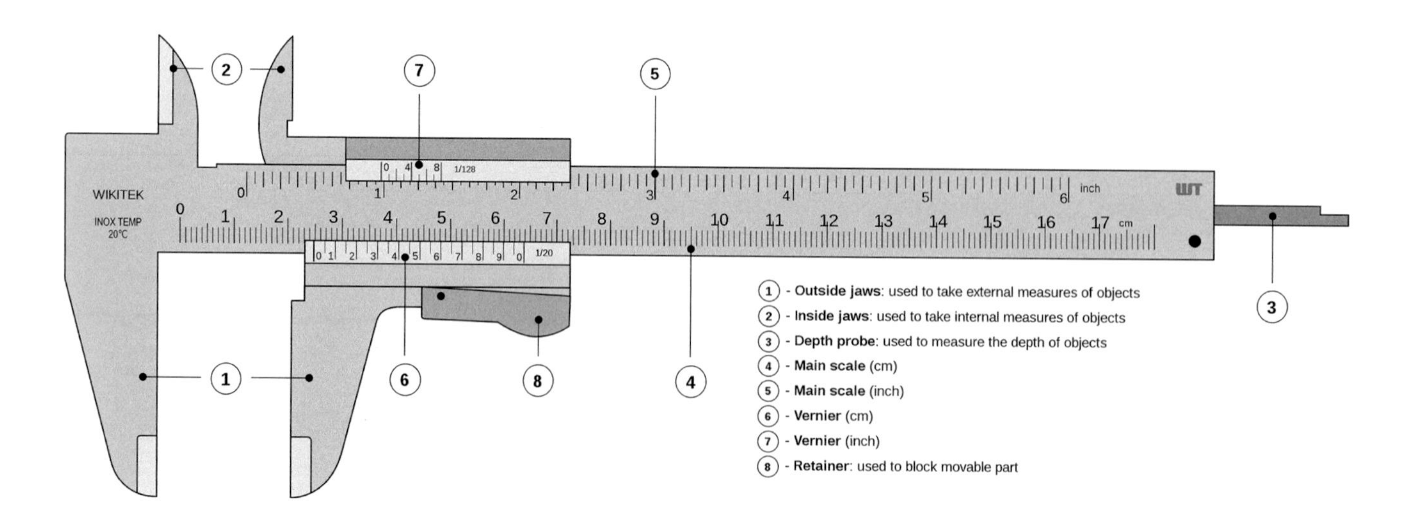

Figure 2.4. Vernier sliding caliper.

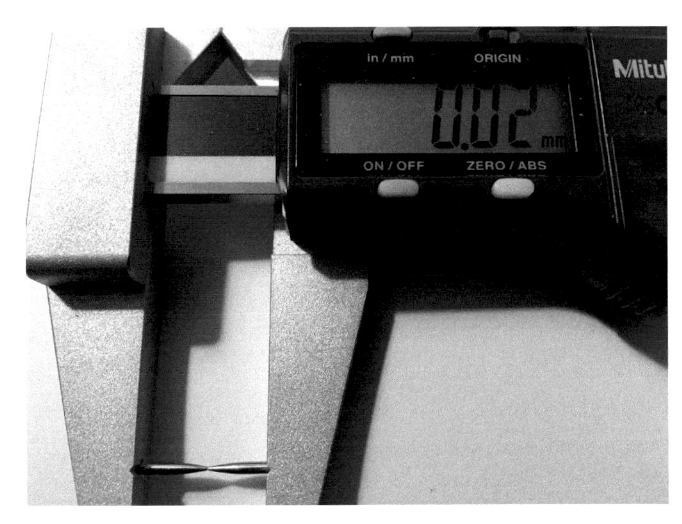

Figure 2.5. Caliper tips that are not correctly calibrated. Note the caliper tips are closed but the digital readout still reads 0.02mm.

Inter-observer Error

Inter-observer error results from variation in the accuracy of data recorded by *different* observers. If two observers record the same measurement (such as head length) differently using different definitions of where to place the caliper tips, then they cannot combine their data sets without accounting for this error. Factors that increase inter-observer error include:

—Observer experience: more experienced researchers are often more accurate and precise than novice researchers.
—Use of different measuring equipment: calipers may be calibrated differently or have slightly bent tips and other mechanical imperfections.
—Use of different measurement definitions: even if two researchers think they are measuring the same measurement by name (cranial length), there could be differences in how they are each defining the measurement landmarks. In addition, some measurement definitions changed through time as the field evolved.
—Differences in how one is trained to record the data.

LEARNING CHECK

Q7. There are many ways to reduce inter-observer and intra-observer error. One is to have the same observer take measurements at the same time of day with the same calipers. What does this accomplish?
A) It reduces inter-observer error.
B) It reduces intra-observer error.

Q8. What if two different observers used the exact same measuring equipment to measure different specimens. This would have what effect?
A) It reduces inter-observer error.
B) It reduces intra-observer error.

Systematic Error

Although random error introduces minor statistical variations in an analysis, other sources of error can be systematic, resulting in biased measurements. Systematic error affects measurements of a variable across the entire sample or subsample. Systematic error skews the sample statistics, making the mean incorrectly high or incorrectly low compared to the true population mean. This affects the *accuracy* of a measurement. Systematic error often results from issues with the equipment one uses to record the measurements. This is why measuring equipment should be calibrated repeatedly during the data collection phase of research using calibration equipment such as measurement blocks of known size. However, even if the equipment is calibrated correctly, inter-observer error can result if one of the observers does not know how to read the calipers correctly.

LEARNING CHECK

Q9. In the following chapters, considerable time will be spent discussing sexual dimorphism (the fact that men tend to be larger than women). Does this apply to the human skull as well? This is a testable hypothesis—males have significantly larger skulls than females (**Figure 2.6**).

To test this, you measure a sample of male skulls and a sample of female skulls and compare the mean values to see which is larger. However, the male skulls included in your sample are from unusually small individuals. Your study suffers from which kind of error? In making your choice, consider whether the error amounts to minor statistical errors (random) or seems insurmountable (bias), meaning it is hard to recover from this mistake.

A) Random error
B) Systematic error (bias)

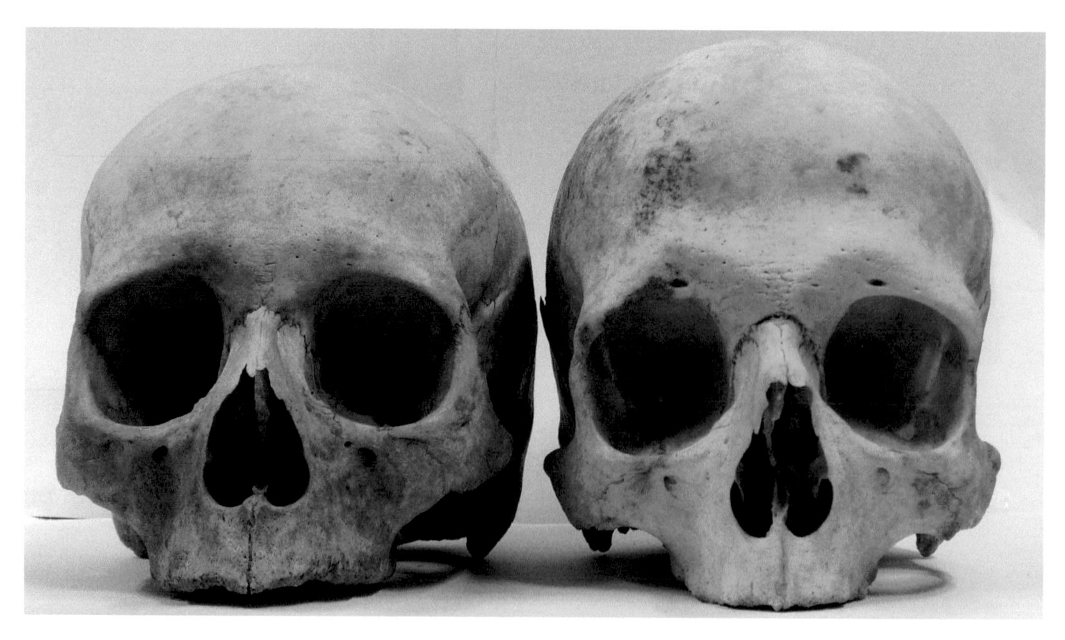

Figure 2.6. Female (*left*) and male (*right*) skulls demonstrating sexual dimorphism.

Q10. During the same research project, you later find out that the calipers you used to measure the female skulls were not calibrated correctly. Your study suffers from which type of error? In making your choice, consider whether the error will result in minor statistical variation (random errors) or serious errors that will significantly affect the study results.
A) Random error
B) Systematic error (bias)

Q11. Furthermore, the lighting in the room was poor. Despite having a window with natural light when you collected the data, there was intermittent cloudiness that week, which affected the natural light coming into the room throughout the day. The study suffers from:
A) Random error
B) Systematic error (bias)

Q12. How could you avoid introducing the kind of error caused by faulty equipment?
A) Use different calipers to measure different subsamples.
B) Check the caliper accuracy at the beginning of each measurement session.
C) Measure all males at once and then all females at once.

Q13. What type of error would you be testing by comparing measurements recorded a couple weeks apart by the same person?
A) Inter-observer error
B) Intra-observer error

Q14. What type of error would you be testing by comparing measurements taken by you with those of a second researcher whose data you wanted to use in your study?
A) Inter-observer error
B) Intra-observer error

Calculating Error Rates

Error rates should be quantified for each variable one uses in a study by comparing two measurements of the *exact* same specimen. The two measurements can be recorded several weeks apart by the same observer, which tests for intra-observer error. Or the two measurements being compared could combine those of two different researchers, which tests for inter-observer error. One common way to assess error rates is to calculate the **technical error of measurement**, or **TEM** (Perini et al., 2005):

$$(1)\ \text{TEM} = \sqrt{\frac{\sum_{i=1}^{n} (X_{1i} - X_{2i})^2}{2n}} = \sqrt{\frac{\sum_{i=1}^{n} d_i^2}{2n}}$$

The TEM expresses how much of the difference between two measurements of the same specimen results from technical error (Harris, 2008). The TEM can be combined with the **variable average value** (VAV) to generate a relative TEM (Perini et al., 2005):

$$(2)\ \text{VAV} = \frac{\sum_{i=1}^{n} \frac{(X_{1i} + X_{2i})}{2}}{n} = \frac{\sum_{i=1}^{n} \overline{X_i}}{n}$$

$$(3)\ \text{Relative TEM} = \frac{TEM}{VAV} \times 100$$

A relative TEM expresses the TEM as a percentage of whatever is being measured. By convention, a relative TEM that is 5 percent or greater is considered problematic.

Although the formulas shown above look complicated, they are describing a process that consists of several simple steps. The part that might look intimidating is the summation notation $\sum_{i=1}^{n}$ but this symbol is just telling you that you will be adding a series of numbers—from the first specimen to the *n*th specimen—together to get a sum. Let's walk through a short example. Say you are measuring the maximum length of tibiae (plural of tibia) and you decide to take a second set of measurements on five tibiae to see how consistent your measurements are. The values you obtain for your first (X_{1i}) and second (X_{2i}) sets of measurements of these five tibiae are shown in **Table 2.4**. To calculate the TEM, we will start with tibia 1 and subtract our second measurement (36.8 cm) from our first measurement (36.5 cm). This gives us the difference between the two measurements, d_1 , which equals -0.3 cm. We multiply this value by itself to get d_1^2, or 0.09 cm. We then find the d_i^2 values for the other four tibiae and add all five numbers together to get $\sum_1^5 d_i^2$, or 0.30. To get the TEM, we divide this number by the number of tibiae that we are remeasuring (*n*, or 5 in this example) and then take the square root. In this case, the TEM is 0.245 cm. To calculate the VAV, we take the average of the average of each pair of measurements. For each tibia, the average measurement, $\bar{X}_i$, is found by adding the two measurements together and dividing by two. For tibia 1, $\bar{X}_1$ is equal to (36.5 + 36.8)/2, or 36.65. We then find the values of $\bar{X}_i$ for the other four tibiae, add all five values together, and divide by $n = 5$ to obtain a VAV of 35.8 cm. Our relative TEM is then 0.245 cm (the TEM) divided by 35.8 cm (the VAV), and multiplied by 100, or 0.68 percent—well below the 5 percent threshold.

Metadata

Metadata refers to data that are recorded about other data. That is, metadata refers to variables that you collect that don't relate to the specific research question (e.g., are male skulls larger than female skulls?) but allow one to assess the quality of the data and their susceptibility to random and systematic error.

Consider a large multi-investigator study. You and three colleagues are going to be collecting data on femur length in the samples that are curated by each of your institutions. You decide over email which measurement definitions you will use. Each investigator will collect data with the equipment available at their university. You have standardized your data collection sheets so that each sheet has the line: *Name of the Investigator.* This *Name* is metadata

Table 2.4. Repeated measurements of maximum tibia length (cm)

Tibia	X_{1i}	X_{2i}	d_i	d_i^2	$\bar{X}_i$
1	36.5	36.8	-0.3	0.09	36.65
2	35.2	35.1	0.1	0.01	35.15
3	35.7	35.7	0.0	0.00	35.7
4	36.2	35.8	-0.4	0.16	36.0
5	35.6	35.4	0.2	0.04	35.5
				$\sum_1^5 d_i^2 = 0.30$	$\sum_1^5 \bar{X}_i = 179.0$

because it is a proxy for a whole host of factors that could affect the results of the study. For example, the *Name* variable tells you the university, which tells you the calipers used, the room the femora were measured in, the weather and temperature at that time of year, etc.

Metadata are not data that you collect that specifically address your research question. For example, if you are interested in creating a new method of age estimation based on the size of one's head, then neither *specimen age* nor *head size* would be considered metadata. These are both *primary variables* directly relevant to your research question. However, the calipers you used to measure head size, the time of day you measured the heads, and whether you had coffee for breakfast that day would all be considered metadata. In forensic anthropology, any basic data sheet should contain at least the following metadata variables: Recorder Name, Date, Caliper Type, and Caliper Serial Number.

LEARNING CHECK

Q15. Based on your understanding of error and how it is measured, what could X_{1i} and X_{2i} represent in the equations for TEM and VAV? Select all that apply. Hint: there are two correct responses.

A) X_{1i} and X_{2i} could reflect the measurement of the *same* specimen by the same observer.

B) X_{1i} and X_{2i} could reflect the measurement of the *same* specimen by different observers.

C) X_{1i} and X_{2i} could reflect the measurement of *different* specimens by the same observer.

D) X_{1i} and X_{2i} could reflect the measurement of *different* specimens by different observers.

Q16. You are interested in male/female differences in overall height and your hypothesis is that males are taller than females, on average. To test this hypothesis, you collect data on femur length using the equipment made available to you at the different museums that house the documented skeletal collections that you visited.

You run your analyses, but the results do not show a systematic difference in femur length in the male and female samples. You suspect there was a problem. Metadata can help ascertain what may have happened during the data collection process that impacted your results.

Select *three* metadata variables from this list that would help reduce error or help you identify error that may have crept into your analysis.

A) Date and time of day specimen was measured

B) Femur length

C) Specimen sex

D) Caliper serial number

E) Date of last calibration of calipers

End-of-Chapter Summary

This chapter discussed rules of evidence, how they have changed over time, and what the effect has been on the practice of forensic anthropology. The *Daubert* ruling occurred in 1993, which established a series of criteria that judges use in their gatekeeping capacity to determine what kinds of testimony are admissible in a court of law. The primary outcome of the

Daubert ruling for the forensic sciences was a greater concern with analytical rigor, creating methods that are repeatable and testable, and establishing error rates for the methodologies used in courtroom testimony. Forensic anthropologists reacted positively to the *Daubert* ruling and began research projects designed to improve the methods used with respect to potential *Daubert* challenges in the courtroom. Estimating known rates of error was a key consideration. This chapter introduced basic concepts that apply to any type of scientific study, not just forensic anthropology. The key concepts include the difference between accuracy and precision, the difference between random error and systematic error (bias), an understanding of inter-observer and intra-observer error and ways to reduce both, how to quantify error and assess whether the error rate is acceptable, and the concept of metadata.

End-of-Chapter Exercises

Exercise 1

Materials Required: A computer with an internet connection, **Figure 2.7,** and **Table 2.5.**

Scenario: You and a colleague are collecting data on the maximum width of male sacra (plural of sacrum). To evaluate your data collection protocol, you both decide to measure a series of specimens with 3D scans available online and compare the values that you get.

Directions: **Table 2.5** contains the specimens that you have chosen to measure as well as the values obtained by your colleague for those specimens. To collect your own measurements, begin by going to www.morphosource.org. In the search bar at the top of the page, enter the specimen number that you wish to measure (including the zeros) and click on "Go." On the next page click on the thumbnail of the scan to open it in the site's web viewer. You will be measuring the sacra following the protocol illustrated in **Figure 2.7A**.

To take a measurement, begin by orienting the scan so that it matches the picture in **Figure 2.7A**. The easiest way to do this is to open the "Tools" bar to the left of the view screen. Here, click on the button that says "Orbit" and it should change to "Rotate." Now you can move the scan by holding the left mouse button and moving your mouse. Once you have the scan in the right orientation, click on the "Nodes" button in the "Tools" bar. Then, switch the units to "millimeters" and click on the button to "Enable Nodes" turning it blue. Notice that now, when you hover the mouse over the scan, the icon is a pointing finger instead of an open hand. You can now collect your measurements. To do so, click on one end of the line that you wish to measure along. Then, holding the shift key, click on the other end of the line. This technique will result in a line of measurement being placed over the image of the scan on your screen. You can left-click the dots at the end of the line to reposition and adjust the measurement. Be sure to place the dots fully on the bone surface. Record the length of the line for this specimen and then follow the same procedure to measure the other specimens listed in **Table 2.5**.

Question 1: The data that you just collected can be considered which of Stevens' data types (see chapter 1)?

Question 2: How do the values that you obtained compare to those obtained by your colleague? Do these differences appear to be the result of random error or systematic error?

Question 3: After consulting with your colleague, you discover that they have collected their measurements following the protocol illustrated in **Figure 2.7B**. Based on this new information, do the differences between your measurements and their measurements represent random or systematic error?

Question 4: After publishing your data, you discover that Specimens 000360979 and

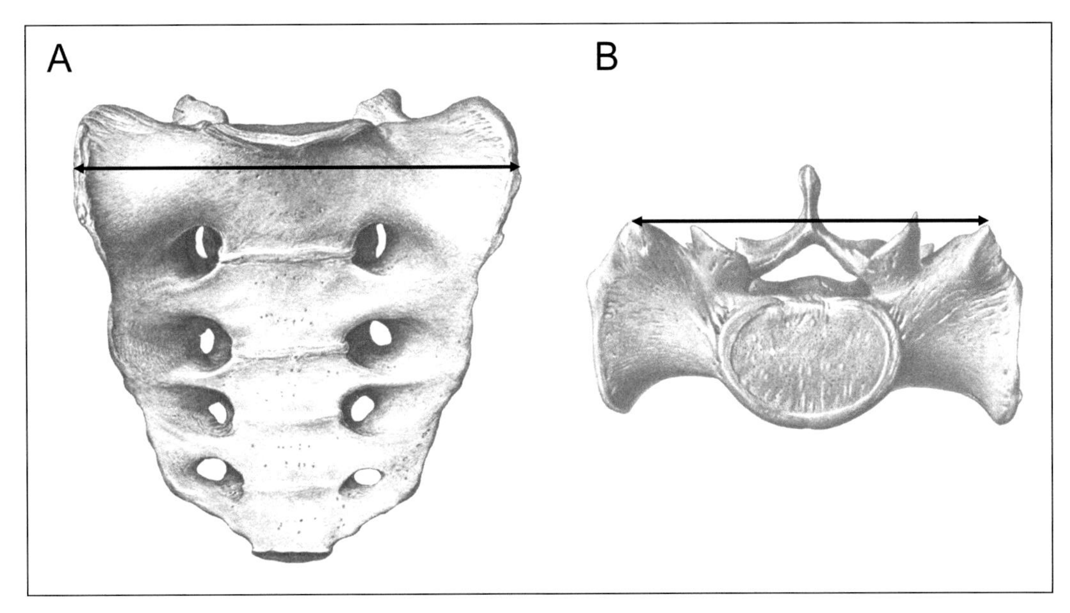

Figure 2.7. Two views of the human sacrum.

Table 2.5. Specimen list and colleague's measurements for a series of male sacra

Specimen	Colleague's Measurement (mm)
000360660	102.843
000360979	100.603
000361018	90.167
000361062	85.316
000361082	96.138
000361096	94.010
000361122	93.725
000361190	90.957
000361336	90.985

000361336 are actually female sacra. Based on the values that you recorded for these specimens, does their inclusion in your research sample seem to represent random or systematic error?

Exercise 2

Materials Required: **Table 2.2** and a computer with an internet connection.

Scenario: You have been asked to serve as an expert witness in a federal murder trial. Your testimony will be concerned with the timing of the fractures observed on a decedent's skeleton. The technique that will form the basis of your testimony was the product of your doctoral dissertation research. As part of your dissertation, you were able to standardize this technique and make it easy for other practitioners to reliably apply. Further, you subjected the technique to extensive testing and determined its error rate. You have not yet published your technique, however, and so other researchers have yet to validate your findings.

Question 1: Based on **Table 2.2,** how many of the *Daubert* criteria does the method that you developed meet?

Question 2: Based on this information, do you think your expert testimony will be admissible in this court? Why or why not?

Question 3: A few months later, you are asked to provide expert testimony on the timing of bone fractures for a murder trial in the state of Pennsylvania. Having still not published any of your dissertation research, you are curious about whether a testimony based on the technique you developed will be admissible in Pennsylvania. Use the internet to research Pennsylvania's laws concerning expert witness testimony. Based on your results, do you think a testimony based on your technique would be admissible in this court? Why or why not?

Exercise 3

Materials Required: **Table 2.6.**

Scenario: A colleague of yours is interested in whether the maximum length of the first metacarpal (the bone at the base of your thumb) can be used to estimate age-at-death. They have taken measurements on 20 metacarpals and asked you to measure the same specimens to see how consistent your measurements are with theirs. The values that each of you recorded are given in **Table 2.6**. Use this information and the equations for TEM, VAV, and relative TEM provided in this chapter to answer the following questions.

Question 1: By comparing these two sets of measurements, are you evaluating intra-observer error or inter-observer error?

Question 2: Based on these values, what is the TEM of maximum length of the first metacarpal?

Question 3: Based on these values, what is the relative TEM?

Question 4: Based on your answer to Question 3, is the maximum length of the first metacarpal being measured with sufficient precision for use in this research?

Table 2.6. Maximum lengths of 20 first metacarpals

Specimen	Your Measurement (mm)	Colleague's Measurement (mm)
1	47.71	41.50
2	45.58	48.27
3	53.63	47.48
4	40.17	50.42
5	47.49	46.80
6	48.28	45.48
7	43.91	49.09
8	46.05	46.08
9	47.50	45.37
10	41.38	52.73
11	43.42	53.96
12	55.79	47.73
13	51.03	49.15
14	53.35	49.32
15	48.25	50.53
16	43.00	45.98
17	44.23	46.82
18	51.58	43.47
19	45.48	45.55
20	47.03	50.35

References

Christensen AM. 2004. The impact of *Daubert:* implications for testimony and research in forensic anthropology (and the use of frontal sinuses in personal identification). *Journal of Forensic Sciences* 49: 427–430.

Christensen AM, Crowder CM. 2009. Evidentiary standards for forensic anthropology. *Journal of Forensic Sciences* 54: 1211–1216.

Grivas CR, Komar DA. 2008. *Kumho, Daubert,* and the nature of scientific inquiry: implications for forensic anthropology. *Journal of Forensic Sciences* 53: 771–776.

Harris EF. 2008. Statistical applications in dental anthropology. In: Irish JD, Nelson GC (Eds.), *Technique and Application in Dental Anthropology.* Cambridge: Cambridge University Press. p. 35–68.

Lesciotto KM. 2015. The impact of *Daubert* on the admissibility of forensic anthropology expert testimony. *Journal of Forensic Sciences* 60: 549–555.

Perini TA, Lameira de Oliveira G, dos Santos Ornellas J, Palha de Oliveira F. 2005. Technical error of measurement in anthropometry. *Revista Brasileira de Medicina do Esporte* 11: 86–90.

Sobotta J. 1909. *Atlas and Text-Book of Human Anatomy.* Philadelphia, PA: W.B. Saunders Co.

3

Basic Bone and Tooth Biology

Learning Goals

By the end of this chapter, the student will be able to:

- Define basic anatomical terminology and directional terms used in forensic anthropology.
- Characterize the molecular structure of bone.
- Differentiate types of bone at the level of gross observation.
- Describe the process of long bone growth, maintenance, and repair.
- Describe the different types of joints in the body and define their movements.
- Define basic terminology of the dental arcade.

Introduction

As a sub-specialization within the broader field of biological anthropology, forensic anthropology employs skeletal analysis for the purposes of aiding death investigations. As such, forensic anthropologists have extensive training in human anatomy, osteology, and odontology. *Osteology* is the study of the structure and form of the bony skeleton, which includes assessments of gross (observable with the naked eye) anatomy as well as different aspects of microscopic structure. *Odontology* is the study of the structure and form of teeth, or the dentition. *Morphology* is the study of the shape and appearance of the skeleton and dentition.

Forensic anthropologists primarily focus on human osteology and odontology; however, more general knowledge of skeletal variation in non-human animals is also important for determining whether a given skeletal element belongs to a human or not (and hence has the potential to be forensically relevant). Both osteology and odontology are branches of anatomy laden with technical language. It is important to understand the meaning of the terminology used throughout this book.

For this book's purposes, a brief overview is sufficient, with a focus on whole bone identification. Those serious about a career in forensic anthropology will need to take a semester-long class in human osteology, which most anthropology departments offer. There, you will focus on identifying small bone fragments, the nuances of how the side (left/right) is determined, and the details of anatomical variation that have been identified over several hundred years of comparative anatomical research. With enough skill and patience, the skilled osteologist can reconstruct the human skeleton even when highly fragmented.

Anatomical Terminology

Before discussing the specifics of the human skeleton, it is critical that a shared, basic vocabulary is established, which will allow the student to orient each skeletal element in the same manner and describe where on the bone a specific feature is located. For example, one might encounter a description of a gunshot wound located on the anterior aspect of the proximal humerus. More specifically, the gunshot may be located on the anterior aspect of the proximal metaphysis of the humerus, with a fracture radiating inferomedially toward the nutrient foramen. This is a lot to take in. By the end of this chapter, you will know exactly what is being described here.

Basic anatomical terminology includes both *directional terms* and *planes of reference*. You can read the definitions of these in **Tables 3.1** and **3.2**. It is important to consider **Standard Anatomical Position** (see **Figure 3.1A**), which is how the human body is oriented with respect to the directions and planes of reference. In Standard Anatomical Position, the body is standing with the feet pointing forward and planted on the ground, the head is facing forward, and the palms are facing forward. This last point may be different than one would

Table 3.1. Directional terms of the human body

Directional Terms (Synonyms)	Definition
Superior (Cranial/Cephalic)	Toward the head, or top, of the body
Inferior (Caudal)	Toward the bottom, away from the head, the opposite of superior. Caudal means toward the tail in quadrupeds.
Anterior (Ventral)	Toward the front. Ventral means toward the stomach in quadrupeds.
Posterior (Dorsal)	Toward the back, the opposite of anterior
Medial	Toward the midline of the body
Lateral	Away from the midline, the opposite of medial
Internal (deep)	Away from the surface of the body, inner
External (superficial)	Toward the surface of the body, outer
Palmar	Toward the palm of the hand. The opposite of palmar is dorsal.
Plantar	Toward the sole of the foot. The opposite of plantar is dorsal.
Proximal	Toward or near the axial skeleton, or main mass of the body
Distal	Away from the axial skeleton, or main mass of the body, the opposite of proximal

Table 3.2. Anatomical planes of the human body

Anatomical Planes	Definition
Sagittal (medial/midsagittal)	A plane through the middle of the body that divides it into left and right halves. These halves are symmetrical. The sagittal plane defines the midline of the body.
Coronal	A plane through the middle of the body that divides it into anterior and posterior halves. This plane is orthogonal (at a right angle) to the sagittal plane.
Transverse	A plane that is perpendicular to *both* the sagittal and coronal planes that is parallel to the ground surface.
Frankfort Horizontal	A plane that connects the lower margin of the orbit with the upper margin of the auditory canal in the dry skull. This plane is used for orienting the cranium for measuring.

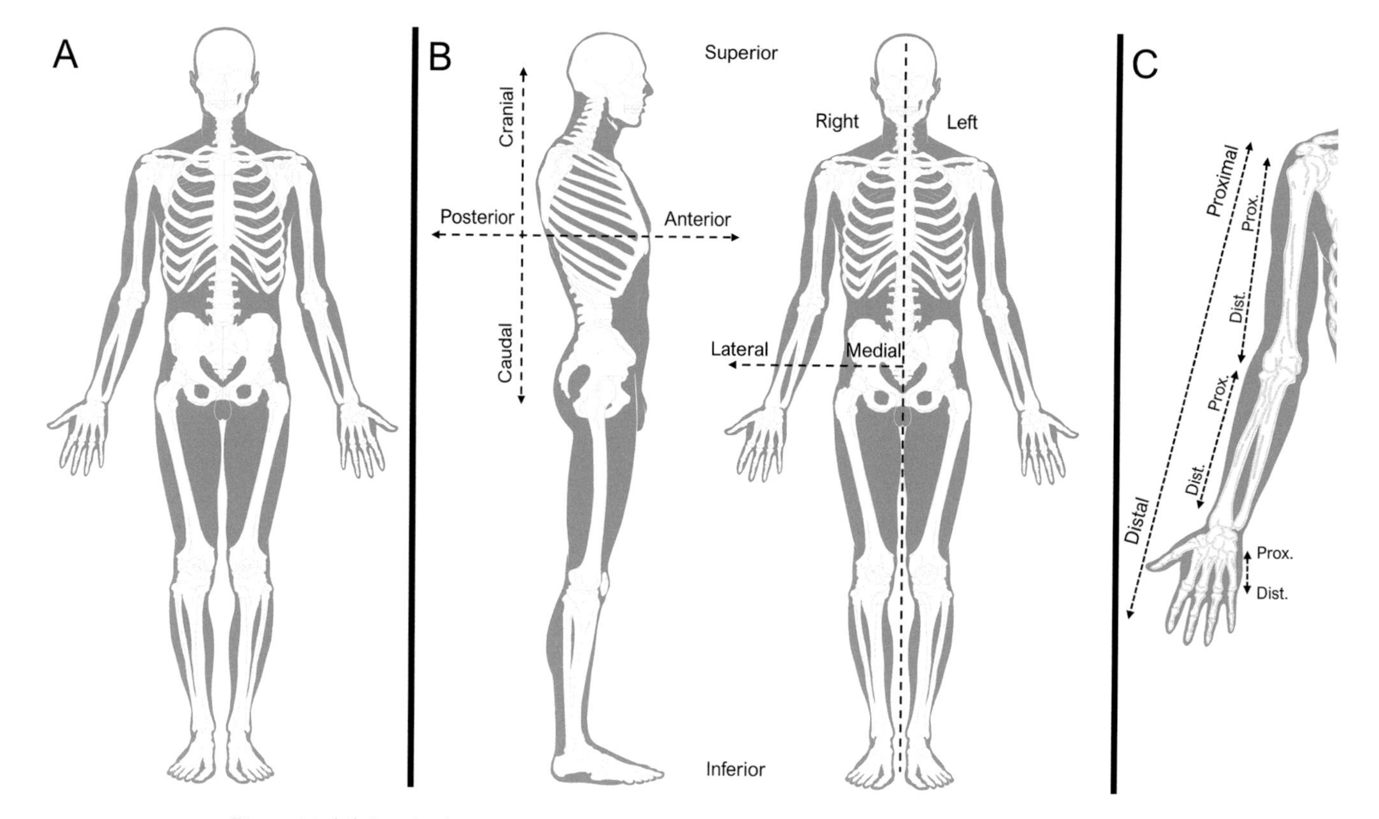

Figure 3.1. (*A*) Standard anatomical position. (*B*) Lateral and anterior views of the body with some directional terms superimposed. (*C*) Schematic of the upper limb showing different applications of the terms "proximal" and "distal."

expect. Because the bones of your forearm cross when standing with your hands at your sides, positioning them in Standard Anatomical Position with the palms turned forward and the thumbs facing out (laterally), your forearm bones will now be "uncrossed."

Most of these terms are straightforward because they also have a common meaning. However, *proximal* and *distal* do not (**Figure 3.1C**). Proximal refers to the aspect of the bone that is toward the articulation with the trunk, while distal means the opposite. Keep in mind that proximal and distal are relative terms. Each bone in the arm, for example, has a proximal half and a distal half. However, it is also the case that the radius is distal to the humerus because the humerus is closer to the shoulder articulation.

In addition to these directional terms, there are standard anatomical planes that are important to know. These are defined in **Table 3.2** and shown in **Figure 3.2**.

Features of Bone

There are numerous terms that describe specific aspects of skeletal anatomy that are generally applied to different parts of the skeleton. For example, there is terminology that refers to *positive* features of a bone. These terms define aspects of a bone that project above the bone surface. One might call these "bumps." However, in osteology you will see terms such as: eminence, boss, process, protuberance, tubercle, tuberosity, torus, trochanter, crest, or spine. Each of these has subtle differences in meaning, but what is relevant here is that these are all positive projections of bone.

To the contrary, osteology also has a specialized terminology that refers to features that are negative. For example, depressed areas can be called a fovea or fossa; grooved features are

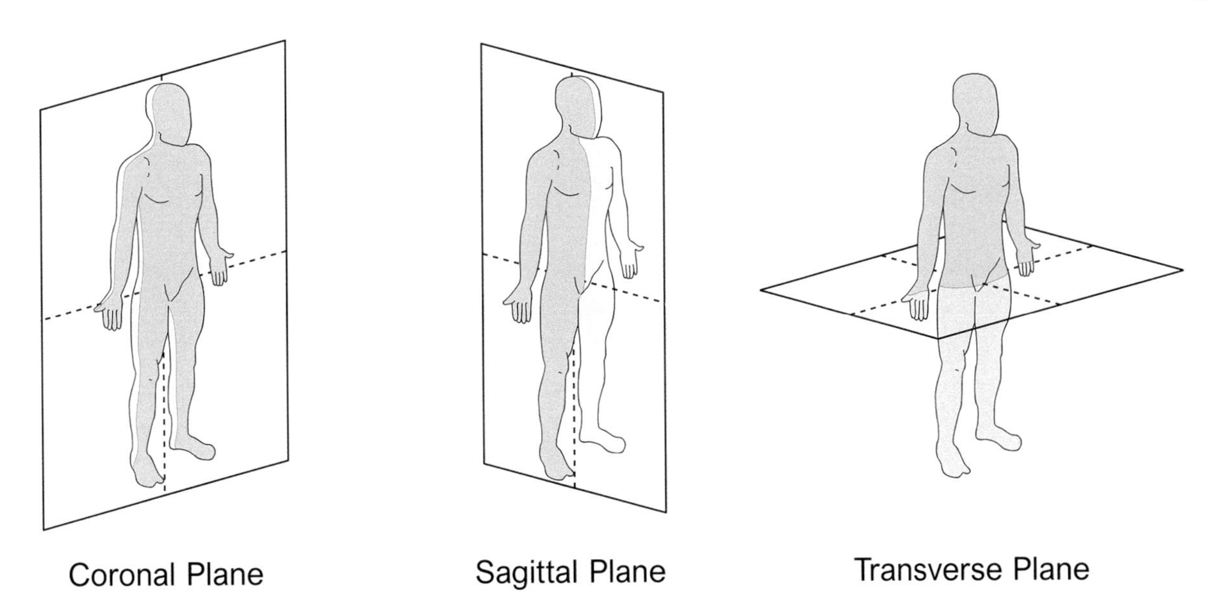

Figure 3.2. Anatomical planes of the human body.

called a sulcus; holes through a bone could be a foramen, fontanelle, aperture, or meatus. Of these, the most relevant terms for forensic anthropology are *meatus* and *foramen* as they are referred to throughout the book (**Figure 3.3**). The external auditory meatus is the hole for the ear, while the foramen magnum, located in the bottom of the cranium, is the hole that allows the spinal cord to connect to the brain. A fontanelle is a cartilage covered hole between the bones of the skull found in infants. Newborns have an anterior fontanelle often referred to as "the soft spot." For the interested reader, please consult the book *Human Osteology* by White, Black, and Folkens (2012) for formal definitions of these terms.

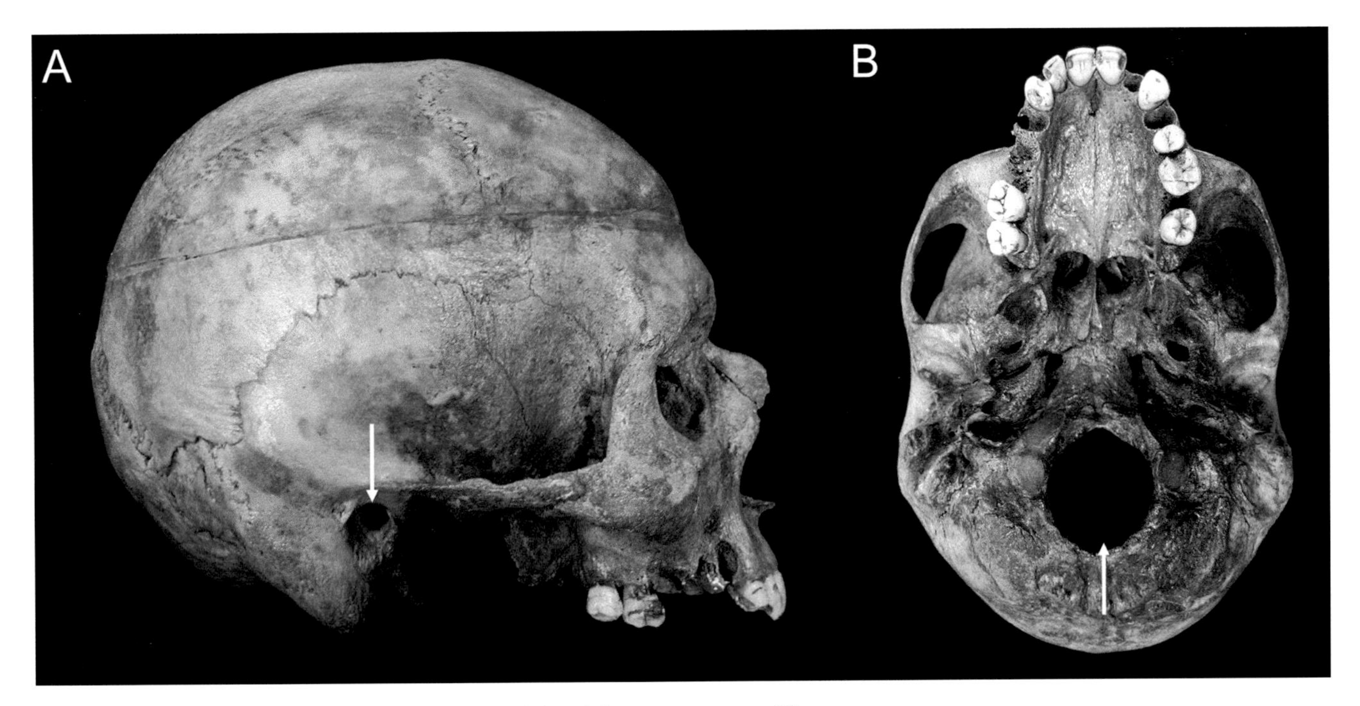

Figure 3.3. Location of the external auditory meatus (*A*) and foramen magnum (*B*).

Odontology

The terminology used to describe anatomical position within your dental arcade is different from that used to describe the skeleton. This is also the terminology your dentist uses when describing the location of cavities. These terms are summarized in **Table 3.3** and shown in **Figure 3.4**.

As noted in **Table 3.3,** the curvature of the dental arcade (it is parabolic in shape) means that different terminology is needed to describe directions in the oral cavity. For example, the labial surface of your front incisors is actually anterior with respect to the body and the lingual surface is posterior. However, the lingual surface of your molars is not posterior, but rather medial. To avoid confusion, the terms *mesial, distal, labial, lingual,* and *buccal* are used. These terms also define axes within the dental arcade. The mesiodistal axis is oriented from the midline (between your central incisors) toward the molars in the back of your mouth, but the axis *follows the curvature of the tooth row.* In other words, it is curved. The buccolingual axis is oriented from side to side, or from the tongue side of a tooth toward the cheek/lips side of a tooth.

Table 3.3. Directional terms of the dental arcade

Directional Terms	Definition
Mesial	Toward the midline when following the dental arcade
Distal	Opposite of mesial, away from the midline when following the dental arcade
Buccal	Toward the cheek when referring to the premolars (bicuspids) and molars
Lingual	Toward the tongue, the opposite of buccal for premolars and molars
Labial	Toward the lips, the opposite of lingual for canines and incisors
Interproximal	The contact points between adjacent teeth, where you floss
Occlusal	The chewing surface of a tooth, especially the premolars and molars
Incisal	The biting surface of a tooth, especially the canines and incisors

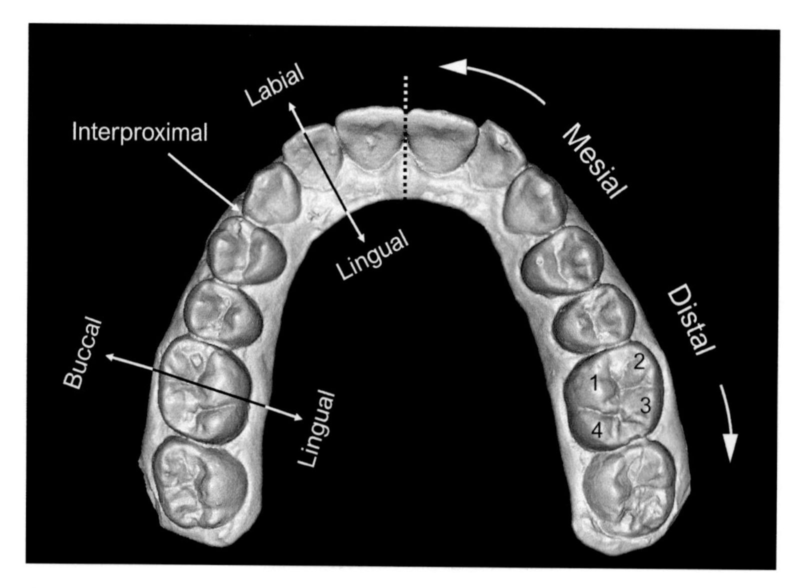

Figure 3.4. Directional terms of the dental arcade.

Note that the occlusal surface of a maxillary molar typically has four cusps, which are labeled in **Figure 3.4** as cusps 1, 2, 3, and 4 (there is a reason the numbering is clockwise, but details will not be discussed here). Cusp 1 is the mesiolingual cusp. Cusp 2 is mesiobuccal, Cusp 3 is distobuccal, and Cusp 4 is distolingual. You bite an apple with the incisal edge of your incisors and chew that apple using the occlusal surfaces of your molars. To remove that piece of apple skin that always gets trapped between your teeth, you floss interproximally. You can see that with very few terms one can combine them to refer to very specific locations within the oral cavity. This becomes useful when using dental records for positively identifying an individual (chapter 11).

LEARNING CHECK

Q1. In **Figure 3.5,** line "a" is _______ relative to the skull.
   A) Lateral
   B) Medial
   C) Superior
   D) Proximal

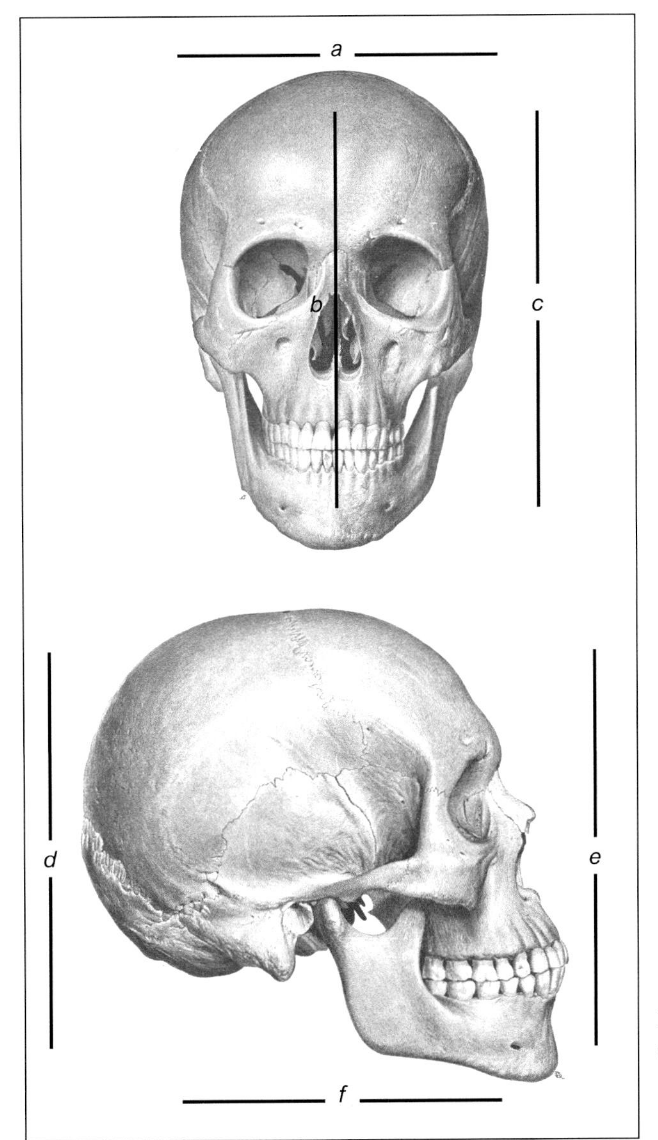

Figure 3.5. Two views of the human skull.

Q2. In **Figure 3.5,** line "b" is _______ relative to the eye orbits.
A) Lateral
B) Medial
C) Superior
D) Proximal

Q3. In **Figure 3.5,** line "c" is _______ relative to the skull.
A) Inferior
B) Medial
C) Superior
D) Lateral

Q4. In **Figure 3.5,** line "d" is _______ relative to the skull.
A) Posterior
B) Medial
C) Superior
D) Lateral

Q5. In **Figure 3.5,** line "e" is _______ relative to the skull.
A) Inferior
B) Proximal
C) Superior
D) Anterior

Q6. In **Figure 3.5,** line "f" is _______ relative to the skull.
A) Inferior
B) Anterior
C) Superior
D) Posterior

Q7. **Figure 3.6** presents a radiograph of a left arm. Both the radius and the ulna are fractured, with the location of the breaks shown by the white arrows. Note that the trunk of the body is to the left of the image—the humerus articulates with the scapula (not pictured) to form the shoulder joint. Based on this information, the radius is _______ to the humerus.
A) Proximal
B) Distal

Q8. In **Figure 3.6,** there is a fracture of the _______ ulna.
A) Proximal
B) Distal

Q9. In **Figure 3.6,** the ulna is _______ to the carpals.
A) Proximal
B) Distal

Q10. The fracture of the radius is _______ to the fracture of the ulna.
A) Proximal
B) Distal

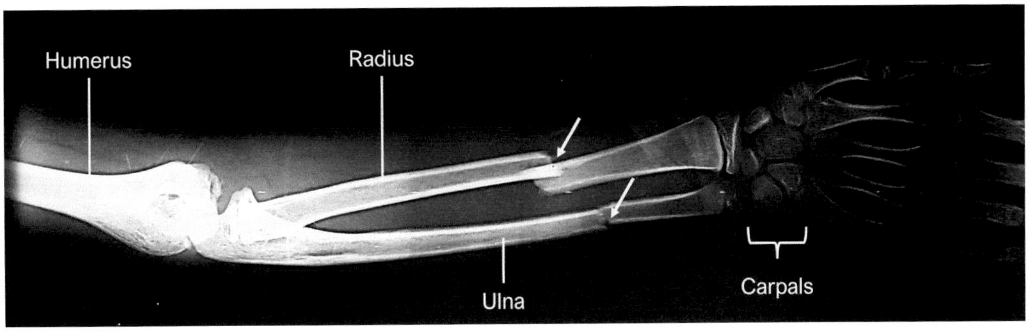

Figure 3.6. Radiograph of the lower arm demonstrating two fractures.

Basic Bone Biology

As shown in **Figure 3.7,** bones vary greatly in shape. There are flat bones, short bones, long bones, and irregular bones that defy easy description. Some individuals have extra bones, often in their hands or feet. When these occur in a tendon, they are called *sesamoids*. Some individuals also have extra bones in the joints of their skull. These are called sutural bones, *ossicles,* or Wormian bones. This variation will be further discussed in the next chapter. What is important here is that you recognize that these bones vary greatly in outward appearance but exhibit tremendous similarity when you consider them at different scales.

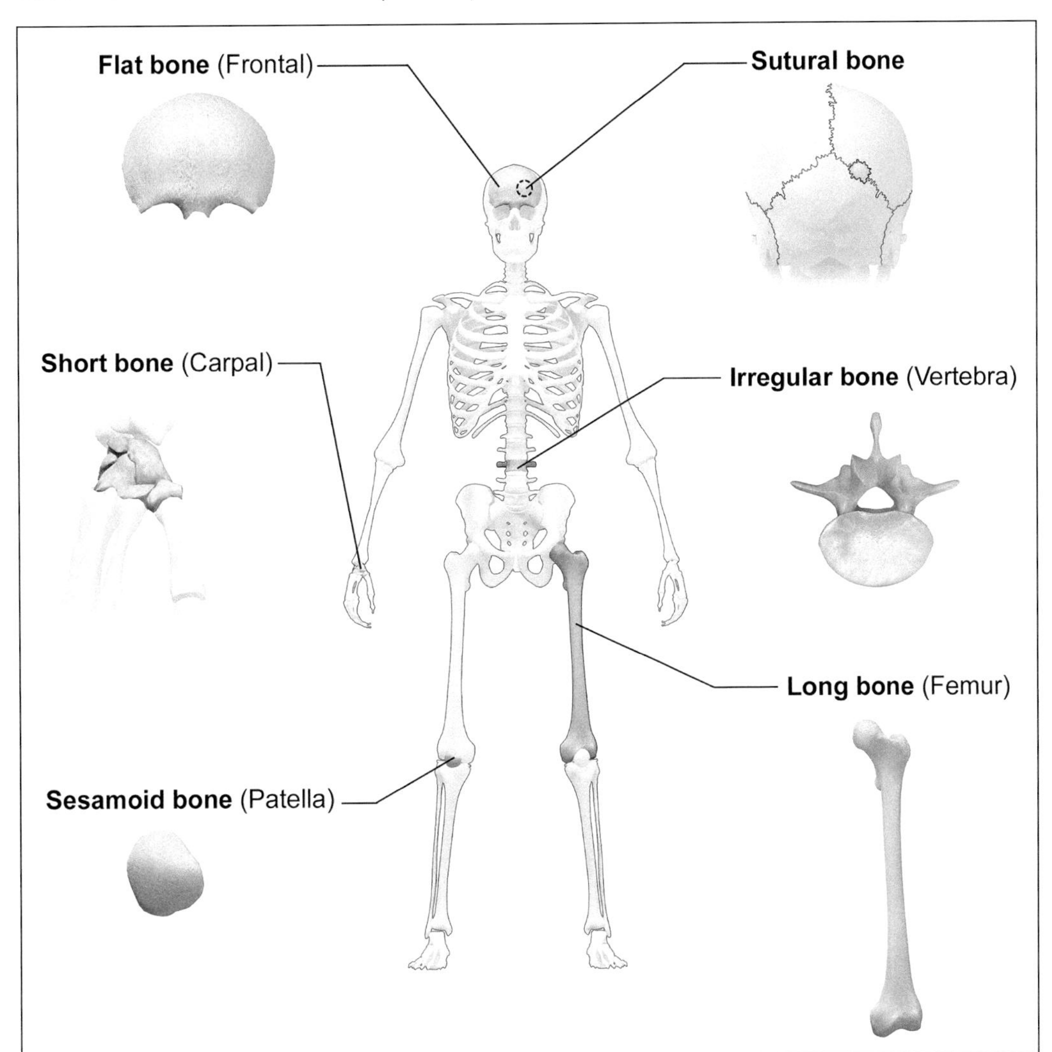

Figure 3.7. Classification of bones according to gross shape.

Molecular Structure of Bone—Collagen and Hydroxyapatite

Bone is a remarkable biological tissue. It is tremendously strong yet also very light given the amount of structural support it provides. This is because bone is a composite tissue that includes both an organic and inorganic (mineral) component. At the molecular level, all bone is composed of two components: collagen and hydroxyapatite. It is critical for our survival that both components are present, because they work in a synergistic fashion to provide both flexibility as well as rigidity to the skeleton. *Collagen* is the primary protein component of bone that provides flexibility. It exists in the form of woven fibers of protein. This matrix is interspersed with *hydroxyapatite*, a calcium phosphate mineral that provides rigidity to bone. It is the combination of both collagen and hydroxyapatite that gives bone its structural properties. Without collagen, bones would be extremely brittle and shatter. Without hydroxyapatite, your bones would have the texture of hard rubber and provide little structural support for your muscles or protection for your vital organs.

Gross Anatomy—Compact and Trabecular Bone

In addition to being similar at the molecular level, all bones in the human skeleton are similar at the level of gross observation (gross means visible to the naked eye without magnification). That is, regardless of external differences in appearance, the bones of the skeleton consist of two different types of bone visible without magnification (**Figure 3.8A**).

The first is called *compact bone* (also known as *cortical bone*), which as the name implies is the smooth, dense bone found on the external surfaces of bones and lining the interior surface of the *medullary cavity* of the long bones. The medullary cavity is the hollow space within long bones where yellow marrow (a fatty reserve located exclusively in the long bones) is stored. Cortical bone that is located on the surfaces of joints is smoother than regular cortical bone and is covered with a thin layer of cartilage (articular cartilage) to protect the joints during movement. This type of compact bone is called *subchondral bone* (subchondral means "below cartilage" in Latin).

The second type of bone is called *spongy bone,* or *trabecular bone,* or *cancellous bone.* Although identical to compact bone at the molecular level (both being composed of collagen and hydroxyapatite), trabecular bone is much more *porous* than compact bone. Trabecular bone is found inside bones and not on the external surface. Its purpose is to provide lightweight structural support. In the flat bones of the skull, this spongy bone is called *diploë.* Osteoporosis is a thinning of the *trabeculae* (spicules) that form the spongy bone in your skeleton. Compression fractures can result if this thinning occurs in the spine. A thinning of the trabeculae in the neck of the femur can result in sudden breakage, often assessed as a broken hip in elderly individuals.

Figure 3.8B presents a sectioned proximal femur showing the location of subchondral bone (dotted white line), cortical bone (arrows), and the internal structure of the neck and head consisting of trabeculae. The lattice-like structure (**Figure 3.8C**) provides support to the bone while minimizing weight.

Figure 3.9A shows a section of flat cranial bone with spongy bone (diploë) sandwiched in between two layers of compact bone. A close-up of the diploë (white arrows) surrounded by compact bone (black arrows) is provided in **Figure 3.9B**, associated with a fatal gunshot wound.

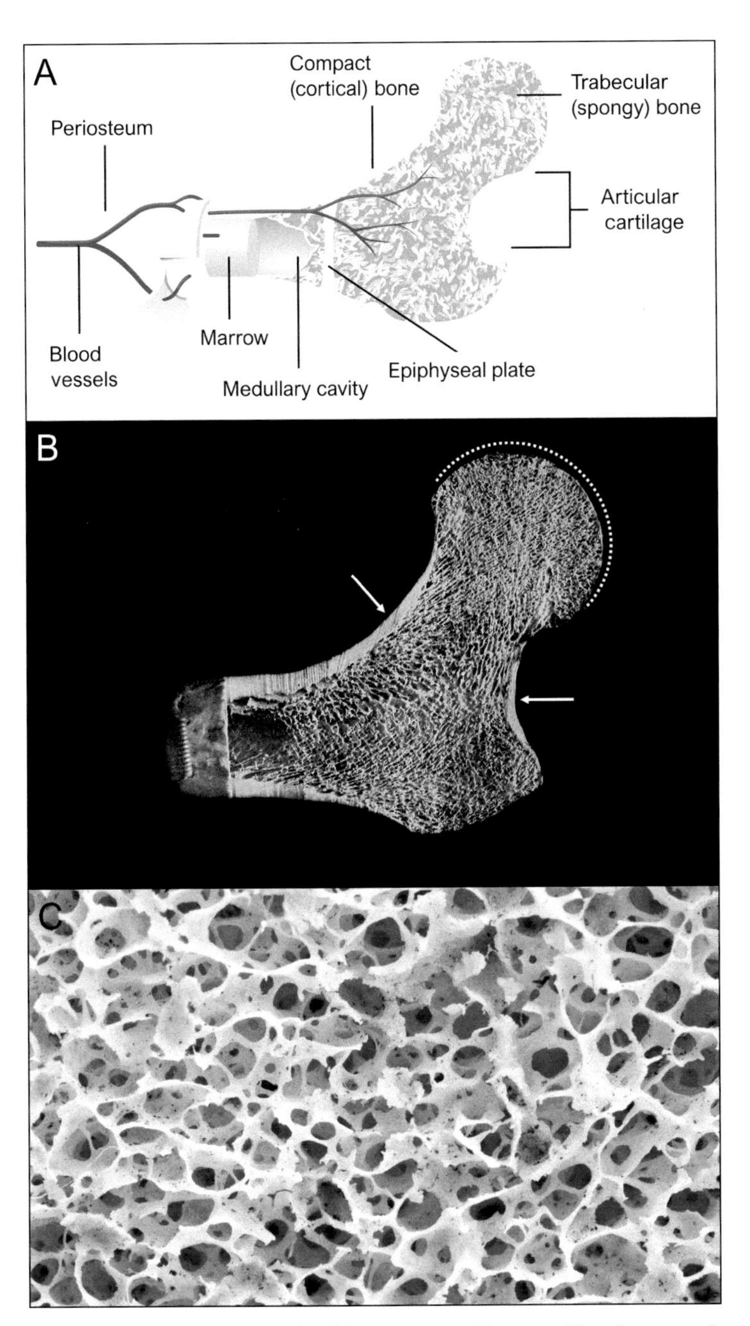

Figure 3.8. (*A*) Locations of the different types of bone and key features of a long bone. (*B*) Location of subchondral (dotted lines) and compact bone (arrows) in a proximal femur. (*C*) Close-up of trabecular bone.

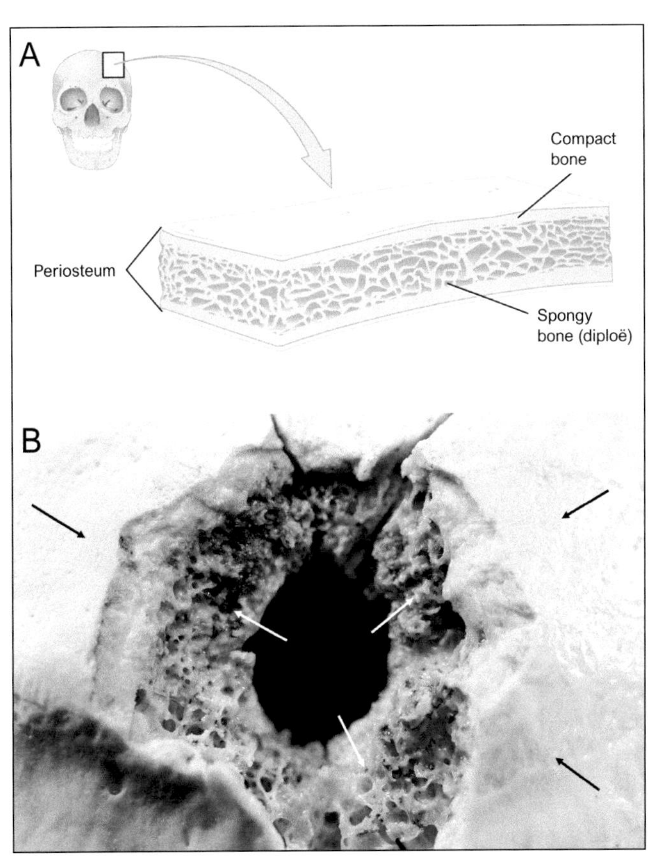

Figure 3.9. (*A*) Flat bones and diploë. (*B*) Compact bone (black arrows) and diploë (white arrows) in a gunshot wound.

The Structure of Long Bones

Long bones are some of the most recognizable in the human body and provide an excellent model for conceptualizing gross skeletal anatomy. **Figure 3.10** summarizes much of this anatomy. Spongy bone, compact bone, the medullary cavity, and the articular cartilage have already been defined. Subchondral bone is located under the articular cartilage (on both the proximal and distal joint surfaces). This figure also demonstrates that, in addition to providing structural support, the spongy bone is where the red marrow is located. Red marrow is responsible for the production of blood cells.

The left side of **Figure 3.10** labels the gross anatomical components of a long bone (*epiphysis, metaphysis, diaphysis*). During early skeletal development, a cartilage precursor forms for

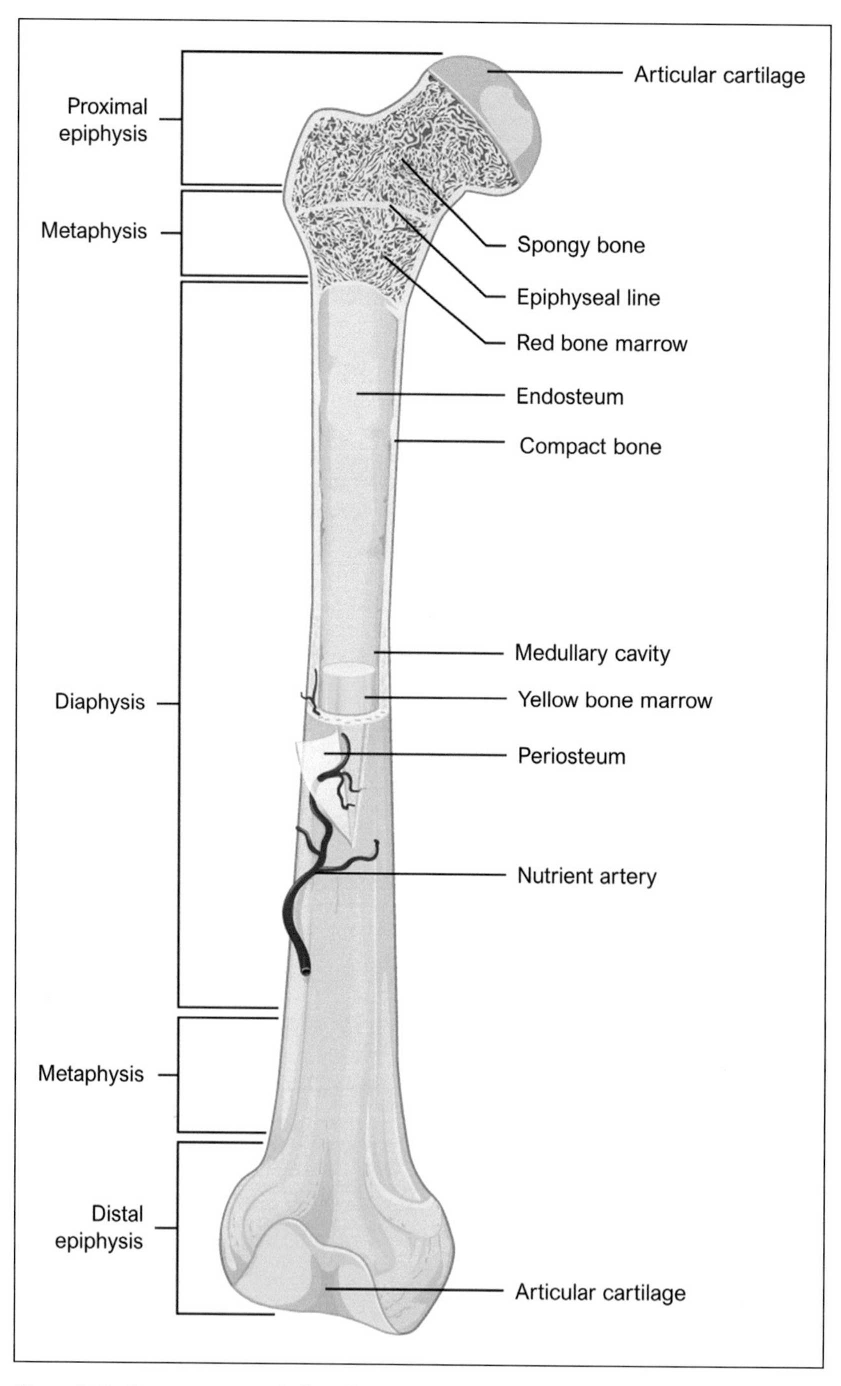

Figure 3.10. Gross anatomy of a long bone.

each long bone. This center of bone development is penetrated by blood vessels that supply nutrients, and the initial site of penetration becomes the *nutrient foramen*. This foramen connects the yellow marrow to the blood supply of the body. The site of penetration is indicated in **Figure 3.10** by the nutrient artery. The cartilage precursor is then replaced with bone, a process that is called *ossification* (to make into bone).

In a long bone, the primary center of ossification is called the *diaphysis,* which forms its shaft. However, all long bones form from multiple distinct centers of ossification. These secondary centers of ossification are called *epiphyses,* which increase in size until they eventually fuse to the diaphysis during epiphyseal union. Epiphyses form the proximal and distal joint surfaces. This is the subchondral bone that was previously described. The *metaphysis* is the segment of the long bone that flares and expands in width to accommodate the growing epiphyses. While the diaphysis and epiphyses are well-defined anatomical structures, and distinct from each other during development, the metaphysis is less clearly demarcated on the adult long bone.

Figure 3.11A shows the humerus of a teenager prior to the fusion of the diaphysis and proximal epiphysis. Note the difference in color of the two pieces of bone. The epiphysis is made of subchondral bone, with exposed trabeculae visible in both images (see, in particular, the small section of damage on the left side of **Figure 3.11A**). The thinner subchondral bone is damaged more easily and also responds to soil conditions differently than compact bone as a result of this difference in density. Epiphyses can be difficult to identify in isolation; however, a skilled osteologist will be able to do so once a certain stage of maturity is reached (**Figure 3.11B**). Other bones also form from distinct primary growth centers. For example, the adult os coxa forms from three distinct primary ossification centers (the ilium, ischium, and pubis) in addition to a number of secondary ossification centers (epiphyses). In

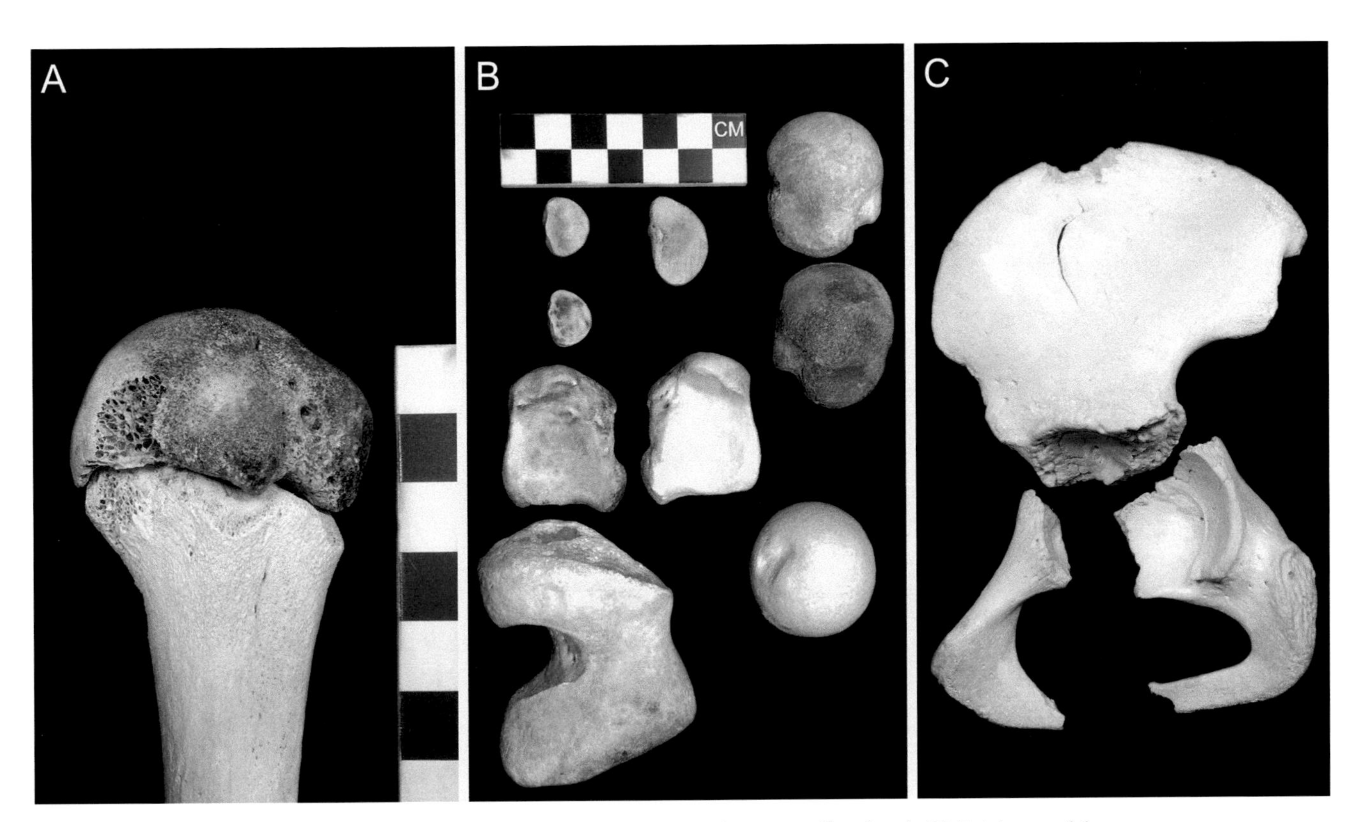

Figure 3.11. (*A*) Humerus of a growing individual with the proximal growth center still unfused. (*B*) Epiphyses of the upper and lower limbs. (*C*) Primary growth centers of the os coxa.

Figure 3.11C, these three ossification centers have yet to fuse; the hip joint (acetabulum) will be formed in the middle once growth has completed.

Thus far, all but two terms labeled in **Figure 3.10** have been discussed: *periosteum* and *endosteum.* In life, bones are covered by a fibrous connective tissue called the periosteum. The periosteum is home to a type of bone cell called an *osteoblast,* whose primary function is to make new bone. On the inner surface of the medullary cavity is another layer of connective tissue called the endosteum. The endosteum is home to a type of bone cell called an *osteoclast,* whose primary function is to remove bone. In adults, both types of bone cells are distributed throughout the element because both are critical to the general maintenance of the skeletal system.

Long Bone Growth

How does a long bone begin as an uncalcified collagen precursor and end as a complete long bone with multiple epiphyses fused to the diaphysis? This process is called *endochondral ossification* (**Figure 3.12**). The collagen precursor begins the process of ossification when hydroxyapatite crystals are deposited within the organic matrix of collagen fibers. Long bones grow in two directions. They increase in length through *longitudinal growth*, and they increase in diameter through *appositional growth.*

Longitudinal growth is easy to conceptualize. Bones increase in length on both ends as osteoblasts produce more bone at the ends of the diaphysis. Epiphyses increase in size as osteoblasts produce more bone in these secondary ossification centers. Nourishment for this process is provided by blood vessels that penetrate the diaphysis at the nutrient foramen and penetrate each of the growing epiphyses (see **Figure 3.12**).

Appositional growth is more complicated. For a bone to increase in diameter, new bone must be added to the outside surface of the bone. However, this process alone would produce a solid bone with no medullary cavity, and the bone would be too heavy. Therefore, during appositional growth, osteoblasts located underneath the periosteum (that covers the external

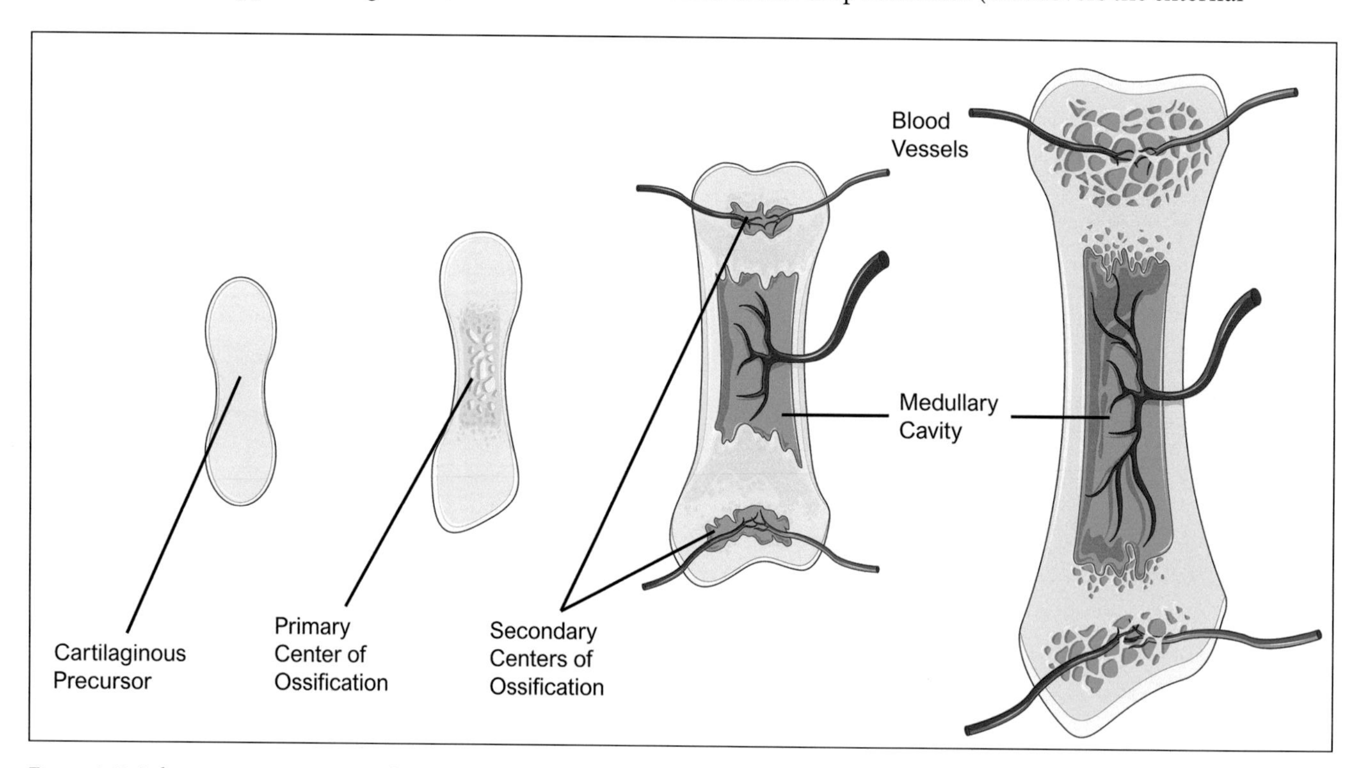

Figure 3.12. Schematic representation of the process of long bone ossification.

surface of the bone) produce new bone in a series of layers. These layers are called *lamellae*. At the same time, the medullary cavity is formed by the activity of osteoclasts (cells that remove bone) located in the endosteum. Thus, the long bone can expand in outside diameter without increasing significantly in weight. This process is guided by a series of genes responsible for new cell formation, bone production, bone removal, and overall regulation.

Bone Maintenance and Repair

Once bone growth has completed, osteoblasts and osteoclasts continue to play a role in the maintenance of the skeletal system. For example, if a bone is fractured, within a few days osteoblasts will produce new bone to heal the site of the fracture. Initially, a *hematoma* forms; this is coagulated blood that attempts to stop any internal bleeding by closing off capillaries. The swelling and bruising one experiences with a fracture are a result of hematoma formation.

Within a few days, a very primitive form of bone, called *woven bone* or *immature bone,* is laid down. The purpose of woven bone is to quickly stabilize the fracture through the formation of a primary fracture *callus*. Woven bone is structurally weak and poorly organized (see **Figure 3.13**). It is only present in the adult skeleton associated with fracture repair or certain diseases. The primary benefit of woven bone is that it can be laid down quickly.

Because woven bone is structurally weak, after several weeks it is replaced with *mature* or *lamellar* bone. The woven bone is resorbed (removed) as mature, compact bone is laid down. This process of iterative addition and removal of bone is called *remodeling,* the goal being to restore the fractured bone to its original state. It is for this reason that bone fractures are "set" at the hospital; the purpose is to align the original contours of the fractured bone edges to aid in the healing process.

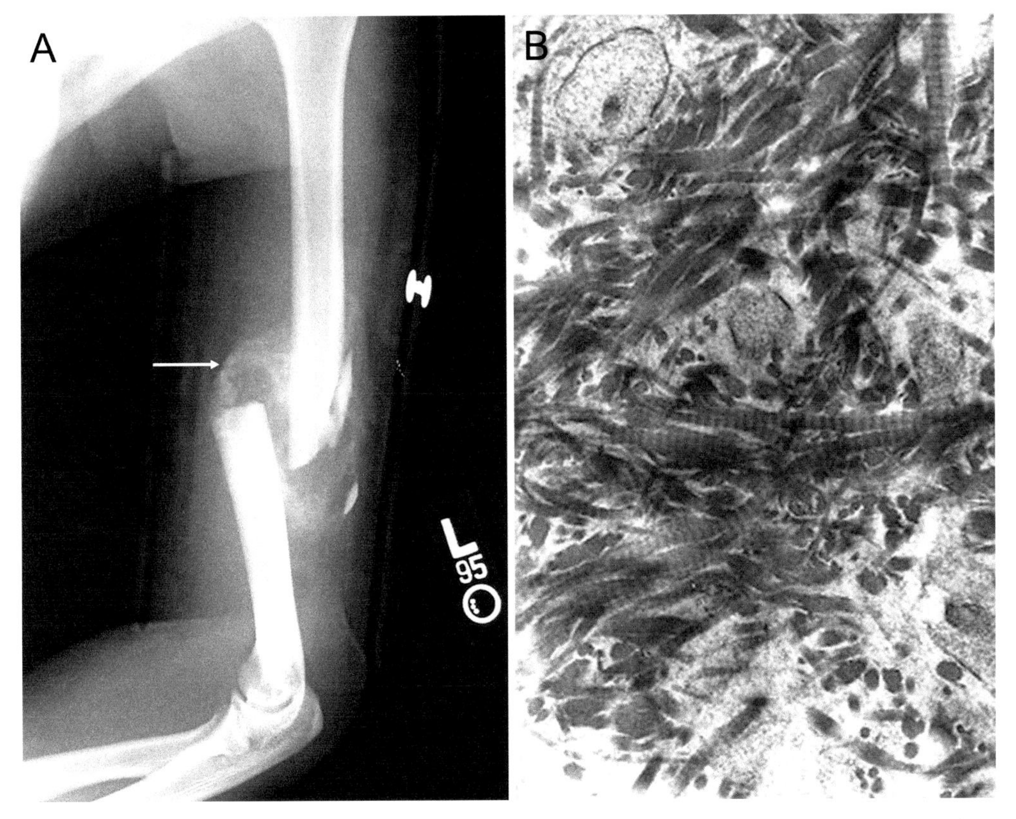

Figure 3.13. (*A*) Fractured humerus with initial callus indicated by arrow. (*B*) Microscopic structure of woven bone, note the disorganized appearance.

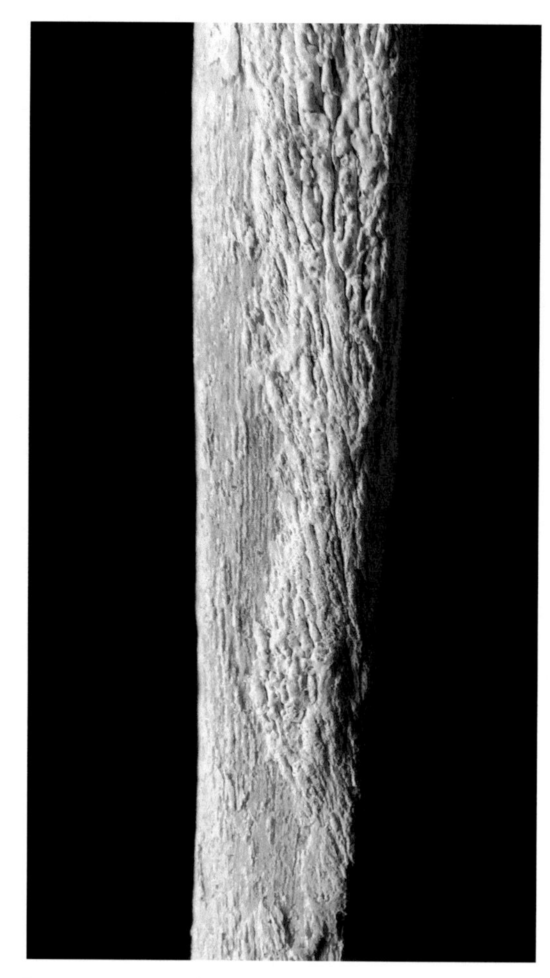

Figure 3.14. Woven bone reflecting a pathological response on a long bone.

Osteoblasts and osteoclasts are also important for more routine maintenance of the skeleton. Even minor injuries to the periosteum (such as being kicked in the shin) can produce a bony response that results in new bone being laid down on the surface (periostitis) (**Figure 3.14**). In addition, the body will remove bone where it is not needed and add new bone where it is needed in response to functional stresses. This is called **Wolff's Law**. Colloquially one might say, "use it or lose it." For example, it is well known that certain types of exercise can change the shape of your long bones. This is in response to the physics of stress being applied to the bone surfaces, which must be shaped in a certain way to withstand injury.

Bone Histology

Histology is the microscopic study of biological tissues with a primary focus on cells, their structure and function. The two types of bone at the microscopic level have already been described: 1) immature or woven bone, and 2) mature or lamellar bone. As noted above, the entire adult skeleton is composed of mature or lamellar bone, which is more highly structured and organized than woven bone. *Note that cortical and trabecular bone are both made of mature bone.* The properties of these different classifications of bone are presented in **Table 3.4**.

Both cortical and trabecular bone are lamellar bone. Bone cells located within the trabecular bone can be supplied with nutrients from the surrounding red marrow. This is because

Table 3.4. Gross differences in bone classification

Woven Bone	Lamellar Bone
Disorganized	Organized (lamellae)
Laid down quickly	Laid down slowly
Structurally weak	Structurally stronger than woven bone
Found near fractures and with certain injuries or diseases	Characterizes the adult skeleton
	Includes both *Cortical* and *Trabecular* bone

Cortical/Compact Bone	Spongy/Trabecular Bone
Dense and compact	Porous
Defined by lamellar appearance	Defined by trabeculae spicules
Found on the external surfaces of bones	Found inside the long bones and flat bones of the skull and pelvis
Osteons	No osteons

trabecular bone is so porous (see **Figure 3.8** above). However, the lamellar bone that constitutes cortical bone is so densely packed that it cannot easily receive the nutrients it needs to survive without a more structured delivery system. The system for delivering blood and nutrients to dense, compact bone is called a *Haversian system.* The key feature of the system is the *osteon,* which is a circular structure of concentric or circumferential lamellae of bone surrounding the nerve and blood supply that runs through the middle.

These osteons are depicted in **Figure 3.15A**. Note the outer surface is located to the left where the *periosteum* is indicated. The right side of the image represents the *medullary cavity,* with *trabecular* bone sandwiched in between the *medullary cavity* and the *compact* bone that

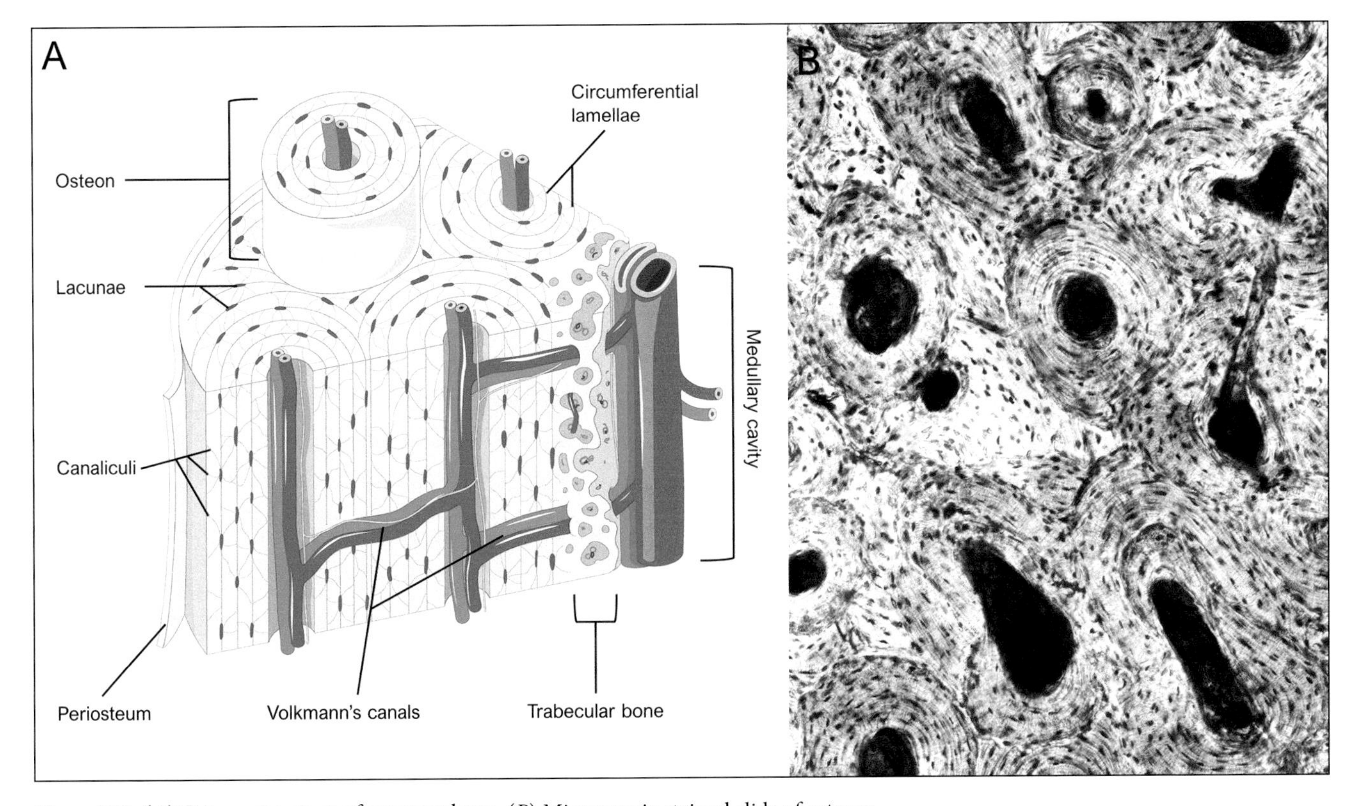

Figure 3.15. (*A*) Osteon structure of compact bone. (*B*) Microscopic stained slide of osteons.

forms the outer bone surface. The tree ring–like appearance of the *circumferential lamellae* is clearly seen in this image. Multiple *osteons* are depicted here (each circular structure surrounding a blood vessel), and the nerve and blood supplies of each are linked by a series of canals that are called *Volkmann's canals.* Two of these are depicted on the right side of the diagram. The *lacunae* are cavities in which the living bone cells reside. The blood is transported from the central vessel in each osteon to the cells in the lacunae through a series of smaller channels called *canaliculi.*

When compact bone from the shaft of a long bone is sectioned and observed under the microscope, it has the appearance of a series of tree trunks (**Figure 3.15B**). The ring-like structures that surround the dark red central canal (where the blood supply is located) are circumferential lamellae. It is important to note that the collagen fibers in the different lamellae are oriented in different directions, which adds strength to the bone. The small red specks are lacunae where the bone cells (osteocytes) are located. Note that as bone remodels, the osteons are replaced. This leaves behind fragmentary, older osteons that are partially intercut by newer osteons. Remnants of remodeled osteons are also visible in **Figure 3.15B**.

The Function of Bone in the Body

One of the primary purposes of the skeleton is to provide attachment points for muscles, which contract to control movement of the body segments to perform life's functions. Although muscle and bone are functionally linked, muscles do not attach directly to the surface of a bone. Instead, *tendons* serve this purpose. These areas of tendon attachment on the outer bone surfaces are often rougher than the surrounding bone surface and can easily be identified as areas of muscle attachment. The *origin* of the muscle is the attachment site that is usually closest to the trunk of the body; it moves less during muscle contraction. The *insertion* of the muscle is the location of muscle attachment where there is more movement during muscle contraction.

Where any two bones meet in the human body, a joint is formed. Another word for a joint is *articulation,* and any two bones that meet are said to articulate. These articulations are often covered by cartilage that helps buffer the stresses related to movement. There is more cartilage in the skeletons of infants, which then gets replaced with bone as one grows and matures. Collagen is a flexible tissue that responds well to compressive forces. *Ligaments* are another type of connective tissue that serve to secure a joint and restrict its movement. The way in which bones move at a joint depends on the muscle origin and insertion locations, the shape of bone at the joint, and the ligaments that surround that joint.

The two basic types of movements caused by muscles contracting are flexion and extension. *Flexion* reduces the angle between two bones. When the bicep flexes, the angle between the upper and lower arm decreases. *Extension,* in contrast, straightens the arm, which opens or increases the angle between the bones of the upper and lower arm. Other muscle movements are presented in **Table 3.5** and shown in **Figure 3.16.** Movement outside the range of tolerance can result in injury.

Types of Joints

When you hear the word "joint," you likely think of the major joints of the limbs: the shoulder, elbow, or knee. However, there are several types of joints that vary in the way the bones articulate and the soft tissue connections in and around the articulations.

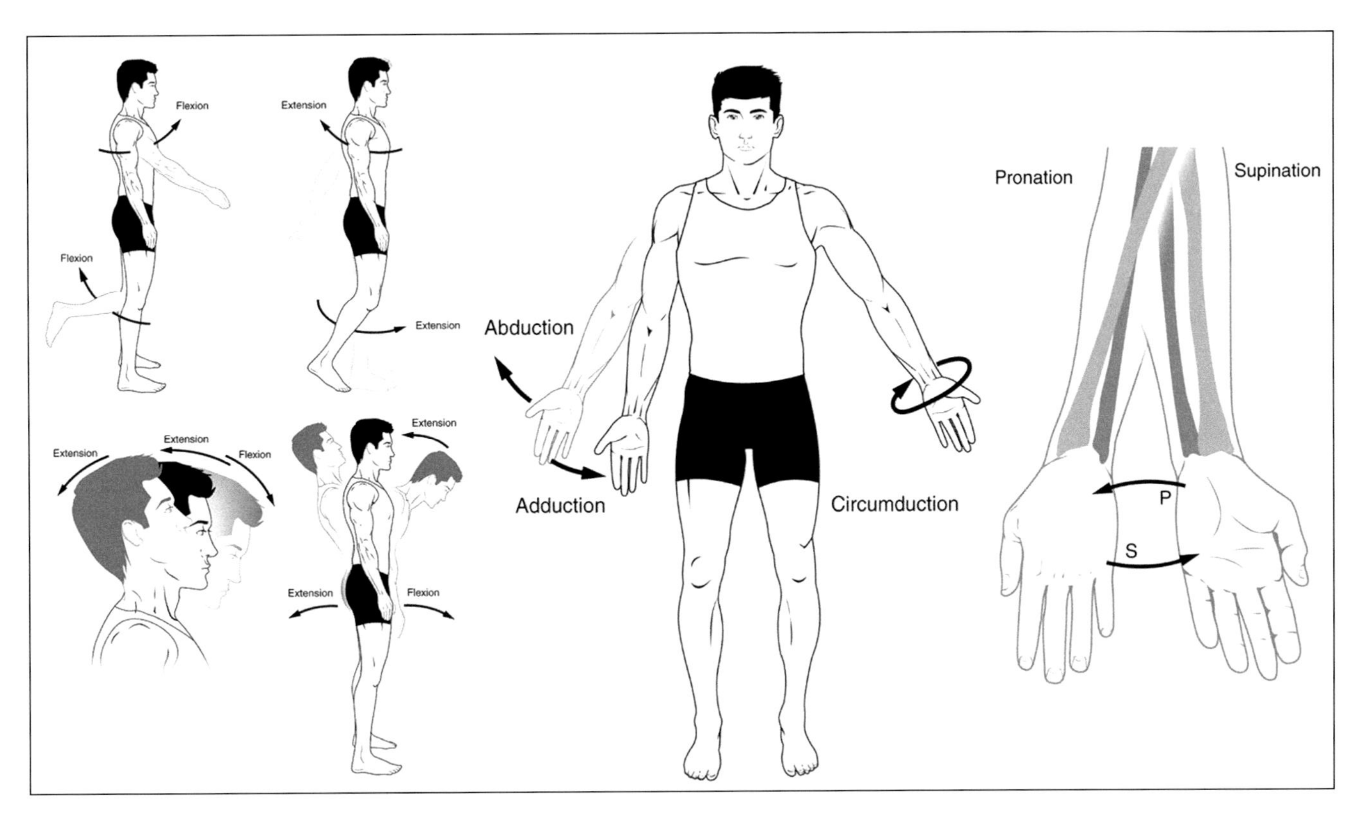

Figure 3.16. Types of movement at the joints.

Table 3.5. Movement of the body via muscle action

Term	Movement
Flexion	Muscle movement that results in a decrease in the angle between bones
Extension	Muscle movement that results in an increase in the angle between bones, the opposite of flexion
Abduction	Muscle movement that results in a bone moving away from the midline
Adduction	Muscle movement that results in a bone moving toward the midline, the opposite of abduction
Circumduction	Muscle movement that combines all four of the above resulting in the body part tracing a cone in its movement
Pronation	Rotation of the forearm from palm forward to palm facing backward; that is the hand is moved from Standard Anatomical Position (thumb is lateral) to the thumb being medial
Supination	The opposite of pronation

Synovial Joints

The most common type of joint in the human skeleton is a synovial joint. The shoulder, elbow, and hip are all synovial joints and allow maximum movement between the skeletal elements that form the joint capsule. The surface of the bones that comprise a synovial joint is covered with cartilage to ease movement, and the entire joint is encapsulated and filled with a lubricating fluid. The cartilage covering the bone surfaces, combined with the synovial fluid, serves to lubricate the joint capsule and reduce friction (and damage) while moving. Because synovial joints are so common and varied in the human skeleton, they are further classified according to their shape, which determines the type of movement allowed at that

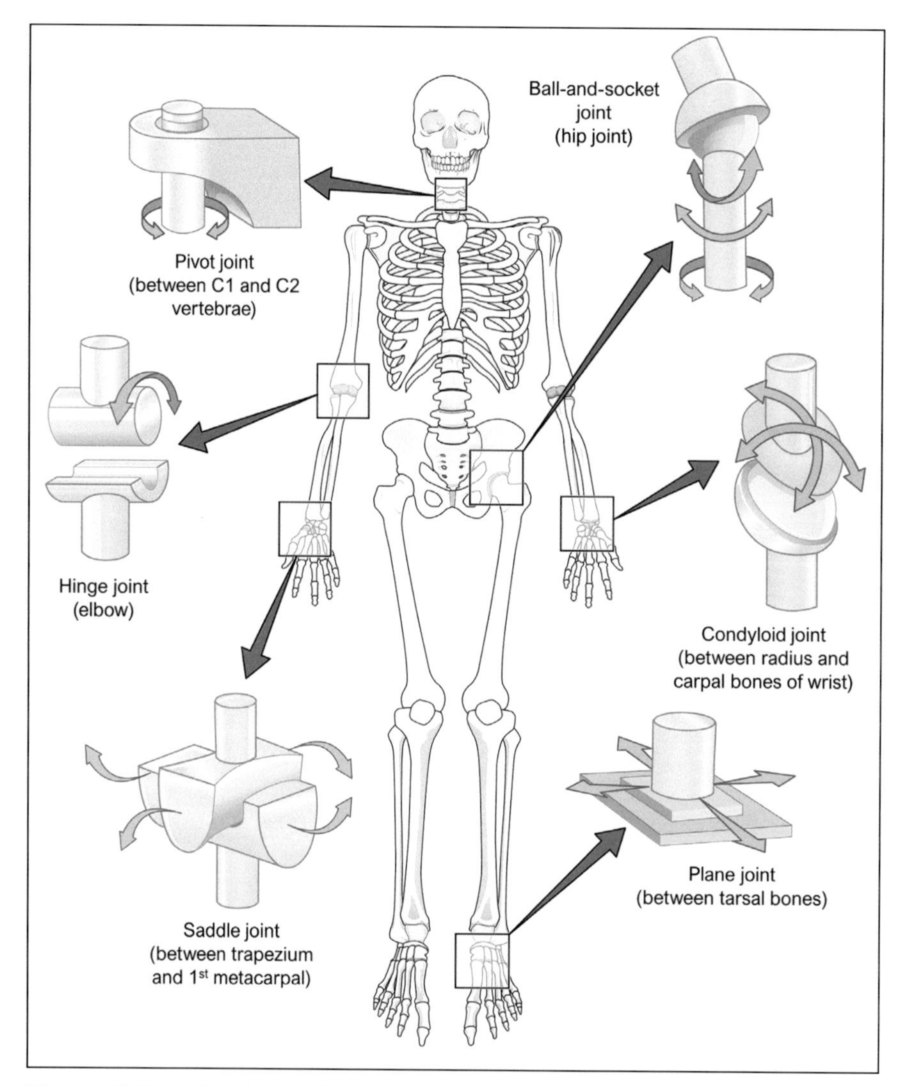

Figure 3.17. Examples of synovial joints.

joint (**Figure 3.17**). Synovial joints are also called *diarthroses* (singular, diarthrosis) because they allow free movement between the bones.

The shoulder and hip are both examples of ball-and-socket joints. The increased movement that is allowed by the shoulder (but not the hip) results from the shape of the joint surface. The shoulder joint is formed by the scapula and humerus, while the hip joint is formed by the os coxa and femur. When you flex your biceps muscle, the bones of the upper and lower arm move in a hinge-like fashion. The elbow is an example of a hinge joint.

When you twist your head from side to side, your cranium is rotating around a projecting piece of bone called the dens, which is part of your second cervical vertebra. This forms a pivot joint.

Cartilaginous Joints

Although synovial joints are the most common type of joint in the body, there are two others that need consideration. Recall that any articulation between two skeletal elements creates a joint. Clearly not all of these are mobile, and indeed some joints have very limited mobility (which can be a good thing). The second type of joint is a cartilaginous joint, also called a *synchondrosis*. Here, the two bone elements articulate but are connected by cartilage, which limits the movement between the bones. Growing long bones form from different growth

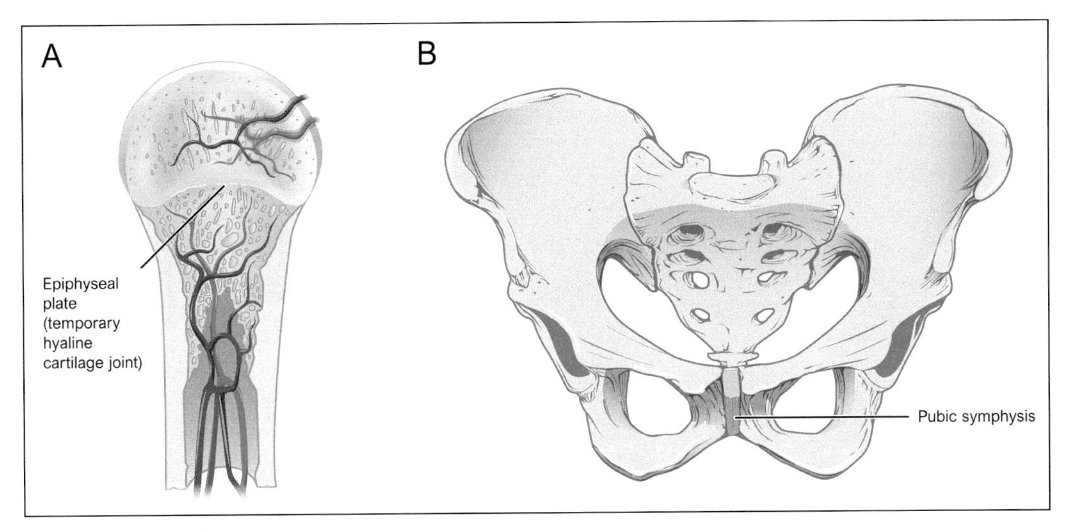

Figure 3.18. Examples of cartilaginous joints.

centers, which are connected by a cartilaginous joint (**Figure 3.18A**). Some cartilaginous joint surfaces are covered by a specific type of cartilage (hyaline) and are called a *symphysis*. This is relevant to forensic anthropology because the two halves of the pelvis articulate anteriorly at what is called the pubic symphysis, which provides critical information on age-at-death in forensic cases (**Figure 3.18B**). Cartilaginous joints are also called *amphiarthroses* (singular, amphiarthrosis) because they allow limited movement.

Fibrous Joints

The final type of joint is called a fibrous joint. Fibrous joints are bone articulations that are connected by dense, fibrous connective tissue that significantly limits the movement between the two skeletal elements. These are also called *syndesmoses,* with examples being the articulation between the radius and ulna in your forearm or between the tibia and fibula in your lower leg (**Figure 3.19A**). A special kind of fibrous joint is found between the bones of the skull. These are called *sutures,* which will be discussed again in chapter 8 (**Figure 3.19B**). Finally, a *gomphosis* is a fibrous joint that is found between the roots of your teeth and the alveolar bone that houses your teeth (**Figure 3.19C**). Fibrous joints are also called *synarthroses* (singular, synarthrosis) because they allow no movement between the bones.

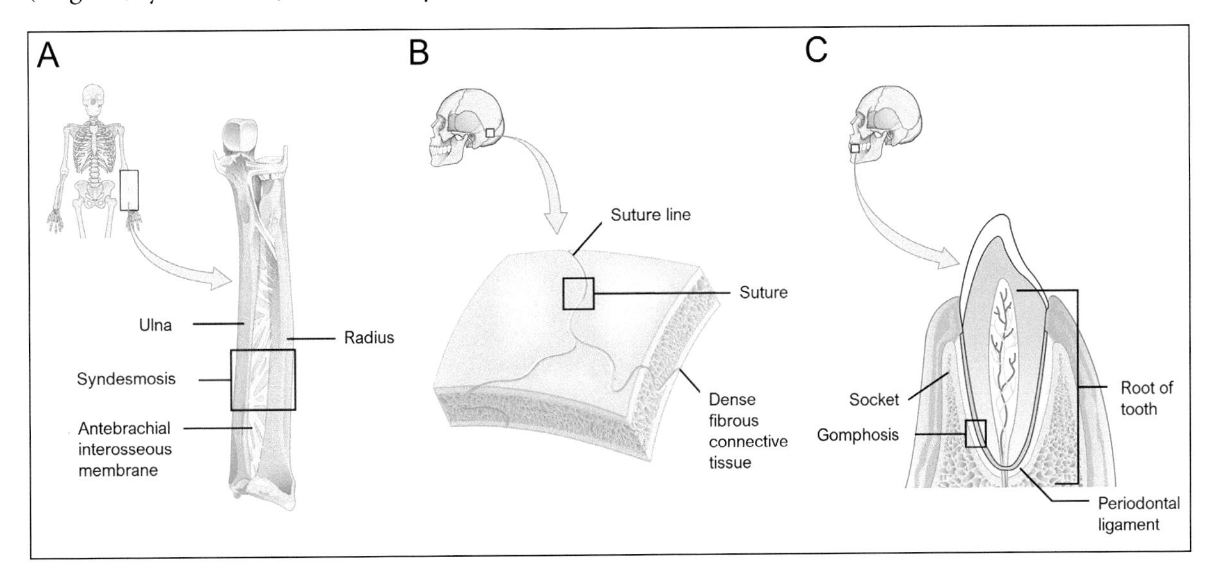

Figure 3.19. Examples of fibrous joints.

LEARNING CHECK

Q11. Please identify the type of joint represented in **Figure 3.20**.
  A) Synovial
  B) Cartilaginous
  C) Fibrous

Q12. How would you describe the joint shape presented in **Figure 3.20**?
  A) Planar
  B) Saddle
  C) Ball and socket
  D) Sutural

Q13. What type of bone is indicated on the image in **Figure 3.20** by the areas labeled 1?
  A) Trabecular
  B) Compact
  C) Subchondral

Q14. What soft tissue type is represented in **Figure 3.20** by the areas labeled 2?
  A) Cartilage
  B) Tendon
  C) Ligament

Q15. What soft tissue type is represented in **Figure 3.20** by the areas labeled 3? This question refers to the white tissue shown here. The red tissues are muscles of the shoulder joint. Note the relationship between the red muscles, the bone, and the white tissue shown in areas labeled 3.
  A) Tendon
  B) Cartilage
  C) Ligament

Q16. What type of bone is located underneath the soft tissue labeled 2 in **Figure 3.20**?
  A) Compact
  B) Trabecular
  C) Subchondral

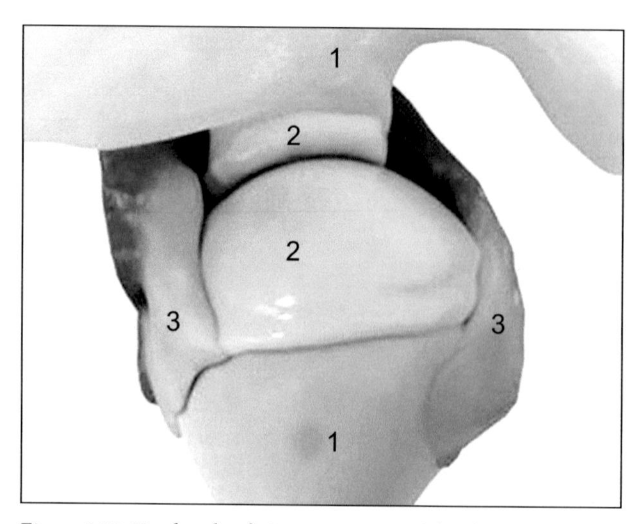

Figure 3.20. Hard and soft tissue anatomy of the shoulder joint.

End-of-Chapter Summary

This introduction to bone biology provided the vocabulary needed to discuss skeletal variation throughout the rest of the book. In addition to defining the terminology used to describe anatomical directions and planes, this chapter has introduced an extensive list of terminology used to describe basic bone biology. The key topics include: the molecular composition of bone (collagen and hydroxyapatite), the gross anatomy of bone (compact/spongy/trabecular vs. compact/cortical OR subchondral), the anatomical features of long bones (diaphysis, epiphysis, metaphysis; medullary cavity; endosteum and periosteum), long bone growth (longitudinal and appositional), the difference between woven/immature and lamellar/mature bone, Haversian systems and their key components (osteons, lacunae, canaliculi, Volkmann's canals, circumferential lamellae), and joint types (synovial, cartilaginous, fibrous).

End-of-Chapter Exercises

Exercise 1

Materials Required: **Table 3.1** and **Figure 3.21.**

Scenario: You have been asked to review a case report written by one of your colleagues for the remains shown in **Figure 3.21**. The following passage is part of your colleague's description of the body and its positioning:

Figure 3.21. An example of a bog body.

> The body is positioned on its right side. The upper extremities [arms] are flexed at the elbow with the forearms positioned anterior to the chest. The left hand is positioned several centimeters inferior to the right hand. The lower extremities [legs] are flexed at both the hip and the knee joints, positioning the knees anterior to the abdomen. The left lower extremity crosses anterior to the right and the right foot is positioned superior and posterior to the left foot. There is a length of cord that has been wrapped around the decedent's neck with two loops of cord positioned anterior to the neck.

Directions: Using the directional terms in **Table 3.1** and **Figure 3.21** as references, rewrite your colleague's description of the body and its positioning, correcting any errors that you may find.

Exercise 2

Materials Required: **Table 3.3** and **Figure 3.22**.

Scenario: The same colleague has asked you to review their description of the maxillary dental arcade from a different case. The maxillary arcade in question is shown in **Figure 3.22** and your colleague's description of it is as follows:

> The left first molar (LM1), left second molar (LM2), and right second molar (RM2) have dental restorations—these are metallic for LM1 and RM2. LM1 has two separate restorations. The larger is located on the mesiobuccal cusp [mesiobuccal = mesial + buccal] and the smaller is distal to this. LM2 also has two separate restorations, both of which are located on the lingual aspect of the crown. RM2 has a single restoration located on the labial aspect of the incisal surface. There is a small, lighter colored bony outgrowth located on the lingual aspect of the alveolar bone distal to LM2.

Directions: Using the directional terms in **Table 3.3** and **Figure 3.22** as references, rewrite your colleague's description of this maxillary arcade, correcting any errors that you may find.

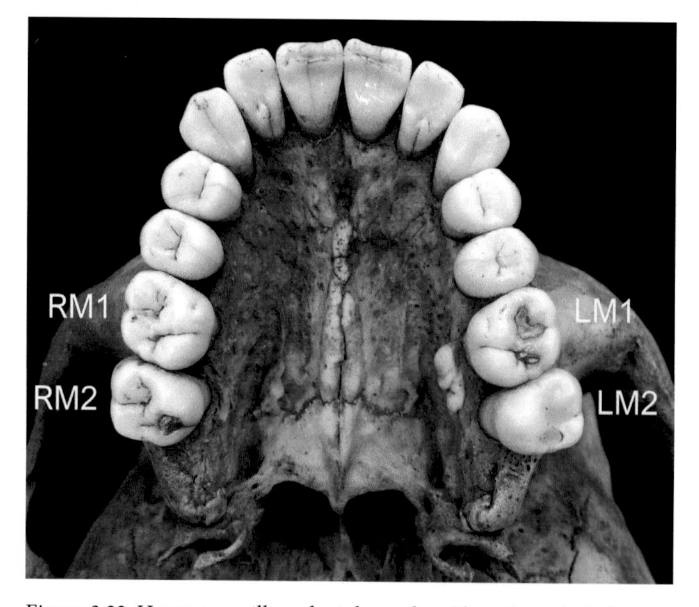

Figure 3.22. Human maxillary dental arcade with molars labeled.

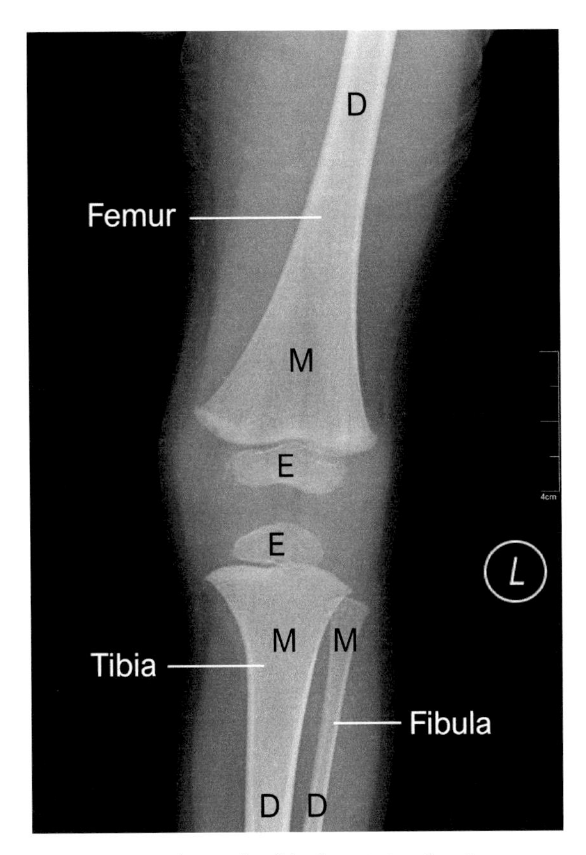

Figure 3.23. Radiograph of the knee joint showing unfused epiphyses.

Exercise 3

Materials Required: **Figures 3.17** and **3.23**.

Directions: **Figure 3.23** shows a radiograph taken of a left knee. The diaphyses (D), metaphyses (M), and epiphyses (E) for each of the bones shown have been labeled. Use the information in **Figure 3.31** to answer the following questions.

Question 1: Is the proximal or distal epiphysis of the tibia included in this radiograph?

Question 2: Is the distal epiphysis of the femur proximal or distal to the metaphysis of the tibia?

Question 3: The knee joint, like the elbow joint, only allows for flexion and extension. Based on this similarity (and using **Figure 3.17** as a reference), how would you describe the shape of the knee joint (i.e., is it a ball-and-socket joint, a pivot joint, etc.)?

Question 4: The proximal epiphysis of the fibula ossifies around the age of four in females and about a year later in males. Based on this information, what can you say about the age of the individual in this radiograph?

References

Sobotta J. 1909. *Atlas and Text-Book of Human Anatomy.* Philadelphia, PA: W.B. Saunders Co.
White TD, Black MT, Folkens PA. 2012. *Human Osteology,* Third Edition. Burlington, MA: Academic Press.

4

Human Osteology

Learning Goals

By the end of this chapter, the student will be able to:

- Identify the major bones of the human body.
- Identify key anatomical structures of the skull and pelvis.
- Apply whole bone osteology to fragmentary remains.
- Describe the ways in which individual skeletons vary from one person to another.

Introduction

The adult human skeleton contains 206 bones (**Table 4.1**) with more than half (106) located in the hands and feet. While this number is daunting, the situation is even more complex when one considers the juvenile skeleton. As discussed in the previous chapter, each adult long bone forms from at least three distinct growth centers (diaphysis, proximal epiphysis, distal epiphysis) that fuse together during adolescence. Some epiphyses may not appear until well after birth, making the process of counting the number of bones in the juvenile skeleton a moving target. An *infant* may have 300–350 bones, for example. However, it is generally stated that the *juvenile* skeleton has 270 bones that fuse together as one grows, a process which reaches completion in the third decade of life. For example, the mandible (lower jaw) forms from two distinct halves that fuse together during the first year of life. In addition, the cranium is composed of many bones joined by sutures (joints) that fuse as one ages (sutural obliteration is one method used to estimate age-at-death).

There are many ways in which these bones, or *elements,* are categorized (**Figure 4.1**). For example, most bones are identified as either belonging to the cranial or postcranial skeleton**.** The cranial bones include the skull (cranium + mandible), while the postcranial skeleton includes everything below the cranium. The hyoid, which is the only bone that does not articulate with another bone, is suspended within the neck and does not neatly fit into either the cranial or postcranial skeleton. In general, all elements *caudal* to the cranium are considered postcranial. This system for organizing the skeleton simply focuses on position within the body.

The skeleton can also be categorized in terms of function. The *axial* skeleton includes the skull, the vertebral column, the sacrum, the ribs, and the sternum. These bones, together, form the midline of the body, or its axis. The *appendicular* skeleton includes the upper limbs, lower limbs, and both the pelvic and shoulder girdles. These comprise the appendages that move about the axial skeleton.

Table 4.1. Bones of the adult skeleton, after Mays (2021: Table 1.1)

Region		Number of Bones
Skull	Cranium	27 Total
		Unpaired: Frontal, Occipital, Vomer, Ethmoid, Sphenoid
		Paired: Parietal, Temporal, Maxilla, Palatine, Malleus, Incus, Stapes, Inferior Nasal Concha, Lacrimal, Nasal, Zygomatic
	Mandible	1
	Hyoid	1 unpaired
Vertebrae and Spine	Vertebrae	24 unpaired (7 Cervical, 12 Thoracic, 5 Lumbar)
		1 Sacrum, 1 Coccyx
Thoracic	Ribs	24, 12 pairs
	Sternum	1, unpaired
Pelvic Girdle	Os Coxa	2, paired
Upper Limb	Humerus	2, paired
	Radius	2, paired
	Ulna	2, paired
	Carpals	16, 8 pairs
		Hamate, Capitate, Trapezoid, Trapezium, Pisiform, Triquetral, Lunate, Scaphoid
	Metacarpals	10, 5 pairs
	Phalanges	28, paired
		10 Proximal, 8 Intermediate, 10 Distal
Lower Limb	Femur	2, paired
	Patella	2, paired
	Tibia	2, paired
	Fibula	2, paired
	Tarsals	14, 7 pairs
		Calcaneus, Talus, Navicular, Cuboid, Medial Cuneiform, Intermediate Cuneiform, Lateral Cuneiform
	Metatarsals	10, 5 pairs
	Phalanges	28, paired
		10 Proximal, 8 Intermediate, 10 Distal

For the purposes of counting elements to ascertain the *Minimum Number of Individuals* in a commingled sample, it is useful to know whether elements are paired or unpaired (**Table 4.1**). An unpaired element is found along the midline of the body. Examples include the sternum, the vertebrae, and the sacrum in the postcranial skeleton, and the frontal bone, the occipital bone, and the vomer in the cranial skeleton. The hyoid is also an unpaired skeletal element. Paired elements occur on both sides of the body, that is, there is a left and a right. Although there is always minor variation, paired bones are generally mirror images of each other in terms of overall appearance.

Finally, bones can be categorized based on their overall appearance and shape. There are flat bones, short bones, long bones, and irregular bones that defy easy description (**Figure 4.2**). Some individuals have extra bones, often in their hands or feet. If these occur in a tendon, they are called sesamoid bones. Some individuals also have extra bones in the joints of their skull. These are called sutural bones, or ossicles, or **Wormian** bones.

Flat bones are found primarily in the skull, the shoulder and pelvic girdles, and the ribs. The flatness results from their cross-sectional shape, rather than having the quality of being

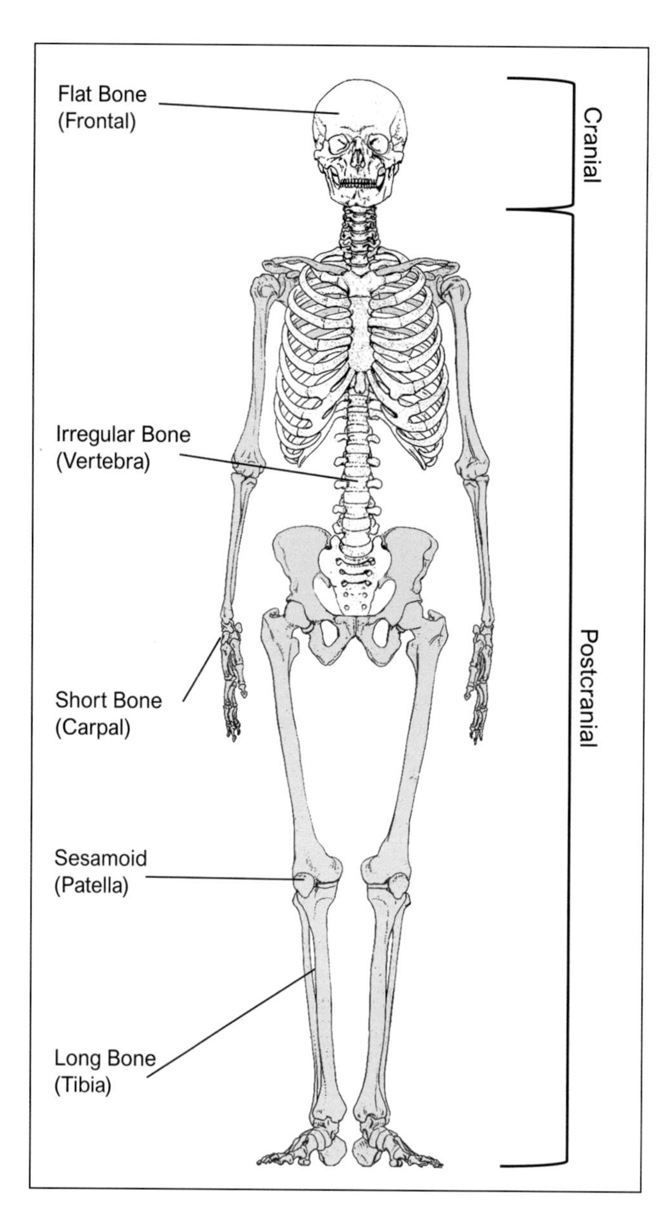

Figure 4.1. Categories of bones of the human skeleton. Elements belonging to the appendicular skeleton are shaded blue; those of the axial skeleton are white.

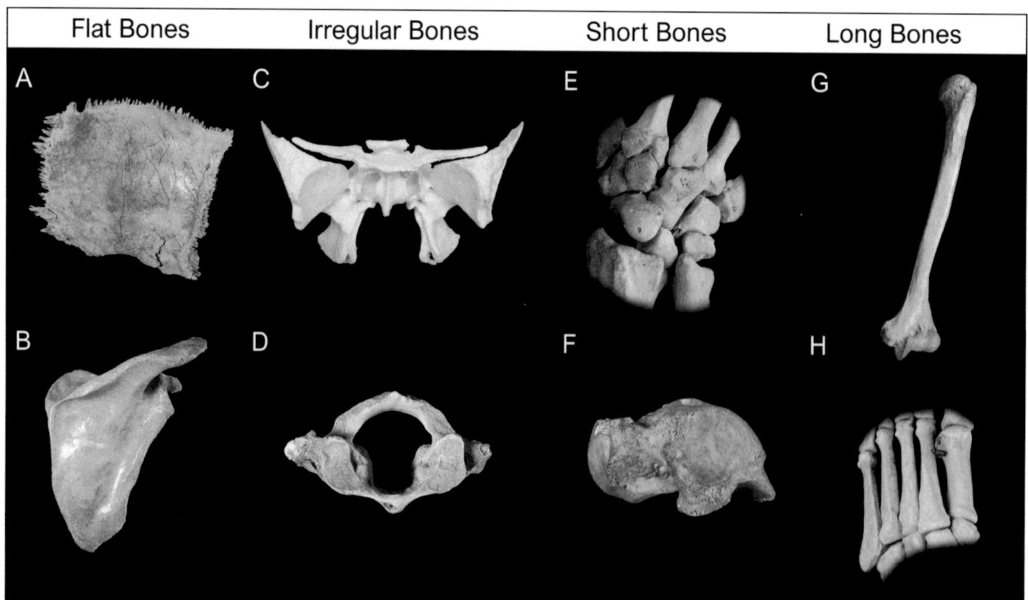

Figure 4.2. Categorizing bones by shape. Examples of flat bones include the parietal (*A*) and scapula (*B*); irregular bones include the sphenoid (*C*) and vertebrae (*D*); short bones include the carpals (*E*) and talus (*F*); and long bones include the humerus (*G*) and the metatarsals (*H*). Note: images are not to scale.

completely flat. This distinction is important for predicting the effects of trauma on patterns of bone damage.

Irregular bones are found in the facial skeleton and spinal column. **Figure 4.2C** shows a sphenoid, a bone of the skull that is buried deep within the braincase and is difficult to see (except in the temporal region). The shape of the sphenoid is highly irregular; it is said to resemble a bird of prey. Note that while the flat and irregular bones form the axial skeleton, the long and short bones of the skeleton comprise the appendicular skeleton.

The Skull

The skull is the most complex skeletal structure in the body. It is also one of the most important in forensic anthropology. The sex of an individual can be assessed through an analysis of their skull. Age-at-death can be roughly estimated based on the closure of the sutures. The facial skeleton contains information on population affinity or ancestry because it is shaped by climatic adaptation and is therefore geographically patterned in humans (to some extent). Because the skull protects the brain, it is also one of the most targeted areas for interpersonal violence. Many blunt force trauma deaths will target the head, and gunshot wounds to the skull are quite common.

Basic Terminology

The terms "skull" and "cranium" are often used interchangeably in everyday language; however, they are not equivalent terms. In fact, there are a number of terms used to describe the various components of the skull depending on what portions are present. These terms are summarized in **Table 4.2**.

Table 4.2. Anatomical terminology of the head, after White et al. (2012)

Term	Definition
Skull (pl. skulls)	The complete skeletal structure of the head
Cranium (pl. crania)	A skull that lacks the mandible
Mandible (pl. mandibles or mandibulae)	The lower jaw bone
Calvaria (pl. calvariae)	A cranium that lacks the facial skeleton
Calotte (pl. calottes)	A calvaria that lacks the base
Splanchnocranium (pl. splanchnocrania)	The facial skeleton
Neurocranium (pl. neurocrania)	The vault or braincase

The Bones of the Skull

The bones of the skull classify as either flat bones (bones of the vault) or irregular bones (bones of the face, or splanchnocranium). The bones of the vault have names that generally correspond with the parts of the brain that they protect. The bones of the skull are summarized in **Table 4.3** with specific notes about their forensic relevance. The location and articulation of each are presented in **Figure 4.3**. Examples of dry bone specimens are presented in **Figures 4.4** and **4.5**.

Table 4.3. Bones of the skull

Bone	Description and Importance
Frontal	Forms the upper portion of the orbits and forehead, provides information on sex, a frequent site of trauma
Parietal (p)	Forms the sides of the skull, a frequent site of trauma
Occipital	Forms the back of the skull, provides information on sex
Temporal (p)	The location of the ear and mastoid process, the latter provides information on sex
Auditory ossicles (p)	Small bones of the inner ear, little forensic relevance
Maxilla (p)	Houses the teeth, forms the nasal and oral cavities
Palatine (p)	Posterior aspect of hard palate, little forensic relevance
Vomer	Forms part of the nasal cavity, little forensic relevance
Inferior nasal concha (p)	Forms part of the nasal cavity, little forensic relevance
Ethmoid	Little forensic relevance
Lacrimal (p)	Inner surface of orbit, little forensic relevance
Zygomatic (p)	Forms part of the cheek, sometimes damaged as the result of trauma
Sphenoid (p)	Complex bone that articulates with many others, little forensic relevance
Mandible	Houses the teeth, forms the oral cavity, sometimes damaged as the result of trauma

Note: A (p) means the element is paired.

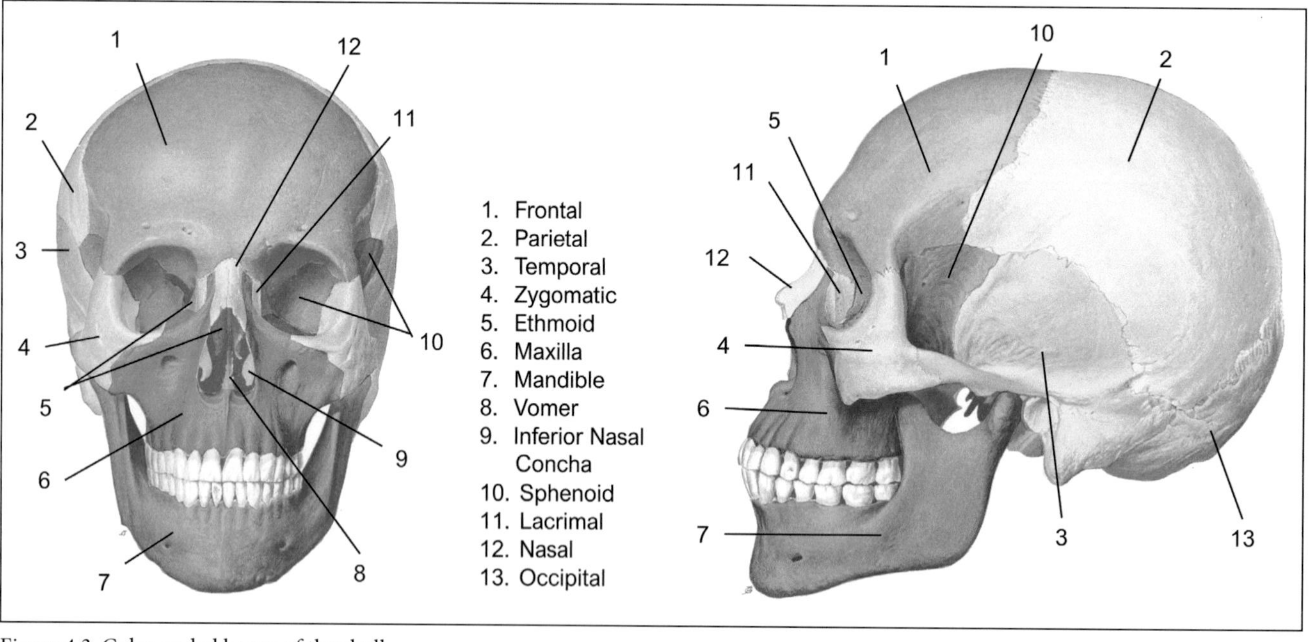

Figure 4.3. Color-coded bones of the skull.

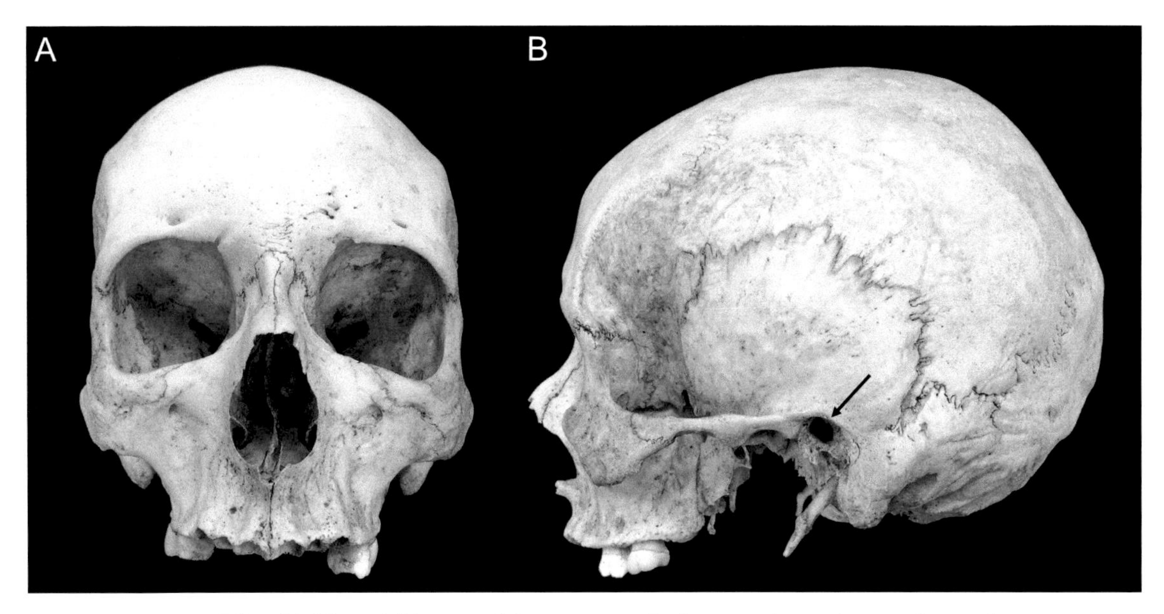

Figure 4.4. Anterior (*A*) and left lateral (*B*) views of the cranium—dry bone specimen. Arrow indicates the external auditory meatus.

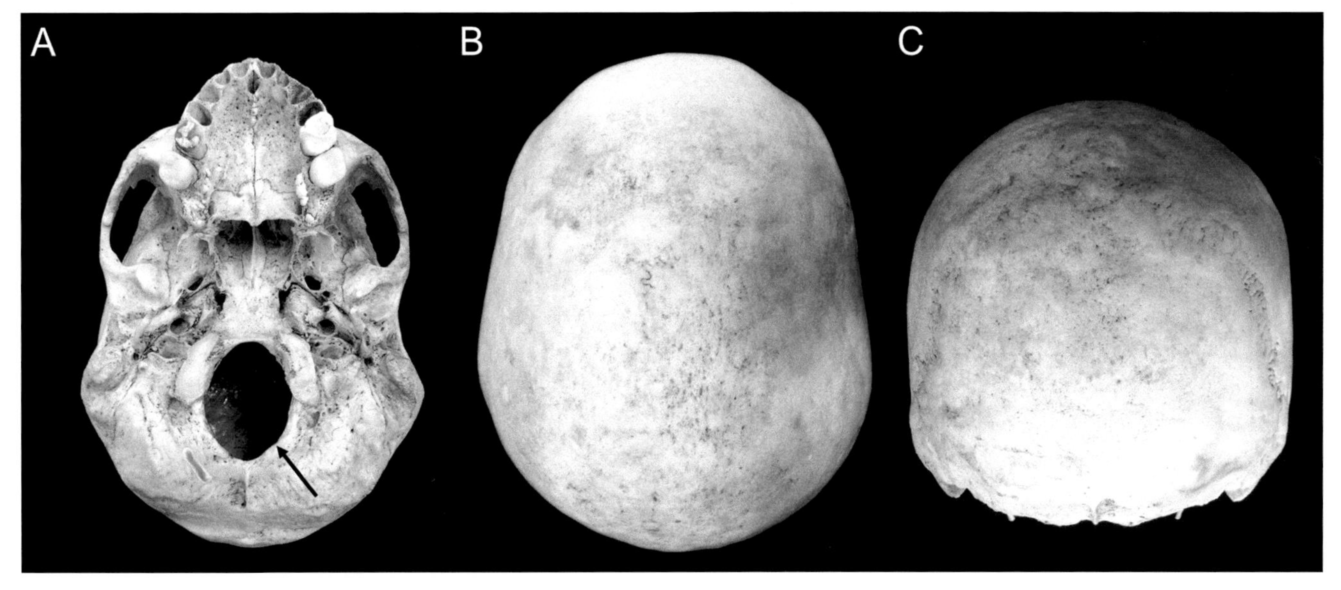

Figure 4.5. Inferior (*A*), superior (*B*), and posterior (*C*) views of the cranium—dry bone specimen. Arrow indicates the foramen magnum.

The Sutures

The articulations between the bones of the skull are fibrous joints that are tightly connected in life. The soft tissue connections between the joints are reinforced with complex bony configurations that vary in complexity from person to person. The sutures obliterate as one ages, due to the continued process of new bone formation connecting the two skeletal elements.

Every point of articulation between different bones defines a suture with a different name. These are labeled in **Figure 4.6**. Although at first glance this seems overly complicated, most sutures are named based on the two bones that articulate there. For example, the occipito-mastoid suture is located near the posterior of the skull where the occipital bone and the mastoid portion of the temporal bone meet.

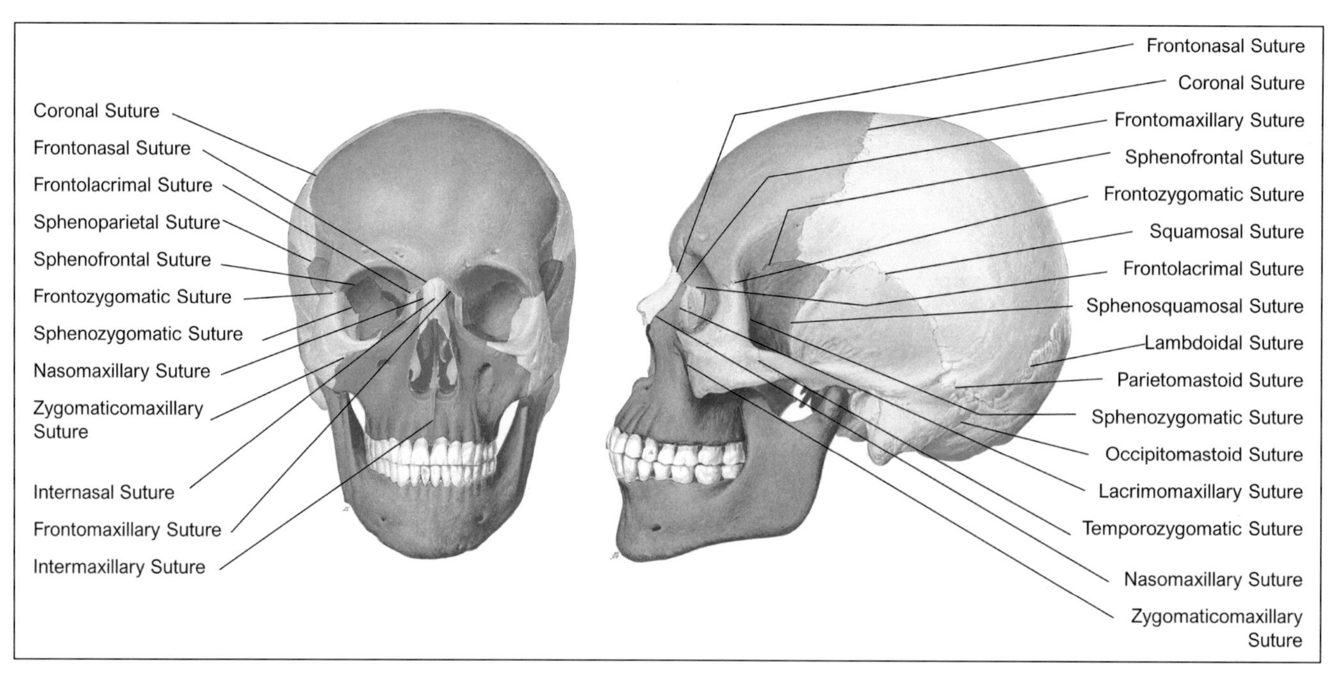

Figure 4.6. Sutures of the skull.

There are four exceptions to this general rule:

1) The *coronal suture*: orients the coronal plane, runs mediolaterally at the articulation between the frontal and parietal bones.
2) The *sagittal suture*: orients the sagittal plane, runs anteroposteriorly at the articulation of the left and right parietal bones.
3) The *squamosal suture*: the articulation between the temporal and parietal bones on the side of the vault.
4) The *lambdoidal suture*: the articulation between the occipital and parietal bones on the posterior of the vault.

The "Holes"

Negative spaces (versus holes) may be the more appropriate term here.

1) The *foramen magnum* is the large foramen in the base of the cranium that allows the spinal cord to connect to the brain (**Figure 4.5,** *left*, see black arrow). It is formed completely by the occipital bone.
2) The *external auditory meatus* is the opening for the ear (**Figure 4.4,** *right*, see black arrow). The skeletal structure funnels sound waves into the inner ear where the skeletal and soft tissue anatomy helps produce the sensation of hearing. The external auditory meatus is located in the *temporal* bone.
3) The *oral cavity* is where food is ingested and chewed, i.e., the mouth. It is formed by the *maxillae, palatines,* and *mandible.* The palatines are the small rectangular bones near the rear of the palate. Importantly, the oral cavity houses the dentition, from which a tremendous amount of relevant forensic data can be generated.
4) The *orbits* house the eyes and are one of the more complex structures of the skull (**Figure 4.7**). Seven bones articulate to form the bony eye socket: *frontal, sphenoid, zygomatic, maxilla, nasal, lacrimal, ethmoid,* and *palatine.* The optic canal is critical

for transmitting sensory data to the brain. The supraorbital and infraorbital foramina open into passages that transmit nerves and blood vessels to the face.

5) The *nasal cavity* is also a complex, composite structure in the skull (**Figure 4.8**). It is formed by the *frontal* bone, the *nasals,* the *maxillae,* the *palatines,* the *ethmoid,* the *vomer,* the *sphenoid,* and the *inferior nasal conchae.*

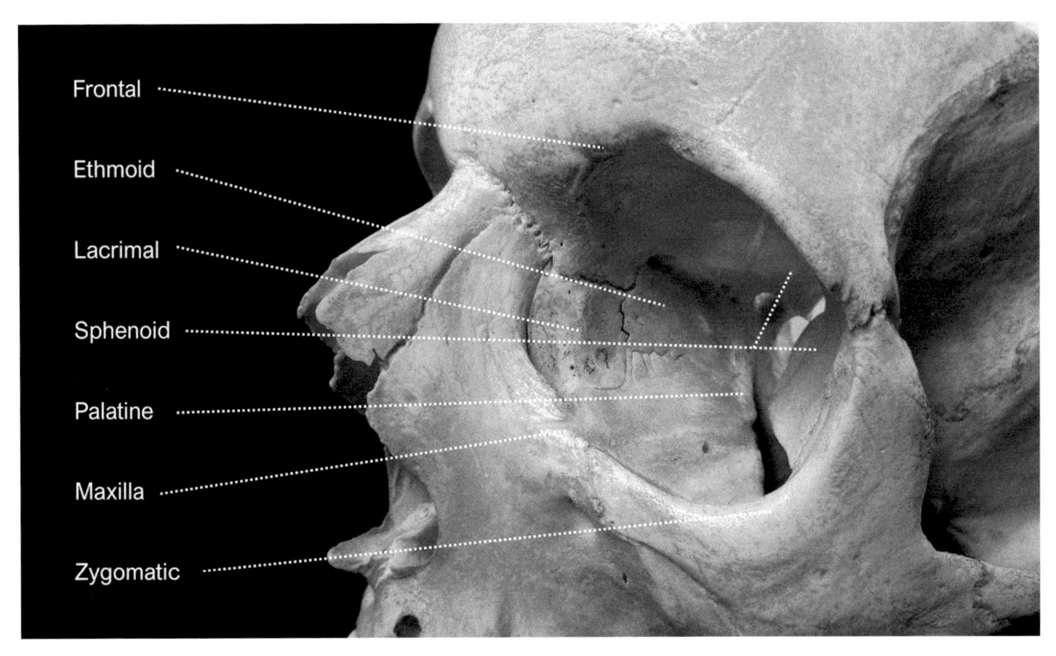

Figure 4.7. Bones of the orbit.

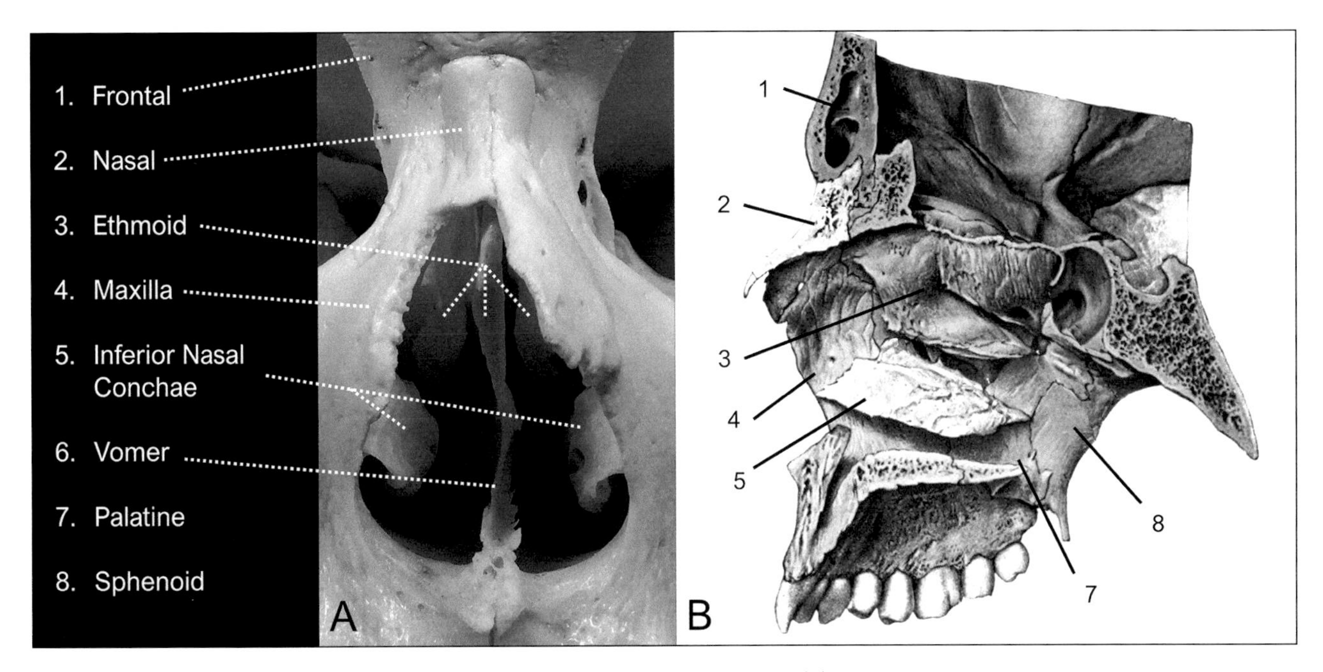

Figure 4.8. Bones of the nasal cavity in dry bone (*A*) and schematic sagittal section (*B*).

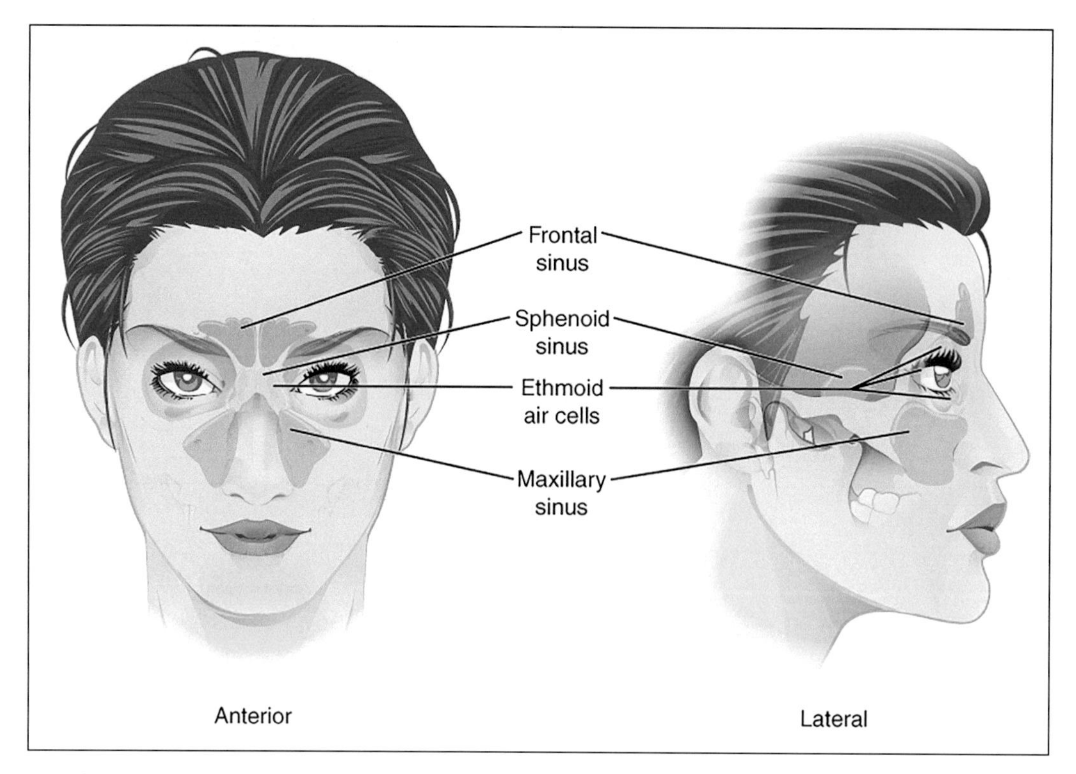

Figure 4.9. Locations of the paranasal sinuses.

The Sinuses

Sinuses are cavities within the skull that are found surrounding the nasal cavity (paranasal sinuses) (**Figure 4.9**). Four are identified here: *frontal sinus, maxillary sinus, ethmoid sinus/sacs,* and the *sphenoidal sinus.* Of these, the frontal sinus is the most significant from a forensic perspective because its shape is thought to be unique among individuals, thus serving as a type of fingerprint that can be seen radiographically. The frontal sinus is discussed in more detail in chapter 11.

LEARNING CHECK

Q1. What is depicted in **Figure 4.10A**?
A) Cranium
B) Calvaria
C) Calotte
D) Skull

Q2. What is depicted in **Figure 4.10B**?
A) Cranium
B) Calvaria
C) Splanchnocranium
D) Skull

Questions 3 through 6 refer to **Figure 4.11**. In answering them, first identify the two bones that are articulating at the indicated suture and then select among the answers provided. Choose your answer carefully and think about the names of the bones articulating at the indicated suture.

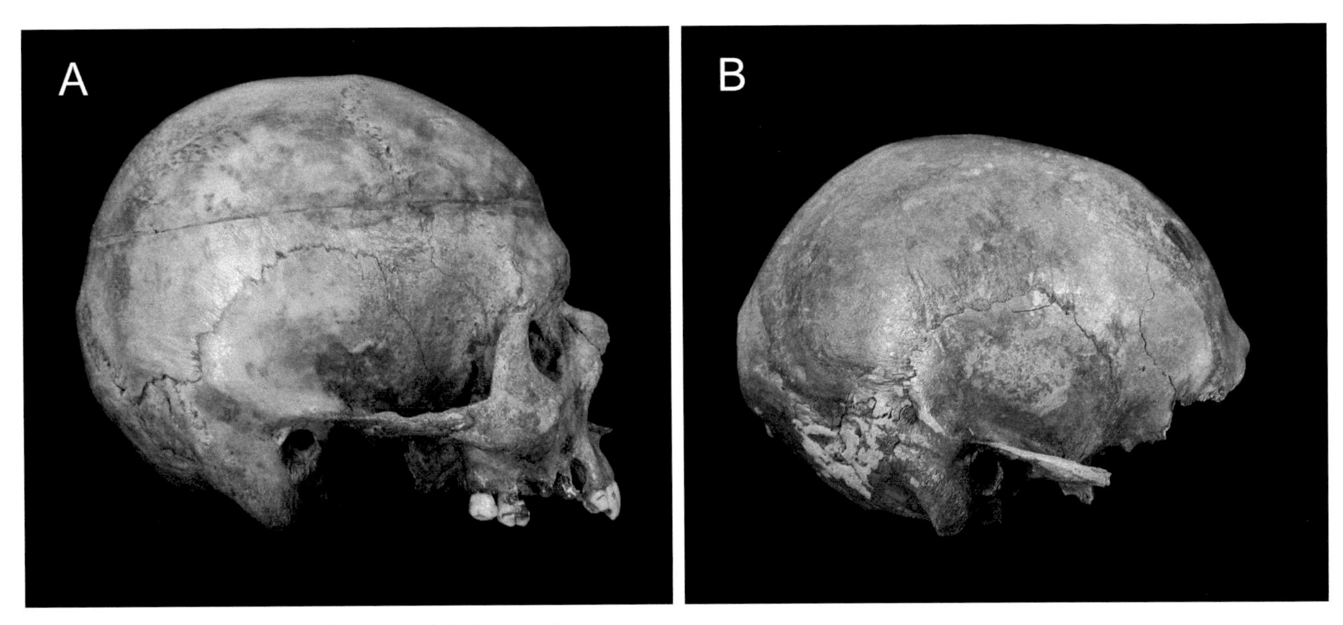

Figure 4.10. Right lateral views of two partial skeletonized elements.

Q3. Identify the suture indicated by the arrow labeled **A**.
A) Fronto-maxillary
B) Fronto-nasal
C) Fronto-zygomatic
D) Zygomatico-maxillary

Q4. Identify the suture indicated by the arrow labeled **B.**
A) Fronto-maxillary
B) Fronto-nasal
C) Fronto-zygomatic
D) Zygomatico-maxillary

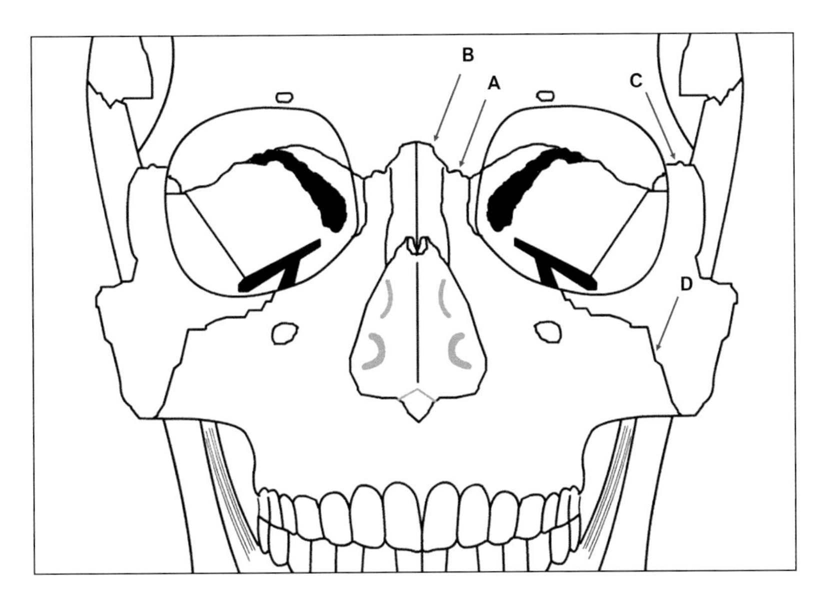

Figure 4.11. Schematic of the human skull, frontal view, with sutures identified by arrows.

Q5. Identify the suture indicated by the arrow labeled **C.**
A) Fronto-zygomatic
B) Fronto-nasal
C) Fronto-maxillary
D) Zygomatico-maxillary

Q6. Identify the suture indicated by the arrow labeled **D.**
A) Fronto-maxillary
B) Fronto-nasal
C) Zygomatico-maxillary
D) Fronto-zygomatic

The Axial Skeleton—Postcranial

The remainder of the axial skeleton includes the postcranial elements of the spinal column and thorax. These elements are summarized in **Table 4.4**. There are 80 bones in the axial skeleton, including the cranial elements. Excluding the cranium, there are 52 postcranial axial elements.

The Sternum and Ribs

The **sternum** and **ribs** form the thorax or rib cage. The *sternum* is a very light skeletal element divided into three parts. The superior portion is called the *manubrium,* the middle portion is called the *corpus sterni,* and the inferior portion is called the *xiphoid process.* All three are visible in **Figure 4.12A**. The manubrium articulates with the shoulder girdle via the clavicle, while the corpus sterni articulates via cartilage with the ribs.

There are 12 *ribs* on each side of the body, which increase in length until around rib 9/10 and then decrease in size from rib 9/10 to rib 12. Rib anatomy consists of the head, tubercle, and shaft. The head articulates with the vertebral body, the tubercle articulates with the transverse process of the vertebra, and the shaft is the long curving segment that terminates in the sternal end that articulates (for most ribs) with the sternum via costal cartilage (**Figure 4.12B**). Because the rib is a flat bone, it can be divided into a cranial (upper) edge and a caudal (lower) edge. Ribs are commonly fractured as a result of trauma.

Table 4.4. Bones of the axial skeleton

Bone	Description and Importance
Sternum	Breastbone, one of the least dense in the body, prone to fluvial transport, articulates with ribs
Vertebral Column	Supports the cranium, protects the spinal cord, and articulates with the ribs. Three types: cervical (neck), thoracic (mid-back), lumbar (lower back). Cutmarks on cervical spine vertebrae may indicate decapitation.
Ribs	Can be used for assessing age-at-death, often damaged via blunt force trauma
Sacrum	Forms the midline of the pelvis, possible sex-related variation in shape
Coccyx	Little forensic relevance

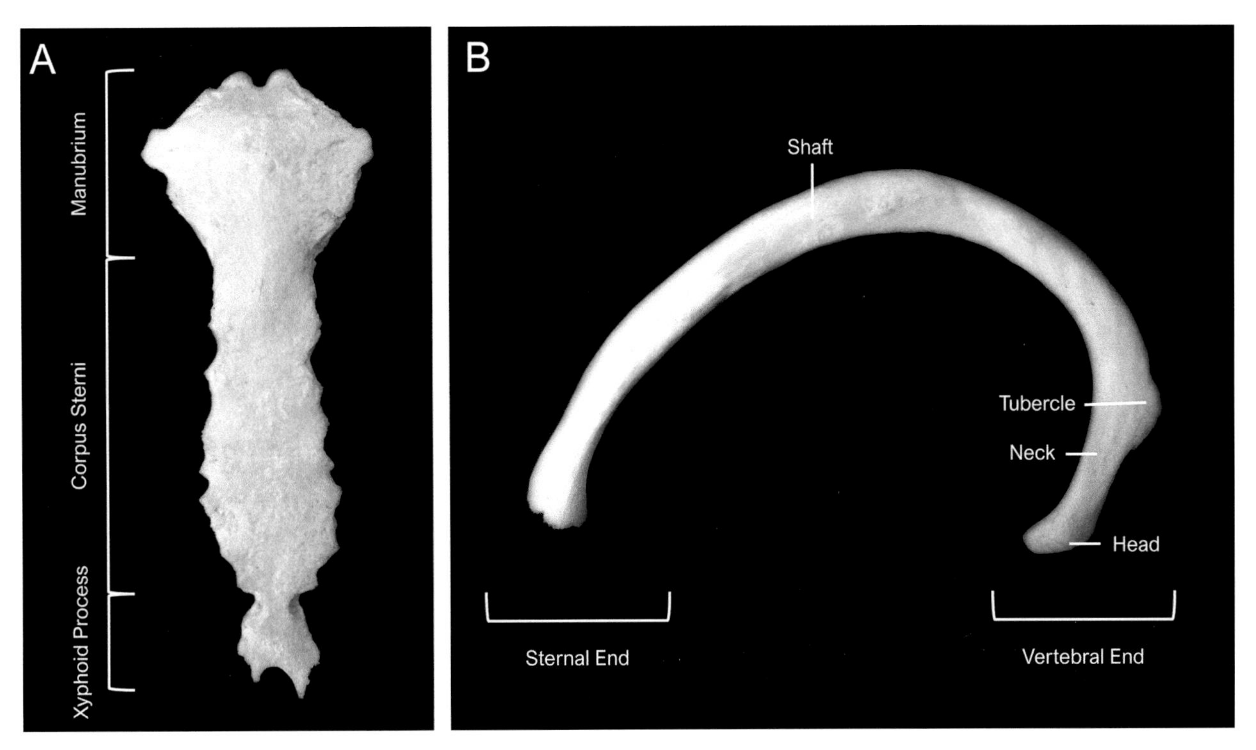

Figure 4.12. Skeletal anatomy of the sternum (*A*) and a typical rib (*B*).

The Spinal Column

The bony spine supports the weight of the cranium and provides structural protection for the spinal cord. Three elements are identified in the spine: the vertebrae, the sacrum, and the coccyx.

Although each **vertebra** is distinct, there are commonalities in their anatomy (**Figure 4.13**). The *body* supports the weight of the cranium and protects the spinal cord anteriorly. The *vertebral foramen* is the hole through which the spinal cord passes and is defined on the posterior side by the *vertebral arch*. In other words, the vertebral body and vertebral arch define the vertebral foramen. The sides of each vertebra have bony protrusions called *transverse processes,* which have articulations for the ribs in the thoracic vertebrae. The posterior aspect of each vertebra has a spinous process. The *spinous process* is what you palpate when

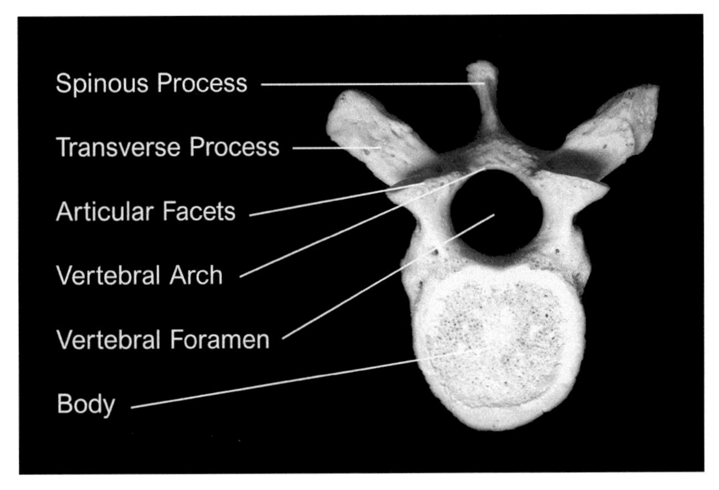

Figure 4.13. Skeletal anatomy of a vertebra.

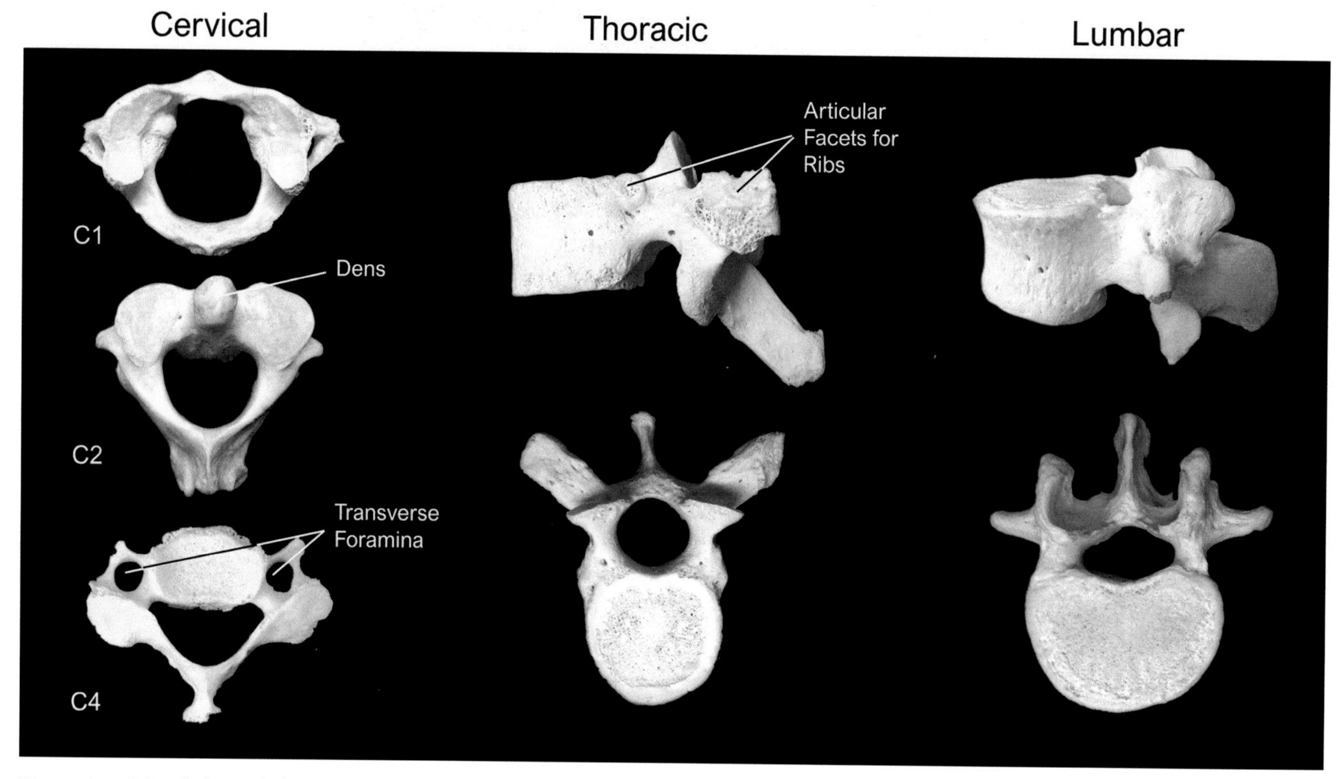

Figure 4.14. Morphological characteristics of cervical, thoracic, and lumbar vertebrae.

you run your finger along your spine. Finally, the *articular facets* are paired joints found on the upper and lower surfaces of the neural arch that allow the vertebrae to articulate with each other. This provides flexibility to the spine. There are three types of vertebrae: cervical, thoracic, and lumbar. The vertebrae are all unique in anatomy and can be identified to the specific position with advanced training in human osteology.

There are seven cervical vertebrae located in the neck. These vertebrae are the smallest because they support the least amount of weight. The first cervical vertebra (C1) lacks a body and articulates directly with the occipital bone. The second cervical vertebra (C2) is noteworthy because of the *dens* or *odontoid process* that provides a means for rotating the skull (**Figure 4.14**). The *dens* is often what is fractured when someone "breaks their neck." Cervical vertebrae also feature *transverse foramina* for the passage of the vertebral arteries.

There are 12 thoracic vertebrae located in the mid-back. These increase in size as one moves inferiorly down the spine. Thoracic vertebrae are distinguished by the fact that they articulate with ribs both at the body and transverse processes (**Figure 4.14**).

There are five lumbar vertebrae located in the lower back (**Figure 4.14**). They are the largest vertebrae with the tallest and widest bodies that support the weight of the upper half of the body. Lumbar vertebrae lack rib articulations and have large, hatchet-shaped spinous processes.

The **sacrum** is formed by four to six vertebral segments that fuse together during development. Failure of these segments to fully fuse in adulthood leads to a condition called spina bifida, a neural tube defect that prenatal (B12) vitamins are meant to protect against. In **Figure 4.15,** the five sacral segments have fused but are still visible. The sacrum is relevant because it forms the posterior aspect of the pelvis, where it articulates with the os coxa ("hip bone") at the auricular surface, a joint that is important for age-at-death assessment. The *coccyx* represents segments of a vestigial tail and is of limited forensic relevance.

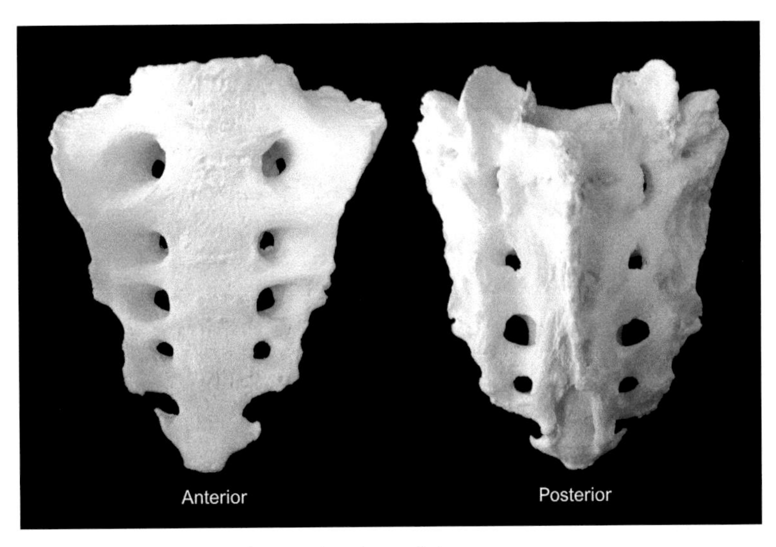

Figure 4.15. Anterior and posterior views of the sacrum.

The Appendicular Skeleton

The appendicular skeleton includes the bones of the shoulder girdle, upper limb, pelvic girdle, and lower limb. These elements are summarized in **Table 4.5**. There are 126 individual elements that comprise the appendicular skeleton.

The Pectoral/Shoulder Girdle and Upper Limb

The bones of the shoulder and upper limb include the clavicle and scapula that articulate with the humerus to form the shoulder joint.

The **clavicle** is an S-shaped long bone that serves as a strut that keeps the arm away from the body and allows for a wide range of movement in the upper limb (**Figure 4.16**). Because

Table 4.5. Bones of the appendicular skeleton

Bone	Description and Importance
Clavicle	Collarbone, the most frequently fractured bone in the body
Scapula	Shoulder blade, thin and translucent, easily damaged, limited forensic relevance
Humerus	Upper arm bone, useful for estimating stature
Radius	Forearm, limited forensic relevance
Ulna	Forearm, fractured during self-defense
Carpals	Bones of the wrist, limited forensic relevance
Metacarpals	Bones of the palm, limited forensic relevance
Phalanges	Fingers, limited forensic relevance
Os Coxa	Paired elements that form the pelvis, most useful for age and sex estimation
Femur	Thigh bone, useful for stature reconstruction
Patella	Kneecap, limited forensic relevance
Tibia	Shin bone, useful for stature reconstruction
Fibula	Lateral bone of the lower leg, limited forensic relevance
Tarsals	Bones of the ankle, limited forensic relevance
Metatarsals	Bones of the foot, limited forensic relevance
Phalanges	Toes, limited forensic relevance

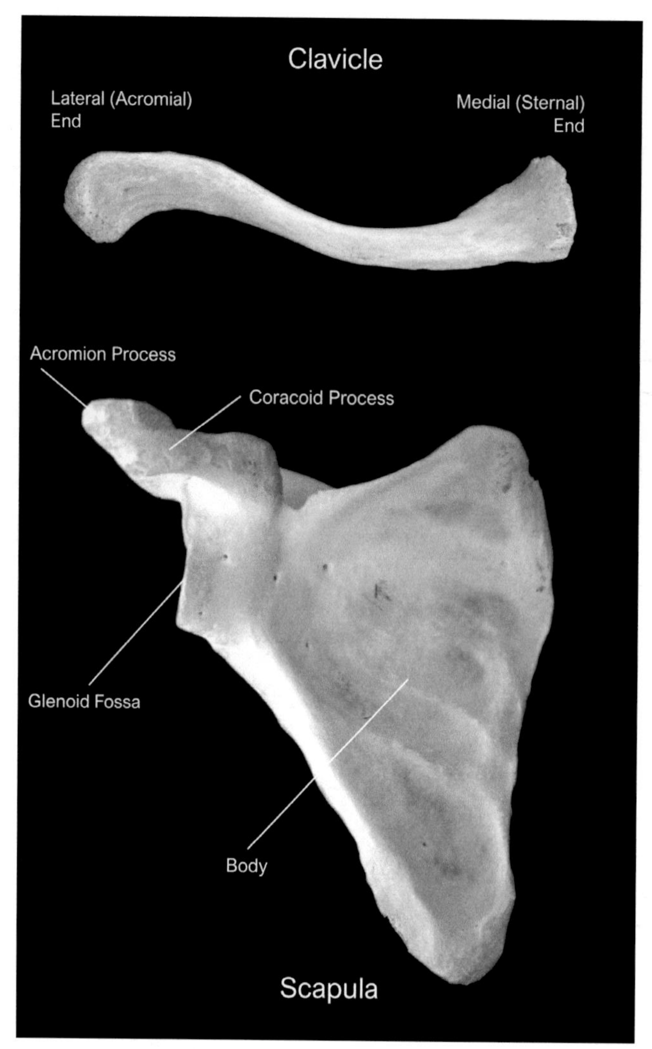

Figure 4.16. Skeletal anatomy of the right clavicle (superior view) and right scapula (anterior view).

of this function, the clavicle is frequently broken in falls as the arm is outstretched to support the body as it hits the ground. There is a medial (sternal) and lateral end to the clavicle; the former is round in cross-section while the latter is flat in cross-section.

The **scapula** is a flat bone that can be palpated by placing your arm behind your back and feeling for the triangular bone that protrudes. The scapula is defined by a thin body and two projections of bone called the *coracoid* and *acromion* processes (**Figure 4.16**). The *glenoid fossa* is the articular surface that forms the shoulder joint with the humerus. It has a bean-shaped appearance and does not constrain the movement of the humerus, hence why shoulders are more mobile than hips.

The **humerus** is the bone of the upper arm. Like all long bones, the proximal and distal epiphyses are diagnostic and can be easily identified in fragmentary remains (**Figure 4.17**). The proximal end has a head that articulates with the scapula and greater and lesser tubercles. The distal end is defined by a spindle-shaped joint with two segments called the *trochlea* and *capitulum,* with a posteriorly placed *olecranon fossa* where the proximal end of the **ulna** articulates. The shaft is round in cross-section and easily identified because of the attachment sites for the muscles of the upper limb.

The **radius** and **ulna** have shafts that are very similar in shape and size. These are easily confused when fragmentary. However, like all long bones, each has distinct proximal and

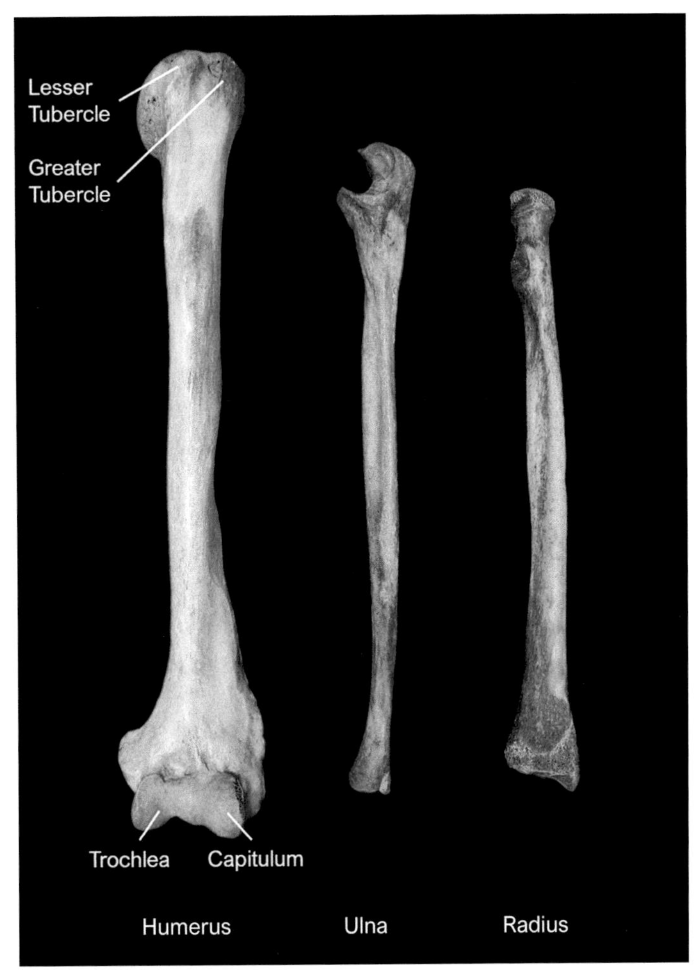

Figure 4.17. Left humerus, ulna, and radius.

distal joint surfaces. The ulna has a U-shaped proximal joint surface that articulates with the humerus. The radius has a round proximal joint surface that articulates with both the humerus and ulna. The distal end of the ulna is thin and rounded, while the distal end of the radius is square and flat (**Figure 4.17**). The articulation of the elbow is shown in **Figure 4.18**. Note how the U-shaped proximal ulna fits perfectly around the distal humerus. The round head of the radius is also visible in this image.

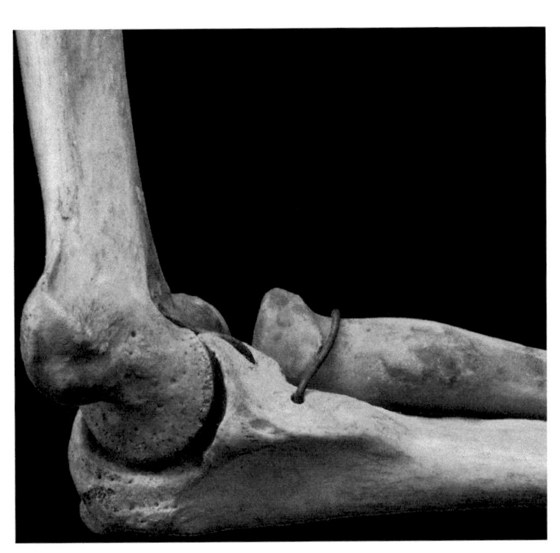

Figure 4.18. Bones of the elbow showing the ulna and radius articulating with the distal humerus.

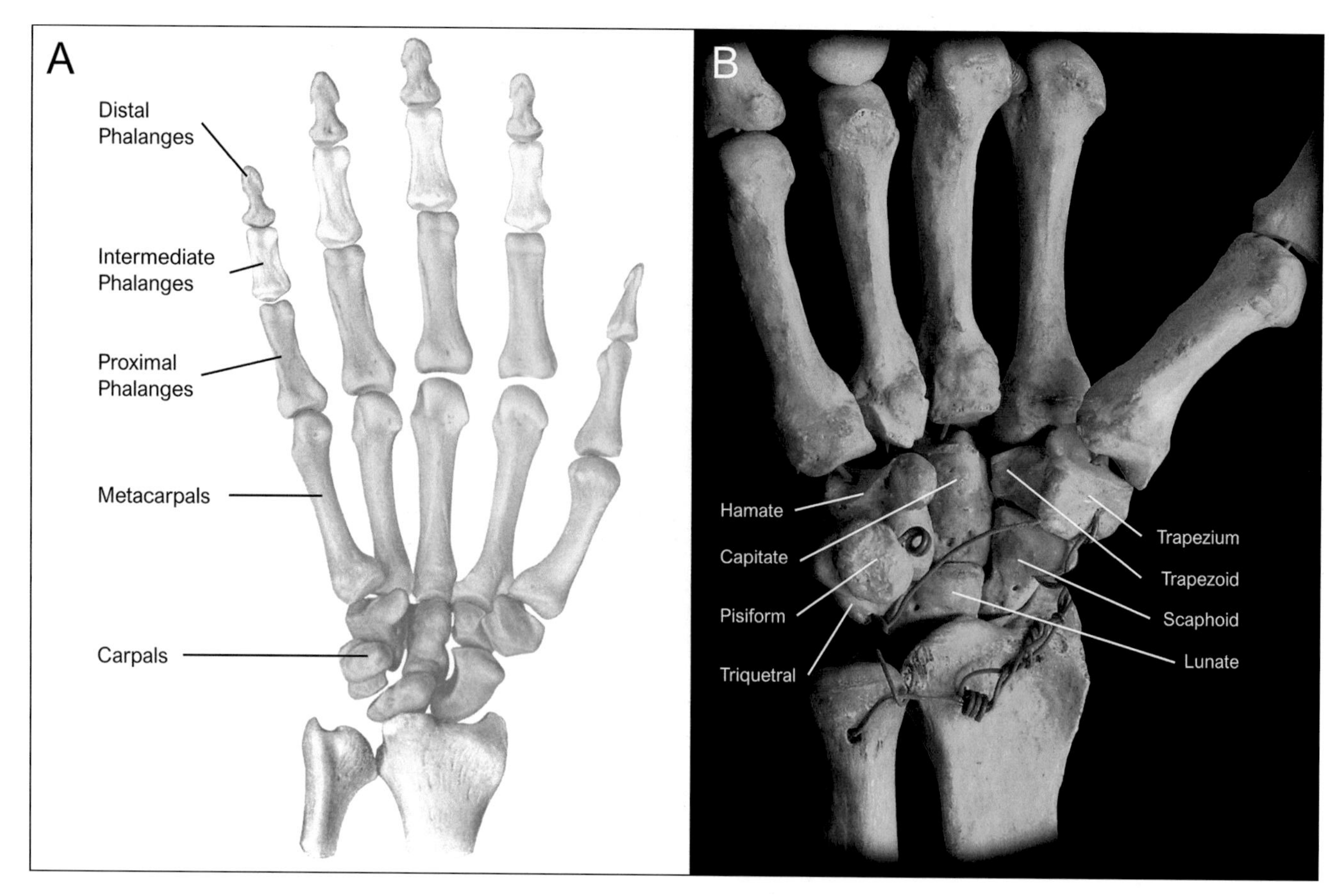

Figure 4.19. Bones of the hand (*A*) with close-up, palmar view of the carpals (*B*).

Figure 4.19A shows the bones of the hand. **Figure 4.19B** shows a close-up image of the carpals taken from the palmar/anterior view. Note the complex series of articulations evident here. These bones are particularly tricky to side. Small accessory bones, called *sesamoids,* can also be found among the bones of the hand.

In addition to the 8 **carpals**, each hand is composed of 5 **metacarpals** and 14 **phalanges (Figure 4.19A)**. The phalanges are divided into *proximal, intermediate,* and *distal* elements. Note that the thumb (digit 1) does not have an intermediate phalanx (singular of phalanges), hence why there are 14 and not 15 phalanges. While each metacarpal is distinct and easy to identify and side, the phalanges can be difficult to identify to their specific position in the body. Siding is also a challenge when found in isolation. Each finger is called a *digit* or a *ray.*

Metacarpals have the general appearance of a long bone: there is a shaft, a proximal joint surface, and a distal joint surface. The distal joints of the metacarpals are hard to distinguish in isolation; however, the proximal joint surfaces are distinct and diagnostic (**Figure 4.20A**). The proximal, intermediate, and distal hand phalanges are easily distinguished from each other based on the appearance of their surfaces (**Figure 4.20B**). Distal phalanges have rough, non-articular distal ends because they form the ends of your fingertips where the nails and fingerprints are located.

The Pelvic Girdle and Lower Limb

The bones of the pelvis and lower limb include the os coxa, femur, patella, tibia, fibula, tarsals, metatarsals, and phalanges.

The **os coxa** ("hip bone") or **innominate** (literally meaning "unnamed," an oxymoron)

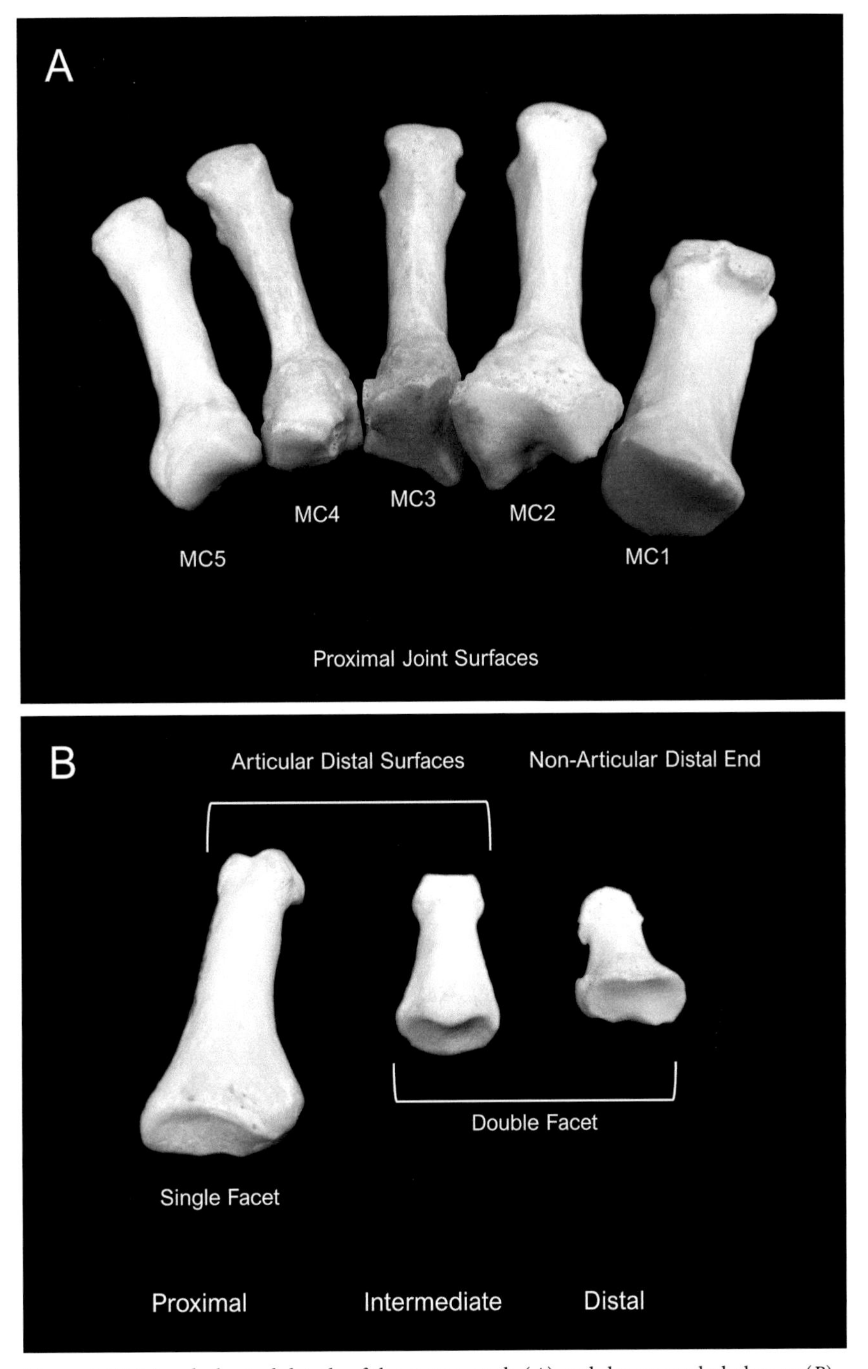

Figure 4.20. Morphological details of the metacarpals (*A*) and the manual phalanges (*B*)

is the most important bone in the postcranial skeleton for the purposes of forensic analyses. This is because the primary sexual characteristics of males and females are observable here. Female pelvic anatomy has been shaped by the necessities of giving birth. Therefore, sex assessment is principally performed using the pelvis. In addition, two joints (the pubic symphysis and auricular surface of the sacroiliac joint) provide critical information on age-at-death. The os coxa forms from three primary growth centers: the *ilium,* the *ischium,* and the *pubis.* In **Figure 4.21,** the ilium (shaded blue) is located superiorly and is blade-like in appearance. The ischium (shaded green) is posterior, and the pubis (shaded red) is anterior. These fuse together in adolescence to form one bone, and the location of fusion forms the socket of the hip called the *acetabulum.* The superior portion of the ilium is called the *iliac*

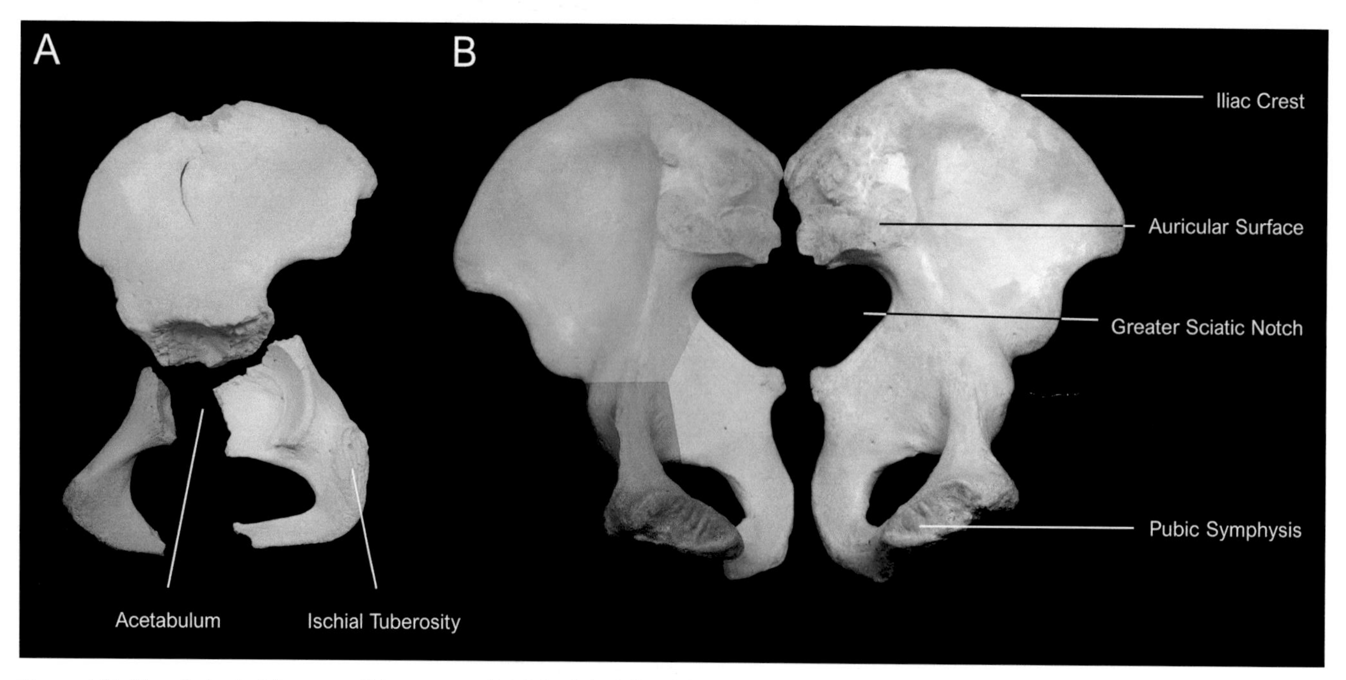

Figure 4.21. Morphological features of the os coxa. (*A*) Subadult, (*B*) Adult.

crest. This is where your pants sit on your hips. The posterior aspect of the ischium is called the *ischial tuberosity,* which you are probably sitting on right now. The two ossa coxae articulate anteriorly at a joint called the pubic symphysis and posteriorly with the sacrum at a joint called the sacroiliac joint. The contribution of the ilium to this joint is called the *auricular surface*.

The **femur** is the thigh bone. Like all long bones, the proximal and distal epiphyses are diagnostic and can be easily identified in fragmentary remains. The proximal end has a head that articulates with the pelvis at the acetabulum. The femur also has a *greater and lesser trochanter* near the proximal end that form from different growth centers (**Figure 4.22**). The distal end is large and has two distinct articular surfaces called *condyles* in addition to a joint surface on the anterior aspect of the shaft for the articulation with the patella. The shaft is round in cross-section, larger and thicker than the humerus, with a distinct raised ridge on the posterior aspect of the shaft called the *linea aspera*.

The **patella** is a small triangular bone that articulates with the femur at its distal end. It forms the joint of the knee with the femur and the tibia.

The **tibia** is the primary weight-bearing bone of the lower leg. It has a shaft that is compressed mediolaterally and triangular in cross-section and a proximal joint surface that is flat and platform-like, with two distinct surfaces for articulation with the distal condyles of the femur. The distal joint is square with a flat surface for articulating with the talus, one of the tarsals in the ankle.

The **fibula** is the other bone in the lower leg. It is angular and thin in cross-section, with round and slightly pointed proximal and distal joints.

In parallel with the hand, the bones of the foot consist of 7 **tarsals** (named in **Table 4.1**), 5 **metatarsals**, and 14 **phalanges**. These phalanges are also divided into proximal, intermediate, and distal types (**Figure 4.23A**). The **calcaneus** and **talus** are two large bones of the ankle that form the ankle joint itself (talus) and the heel of the foot (calcaneus) (**Figure 4.23B**).

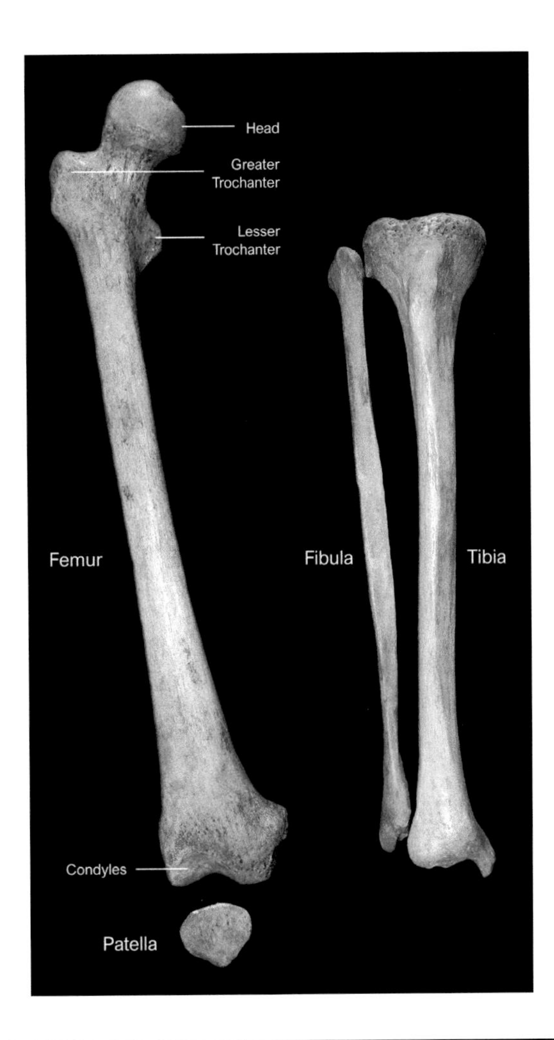

Left: Figure 4.22. Bones of the lower extremity.

Below: Figure 4.23. Bones of the feet (*A*) with close-up, superior view of the calcaneus and talus (*B*).

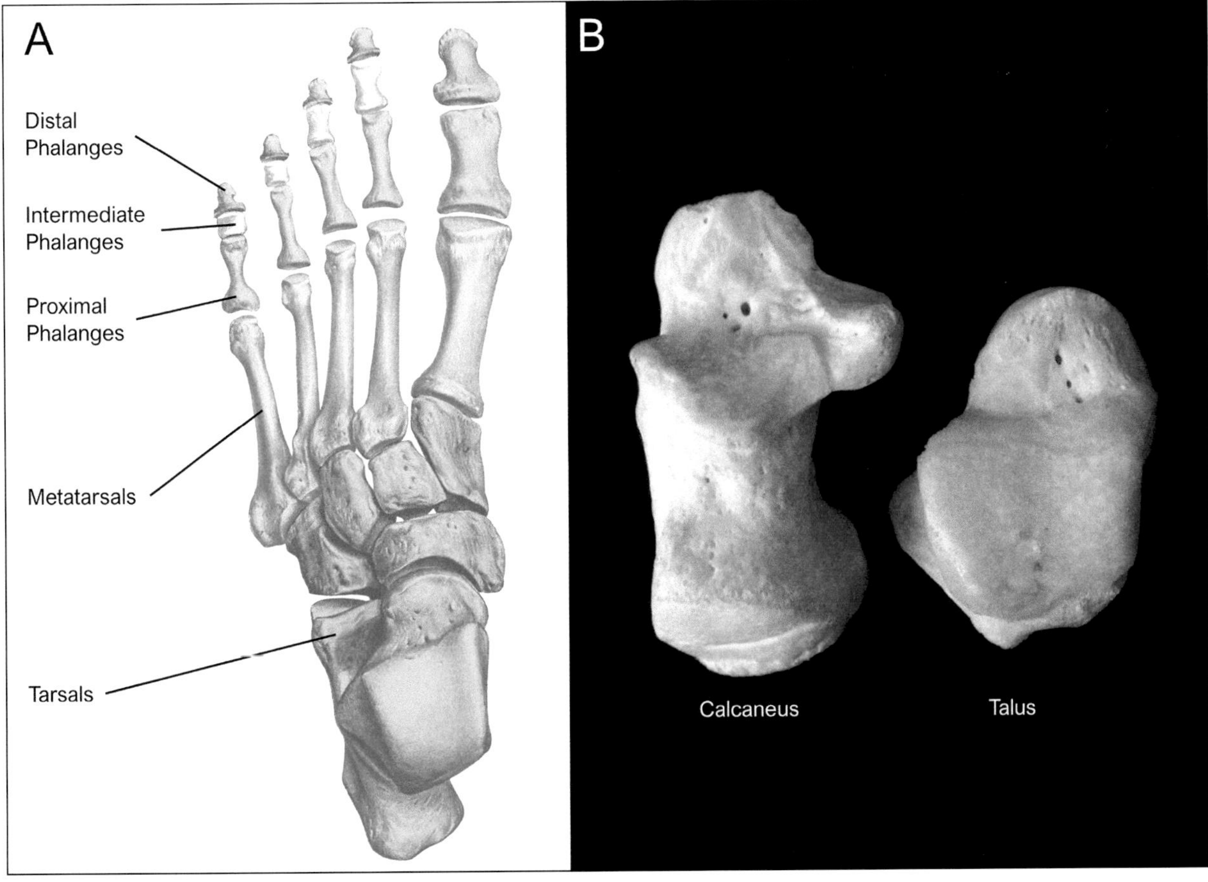

LEARNING CHECK

The next series of questions refer to the bones identified in **Figure 4.24**. You will use all available resources to answer them. Please select all answers that apply.

Q7. Which letter(s) indicate bones that are unpaired?
Q8. Which letter(s) indicate bones that are part of the appendicular skeleton?
Q9. Which letter(s) indicate the phalanges?
Q10. Which letter(s) indicate the zygomatic?
Q11. Which letter(s) indicate flat bones?
Q12. Which letter(s) indicate the bone that is immediately lateral to the tibia?
Q13. Which bone is indicated by the letter C?
Q14. Which bone is indicated by the letter L?
Q15. Which bone is indicated by the letter I?
Q16. Which bone is indicated by the letter E?
Q17. Which bone is indicated by the letter B?
Q18. Which letter(s) indicate a bone that articulates with the frontal bone?
Q19. Which letter(s) indicate a bone that articulates with the scapula?
Q20. Which letter(s) indicate a bone that articulates with the talus?

Sources of Skeletal Variation

The preceding sections have focused on general anatomical features of the human skeleton. However, even paired elements in the same skeleton are not identical. There are always minor variations in size and anatomical detail because the skeleton that is observed in forensic cases is the product of both genetic and epigenetic factors. Researchers have spent considerable time trying to identify the genes and proteins involved in osteogenesis (bone formation). However, while genes provide the code for skeletal gross anatomy, the specific appearance of a bone in an individual also depends on a number of other factors. The factors affecting skeletal morphology that are most relevant within a forensic setting are discussed in later chapters of this book. These include sex (chapter 6), age (chapters 7 and 8), geographic or population variation (chapter 9), and idiosyncratic skeletal changes related to an individual's behaviors, health, and disease (chapter 11).

End-of-Chapter Summary

This chapter presented a basic introduction to human osteology. The goal was to define the terminology used to describe the skeleton, identify the bony elements of the skeleton, and identify the landmarks and features that are most relevant to forensic anthropology. The skull is a complex structure that includes over two dozen individual elements that articulate at sutures. These elements combine to form complex anatomical structures such as the orbits, the nasal cavity, the oral cavity, and the ear. The axial skeleton includes elements of the thorax and spinal column. The ribs, vertebrae, sternum, sacrum, and coccyx comprise the postcranial portion of the axial skeleton. The appendicular skeleton includes the bones that form the shoulder girdle, pelvic girdle, upper limbs, and lower limbs. Most skeletal elements are found in the appendicular skeleton, while the skull is by far the most complex anatomical structure in the skeletal system. Bones can be paired or unpaired. Bones can also be categorized by shape (flat, irregular, long, short). Although osteology focuses on how humans are similar, there are also a number of factors that make each person's skeleton unique. It is important

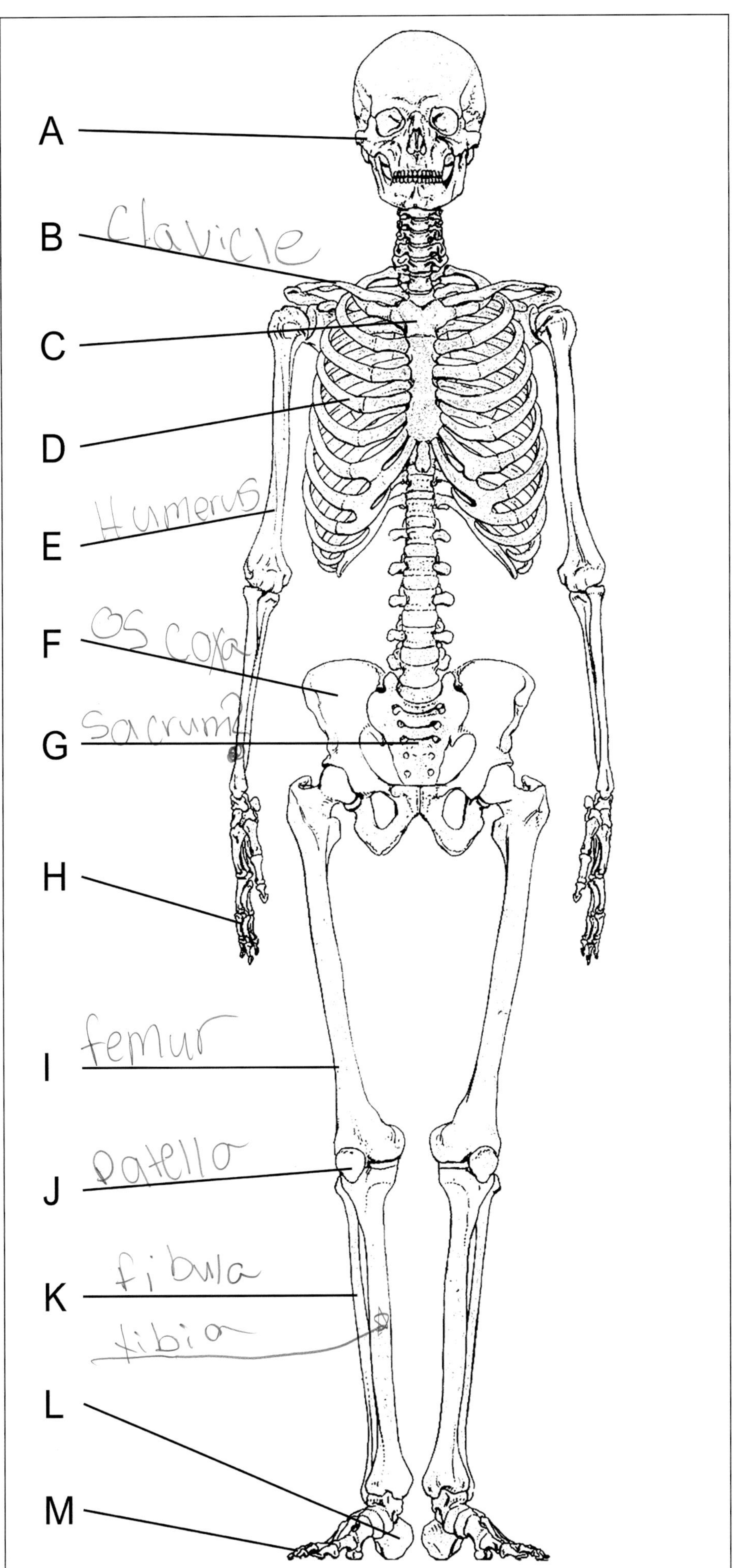

Figure 4.24. Schematic of the human skeleton.

that the forensic anthropologist is aware of these factors as possible explanations for what is observed when analyzing skeletal remains.

End-of-Chapter Exercises

Exercise 1

Materials Required: **Figures 4.25** and **4.26**

Directions: It's time to practice whole bone identification. Use **Figures 4.25** and **4.26** and any other available materials to answer the following questions. The images are not to scale, so pay close attention to the shape of the bone as well as its specific features (e.g., foramina or articular surfaces). For each of the following questions, in addition to identifying the bone that is depicted in each image, please indicate whether the bone is paired or unpaired and whether it is part of the axial skeleton or the appendicular skeleton. If a bone is paired, please indicate the side of the body that it comes from.

Question 1: Which bone is depicted in **Figure 4.25A**?

Question 2: Which bone is depicted in **Figure 4.25B**?

Question 3: Which bone is depicted in **Figure 4.25C**?

Question 4: Which bone is depicted in **Figure 4.26A**?

Question 5: Which bone is depicted in **Figure 4.26B**?

Question 6: Which bone is depicted in **Figure 4.25C**? Which portion (i.e., proximal or distal) of the bone is shown?

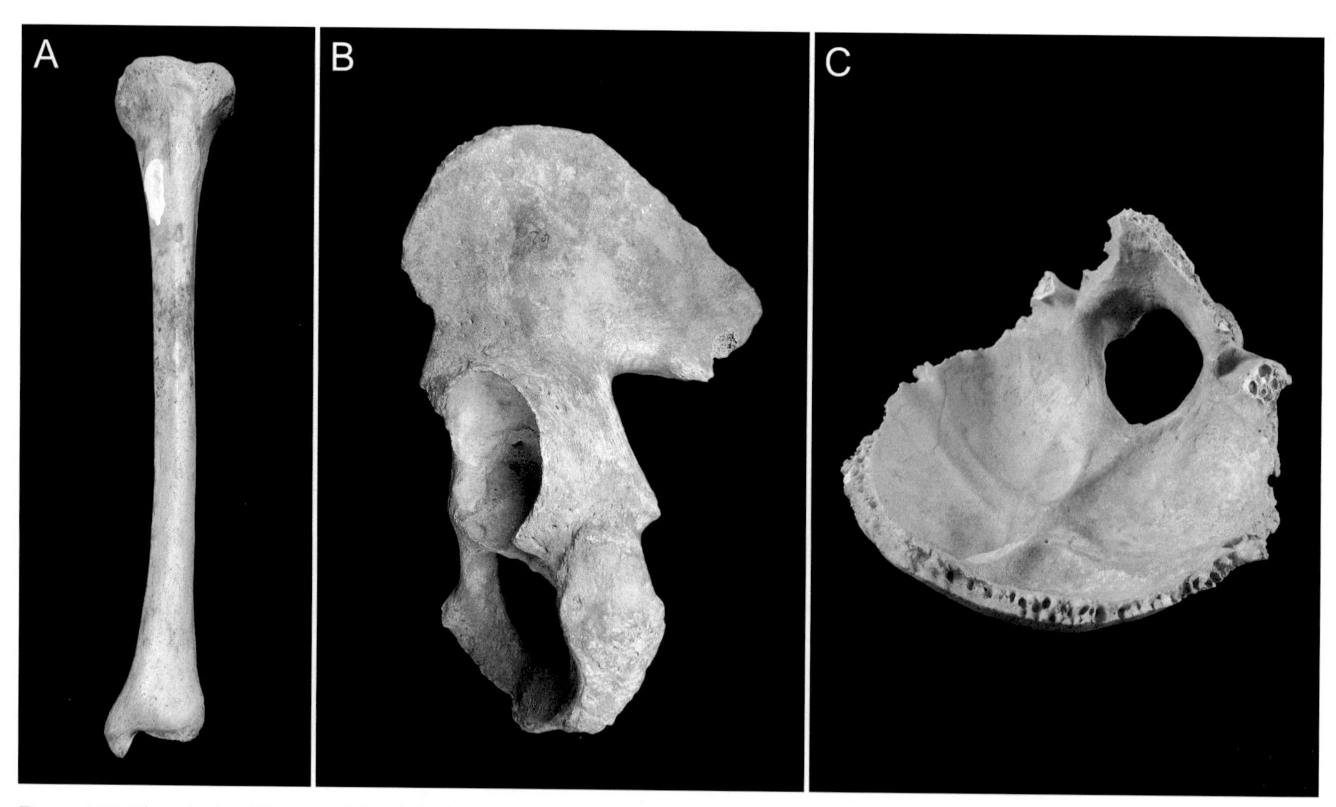

Figure 4.25. Three isolated human skeletal elements.

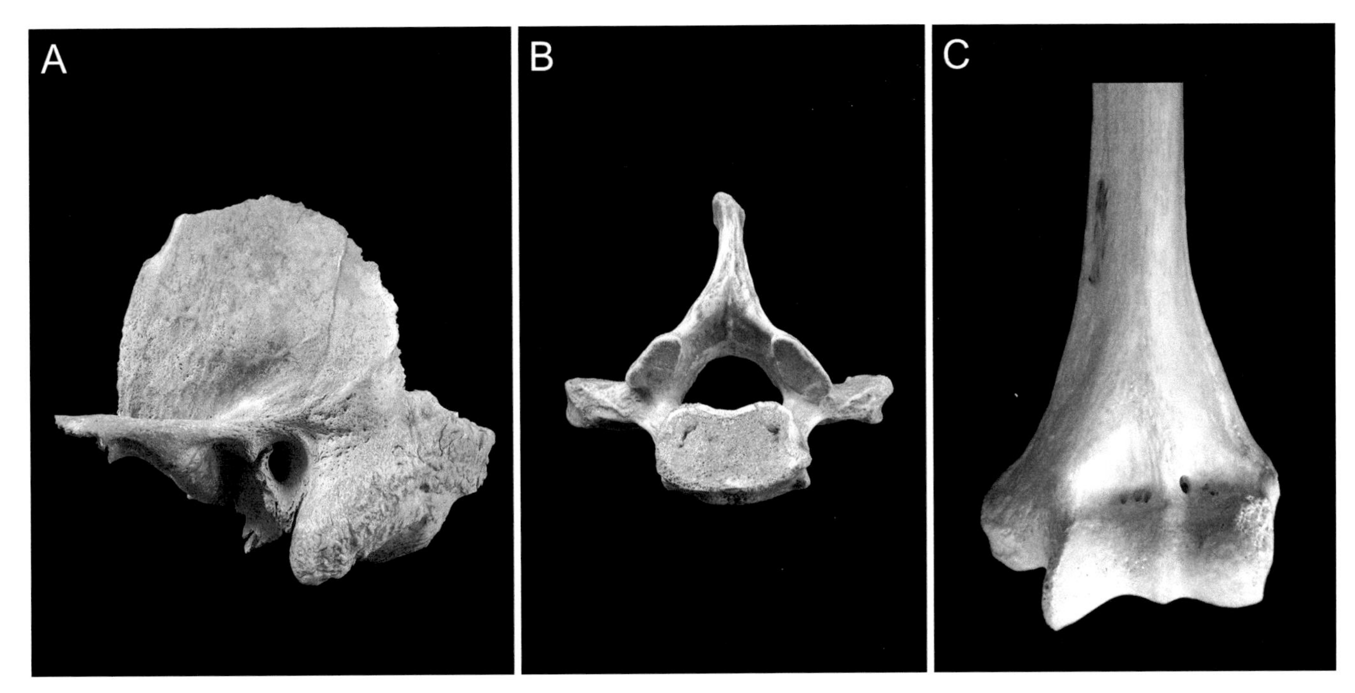

Figure 4.26. Three isolated human skeletal elements.

Exercise 2

Materials Required: **Figure 4.27**

Directions: A firm understanding of human osteology allows forensic anthropologists to create a skeletal inventory. This is the first step in determining the minimum number of individuals (MNI) that a collection of bones is likely to represent. A crude approximation of MNI can be made by simply counting the number of a given skeletal element and dividing it by how many of those bones exist in a single skeleton (rounding the answer up). For example, if a collection of bones has seven radii (plural of radius), then you know that there are at least four individuals represented (7 divided by 2—each skeleton has 2 radii—is 3.5, which rounds up to 4). An accurate calculation of MNI will depend on the side of the body that the bones come from, the age of the individuals, the sex of the individuals, as well as other characteristics, but this requires advanced training in osteology. **Figure 4.27** shows part of one of the arrangements of human bones at the Sedlec Ossuary. Use **Figure 4.27** and all available resources to answer the following questions.

Question 1: What are the different kinds of bones that are present in this arrangement (you may use "cranium" instead of listing the individual bones of the cranium)?

Question 2: How many sacra (plural of sacrum) do you see?

Question 3: How many ossa coxae (plural of os coxa) do you see?

Question 4: Based on your answers to Questions 2 and 3, what is an approximation of the MNI represented in **Figure 4.27** (i.e., you do not have to consider the side of the bone in your calculation)?

Question 5: Those round bones along the borders and in the center of the assemblage are femoral heads. Likewise, that upside-down U-shaped bone under the central skull is a distal femur. Based on this information, what is an approximation of the MNI represented in **Figure 4.27**?

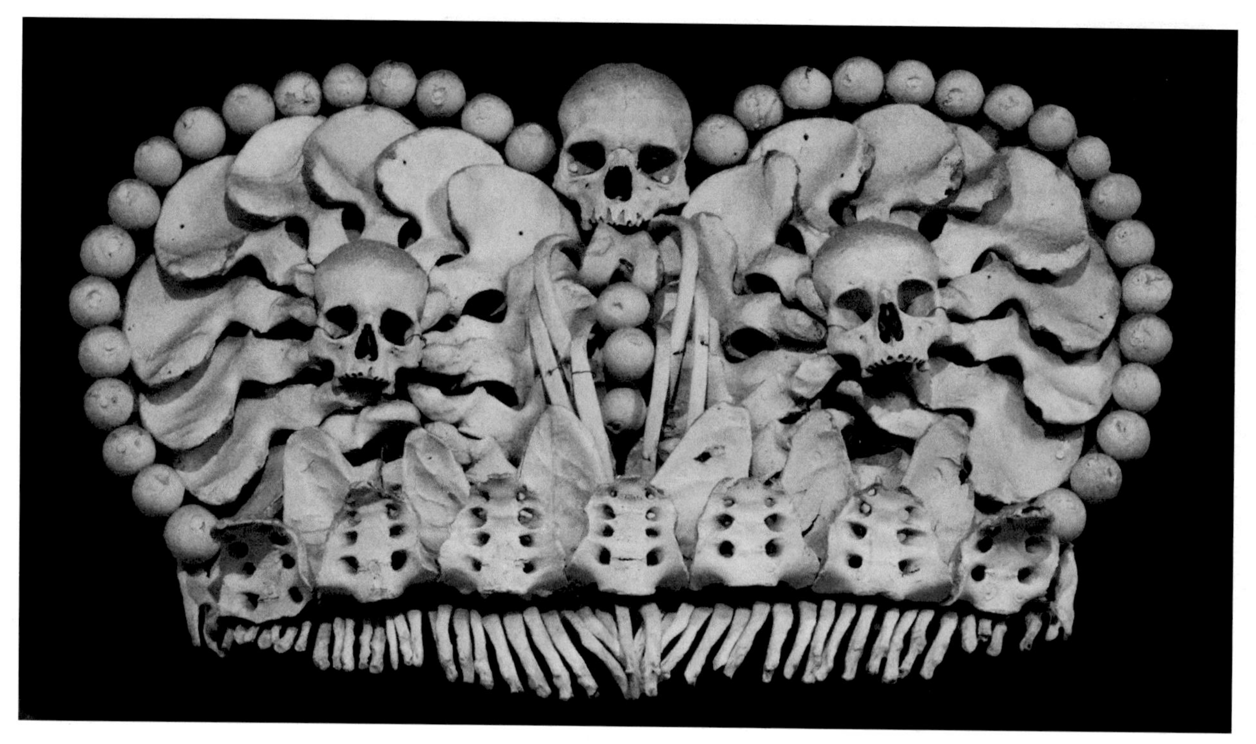

Figure 4.27. Artistic display of human remains using various skeletal elements.

Exercise 3

Materials Required: **Figures 4.28–4.30.**

Directions: While this chapter has focused on whole bone osteology, forensic anthropologists are frequently required to identify pieces of bone that have been fragmented as a result of trauma (chapters 12, 13, and 14) or taphonomic processes such as fire, animal scavenging, or weathering (chapter 15). Accurate fragment identification requires years of experience and advanced training in human osteology, but the diagnostic shape of some bones and/or their specific features and joint surfaces makes identifying some fragments easier than others. Using all available resources, can you identify the following bone fragments?

Question 1: How many bones are represented by the fragment shown in **Figure 4.28**? Which bones are present? The best way to answer this question is to first orient the fragment to the general location in the skeleton. Then, pay attention to the shapes of the bones, sutures/joints that may be present, and the negative spaces created by the elements.

Question 2: **Figure 4.29** depicts a fragment of which bone(s)? Pay close attention to the unbroken contours of the bone and any articular surfaces that are visible. Note the upper portion is highly weathered, which may affect the appearance of any joint surfaces present.

Question 3: **Figure 4.30** depicts burned fragments of two bones. Can you identify which bones are shown? Hint: the distal ends of the bones are pictured and the photographs are at the same scale (a bigger hint: these bones articulate in real life).

References

Mays S. 2021. *The Archaeology of Human Bones,* Third Edition. London: Routledge.

Sobotta J. 1909. *Atlas and Text-Book of Human Anatomy.* Philadelphia, PA: W.B. Saunders Co.

White TD, Black MT, Folkens PA. 2012. *Human Osteology,* Third Edition. Burlington, MA: Academic Press.

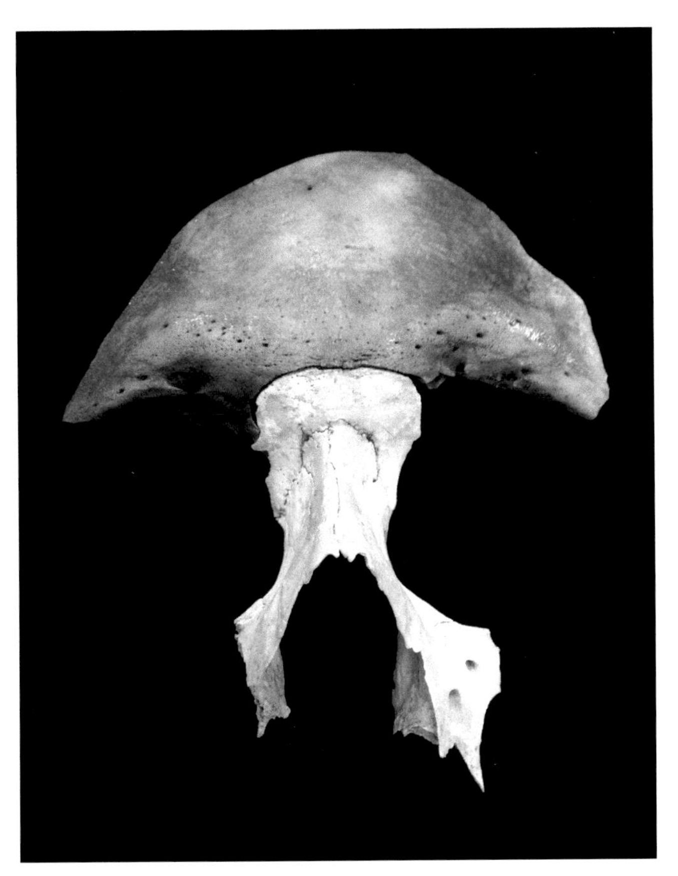

Figure 4.28. A fragment of the human skeleton.

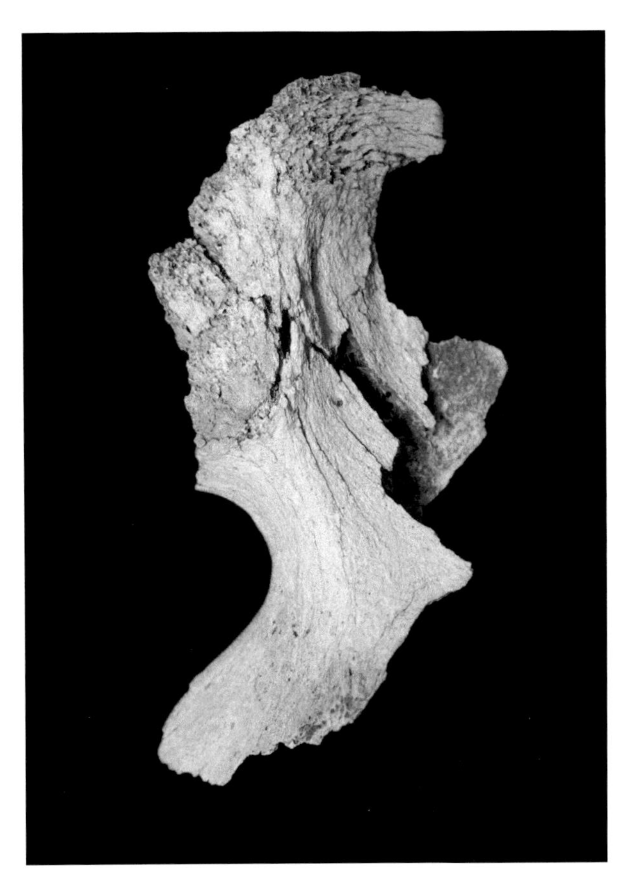

Figure 4.29. A fragment of the human skeleton with severe weathering.

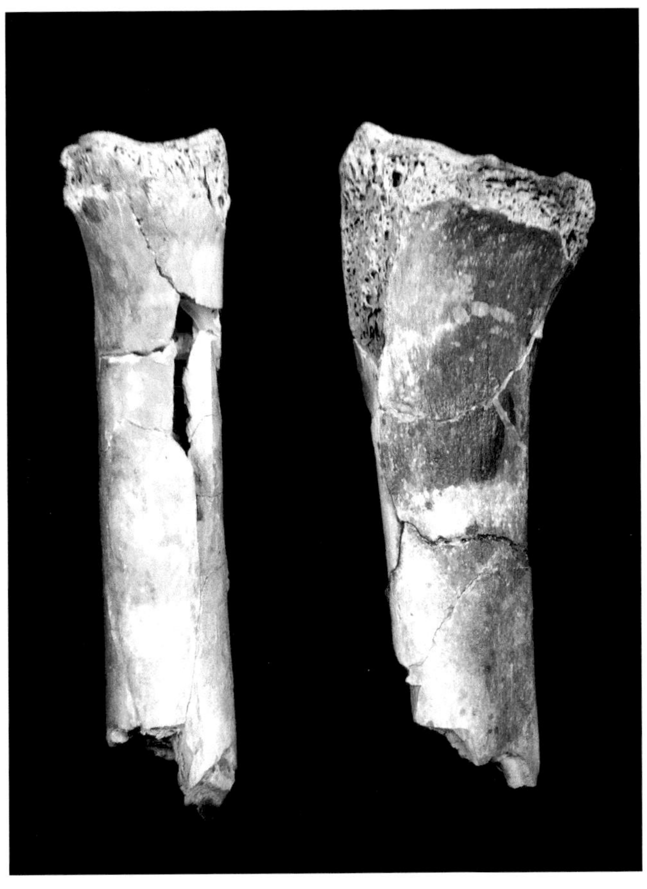

Figure 4.30. Two fragments of the human skeleton with burn marks.

5

Establishing Medicolegal Significance

Learning Goals

By the end of this chapter, the student will be able to:

Differentiate remains that are of medicolegal significance versus general forensic interest.
Describe techniques for determining whether material is skeletal in nature.
Describe techniques for determining whether skeletal material is human or non-human.
Describe observations for determining whether human skeletal material is contemporary or non-contemporary in age.

Introduction

Not all remains that are brought to a forensic anthropologist are of medicolegal significance. In fact, not all specimens brought to a forensic anthropologist are even biological remains, that is, osseous or skeletal, in nature. While this statement may seem confusing given the focus on whole bone osteology and complete skeletons throughout this book, one must also consider very fragmentary materials for which it may be difficult to determine whether the material is even bone. If the material is bone then it may not be human, and if human then it may not have medicolegal significance. Remains that are of medicolegal significance are recent in age, generally individuals who died less than 50 years ago, for whom a death certificate would be issued once the identity of the decedent has been established. Even though a fairly high percentage of "cases" brought to the attention of a forensic anthropologist do *not* result in a criminal investigation, it is best practice to consider all materials *potentially* of significance and to follow legal protocols such as chain of custody and proper care and handling of the materials until a final determination has been made.

In determining medicolegal significance three questions must be addressed:

1) Is the material skeletal or dental in nature—i.e., is it bone, or is it some other material? If the specimen is not bone, then it is not of medicolegal significance to a forensic anthropologist.
2) If the material is bone, then is it human or from some other animal? If the skeletal material is non-human, then it is not of medicolegal significance to a forensic anthropologist.
3) If the material is from a human skeleton, then is it contemporary in age, generally considered to be less than 50 years old? If the skeletal material is not contemporary, then it is not of medicolegal significance to a forensic anthropologist.

It is important to note the difference between **medicolegal significance** and general forensic interest. For remains with medicolegal significance, there will generally be an attempt to positively identify the decedent and/or document trauma to speak to cause and manner of death and aid in the death investigation. For remains of general **forensic interest**, the remains may be part of a criminal investigation, but the specific identity of the decedent is not relevant to that investigation and may be unknowable.

Perhaps the best illustration of this difference involves ancient or non-contemporary skeletal material. The recovery of a skull that had been taken from an archaeological site, for example, provides evidence of illegal looting, but the specific identity of the individual would not need to be established, is likely impossible to know, and is not relevant to prosecuting the case. Under the *Native American Graves Protection and Repatriation Act,* it is illegal to acquire, sell, or trade *any* skeletal remains of Native Americans or Hawaiians. This is a federal law that applies in every state. Individual states may have other laws that allow for the prosecution of the possession and sale of human remains from non-Native Americans—for example, the looting of nineteenth-century graves from New Orleans cemeteries for sale on the collecting market. Recovery of trophy skulls brought home from foreign wars may be evidence of a war crime. Trophy skulls can also be evidence of participation in the illegal trade in antiquities where decorated skulls are sold for their aesthetic appeal, sometimes legally, sometimes not. None of these examples would lead to the identification of the decedent and the issuance of a death certificate. They are potentially of forensic interest but not medicolegal significance. Even non-human skeletal remains can be of forensic interest, such as prosecuting the illegal trade in endangered species under CITES (*Convention on International Trade in Endangered Species*).

Is the Material Skeletal?

Most forensic anthropologists and osteologists keep a collection of materials that, due to chance, look very much like human remains (**Figure 5.1**). These collections form over the course of one's career and make a great reference library for comparing against future cases. They also make great questions on bone quizzes! The number of different materials that can look like bone is nearly endless—PVC pipe, plastics of all kinds, shells, tubing, foam, various types of rocks, twigs and branches, ceramics, metal, etc. How these different materials resemble human bone is idiosyncratic, which makes it very hard to provide hard and fast rules. A pebble, for example, can look like a human tooth, just by chance. A pine nut bitten in half can look very much like a broken canine or incisor root, and one of us (C. Stojanowski) at one time had a twig that looked exactly like a human clavicle. Prehistoric pottery, especially when dirty, can look very similar to fragments of the cranial vault (the presence of diploë is key to identifying cranial fragments) (**Figure 5.2**). As with most things in forensic anthropology, the key is to master human osteology, to know the subtle details of bone anatomy, and to expose oneself to a variety of archaeological and anthropological contexts to experience variation in taphonomic signatures of human bone. Handling skeletal material from a variety of contexts is also important for developing "an eye" for what is bone and what is not. One gains a sense of how human remains *feel* in the hand, the weight given the fragment size (density), the texture of the surface, etc. These subtle textural differences are difficult to describe and come with training and experience.

While it would be difficult to confuse large fragments with human remains, cases involving mass fatality incidents such as plane crashes not only have highly fragmented remains but fragments of many other kinds of materials, much of which is burned and mixed together. In such cases, sorting the human remains from non-skeletal materials is critical for positive identification and repatriating the remains to relatives. Cremations are another context in

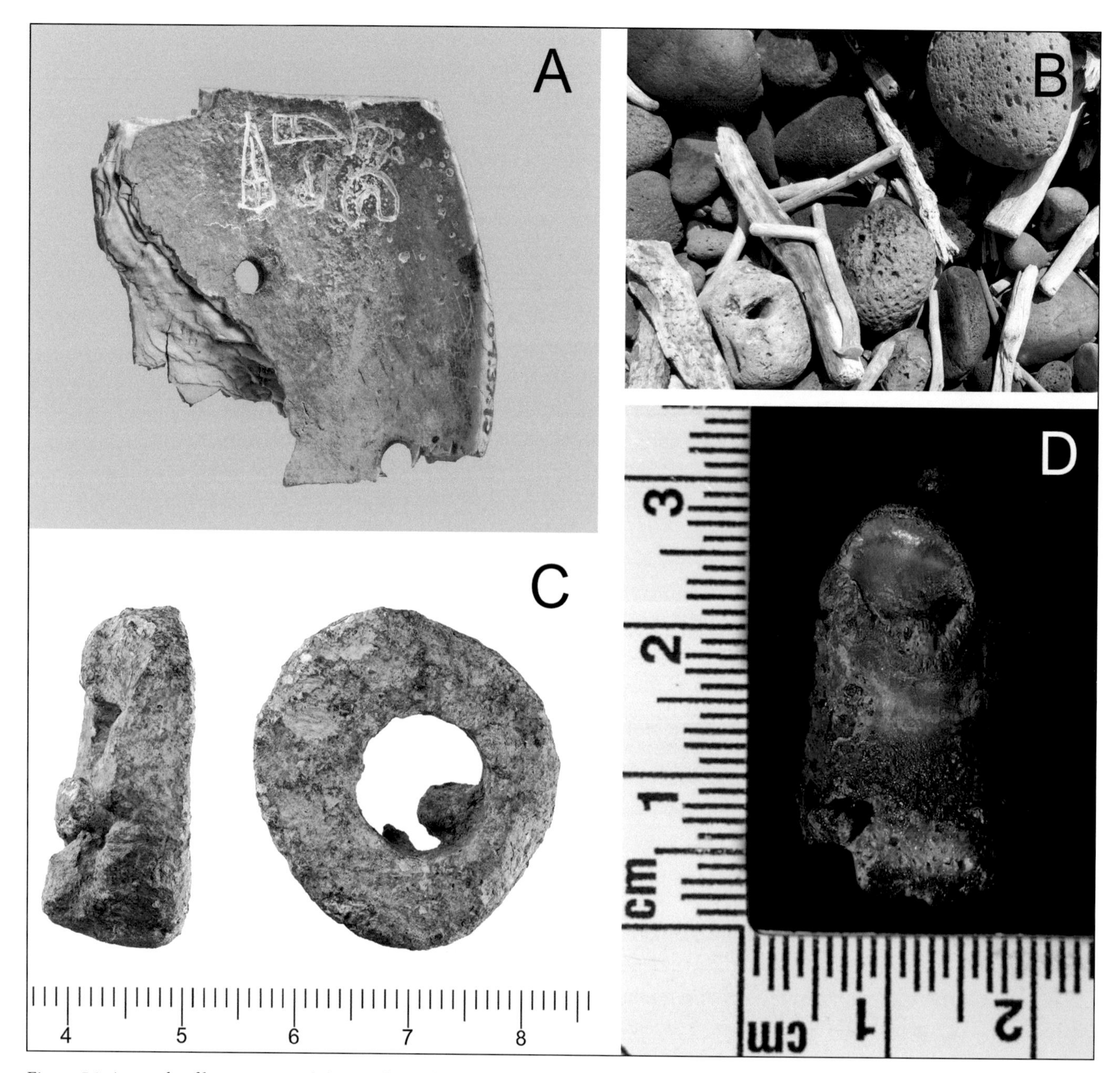

Figure 5.1. A sample of bone mimics: (*A*) carved and drilled shell fragment; (*B*) bleached driftwood; (*C*) lead spindle whorl; (*D*) concretion from inside a pipe.

which it is common to find a mixing of osseous and non-osseous materials, sometimes by design of the perpetrator of the crime or sometimes in cases of fraud where other materials are returned to loved ones as a fraudulent memento. With highly fragmented and burned materials, more labor-intensive methods may need to be used. These methods are summarized by Christensen et al. (2014) and include: *radiography, histology* and *microscopy,* and *elemental analysis.*

Radiography is the use of X-ray or CT scan technology to visualize the density of material and provide a view of its internal structure. The radiodensity of a specimen can be quantified using the Hounsfield scale, where air has a value of -1,000 (radiolucent) and distilled water has a value of 0. Materials with a positive value on this scale are said to be radiopaque, which includes bone. Trabecular bone is less dense than cortical bone, and this is reflected in their

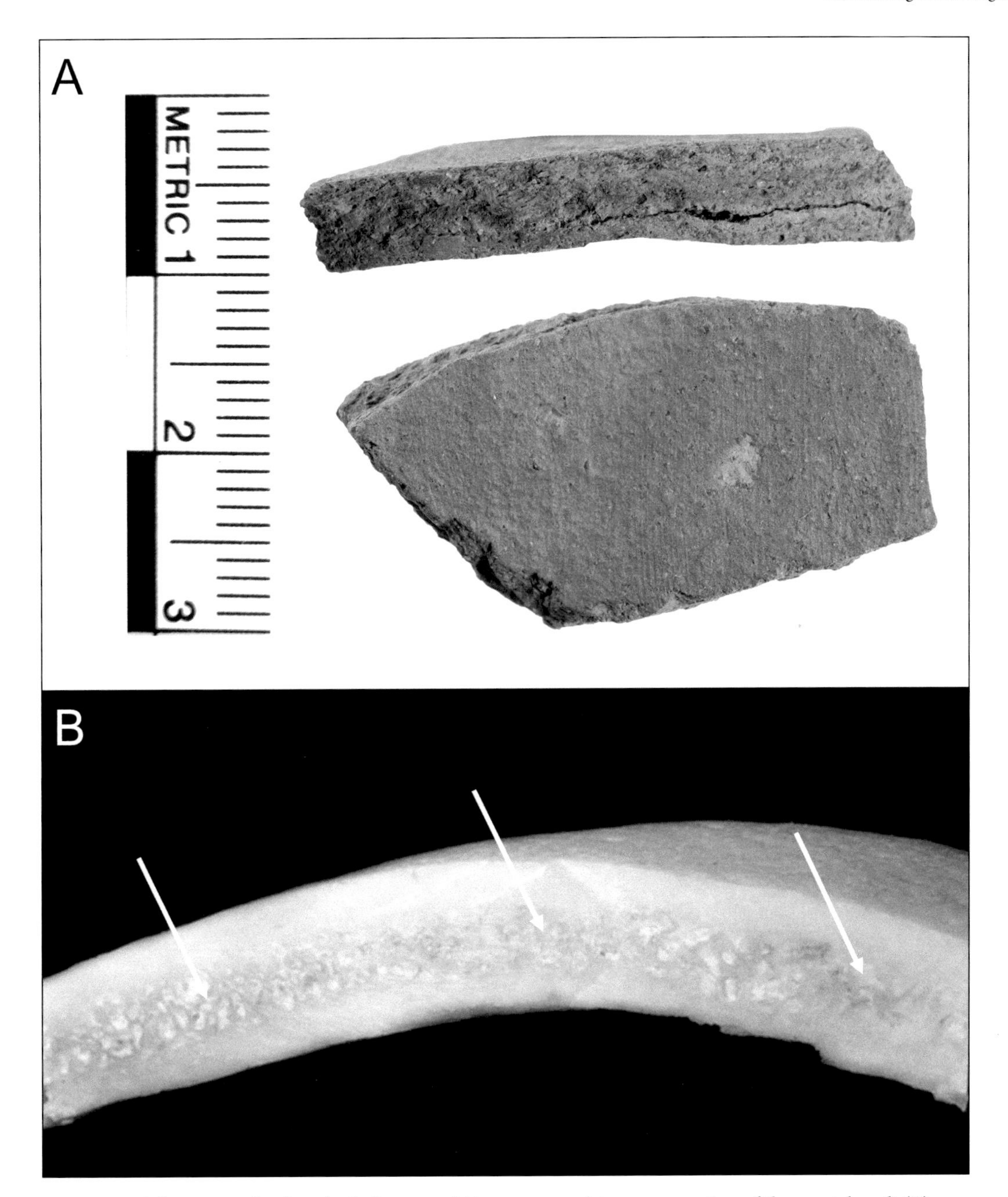

Figure 5.2. A fragment of archaeological pottery (*A*) as compared to a cross section of the cranial vault (*B*). White arrows indicate the layer of diploë that can help differentiate between bone and ceramics.

Hounsfield values (300–400 for trabecular bone, 500–2,000 for cortical bone). Metals and rocks are more radiopaque than bone (values of 3,000 to 20,000), plastics are less radiopaque than bone (values of around 200), while different types of glass have values that overlap those of bone. Radiography can also reveal internal anatomical structures, such as the presence of trabecular bone, a medullary cavity, or in the case of a tooth, the pulp chamber that you wouldn't expect to see in non-organic materials (see chapter 3). For example, a long tubular specimen that is uniformly radiodense would *not* be bone because it lacks the presence of distinct structures such as the cortical periosteal surface, the underlying trabecular bone, and the internal medullary cavity that is seen in long bones (**Figure 5.3**).

Histology is the study of tissue at the microscopic level. As discussed in chapter 3, bone has specific structures when viewed microscopically, including the presence of osteons in

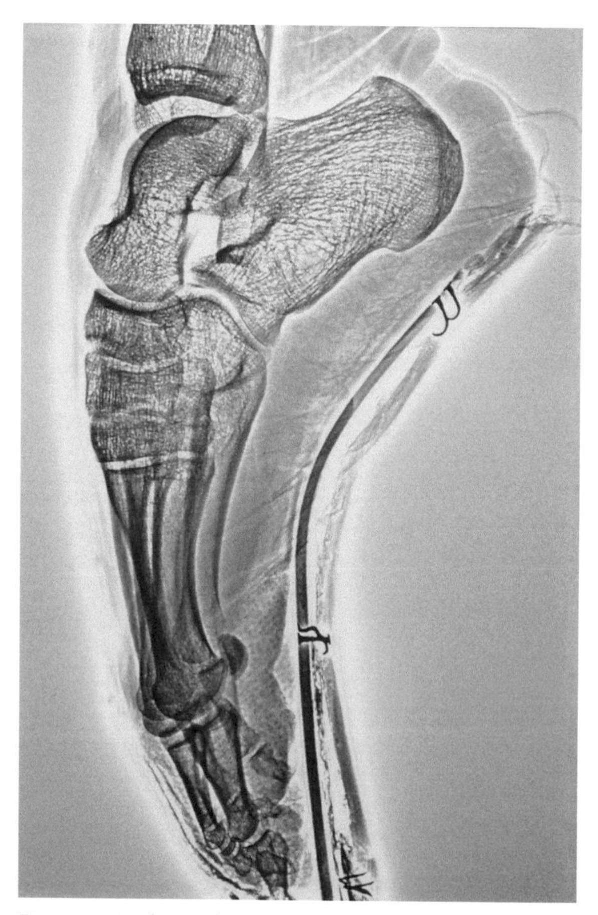

Figure 5.3. Radiograph of a dancer's foot in a ballet slipper showing differences in texture and radiodensity between the bones and the surrounding materials of the soft tissue and shoe.

cortical bone or growth structures preserved in teeth that appear as incremental growth lines of enamel, dentin, and cementum (**Figure 5.4**). None of these features would be present in inorganic materials. Lower resolution microscopic analysis can also be useful in determining whether a specimen was bone. Christensen et al. (2014:94) describe bone as having "a compact and sometimes grainy surface." This appearance is contrasted against other types of materials. For example, Ubelaker (1998) describes cases in which a non-bone determination could be made based on low resolution microscopy indicating a fibrous composition, an appearance that "varied from irregular layers to a mass of irregularly arranged cells (Ubelaker, 1998:516)," or a chalky appearance.

For *elemental analysis,* a variety of methods have been proposed for ascertaining the elemental composition of a fragmentary specimen (Christensen et al., 2012; Ubelaker, et al., 2002). These methods rely on identifying the specific composition of bone, in comparison to other materials, at the elemental level. Since bone is organic it contains carbon, calcium, and distinct signatures of phosphorus, making it easy to determine the presence of these elements that are lacking in metals, glass, and plastics. The presence of phosphorus also allows for the differentiation of coral, shell, and some woods from osseous or dental material (Christensen et al., 2012). Bone's organic nature also gives it properties that make it fluoresce when exposed to short wave light sources, making bone appear brighter than other non-osseous materials. However, the fluorescent properties of bone are a result of its collagen content, and ancient remains or poorly preserved remains with little collagen may not fluoresce, thus giving the false appearance of being non-osseous in nature.

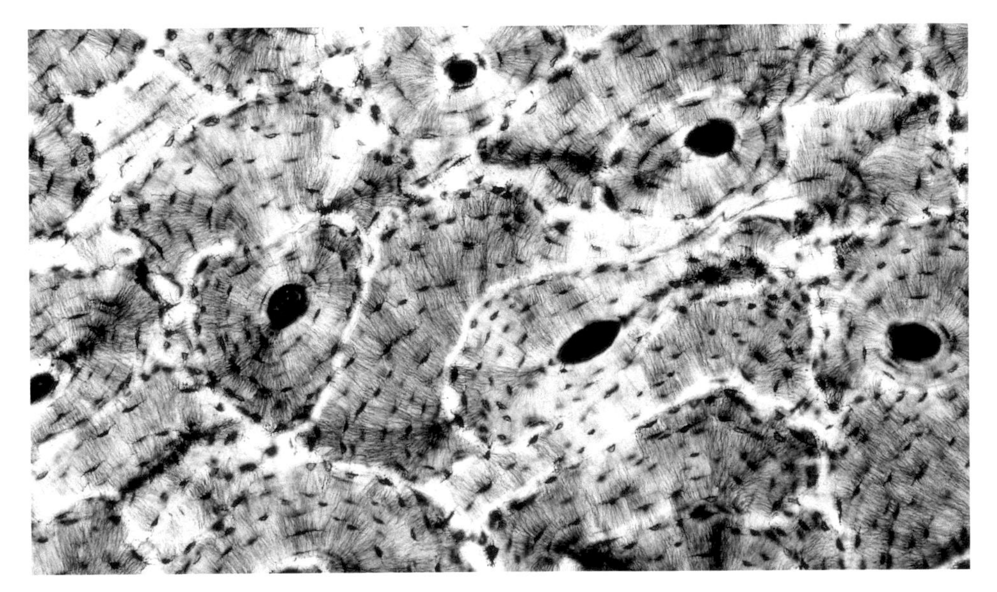

Figure 5.4. Osteons in human cortical bone.

LEARNING CHECK

Q1. Law enforcement has received an anonymous tip that an artist is concealing human remains inside of various sculptures. Radiographs are obtained of these works of art and submitted to you for examination (**Figure 5.5**). Which of the following radiographs contains material consistent with skeletal and/or dental remains?

A) Radiograph A
B) Radiograph B
C) Radiograph C
D) None of these radiographs include skeletal or dental remains.

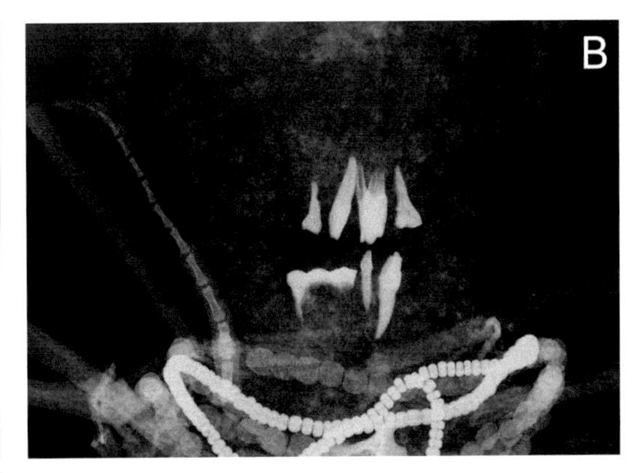

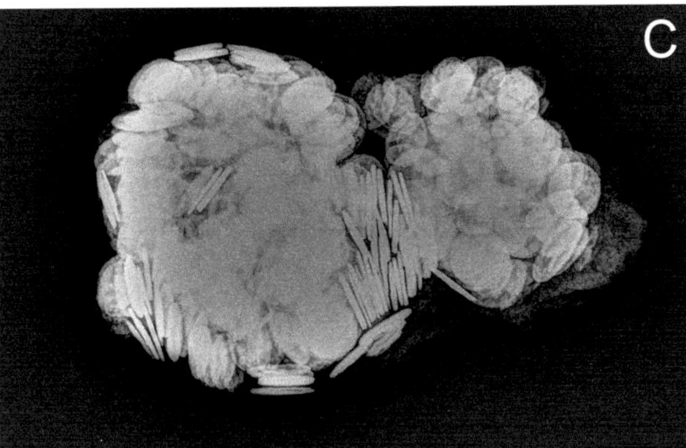

Figure 5.5. Radiographs of three objects of uncertain composition.

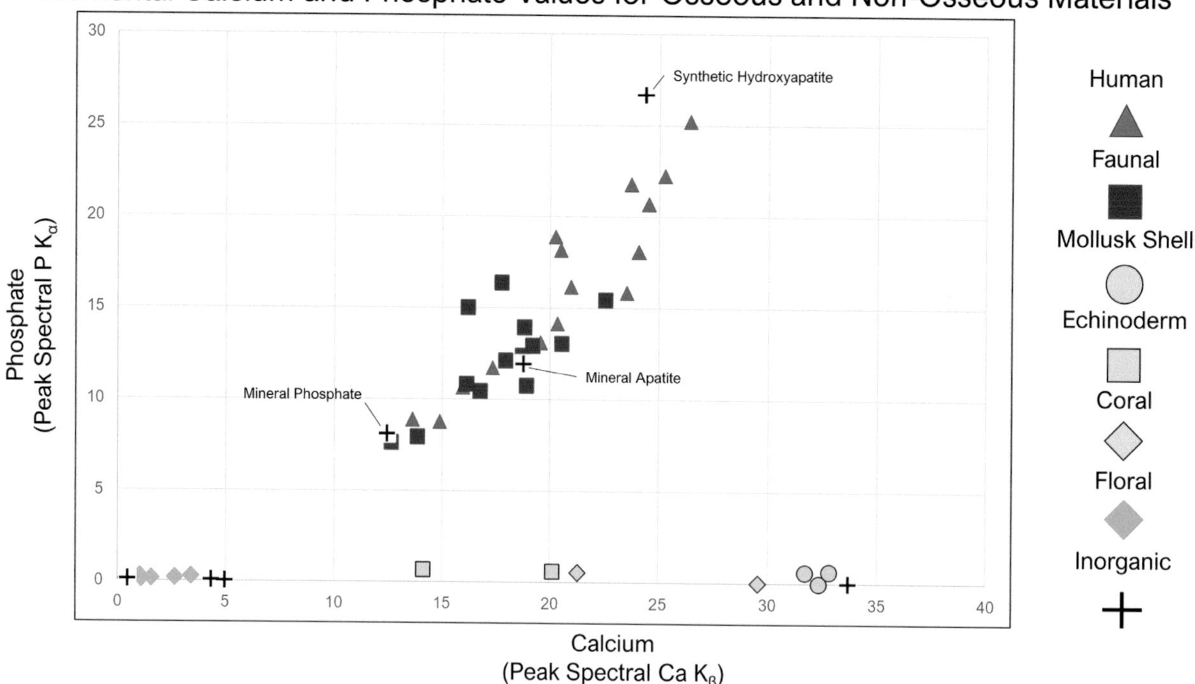

Figure 5.6. Scatterplot of calcium and phosphate values for various materials.

Q2. **Figure 5.6** illustrates the range of elemental calcium (Ca) and phosphate (P) values obtained using a handheld X-ray fluorescence spectrometer to assess various osseous and non-osseous materials. Based on these values, which of the following is an accurate statement?

A) The elemental Ca and P values for mineral apatite are very similar to those of echinoderms (e.g., starfish and sand dollars).

B) Floral (i.e., plant) remains can be easily confused with human bones based on their elemental Ca and P values.

C) Fauna and human bones have similar ranges of elemental Ca and P values, with fauna showing less overall variation.

D) The floral remains tested in this study all exhibit very high elemental phosphate.

Q3. While building a new tool shed, a homeowner found what they thought might be pieces of human bone in their yard (**Figure 5.7**). Which of these, if any, are consistent with skeletal and/or dental remains?

A) The fragment in A is consistent with skeletal and/or dental material.

B) The fragment in B is consistent with skeletal and/or dental material.

C) The fragment in C is consistent with skeletal and/or dental material.

D) None of these fragments are consistent with skeletal or dental material.

Is the Bone Specimen Human or Non-Human?

Once the forensic anthropologist has determined that the material is bone the next question is whether the bone is human or non-human. Methods for identifying whether bone is human or not fall into two categories. The first, which is the most often used, is simple visual inspection of the specimen (macroscopic). The second, which is more labor intensive, uses

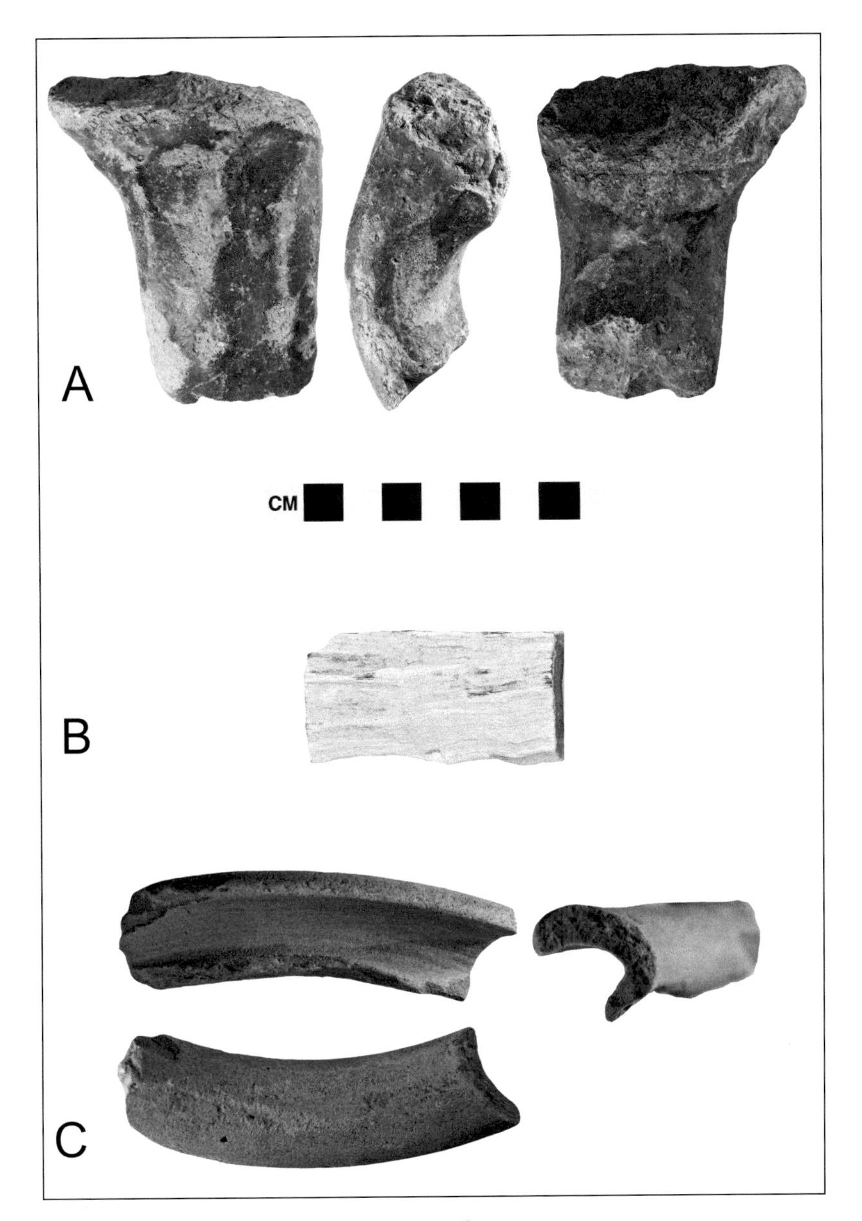

Figure 5.7. Three objects of uncertain material origin.

microscopic methods and histology to determine the species of origin. Because the latter are more labor intensive, they are used only for fragmentary or highly degraded samples.

With macroscopic methods there are two complementary goals. The first is positive identification of a fragment as belonging to the human skeleton. This is done using details of anatomy and allocating the unknown fragment to a specific part of the human skeleton (i.e., the determination is: left lateral proximal condyle of the tibia). Here, the identification is done with confidence because the anthropologist not only knows the bone is human but knows specifically which bone it is and where in the body it is located. A strong foundation in human skeletal anatomy is key. The second approach is exclusionary. The anthropologist may *not* be able to positively identify the sample as belonging to a human but *can* ascertain whether the appearance of the specimen is inconsistent with known variation in the human skeleton (e.g., the determination is: I am not sure what this is, but it is not human). Ideally, the identification can be strengthened by determining which species the fragment belongs to (e.g., the determination is: this is a cow humerus), which requires training in comparative vertebrate osteology. Because such specialties involve mastery of the anatomy of hundreds of different species, it isn't reasonable to expect a forensic anthropologist to know the differences in skeletal anatomy of dozens of fish species, birds, or mammals. In such cases, consulting with an expert in those areas is recommended and numerous books have been published that provide a suitable reference library for comparison (Adams and Crabtree, 2008; France, 2009, 2011). Nonetheless, certain mammals *are* more likely to be confused for human remains due to similarities in size and how common they are in our environment, including wild animals (deer, raccoon), domestic pets, and farm animals (chickens, cows, pigs, sheep, goats) (**Figure 5.8**).

The human skeleton has been shaped from head to toe by evolutionary mechanisms to accommodate one of the things that makes us unique among mammals—we are obligate

Figure 5.8. Examples of commonly encountered animal remains: (*A*) White-tailed deer, (*B*) Chicken, (*C*) Sheep.

bipeds. To the contrary, all other mammals one is likely to encounter in a forensics consultation are quadrupeds. The differences between a bipedal and quadrupedal skeleton reflect the demands of locomotion. Animals need to move about the environment using the least amount of energy. Bones must be heavy enough to support the body and muscles but not too heavy because that would require more energy (calories) to move. Muscles need to insert on the bone in ways that optimize the energetic cost of movement. Joints need to be shaped in such a way as to support the degree of movement needed, but also provide stability to avoid injuries. Regarding our teeth, humans are omnivores and lack a hyper-specialized dentition, meaning evolution has not shaped our teeth to perform a specific function such as exclusively processing meat or grasses. The teeth of carnivores are designed to shear off and process raw meat, while the teeth of herbivores are designed to process leaves and grasses (**Figure 5.9**).

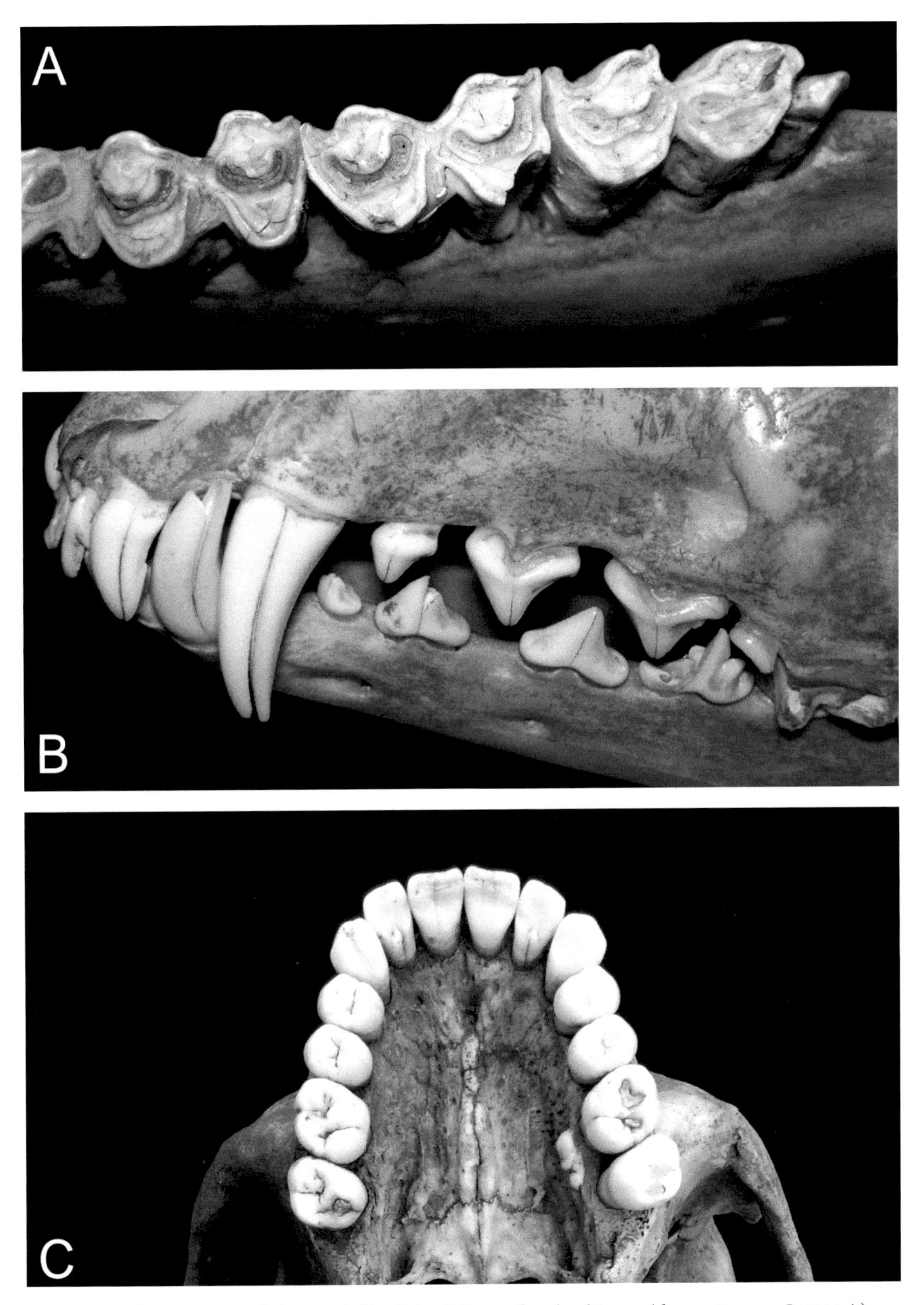

Figure 5.9. Comparison of the specialized dentitions of an herbivore (domestic cow, Image A) and a carnivore (coyote, Image B) with the generalized dentition of humans (Image C).

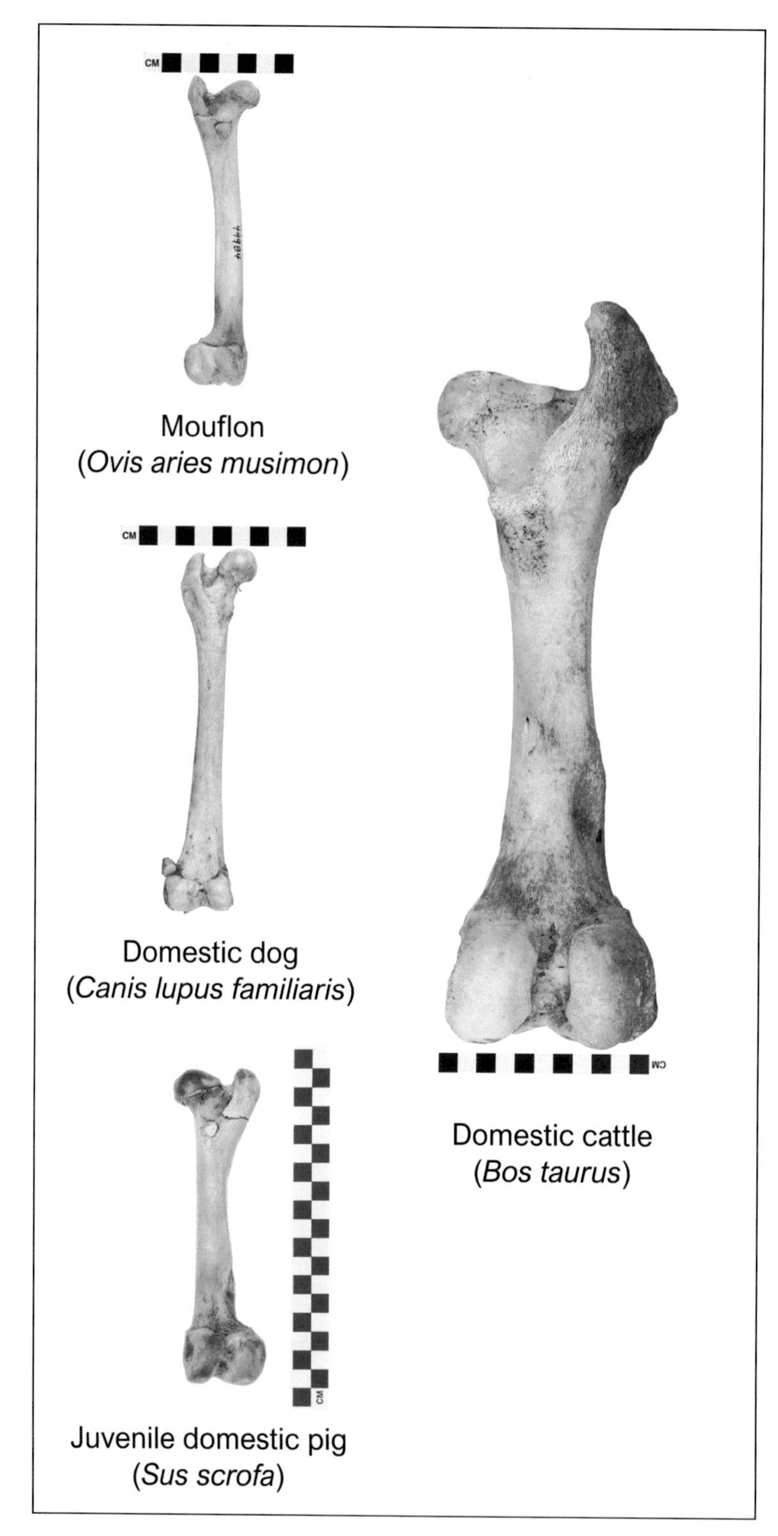

Figure 5.10. Variations in femur size and shape across mammal species.

While evolution is the mechanism that shapes our skeletons and teeth, it is also the mechanism that explains the broad similarities in the skeletal system of most mammals, which is the source of confusion. The principle of *homology* means that we share a similar skeletal structure and that most mammals have the same bones in their bodies. That is, dogs, cats, pigs, beavers, raccoons, cows, horses, and humans all have a humerus, a femur, ribs, vertebrae, frontal bones of the skull, etc. (**Figure 5.10**). The differences we see among species result from adaptations to different patterns of locomotion. That said, there are some bones that *only* exist in other mammals and not in humans, and their identification could be diagnostic of a non-human skeleton. For example, many mammals have a bone in their penis called a *baculum,* and for other animals there are specialized bones in their hands and feet called *metapodials* that have no homologue in the human skeleton (**Figure 5.11**). The presence of horns, claws, hooves, and antlers would also be diagnostic of a non-human skeleton.

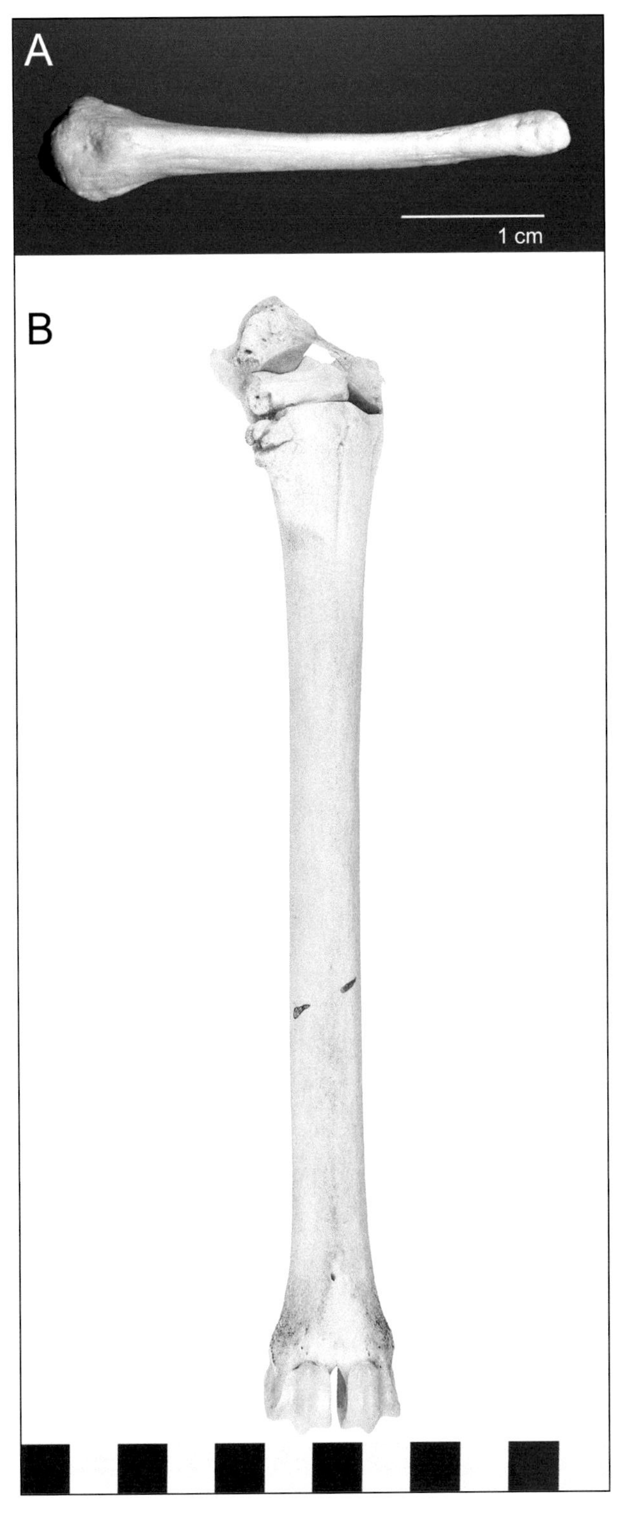

Figure 5.11. Baculum of a Eurasian beaver (*Castor fiber*) (*A*) and metapodial of a pronghorn antelope (*Antilocapra americana*) (*B*). Images on different scales.

To the contrary, evolution has also produced convergences where unrelated species come to resemble each other skeletally in certain ways (Angel, 1974). The two most often cited examples of this in forensics are the similarities in the appearance of bear paws to human hands and feet (Hoffman, 1984; Sims, 2007; Stewart, 1959) and the grossly similar appearance of pig premolars to human molars (**Figure 5.12A**). Stewart (1979, Figures 7 and 8) provides excellent reference images for bear paws (see also **Figure 5.12B**) and describes three other cases where confusion existed about species identification. These involved the rib cage of a sheep (possibly mistaken for a human thorax), the foot of a polydactylous pig (possibly mistaken for a human foot), and the terminal vertebrae of a horse's tail (possibly mistaken for human hand phalanges). Superficial similarities will be easily detected by a skilled osteologist, however, by carefully considering details of joint anatomy.

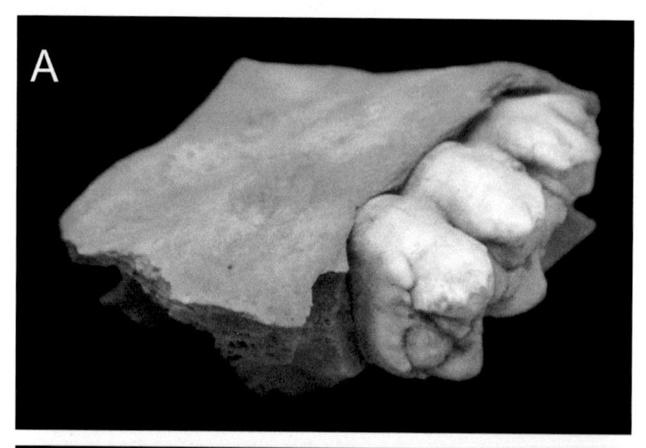

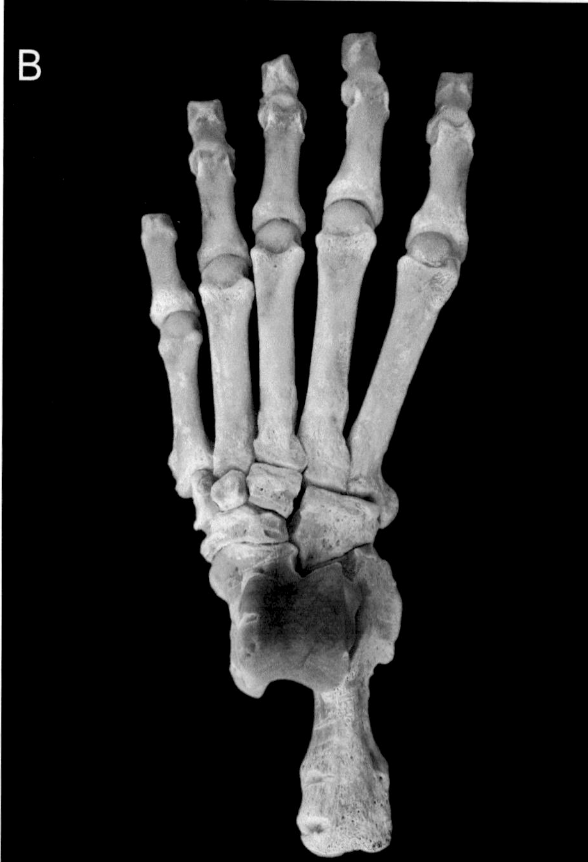

Figure 5.12. (*A*) Archaeological pig teeth. Note how similar the molars look to a human's from this angle. (*B*) Articulated bear paw.

In considering whether an unknown bone belongs to a human, the forensic anthropologist must consider the *size of the bone*, the *shape of the bone*, and the *architectural details of the bone surface*. With regard to size, there is a known range of variation in each bone element's size for humans. A bone being too large or too small would suggest a non-human identification. However, a key consideration is the development status of the individual represented by the unknown specimen. Bones from individuals who were still growing at the time of death will *not* be within the range of adult human sizes. In addition, certain anatomical features of the skeleton may not yet be visible, epiphyses may be separate from the diaphysis or absent entirely, and in fetal remains the overall shape of the element may only roughly resemble the adult bone. For this reason, it is critical that the development status be ascertained for remains that appear to be too small to be human.

For cranial remains, key indicators of subadult status include: 1) the thinness of the cranial bones, from a potato-chip thickness in fetal remains to about ¼-inch thickness in adults, 2) the presence of fontanelles (soft spots) between cranial bones, 3) unfused bones that are normally fused in adults such as the frontal bone and mandible, and 4) very distinct and open cranial sutures. For postcranial remains, key indicators of subadult status include: 1) separate or absent epiphyses, and 2) the presence of active growth "plates" that are usually identifiable based on their denser (and often more darkly stained) texture and billowed appearance (**Figure 5.13**). Training in juvenile human osteology is more limited than adult osteology because of the lack of comparative subadult material in osteological collections. Textbooks and field manuals should be consulted (e.g., Schaefer et al., 2009; Scheuer and Black, 2000). Common areas of misdiagnosis involve rodents and birds because of the similar size of fetal and domesticated bird remains. Because of the demands of flight, however, birds have a very thin cortical bone with a limited trabecular bone and large gaps between trabeculae. As a result, they are lighter than similarly sized juvenile human remains. Fragmented juvenile cranial bones can also be misdiagnosed as turtle shells; however, the latter are often much flatter than one would ever find in a human juvenile or fetal cranium.

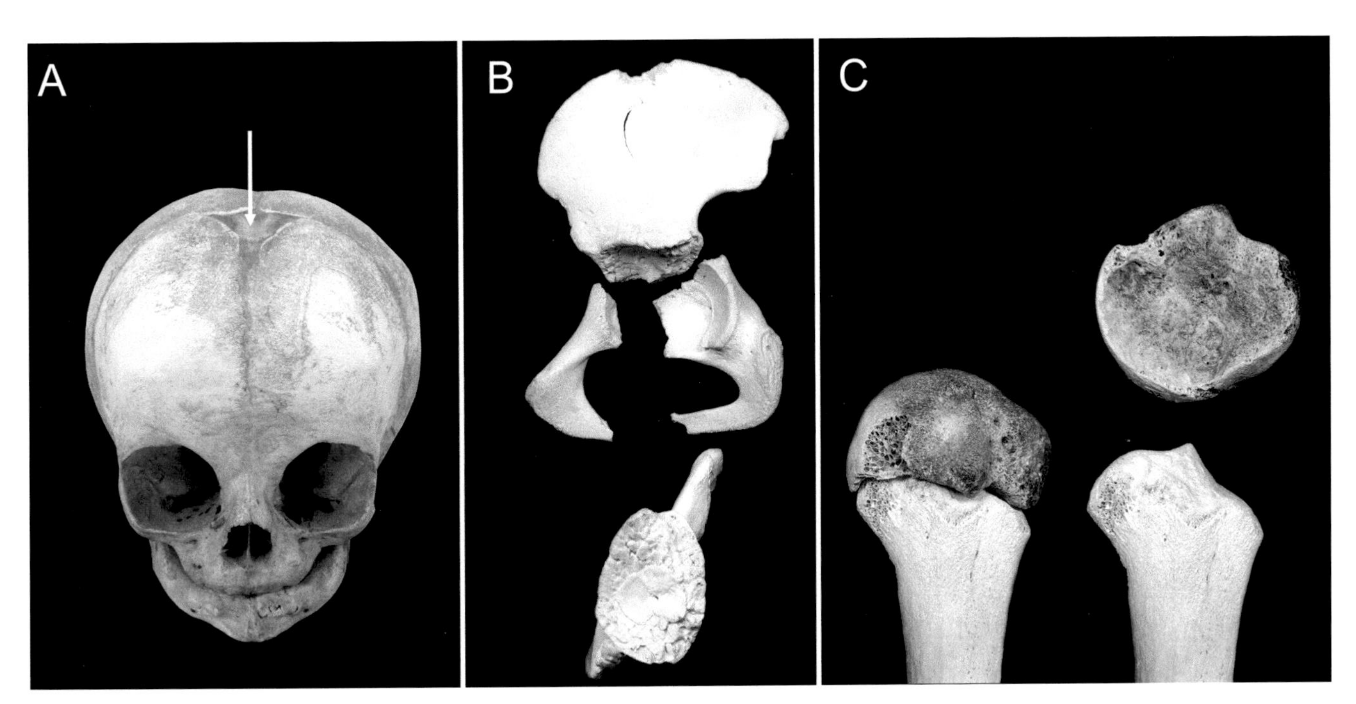

Figure 5.13. Some osteological indicators of subadult status. (*A*) Cranium of an infant with arrow indicating anterior fontanelle; (*B*) Left innominate of a 13-year-old, note billowing on unfused surfaces; (*C*) Proximal left humerus of a 13-year-old with unfused epiphysis.

Table 5.1. Key differences between human and non-human skeletal remains—skull and dentition

Region	Human	Non-human
Skull	Round, gently curving, rarely flat sections	Angular, can have flat sections
	Thick cortical bone relative to diploë, clear layering	Thin cortical bone relative to diploë, layering unclear
	Vault ~¼ inch thick	Thickness can vary
	Weak muscle markings	Stronger muscle markings
	No crests of bone on vault	Often has crests of bone on vault
	Small face, large vault	Large face, small vault
	Face does not project anterior of braincase	Face projects anterior of braincase, presence of snout
	Foramen magnum is anterior and underneath the brain	Foramen magnum is posterior and behind the brain
	Internal surface of vault smooth, presence of meningeal grooves	Internal surface of vault more complex
	Mandible is U-shaped	Mandible often V-shaped
	Chin is present, fused at the midline	No chin present, unfused at the midline
Dentition	2:1:2:3 dental formula	Most do not have 2:1:2:3 dental formula, 3:1:4:3 common
	Incisors vertically implanted in jaw	Incisors can be forward projecting, not vertical
	Molars have low rounded cusps (omnivores)	Molars vary, usually sharp and pointed (carnivores) or with multiple W-shaped crests (herbivores)
	No gaps between teeth (diastema)	Can have gaps between teeth, especially around the canine (canine diastema)
	Canines small, similar in size to premolars	Canines can be much larger than incisors and premolars

If the unknown specimen is from an adult and within the size range for humans, then comparative analyses of shape and the anatomical details of bone surfaces (architecture) should be considered. Key differences between human and non-human cranial remains are summarized in **Table 5.1**. For a complete cranium there is little chance of misdiagnosis. However, cranial vault fragments can provide challenges. Key differences between human and non-human postcranial remains of the thorax, shoulder girdle, and pelvic girdle are presented in **Table 5.2**. As with the cranium, analysis of complete specimens provides little source of confusion; however, highly fragmented specimens may require microscopic analysis (see below).

Guidelines for identifying the long bones and bones of the hands and feet are presented in **Table 5.3**. Because the joint surfaces are shaped by evolution to meet locomotor demands, they are highly diagnostic of the species. Even fragmented articular surfaces can be identified as human or not with relative ease using a comparative collection. The biggest challenge is identifying isolated fragments of long bone diaphyses. Size, shape, and curvature are key considerations. The locations of muscle attachment sites, their size, and course across the diaphysis are also specific to different species and can be used to compare against reference samples for identification.

Table 5.2. Key differences between human and non-human skeletal remains—thorax, shoulder girdle, pelvic girdle

Region	Human	Non-human
Vertebrae	Column is S-shaped with four distinct curves	Column lacks curvatures, or has a single curve
	Vertebrae vary in size throughout the spine due to greater weight bearing in inferior vertebrae	Vertebrae less variable in size throughout spine, weight bearing not cumulative in quadrupeds
	Short spinous processes, all relatively the same length	Longer spinous processes, can vary in length along the spine; thoracic very long
	Bodies broad and flat, wedge-shaped; large relative to spinous processes	Bodies can be smaller overall with distinct cup-shaped appearance
	Bodies never taller than wide	Bodies can be taller than wide, forming cylinders
	Sacrum triangular, 5 fused vertebrae	Sacrum long and narrows, 3–4 fused vertebrae
Ribs	More curved along their length	Straighter along their length, less curved
	Costal groove present on inferior surface	Costal groove variable or absent
Shoulder	Clavicle present with distinct S-shape	Many animals lack a clavicle or it is reduced in size
	Scapula is taller than wide, triangular	Scapula is wider than tall
	Acromion and coracoid well-defined	Acromion and coracoid minimized, less projecting
Pelvis	Broad and short	Long and narrow
	S-curve profile of ilium blades	Straight profile of ilium blades
	Anterior joint not fused	Anterior joint can be fused

However, even relatively large sections of long bone shaft can lack any diagnostic criteria. In such cases the following guidelines can be applied. First, the outer cortical surface of non-human bones tends to be smoother and denser in appearance. Human long bones tend to look grainier with a more porous texture (**Figure 5.14A**). Second, if the shaft is broken, exposing the medullary cavity, then the thickness of the cortical bone can be observed. In human long bones, the cortical bone is about ¼ the thickness of the entire diameter. In many non-human animals, the cortical bone is much thicker, approximating ½ the total diameter. Third, human long bones have trabecular bone in the diaphysis with a poorly defined border between the cortical and trabecular bone (**Figure 5.14B**). In non-human animals there is less trabecular bone overall. Those species with trabecular bone show a more homogeneous appearance and there is often a very clear delineation between the cortical and trabecular bone at the endosteal surface.

For highly degraded or fragmented samples, microscopy can be used to examine the cellular structure of the sample. As discussed in chapter 3, at the microscopic level human compact bone is composed of a series of structures called osteons that appear as a series of circular rings when viewed in cross-section. These osteons are randomly spaced and not organized into rows or clusters. Because of close packing and bone remodeling processes, human bone samples will often show both primary osteons (partially replaced) and secondary osteons (Haversian systems). In humans, about 50 percent of the bone area is composed

Table 5.3. Key differences between human and non-human skeletal remains—long bones, hands, and feet

Region	Human	Non-human
Long Bones	Cortical bone about ¼ the diameter of shaft	Cortical bone about ½ the diameter of shaft
	Less robust, smaller muscle attachment sites	More robust, larger muscle attachment sites
	Joints have less complex shape	Joints have more complex shape
	Humerus head larger than greater tubercle	For ungulates, head smaller than greater tubercle
	Radius and ulna never fused	Radius and ulna often fused, with complex, distal locking joints
	Femur and humerus heads large relative to bone	Femur and humerus heads small relative to bone size
	Femur has long neck	Femur has short neck
	Distal femoral epiphysis angled	Distal femoral epiphysis straight
	Femur has single linea aspera, raised and distinct	Linea aspera varies, not single, often double
	Tibia and fibula never fused	Tibia and fibula often fused
	Fibula always present	Fibula can be reduced or absent
Hands/Feet	Nails	Claws
	Flat and straight metacarpals, and hand phalanges	Specialized forelimb and hindlimb, often for stability for running and speed
	Divergent thumb on hand	No divergent thumb
	Foot elongated for bipedalism	Foot rarely as elongated
	Double arch in foot	No arch in foot

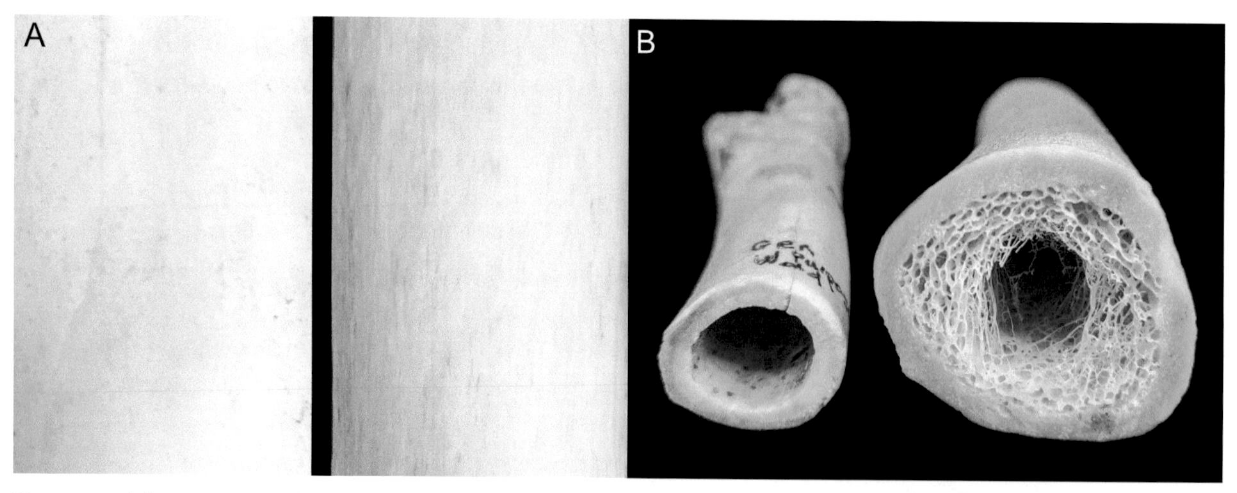

Figure 5.14. (*A*) Comparison of cortical bone from the femur of a dog (*left*) and a human femur (*right*). Note differences in surface texture. (*B*) Comparison of the cross-section of a dog femur (*left*) and a human femur (*right*). Notice the more abrupt transition from the cortical bone to the medullary cavity in the non-human bone as well as the greater thickness of the cortical bone relative to the overall size of the element.

of primary osteons or secondary Haversian systems with the other 50 percent comprising interstitial lamellar bone (**Figure 5.15A**). To the contrary, in many non-human animals, circular osteons are not present or are found only in specific parts of the bone. Instead, the bone consists of organized rectangular structures that look like stacks of bricks, which is called *plexiform bone,* and most commonly found in large, fast-growing mammals such as cow, pigs, and horses. Where osteons are found they tend to occur in organized bands (*osteon banding*) or as alternating bands of osteons and plexiform bone (**Figure 5.15B**). Osteon banding is particularly diagnostic because it is rarely found in human samples (Mulhern and Ubelaker, 2001). *Fibrolamellar bone* is another type of bone organization found in some

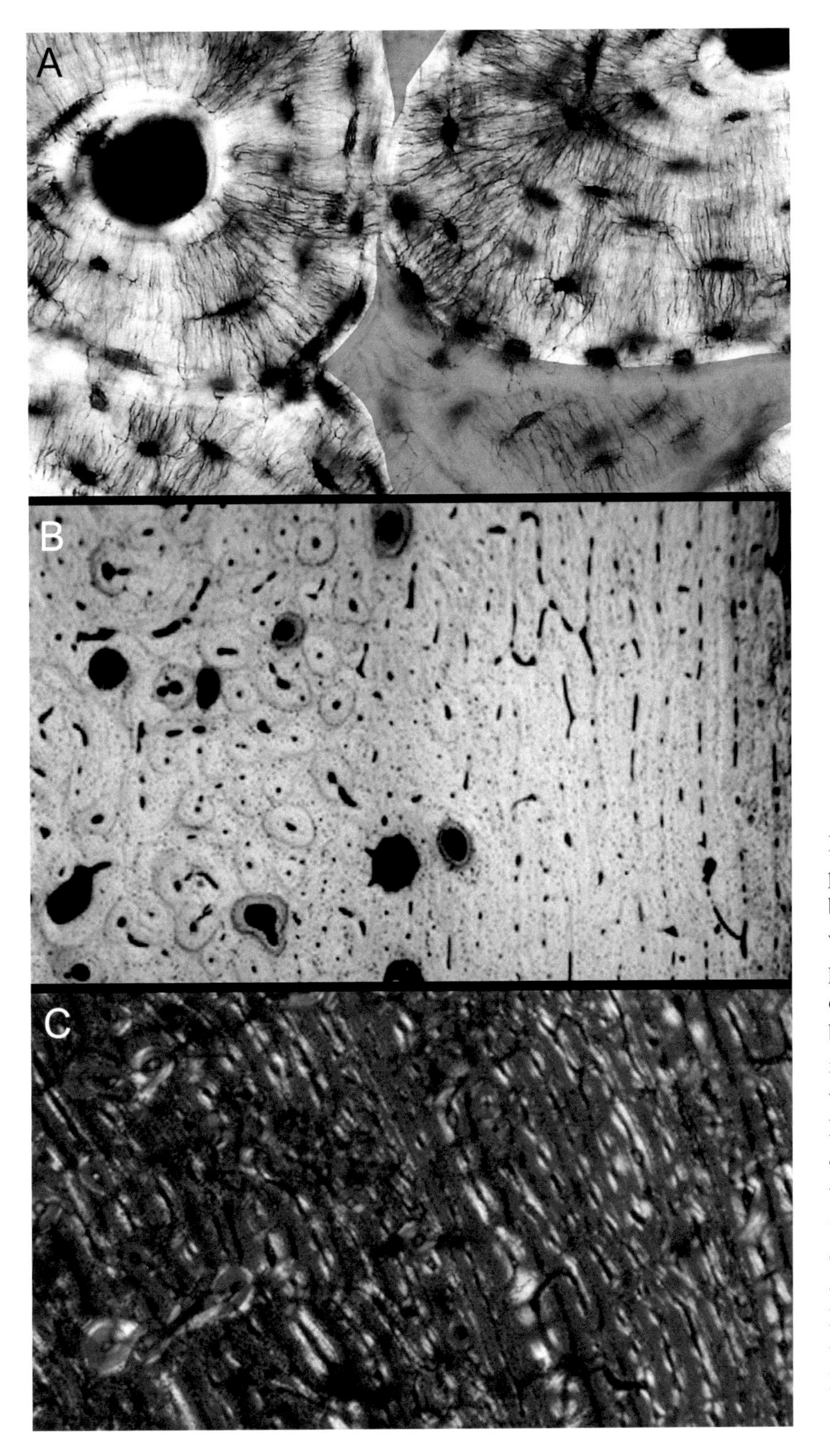

Figure 5.15. Histological comparison of human and animal bone. (*A*) Human compact bone with interstitial lamellae tinted purple. (*B*) Histological section of a metapodial with plexiform bony tissue (brick-like arrangement) on the right transitioning to a Haversian system on the left (circles). Note the vertical banding of the osteons in the plexiform arrangement. (*C*) Histological section of a cave bear femur with poorly organized fibrous bone to the left and parallel layers of bone to the right, creating a fibrolamellar complex.

fast-growing non-human animals that appears as layers of relatively disorganized, fibrous bone alternating with bands of lamellar bone (**Figure 5.15C**). Owsley et al. (1985) presented an analysis of deer and human bone samples that provides a good model for how histology can be used to differentiate human and non-human bone samples. In this case, they found that deer, which are commonly found in wooded areas and mistaken for human remains due to their similar size, had primarily plexiform bone, no secondary Haversian systems, and, where present, osteons were tightly packed and uniform in size and had smaller Haversian canals in comparison to humans.

While it is tempting to associate the presence of circular osteons with humans, and plexiform or fibrolamellar bone with non-humans, the distinctions are not always so clear. For example, plexiform bone and fibrolamellar bone can be found in juvenile or pathological human samples (Caccia et al., 2016), and non-human animals can have osteons present often near large muscle attachment sites (Hillier and Bell, 2007; Mulhern and Ubelaker, 2012). There is also considerable variation in bone histological structure among the species most misidentified in a forensic setting. As Hillier and Bell (2007:250) noted, "factors such as specific bone, bone portion sampled, age, sex, and pathological conditions all affect the 'normal' appearance of bone tissue, resulting in significant variation in bone tissue appearance throughout the skeleton." The histological structure of a sample may also be affected by burning and consumption (i.e., being eaten by other animals), as well as other taphonomic alterations leading to bone diagenesis (the breakdown or change of the chemical composition of bone).

Because of this variation in bone form even within the same element of an individual, it is important to consider more quantitative approaches that also attempt to satisfy the *Daubert* criteria. Potential variables of interest include: the number of osteons for a given area, the number of secondary osteons, osteon size, Haversian canal diameter (area, minimum, and maximum), and osteon shape (oval or round) (Mulhern and Ubelaker, 2012). Several studies have presented equations for estimating whether a sample is human or not based on quantitative measures of the Haversian canal, osteon density, and osteon size (Cattaneo et al., 2009; Martiniaková et al., 2006; Urbanová and Novotný, 2005), with varying degrees of success. If the species cannot be identified using macroscopic or microscopic methods, then more labor-intensive and costly techniques can be used, including protein-based methods and DNA analysis (Ubelaker et al., 2004; Lowenstein et al., 2006).

LEARNING CHECK

Q4. While remodeling their home, a couple found skeletal remains within a wall. Concerned that they might represent the bones of an infant, investigators sent you the following image of the remains, in which the white lines measure 5 mm (~0.2 inches) in length (**Figure 5.16**). Are these remains human in origin?

A) These remains are likely from a human.

B) These remains are not from a human.

C) There is not enough information in this image to determine if these remains are human or non-human.

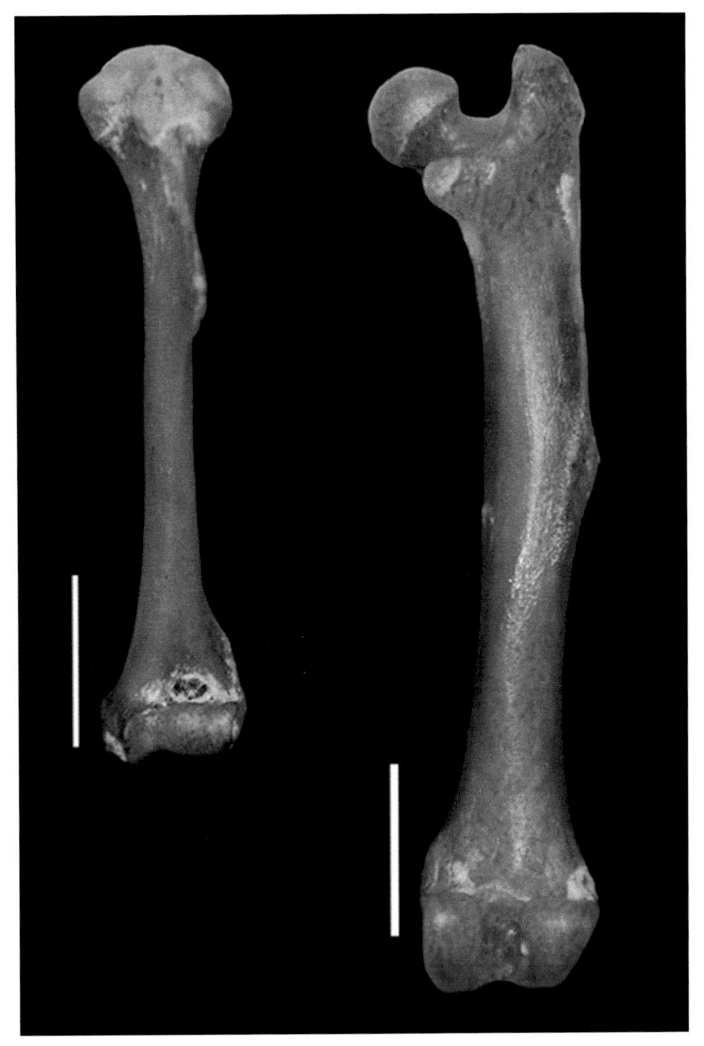

Figure 5.16. Two skeletal elements of uncertain origin.

Q5. Investigating reports of a clandestine grave located near a lake, law enforcement has discovered some remains that look as though they might be cranial fragments. Based on the appearance of the remains shown in **Figure 5.17A**, are they human in origin? (Hint: pay attention to characteristics like surface texture and the presence of vascular grooves, sutures, and/or diploë when making your assessment.)

A) These remains are likely from a human.
B) These remains are not from a human.
C) There is not enough information in this image to determine if these remains are human or non-human.

Q6. Farther along the lakeshore from the remains described in Question 5, investigators discovered the bone fragments shown in **Figure 5.17B**. Based on the appearance of these remains, are they human in origin? (Hint: pay attention to characteristics like surface texture and the presence of vascular grooves, sutures, and/or diploë when making your assessment.)

A) These remains are likely from a human.
B) These remains are not from a human.
C) There is not enough information in this image to determine if these remains are human or non-human.

Figure 5.17. Skeletal element fragments of uncertain origin.

Q7. After an apartment complex burned to the ground, several fragments of potential human remains were submitted to you for examination. Macroscopic examination was inconclusive, so you had histological sections of the fragments prepared. Under a microscope, the fragments submitted to you appear as shown in **Figure 5.18**. Based on this observation, what is your assessment of whether these are human remains?

A) These remains are likely from a human.

B) These remains are not from a human.

C) There is not enough information in this image to determine if these remains are human or non-human.

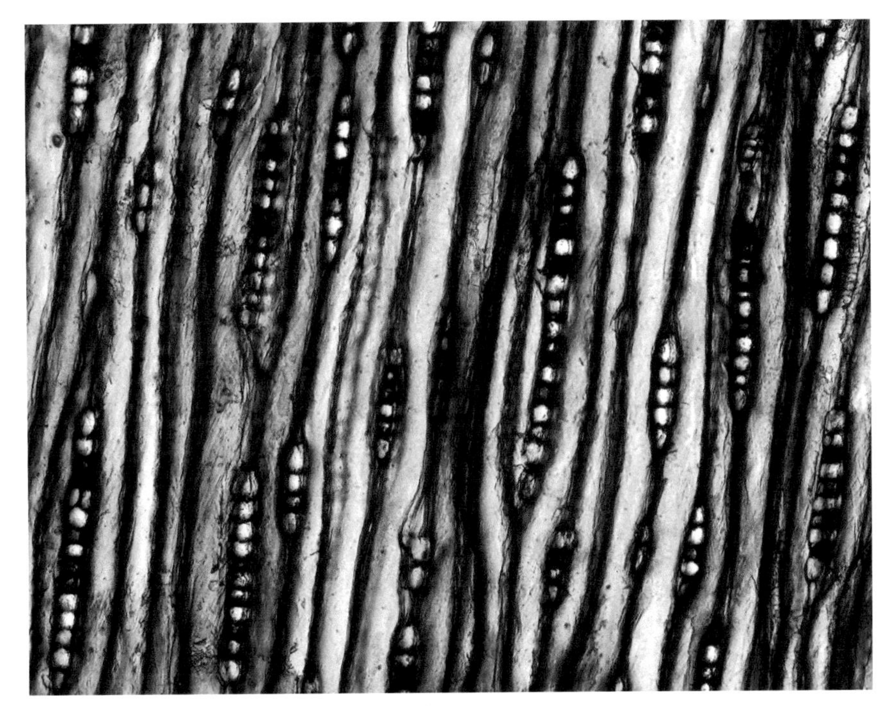

Figure 5.18. Histological section for use with Q7.

Is the Human Skeletal Material Contemporary or Non-Contemporary?

Once the specimen has been determined to be human bone, the final question is whether the remains are contemporary in age, and therefore potentially of medicolegal significance. These decisions are often made based upon simple observations by the examiner, noting the context in which the remains were found and having a strong understanding of the range of ways in which human remains could come to one's attention and *not* be associated with a crime scene. Therefore, there are few true methods to discuss in the section. Examples of non-medicolegal remains include: prehistoric burials, burials associated with a known or historic cemetery, burials associated with an unmarked or family cemetery (i.e., a non-municipal cemetery), trophy skulls collected from a foreign war, tribal art skulls traded in the art market, and skeletons used as teaching specimens or in doctors' or dentists' offices.

For particularly important cases, the examiner does have the option to use radiometric data methods to ascertain how long ago the individual died. For example, the *bomb curve radiocarbon method* can determine, based on changes in atmospheric radioactive carbon levels caused by atomic bomb testing between 1950 and 1963, whether someone was born prior to 1950 (Ubelaker, 2014). However, the methods are still in development and not practical to use on a daily basis.

For most forensic anthropologists, ascertaining the age of the remains combines observations of the following: 1) the context in which the remains were located and recovered, 2) personal effects and material culture recovered with the remains, 3) observations of bone condition and appearance, and 4) biocultural body modifications that may be culturally and temporally specific.

Context refers to the surrounding environment in which the remains were recovered. Observing remains in situ, that is, in place, is critical for making this assessment. This is why it is recommended that human remains are not touched or removed by law enforcement prior to a forensic anthropologist visiting the scene. Certain contexts are more likely to have remains with medicolegal significance. For example, remains found in the woods, on the side of the road, in fields, or in houses are more likely to be significant to the medicolegal community. Remains found associated with cemeteries, known archaeological sites, and university or educational institutions are less likely to be of medicolegal significance. Context also refers to the disposition of the body. Remains scattered on the surface or buried in hastily dug graves are more likely to be significant to the medicolegal community. Bodies that are forced into holes in unusual positions that contradict the prevailing norms of burial (coffins, laid on one's back, hands folded across the stomach) are also likely to be forensically significant. Here, it is important to consider the archaeological history of the region because burial styles and norms do change. For example, it was more common to be buried in a flexed or semi-flexed position on one's side in many precontact Indigenous societies in North America. Burial position includes several descriptors: flexed, semi-flexed, or extended; supine (face up) or prone (face down); head orientation; and hand placement. Consultation with the state archaeologist may help better contextualize remains buried in an atypical manner given the standards of today.

Another consideration is the *material culture* that was found associated with the remains. Obviously, modern clothing, watches, jewelry, and other accessories would suggest the remains are of medicolegal significance. Historic period graves might contain clay pipes, porcelain ceramics, buttons, pins, and antique coffin hardware that provides evidence that can be used to accurately date the body and indicate it is not of medicolegal significance. Prehistoric burials may contain ceramics that will normally be thicker and coarser than post-contact and modern pottery. It will also be unglazed. Surface decorations would be incised or paddled onto the raw clay surface and the entire manufacturing process is done by hand. Stone tools or stone tool flakes (debitage) are another indicator of a precontact date of burial. Precontact material culture will vary considerably across the country and, if in doubt, the state archaeologist should be consulted.

Turning to the bone itself, there are several crude indicators of whether the remains are contemporary in age. Bone undergoes a progressive change in appearance as time since death increases. Fresh bone will be off-white to light yellow in color, feel dense and heavy in hand, have a smooth and dense appearance, a greasy feel due to the presence of lipids, may have soft tissue still adhering, and may present *adipocere* (**Figure 5.19**). As time since death increases, there is a progressive loss of collagen and fat. Bone color changes toward light brown if buried, and white and bleached if exposed to the sun. However, soil composition causes variation in bone coloration, and bone will begin to stain relatively quickly after burial. Color alone does not determine whether a body is contemporary. As time increases, bone texture also deteriorates. Smooth outer surfaces become coarser and grainier. The loss of moisture and collagen decreases bone weight and increases fragmentation, a condition referred to as being friable.

Biocultural body modifications can also provide evidence of whether the remains are contemporary. These include a wide variety of changes observed on the skeleton that provide evidence of the age of the body. Postmortem treatments such as embalming or processing through the funeral industry suggest contemporary but not medicolegal remains. Anatomical preparation hardware including screws, hinges, modern drill holes, and cuts with modern saws (autopsy cuts) may indicate the body is an anatomical specimen and not of medicolegal significance. Such teaching specimens may also show shelf wear and writing on the surfaces if used in classroom settings (**Figure 5.20**). However, other types of surgical interventions such as screws, plates, and other surgical hardware could indicate a contemporary body that is of medicolegal significance. Modern dental restorations (fillings, bridges, crowns, dentures) might suggest a contemporary date depending on the materials used. Dentures

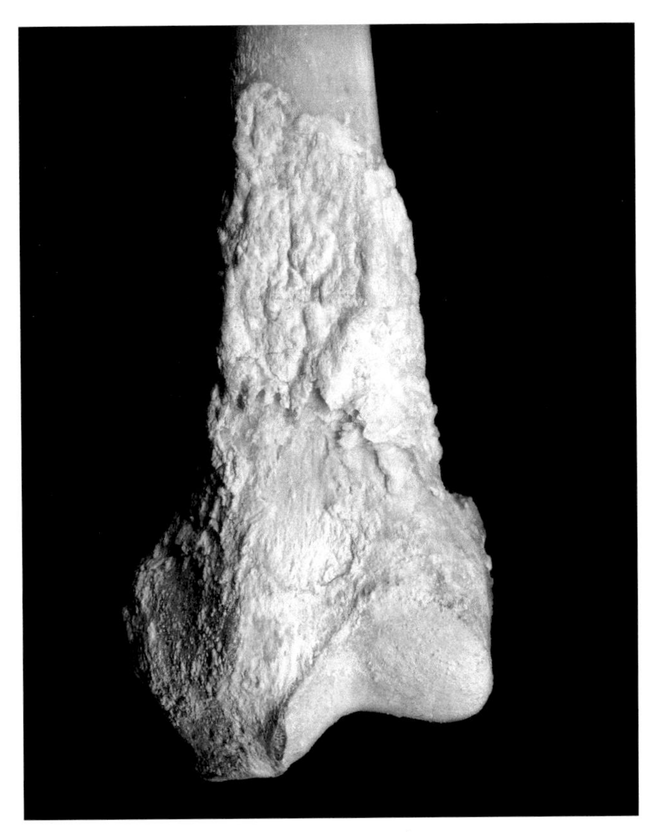

Figure 5.19. Distal femur with adipocere formation.

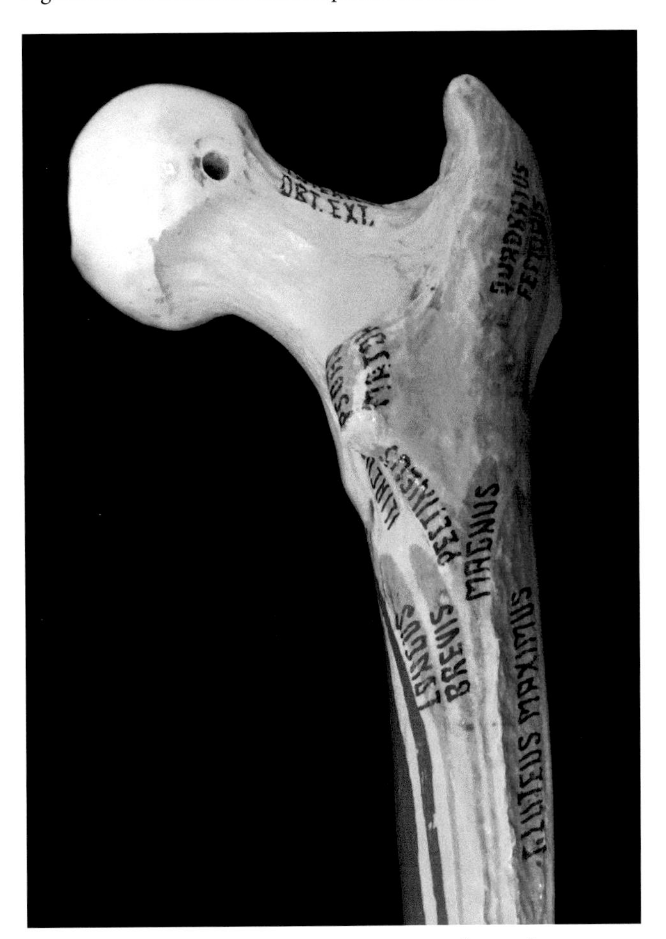

Figure 5.20. Anatomical teaching specimen with muscle origins and insertions labeled and drill hole through femoral head.

constructed from precious metals and animal teeth, for example, were once fashionable but modern dentures are typically made from acrylic resin. To the contrary, some surgical interventions are indicative of non-contemporary remains. Trephination, for instance, is the intentional removal of pieces of the skull using stone tools. It is an ancient practice in the Americas and would suggest an archaeological burial.

Other biocultural modifications may also be encountered that suggest the remains are very old. Certain body modification practices, for example, can leave their marks on the skeleton. Most notable is the practice of intentional cranial modification resulting in elongated skulls or vertically erect skulls with flat planes on the frontal and occipital bones (**Figure 5.21A**). This was practiced widely for a variety of reasons. Cradle boarding describes a child rearing practice that also leaves a flat plane on the occipital and is not typically associated with medicolegal remains. The dentition was also modified for biocultural reasons, including intentional chipping and filing of the teeth and the insertion of inlays into the front surface of the incisors (**Figure 5.21B**). This work is akin to the modern dental beauty industry (whitening, straightening, veneers, and decorative appliances). Pipe wear facets were also common in premodern times as pipes were made of clay that wore down the teeth during repeated clenching over the course of one's life (**Figure 5.21C**). Some societies collected the skulls of ancestors or enemies and modified them for display. This could include drilling holes for inserting a cord for hanging the skull, or the addition of decorative paint, modeling of the face with clay, or the attachment of beads and shells (**Figure 5.21D**). These modifications are so specific that they can be identified as belonging to a specific society.

Finally, some of the more notable differences between contemporary society and the past are changes in diet and food preparation techniques. Our soft diets today lead to high rates of cavities, tooth misalignment, and very low rates of tooth wear. To the contrary, in the past diets were much coarser and prepared in ways that incorporated more grit into the diet. As a result, most archaeological burials (whether prehistoric or historic) show high rates of tooth wear even at relatively young ages (**Figure 5.21E**). This is represented by exposed dentin on the tooth crowns, which will usually be stained dark brown in comparison to the white enamel. Moderate or excessive tooth wear suggests the remains are not contemporary.

LEARNING CHECK

Q8. A family called law enforcement after their dog returned home with a bone in its mouth. After determining that the bone was a human ulna, an investigator was sent to the scene. The investigator found that the dog had been digging in an historic cemetery dating to the early 1800s. Based on this context, which of the following is most likely?

A) These remains are of medicolegal significance.

B) These remains are not of medicolegal significance.

Q9. While digging the foundation for a home in a new housing development, skeletal remains were discovered lying extended and face up. Associated with the remains were the dentures shown in **Figure 5.22**. Based on this information, which of the following is most likely?

A) These remains are of medicolegal significance.

B) These remains are not of medicolegal significance.

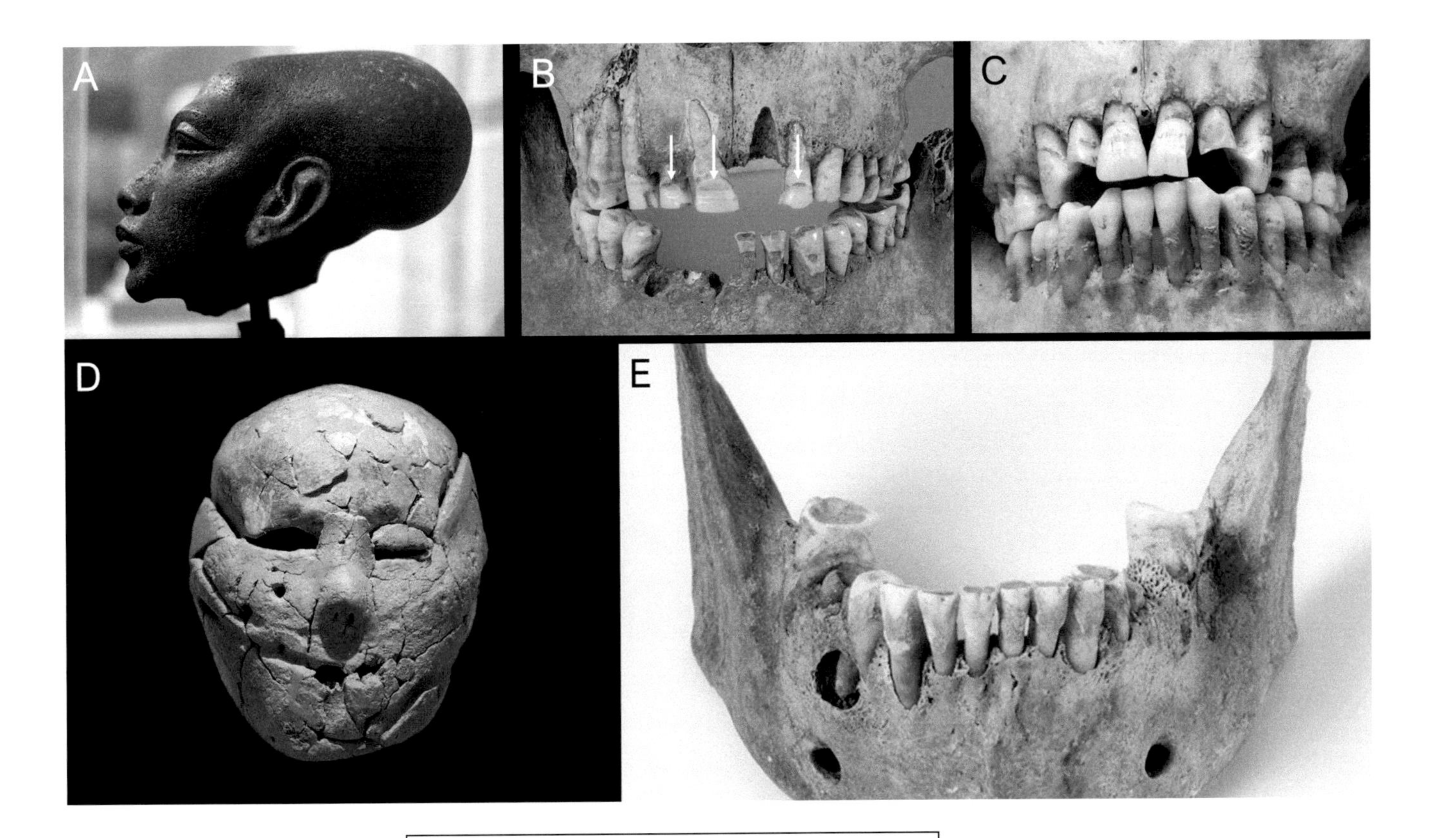

Above: Figure 5.21. Biocultural modifications of the cranium and dentition. (*A*) Cranial modification. Note the elongated head shape. (*B*) Intentional dental modification (arrows indicate grooves carved into the maxillary incisors). (*C*) Unintentional dental modification from habitual pipe smoking. (*D*) Plastered skull. (*E*) Heavy dental wear on an archaeological mandible.

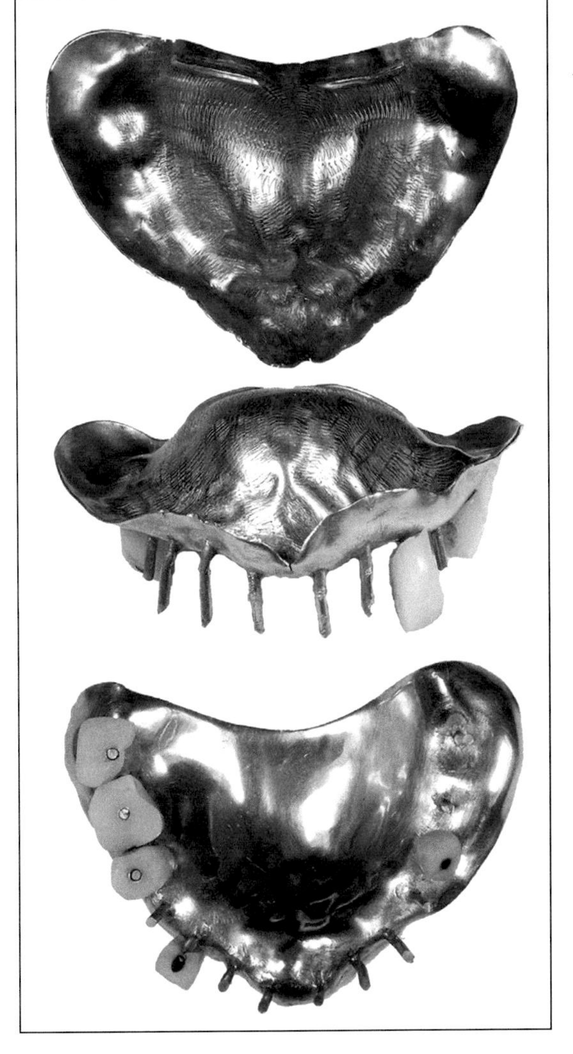

Left: Figure 5.22. Question 9.

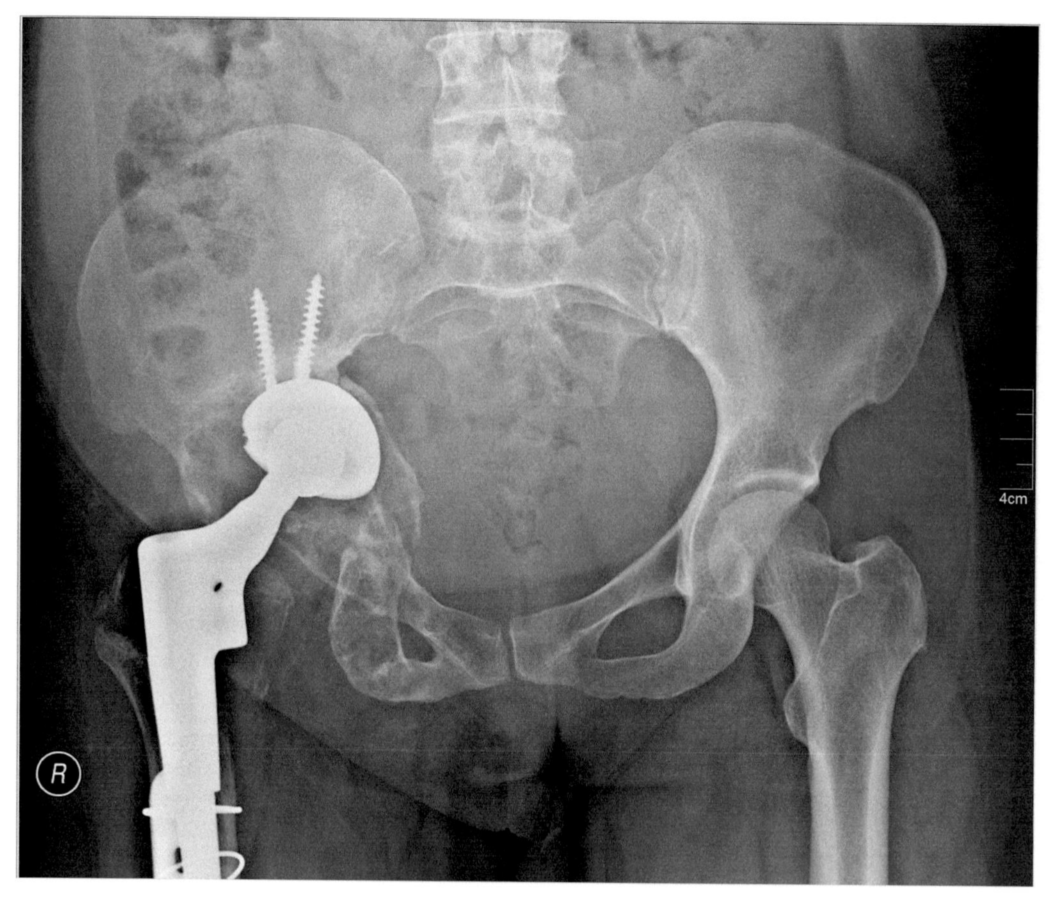

Figure 5.23. Radiograph of a human pelvis and long bones.

Q10. Mummified remains, concealed by a layer of rocks, were recovered from a desert setting. **Figure 5.23** presents one of the X-rays taken of the remains. Based on this information, which of the following is most likely?
A) These remains are of medicolegal significance.
B) These remains are not of medicolegal significance.

End-of-Chapter Summary

The forensic anthropologist must answer three questions when presented with a specimen of unknown origin: Is it bone, is it human bone, and is it contemporary in age and therefore of medicolegal significance? Understanding the macroscopic and microscopic qualities of bone is key to determining whether a fragmentary specimen is bone or some other material. Mastery of human and non-human osteology is critical for being able to determine whether a bone is human or not. Using observations of the context in which the remains were found, the preservation of the bone, and any time-specific biocultural signature on the bone is key to determining whether the remains are contemporary in age. For most cases, if the remains are well preserved and complete, then determining the osseous nature of the sample and its human identity is straightforward.

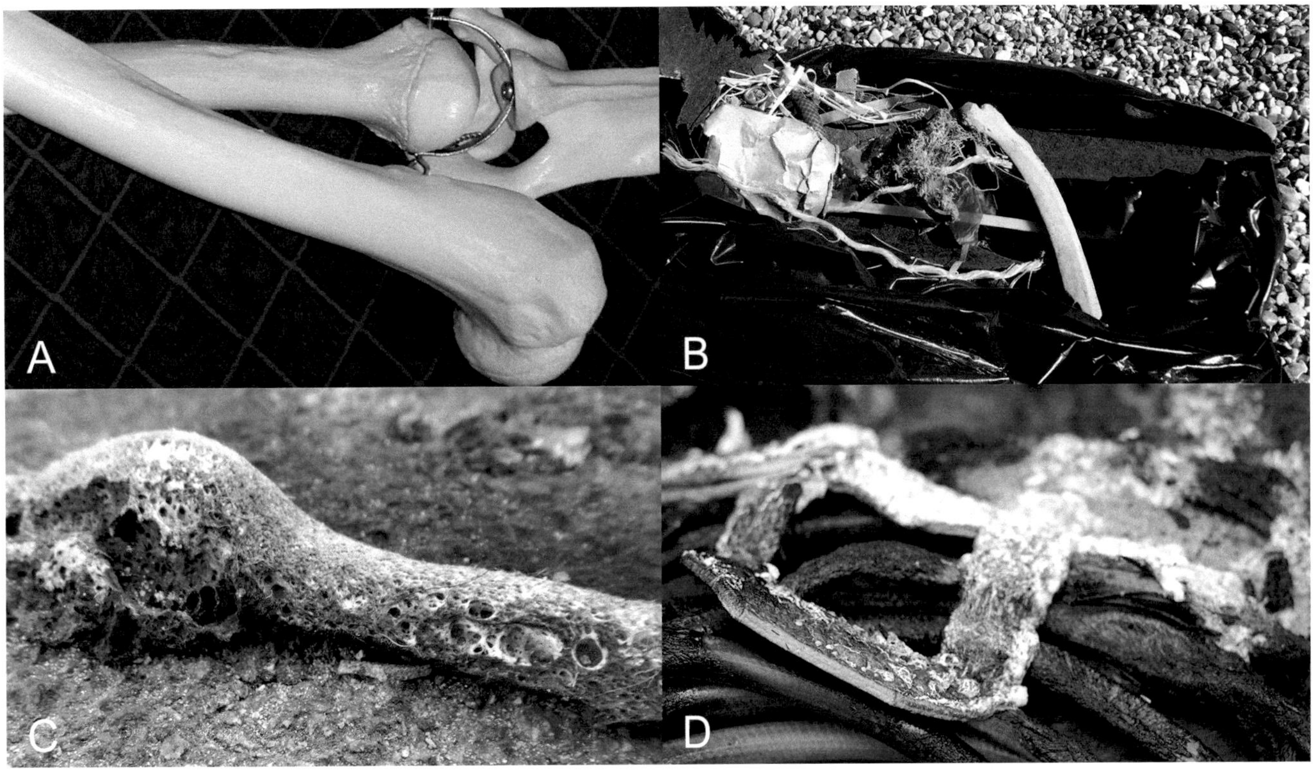

Figure 5.24. Four images showing objects of uncertain material and origin.

End-of-Chapter Exercises

Exercise 1

Materials Required: **Figure 5.6** and **Figure 5.24.**

Scenario: Law enforcement has received an anonymous tip that a missing woman has been murdered and that her body has been concealed in a landfill. They are currently searching the landfill for her remains and have sent you several photographs for evaluation. Four of the photos that have been sent to you are shown in **Figure 5.24**.

Question 1: Based on these photographs, which specimens appear to be osseous or dental tissue?

Question 2: To be on the safe side, you send a colleague who is trained in the use of a handheld XRF (HHXRF) spectrometer to assess the specimens. Your colleague provides you with the following values for elemental calcium (peak spectral Ca K_β) and phosphate (peak spectral P K_α):

Specimen A: Ca–0.49; P–0.11
Specimen B: Ca–18.73; P–12.39
Specimen C: Ca–0.41; P–0.18
Specimen D: Ca–0.44; P–0.06

Compare these values to those shown in **Figure 5.6**. Based on this information, which specimens are consistent with osseous or dental tissue? Does this answer match your answer to Question 1?

Figure 5.25. Three fragmented objects of uncertain origin.

Question 3: Based on the landfill context, the photographs, and the HHXRF results, what can you say about each of the specimens in terms of medicolegal significance? Try to frame your answers in terms of the three main questions discussed in this chapter (i.e., is the specimen skeletal material? Is the specimen human or non-human? Is the specimen recent?) and remember that you may not always be able to determine the answer based on the available information.

Exercise 2

Materials Required: **Figure 5.6, Figure 5.25,** and **Figure 5.26**

Scenario: A hiker reported finding a piece of a human mandible along a riverbank. After searching the area, investigators have located several other specimens that they think may represent human remains and submitted them to you for evaluation. The specimens submitted are shown in **Figure 5.25** (note: images are not to scale).

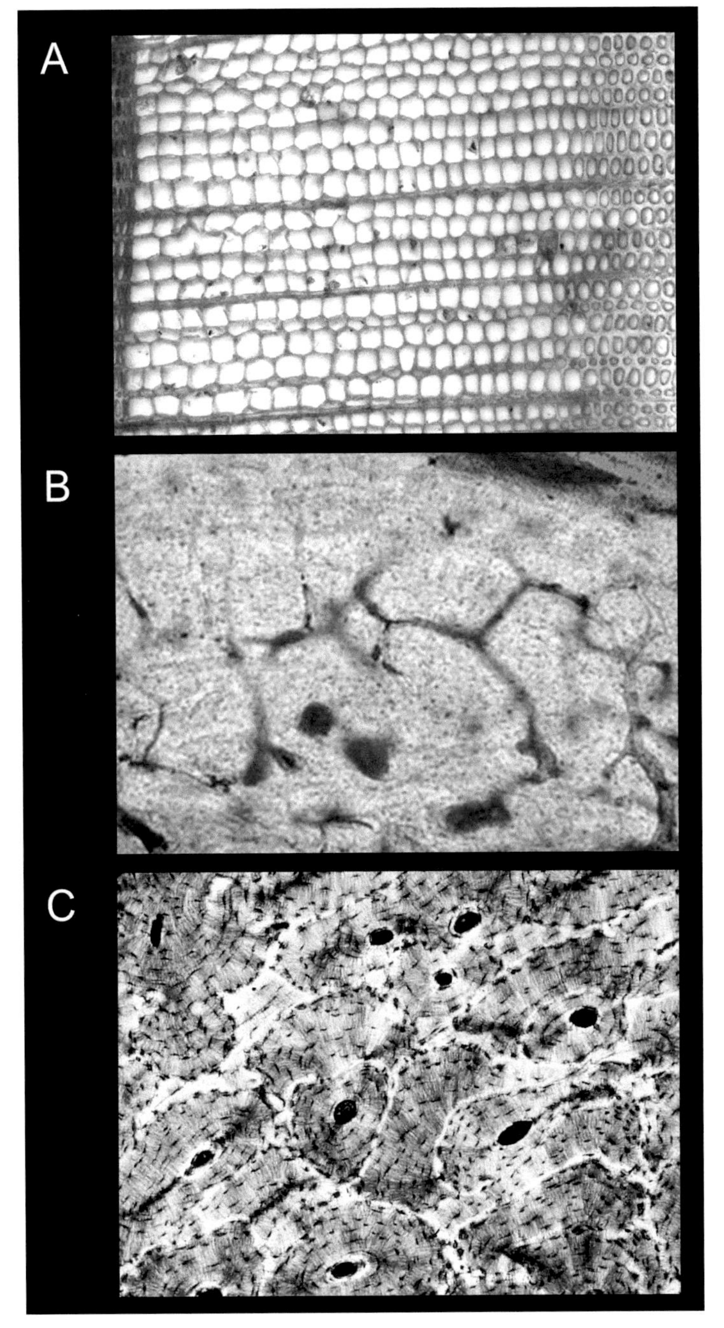

Figure 5.26. Three histological bone sections.

Question 1: Based on the morphology alone, do any of these specimens appear to be consistent with human skeletal material?

Question 2: You have your colleague use a handheld XRF spectrometer to assess the specimens shown in **Figure 5.25,** and they have obtained the following measures of peak spectral calcium (Ca) and phosphate (P):

Specimen A: Ca–2.38; P–0.13

Specimen B: Ca–20.49; P–13.11

Specimen C: Ca–20.92; P–16.19

Comparing these values to those shown in **Figure 5.6,** are any of these specimens consistent with being skeletal material?

Question 3: You ask a different colleague to complete a histological analysis of the specimens. Their slides are shown in **Figure 5.26**. Based on this information, are any of these specimens consistent with human skeletal material?

Question 4: Given your answers to the questions above, which, if any, of these specimens are potentially of medicolegal significance?

Exercise 3

Materials Required: **Figure 5.27** or suitable substitute images or specimens.

Scenario: In 2003, after extensive genealogical research, a family thought they had traced the final resting place of one of their ancestors to a potter's field in Massachusetts—a location that had been used as a public cemetery since 1869. The ancestor in question had died and been buried in 1901. Once the putative grave was located, the family received permission to have the remains exhumed and tested for DNA. After several days, the results of the analysis were received and indicated that the remains were, indeed, those of their relative.

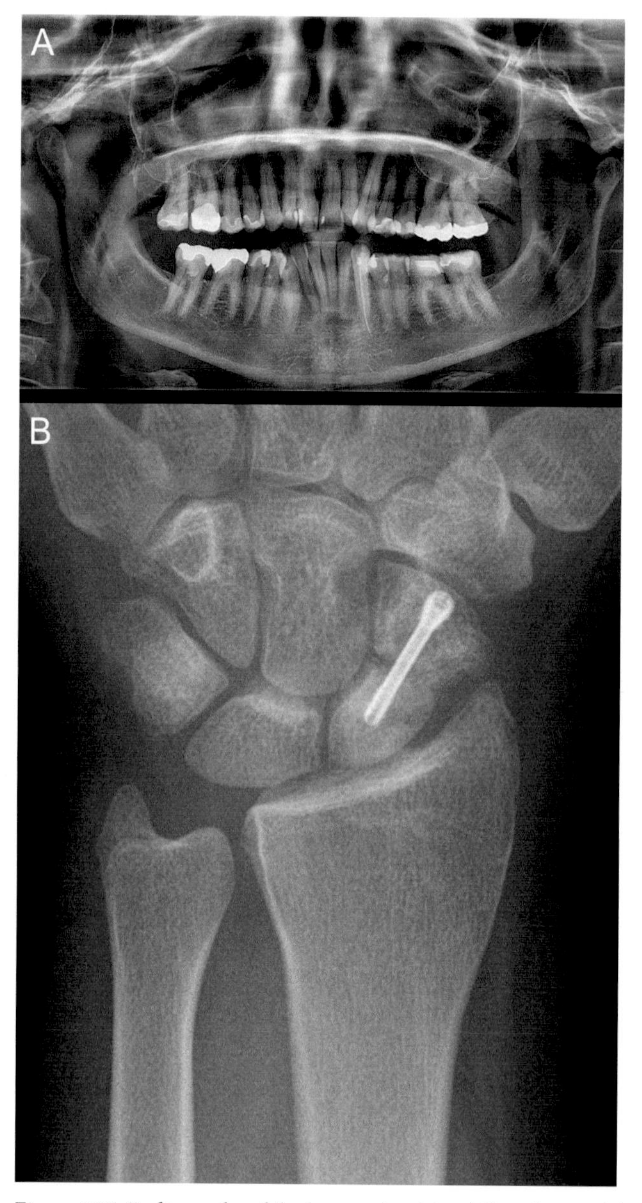

Figure 5.27. Radiographs of the human dentition (*A*) and wrist (*B*).

The skeleton was then returned to its grave in the potter's field. In 2018, the family decided to move the remains of their ancestor to a family plot in a different cemetery. Upon re-exhumation of the grave, a second set of skeletal remains was discovered lying deeper in the grave.

Question 1: Based on this context, would you conclude that the second set of human remains is recent enough to be of medicolegal significance? Why or why not?

Question 2: An historical archaeologist is conducting a research project on this cemetery and has requested permission to analyze this second set of remains. As part of their analyses, radiographs were obtained. Some of the observations are shown in **Figure 5.27**. Does the information contained in these radiographs affect your conclusion as to whether these remains are of medicolegal significance?

References

Adams BJ, Crabtree PJ. 2008. *Comparative Skeletal Anatomy: A Photographic Atlas for Medical Examiners, Coroners, Forensic Anthropologists, and Archeologists.* Totowa, NJ: Humana Press.

Angel JL. 1974. Bones can fool people. *FBI Law Enforcement Bulletin* 43: 16–20.

Bradfield J. 2018. Identifying animal taxa used to manufacture bone tools during the Middle Stone Age at Sibudu, South Africa: Results of a CT-rendered histological analysis. *PLoS ONE* 13: e0208319. Available from: https://doi.org/10.1371/journal.pone.0208319.

Caccia G, Magli F, Tagi VM, Porta DGA, Cummaudo M, Márquez-Grant N, Cattaneo C. 2016. Histological determination of the human origin from dry bone: a cautionary note for subadults. *International Journal of Legal Medicine* 130: 299–307.

Cattaneo C, Porta D, Gibelli D, Gamba C. 2009. Histological determination of the human origin of bone fragments. *Journal of Forensic Sciences* 54: 531–533.

Christensen AM, Passalacqua NV, Bartelink EJ. 2014. *Forensic Anthropology. Current Methods and Practice.* Oxford: Elsevier.

Christensen AM, Smith MA, Thomas RM. 2012. Validation of X-ray fluorescence spectrometry for determining osseous or dental origin of unknown material. *Journal of Forensic Sciences* 57: 47–51.

France DL. 2009. *Human and Non-Human Bone Identification: A Color Atlas.* Boca Raton, FL: CRC Press.

France DL. 2011. *Human and Non-Human Bone Identification: A Concise Field Guide.* Boca Raton, FL: CRC Press.

Hillier ML, Bell LS. 2007. Differentiating human bone from animal bone: a review of histological methods. *Journal of Forensic Sciences* 52: 249–263.

Hoffman JM. 1984. Identification of nonskeletonized bear paws and human feet. In: Rathbun TA, Buikstra JE (Eds.), *Human Identification: Case Studies in Forensic Anthropology.* Springfield, IL: Charles C. Thomas. p. 96–106.

Lowenstein JM, Reuther JD, Hood DG, Scheuenstuhl G, Gerlach SC, Ubelaker DH. 2006. Identification of animal species by protein radioimmunoassay of bone fragments and bloodstained stone tools. *Forensic Science International* 159: 182–188.

Martiniaková M, Grosskopf B, Omelka R. Vondráková M, Bauerová M. 2006. Differences among species in compact bone tissue microstructure of mammalian skeleton: use of a discriminant function analysis for species identification. *Journal of Forensic Sciences* 51: 1235–1239.

Mulhern DM, Ubelaker DH. 2001. Differences in osteon banding between human and nonhuman bone. *Journal of Forensic Sciences* 46: 220–222.

Mulhern DM, Ubelaker DH. 2012. Differentiating human from nonhuman bone microstructure. In: Crowder C, Stout S (Eds.), *Bone Histology. An Anthropological Perspective.* Boca Raton, FL: CRC Press. p. 109–134.

Owsley DW, Mires AM, Keith MS. 1985. Case involving differentiation of deer and human bone fragments. *Journal of Forensic Sciences* 30: 572–578.

Pérez MJ, Barquez RM, Díaz MM. 2017. Morphology of the limbs in the semi-fossorial desert rodent species of *Tympanoctomys* (Octodontidae, Rodentia). *ZooKeys* 710: 77–96.

Schaefer M, Black S, Scheuer L. 2009. *Juvenile Osteology: A Laboratory and Field Manual.* San Diego: Academic Press.

Scheuer L, Black S. 2000. *Developmental Juvenile Osteology.* San Diego, CA: Academic Press.
Scheyer TM. 2009. Conserved bone microstructure in the shells of long-necked and short-necked chelid turtles (Testudinata, Pleurodira). *Fossil Record* 12: 47–57.
Sims ME. 2007. Comparison of black bear paws to human hands and feet. Identification Guides for Wildlife Law Enforcement No. 11. USFWS, National Fish and Wildlife Forensics Laboratory, Ashland, OR.
Stagno V, Mailhiot S, Capuani S, Galotta G, Telkki V-V. 2021. Testing 1D and 2D single-sided NMR on Roman age waterlogged woods. *Journal of Cultural Heritage* 50: 95–105.
Stewart TD. 1959. Bear paw remains closely resemble human bones. *FBI Law Enforcement Bulletin* 28: 18–22.
Stewart TD. 1979. *Essentials of Forensic Anthropology, Especially as Developed in the United States.* Springfield, IL: Charles C. Thomas.
Ubelaker DH. 1998. The evolving role of the microscope in forensic anthropology. In: Reichs KJ (Ed.) *Forensic Osteology, Advances in the Identification of Human Remains,* Second Edition. Springfield, IL: Charles C Thomas. p. 514–532.
Ubelaker DH. 2014. Radiocarbon analysis of human remains: a review of forensic applications. *Journal of Forensic Sciences* 59: 1466–1472.
Ubelaker DH, Ward DC, Braz VS, Stewart J. 2002. The use of SEM/EDS analysis to distinguish dental and osseous from other materials. *Journal of Forensic Sciences* 47: 940–943.
Ubelaker DH, Lowenstein JM, Hood DG. 2004. Use of solid-phase double-antibody radioimmunoassay to identify species from small skeletal fragments. *Journal of Forensic Sciences* 49: 924–929.
Urbanová P, Novotný V. 2005. Distinguishing between human and non-human bones: histometric method for forensic anthropology. *Anthropologie* 43: 77–85.
Veitschegger K, Kolb C, Amson E, Scheyer TM, Sánchez-Villagra MR. 2018. Palaeohistology and life history evolution in cave bears, *Ursus spelaeus* sensu lato. *PLoS ONE* 13: e0206791. Available from: https://doi.org/10.1371/journal.pone.0206791.
Zimmerman HA. 2013. *Preliminary Validation of Handheld X-Ray Fluorescence (HHXRF) Spectrometry: Distinguishing Osseous and Dental Tissue from Non-Bone Material of Similar Chemical Composition.* MA Thesis, University of Central Florida, Orlando.

6

Sex Estimation

Learning Goals

By the end of this chapter, the student will be able to:

Define sexual dimorphism.
Describe the relationship between the mean and standard deviation of a measurement and how to use these statistics to differentiate males from females.
Identify elements of pelvic and skull anatomy used for estimating sex.
Apply methods of sex estimation to a forensics case.

Introduction

After human remains have been recovered, the first task of the forensic anthropologist is to develop the biological profile of the decedent. The **biological profile** refers to the basic demographic characteristics of a person that helps establish their identity, which ultimately leads to the return of the victim's remains to their family and contributes to solving a crime if one has been committed. Estimating the biological profile typically entails an assessment of *sex, age, stature,* and *ancestry.* Such information turns a *John Doe/Jane Roe* (victim identity unknown) into an identity such as "30–40-year-old white male, approximately 5.5 to 6 feet tall." Of these four basic identifying features of a person, biological sex is likely the most important. Here, *biological sex* refers to the individual's genotype as determined by their sex chromosomes (XX = female, XY = male). In forensic anthropology, sex tends to be treated as a binary (male/female), even though biological reality is more complex. Furthermore, sex is not gender (which is a culturally constructed pattern of behavior), and the two terms should not be used interchangeably.

By providing information on the decedent's sex, the forensic anthropologist helps law enforcement narrow the list of possible victim identities by around 50 percent (if you know the decedent was male then you can eliminate female missing persons, and vice versa). While DNA can provide highly accurate estimates of sex, it is costly, destructive, and time consuming. With high caseloads, it is unlikely that most medical examiners' offices around the country will have the financial resources to perform DNA analysis for most cases. In addition, anthropological sex estimation for a complete skeleton is accurate, precise, and expedient, thus making expensive and destructive DNA testing unnecessary in many cases. That said, DNA methods do have their place. For example, there are currently no reliable means of sex estimation for individuals younger than 18 years of age, despite many years of research on this topic (Holcomb and Konigsberg, 1995; Loth and Henneberg, 2001; Mittler and Sheridan, 1992; Schutkowski, 1993; Vlak et al., 2008; Weaver, 1980). This is because many of the skeletal features that differentiate males from females are still developing during adolescence, leading to calls to identify novel features of the subadult skeleton that may relate to sex and to develop more flexible statistical analyses (i.e., KidStats; Stull et al., 2020). For now, DNA may be the only way to ascertain sex in subadults with any degree of certainty.

Dozens of methods have been developed for estimating sex from the skeleton. All are based on the same fundamental biological processes that reflect our evolutionary history as "walking apes"—males are larger and more robust (heavily muscled) than females. This is called **sexual dimorphism** (dimorphism means "two forms"). Sexual dimorphism is common throughout the animal kingdom and refers to patterned differences between males and females with respect to size and shape (and for other animals, coloration and behavior). This is true for humans in nearly every measurable aspect of the skeleton, from the incisors to the smallest bones in your body, located in your middle ear (malleus, incus, stapes), although these differences can be quite subtle and not obvious upon simple examination. One cannot just "eyeball" a rib or vertebra and say whether the skeleton is of a male or female. But simple observational methods are possible with the skeleton, which is the primary focus of this chapter.

Forensic anthropologists should be trained in the use of both **metric** (measurement-based) and **morphoscopic** (ordinal-scale observational) methods of sex estimation. Metric methods are objective, replicable, less prone to observer error, and easily analyzed with statistics that satisfy the *Daubert* criteria. However, metric methods are also more time consuming, require specialized equipment, and assume that there are easy-to-identify landmarks to define measurements, which is not true for the most useful parts of the skeleton, such as the pelvis. In addition, metric measurements are population-specific; what is a male-size range in one population may overlap considerably with a female-size range in another population. Morphoscopic traits are easier to score and provide a simple, visual observational method that generates accurate estimates of sex, and they do so quickly and reliably. In recent years, forensic anthropologists have also developed sophisticated analyses of morphoscopic traits that are statistically sound (Klales et al., 2012).

Here we consider pelvic and cranial traits because it has been shown that some skeletal features are more effective at sex estimation than others (**Table 6.1**). Better performing features reflect differences in human reproductive anatomy or differences in body mass, often the weight-bearing joints and other measures of body width. Widths are a better measure of body mass, whereas lengths are correlated with body height. The teeth, smaller bones of the hands, more distal limb elements, and non-weight-bearing elements tend to perform worse. When forensic anthropologists are presented with a completely skeletonized individual, the first area they will look at is the pelvis because it is generally viewed as the most accurate for sex estimation. In many texts the cranium is considered the second-best region of the

Table 6.1. Sexual dimorphism in the human skeleton

Features That Perform Well	Features That Perform Poorly
Canine tooth crown size	All other tooth crown sizes
Femoral head size	Phalanges
Femoral neck size	Most measurements of bone lengths
Pelvic size and shape	
Cranial dimensions	
Weight-bearing bone width measures	

skeleton for sex estimation; however, several long bone measurements perform better than visual observations of the cranium (Spradley and Jantz, 2011).

Regardless of whether the data are metric or morphoscopic, the same principles of sexual dimorphism apply in determining which techniques are successful and which are not. Consider **Figure 6.1**. This figure shows a hypothetical sexually dimorphic feature plotted along the horizontal (or x) axis, which varies in this sample from a low value of 19 to a high value of 45. The vertical (or y) axis shows how commonly that measurement is observed for males (blue) and for females (red).

Note the following:

1) Sexually dimorphic features have a bimodal distribution. That means there are two peaks—for females the peak is around 29 and for males the peak is around 35.
2) Both males and females show a range of variation in their values. For females, the range is from 19 to around 39. For males, the range is from around 25 to 45.
3) Looking at **Figure 6.1,** you can be reasonably certain that a skeleton with a measured value <25 is female, and you can be reasonably certain that a skeleton with a measured value >39 is male.
4) It is more likely that an individual with a measured value between 25 and 30 is female. Although some males fall within this range (part of the blue line on the chart), most individuals in this range are female. To the contrary, individuals with a measurement between 34 and 39 are more likely male, despite some females falling in this measurement range.
5) Assessments of male or female are not absolute; they are stated in terms of probabilities. Individuals that plot to the extreme lower end of the distribution (the left side of the horizontal axis) are likely female. As you move right along the horizontal axis and measurement values increase, the scale moves from Definitely Female, to Likely Female, to Uncertain, to Likely Male, to Definitely Male.
6) Features that have minimal overlap in the distributions of males and females are ideal for sex determination. The more overlap between the sexes (the gray area in **Figure 6.1**), the more ambiguous the sex estimation is going to be; accuracy will be lower and erroneous sex estimates will be higher.

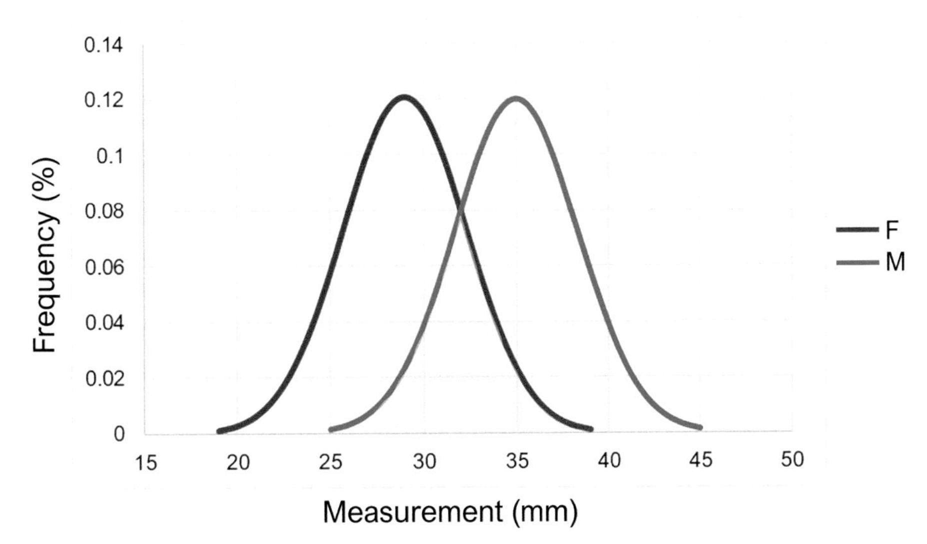

Figure 6.1. Plot of the distribution of a hypothetical trait for males and females. The gray area is the area of overlap, and it would be difficult to assess if an individual with a measurement in this range was male or female.

LEARNING CHECK

Q1. You measure the subpubic angle of a pelvis and determine it to be 75 degrees. Given the distribution of this feature depicted in **Figure 6.2A**, does this pelvis likely belong to a male or female?
A) Male
B) Female
C) Cannot determine from the information provided

Q2. You measure the height of a canine crown and determine that its height is 18 mm. Given the distribution of crown height depicted in **Figure 6.2B**, does the canine belong to a male or female?
A) Male
B) Female
C) Cannot determine from the information provided

Q3. You measure the height of another canine crown and determine that its height is 15 mm. Given the distribution of canine crown height depicted in **Figure 6.2B**, does the canine belong to a male or female?
A) Male
B) Female
C) Cannot determine from the information provided

Q4. You measure the length of a femur and determine that its length is 40 cm. Given the distribution of femoral lengths depicted in **Figure 6.2C**, does the femur belong to a male or female?
A) Male
B) Female

Q5. You measure the length of a femur and determine that its length is 35 cm. Given the distribution of femoral length depicted in **Figure 6.2C**, does the femur belong to a male or female?
A) Male
B) Female
C) Cannot determine from the information provided

Sex Estimation Using the Pelvis

It should come as no surprise that the pelvis shows consistent differences between males and females. The anatomy of the pelvis is reflective of the fundamental difference between the sexes; females give birth while males do not. Females have larger than expected pelvic girdles given their body size, whereas the opposite is true for males. Researchers have identified nearly two dozen features that differentiate the male and female pelvis (Rogers and Saunders, 1994). Many of these focus on the pubic bone and the anatomical differences that reflect the soft tissue anatomy of the genitals. However, there are overall shape differences in the pelvis that provide useful information for sex estimation.

Pelvic Anatomy

The human pelvis is composed of two halves, called the ossa coxae. The ossa coxae (plural) articulate with the sacrum posteriorly and with each other anteriorly at a joint called the

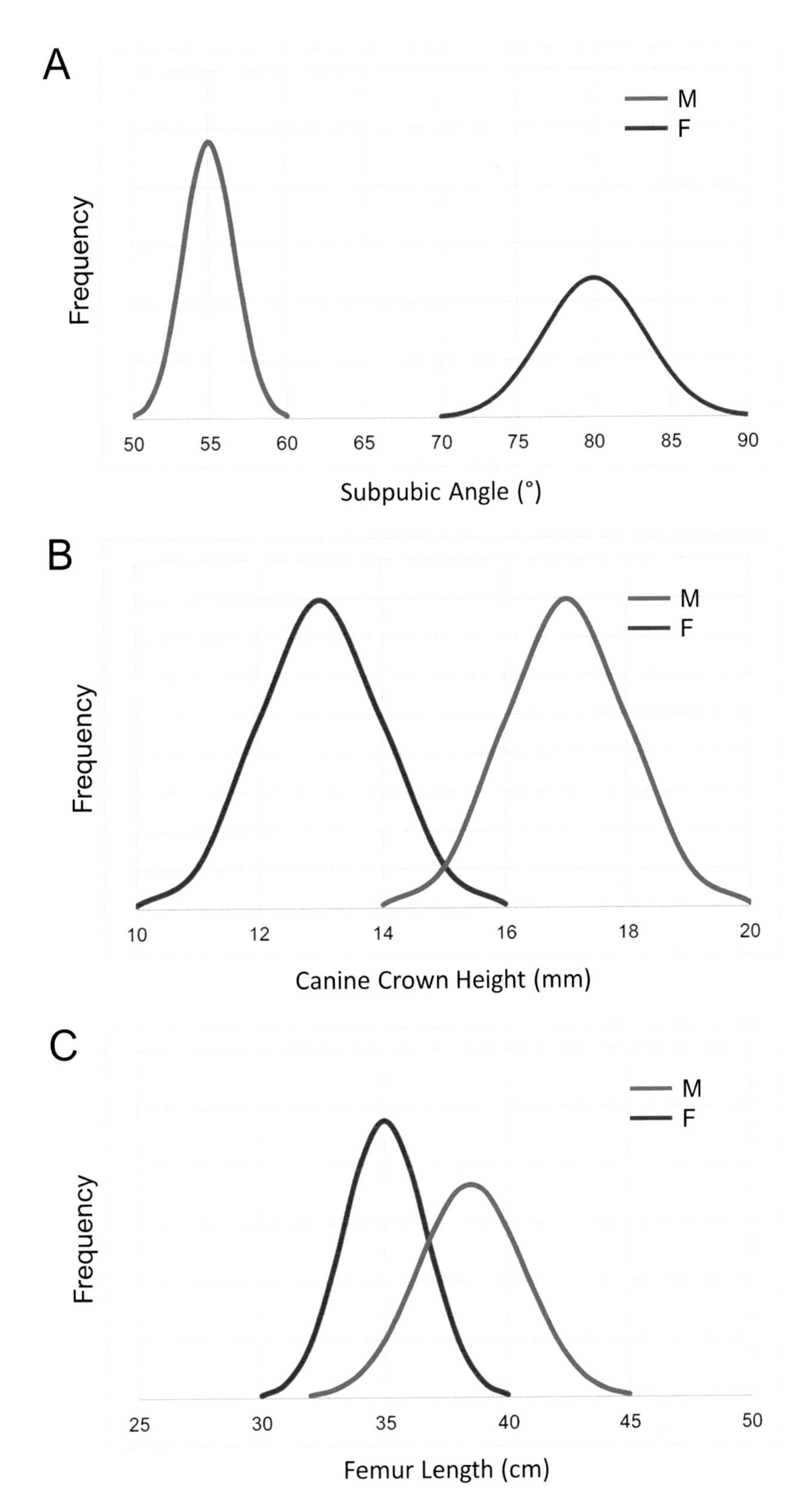

Figure 6.2. Sex-specific distributions of three osteological markers of sex.

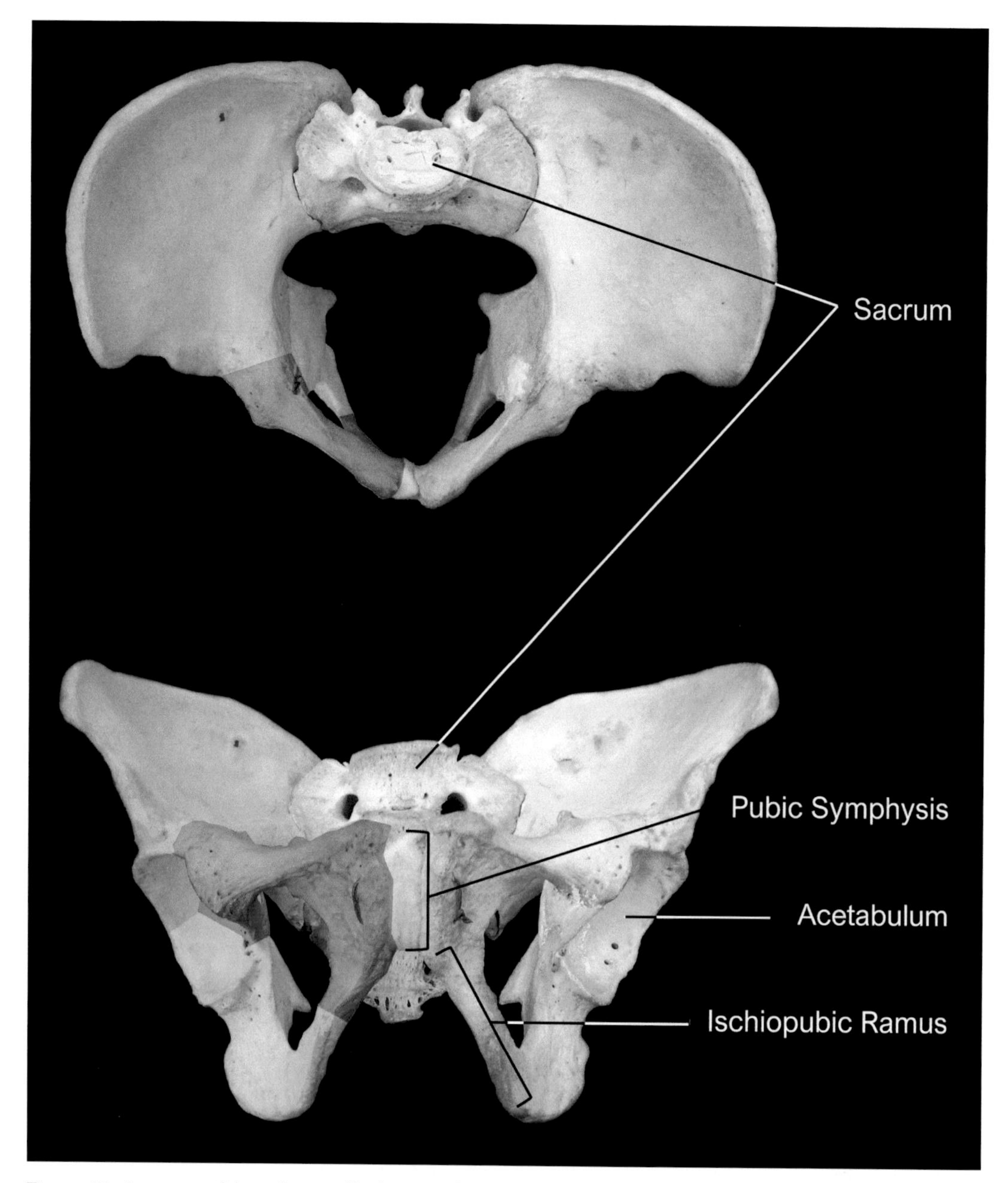

Figure 6.3. Anatomy of the pelvic girdle showing the ilium (blue), ischium (green), and pubis (red).

pubic symphysis. Each os coxa (singular) is constructed from three distinct growth centers, which are called the ilium (superior aspect, the blades of the pelvis that form the hips), the ischium (the roughened posterior part that you sit on), and the pubis (the anterior/ventral part).

Figure 6.3 shows how the pelvic girdle is constructed. The sacrum sits in between the ossa coxae. The coccyx, although not pictured, articulates with the sacrum inferiorly. This chapter will focus primarily on the anatomy of the pubis because there is considerable information about an individual's sex represented here. In particular, the shape of the pubic bone, the shape of the ischiopubic ramus, and the presence or absence of a ventral arc (see below) are most critical. Note that the pubic symphysis will be crucial when discussing age-at-death estimation in chapter 8. The acetabulum is where the femur articulates with the pelvis, which is useful for orienting the os coxa for observation.

Orienting an isolated os coxa can be difficult. The long axis of the bone should be oriented vertically (ilium is superior, ischium is inferior). The pubis is attached to the rest of the os coxa by two struts of bone and is anterior. The acetabulum, which articulates with the femur, is lateral. The ear-shaped auricular surface faces medially. Orienting the isolated os coxa can be challenging at first but must be mastered to apply the techniques discussed in this chapter and the following chapter on age estimation.

LEARNING CHECK

Q6. Label the three primary growth centers of this adult os coxa as shown in **Figure 6.4**. Note that the part of the bone labeled C is closer to the camera and elevated off the table.
A) A = pubis, B = ischium, C = ilium
B) A = ischium, B = pubis, C = ilium
C) A = ilium, B = ischium, C = pubis
D) A = ilium, B = pubis, C = ischium

Q7. Identify the side of the os coxa in **Figure 6.4**.
A) Left side
B) Right side

Q8. Identify what aspect of the os coxa is seen in **Figure 6.4**.
A) Lateral side
B) Medial side

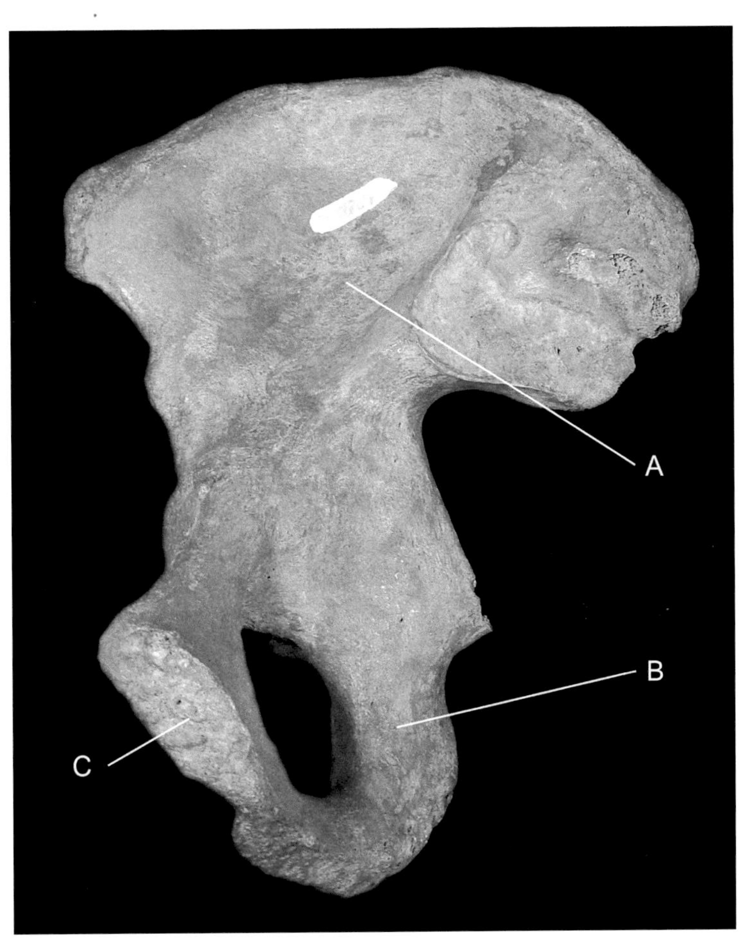

Figure 6.4. Human os coxa.

Features of Pelvic Sex Assessment

Figure 6.5 provides an illustration of the differences between male and female pelvic morphology. The *subpubic angle* is formed by the two pubic bones and is much wider in females than in males. For males, the subpubic angle is V-shaped, measuring approximately 70 degrees or less. For females, the subpubic angle is more U-shaped and measures between 70 and 110 degrees.

The *blades of the ilia* tend to be more open and flared in females, reflecting the wider pelvic girdle. The female pelvis is low, wide and shallow, while the male pelvis is narrow and tall. The *pelvic inlet* tends to be more oval in females and more heart-shaped in males. The *obturator foramen* is smaller and more triangular in females and larger and more oval in males. Note how the obturator foramen is more visible in the female pelvis. This is because the pubic bones are wider and the struts of bone connecting the pubic bones to the acetabulum are longer in the female pelvis.

If you have a well-preserved skeleton, it is very easy to assess the sex of the individual. However, even if you do not have a complete pelvis, you can still accurately estimate the sex of an individual using only a handful of features. Here, we discuss five features of the os coxa that may be observable in fragmented remains.

Feature 1: Ventral Arc

The ventral arc is one of the most reliable indicators of sex. It is a raised ridge of bone *only* found on the ventral (anterior) surface of the pubic bone in females (Phenice, 1969). That

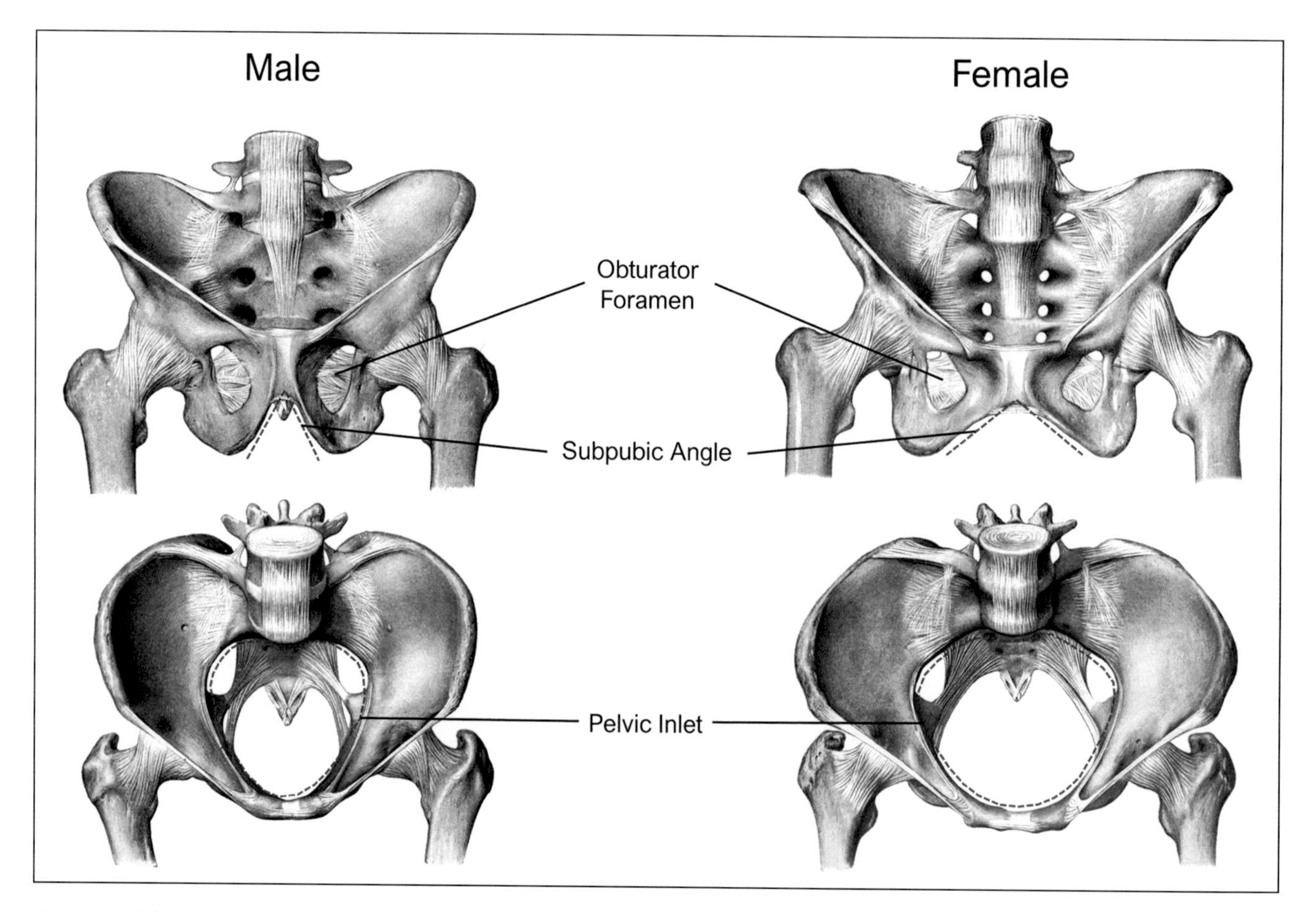

Figure 6.5. Schematic drawings of male and female pelves.

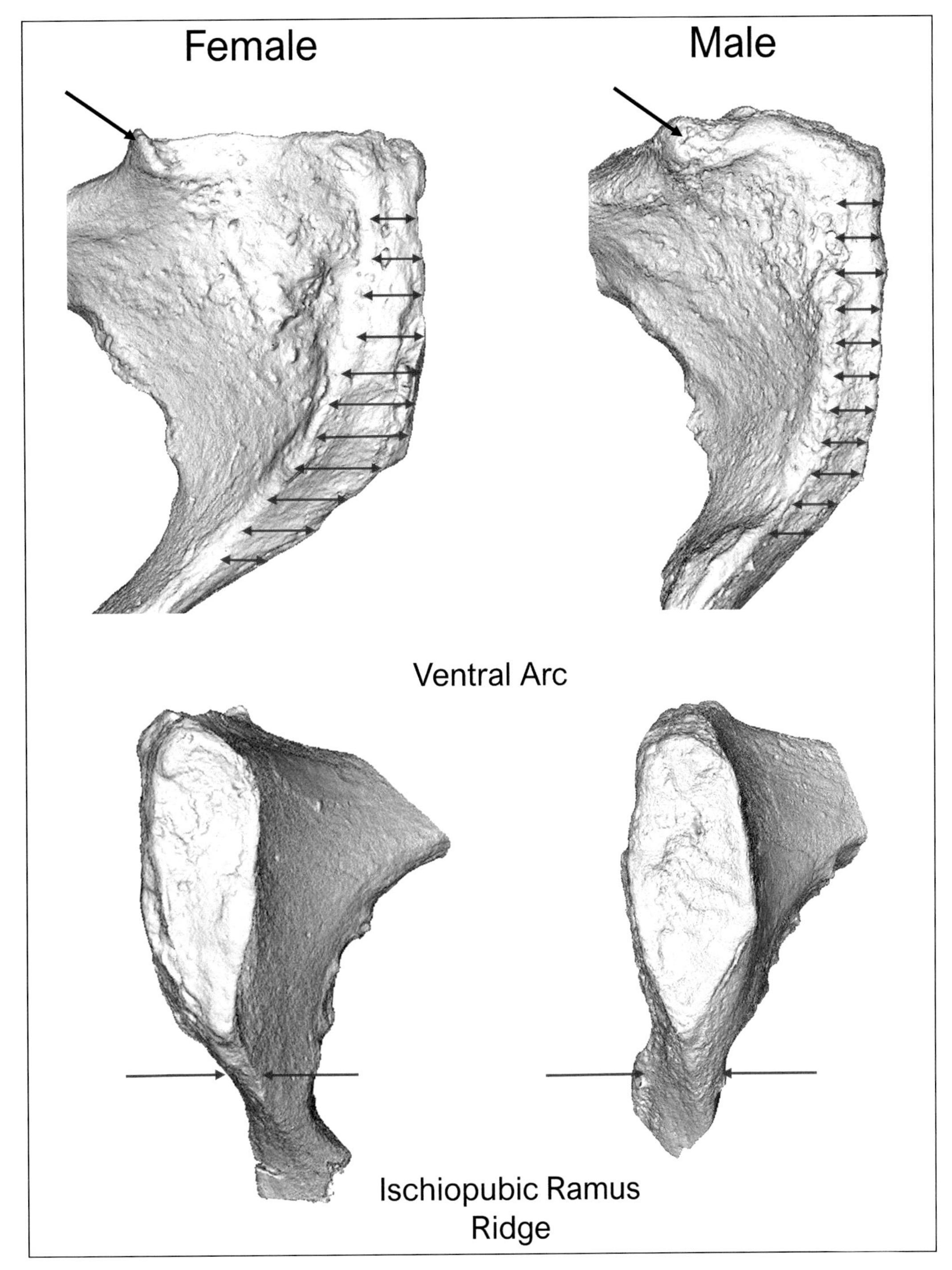

Figure 6.6. Male and female pubic bones exhibit differences in the expression of the ventral arc and the ischiopubic ramus ridge. Black arrows indicate the pubic tubercle.

is, the presence of a ventral arc indicates a female sex. The arc arises during puberty due to growth differences and variation in male and female musculature (Klales et al., 2012).

Consider the two bones in **Figure 6.6** (*top*) with the female on the left and the male on the right. The symphyseal faces, in this case, are oriented toward the right of the image such that the view shown is of the ventral (anterior) surface of the bone. Note that both bones have a raised ridge of bone on their ventral surface and that this ridge is somewhat more pronounced in the female (often, no such ridge is even visible on male pubic bones). Now, look closely at the red arrows that connect the symphyseal face and the inferior border of the

Table 6.2. Klales and colleagues' (2012) trait score descriptions for the ventral arc, ischiopubic ramus ridge, and subpubic contour

Trait	Score	Description[a]
Ventral Arc	1	Arc present at approximately or at least a 40° angle in relation to the symphyseal face with a large triangular portion of bone inferiorly placed to arc
	2	Arc present at approximately a 25–40° angle in relation to the symphyseal face with a small triangular portion of bone inferiorly placed to arc
	3	Arc present at a slight angle (less than 25°) to the symphyseal face with a slight, nontriangular portion of bone inferiorly placed to arc
	4	Arc present approximately parallel to the symphyseal face with hardly any additional bone present inferior to arc
	5	No arc present
Medial Aspect of Ischio-Pubic Ramus	1	Ascending ramus is narrow dorso-ventrally with a sharp ridge of bone present below the symphyseal face
	2	Ascending ramus is narrow dorso-ventrally with a plateau/rounded ridge of bone present below the symphyseal face
	3	Ascending ramus is narrow dorso-ventrally with no ridge present
	4	Ascending ramus is medium width dorso-ventrally with no ridge present
	5	Ascending ramus is very broad dorso-ventrally with no ridge present
Subpubic Contour	1	Well-developed concavity present inferior to symphyseal face along length of inferior ramus
	2	Slight concavity present inferior to face extended partially down inferior ramus
	3	No concavity present, bone is nearly straight (may be a very slight indentation just below the symphyseal face)
	4	Small convexity, especially pronounced along inferior pubic ramus
	5	Large convexity, especially pronounced along inferior pubic ramus

[a] Trait score descriptions from Klales et al. (2012: 109).

ischiopubic ramus to the raised ridge of bone. In the male pubic bone, these arrows are all approximately the same length, meaning that this ridge (when it occurs) parallels the contour of the bone. In contrast, the arrows on the female pubic bone first increase and then decrease in length, illustrating how the ridge of bone curves away from the contour of the bone. This curve, or arc, is the ventral arc.

Directions: View the pubic bone from the anterior/ventral aspect. The ventral surface of the pubic bone is rougher than the dorsal; this is important for orienting the pubic bone. In addition, the ventral surface has a pubic tubercle on the superior aspect. This is a large projection of bone that can be seen in **Figure 6.6** to the left of (lateral to) the symphyseal face (indicated by the black arrows). Klales and colleagues (2012) have presented a revised scoring system for this trait (reproduced in **Table 6.2**). In general, lower scores (e.g., 1 and 2) are more frequently observed among females and rarely observed among males. In contrast, higher scores (e.g., 4 and 5) are more frequently observed among males and rare among females. Intermediate scores (score = 3) are observed in both males and females.

Feature 2: Ischiopubic Ramus Ridge

The ischiopubic ramus is the strut of bone that connects the pubis and the ischium. In females, the ramus is thinner and will have a sharp, elevated ridge just below the pubic symphysis that gives it a pinched appearance. In males, the ramus is broad and rounded and there is no ridge present right below the pubic face (Phenice, 1969). The reasons for these differences are not well understood but are thought to be related to differential growth processes during puberty related to sexual maturation (Klales et al., 2012). Compare the two pubic bones in **Figure 6.6** (bottom). These images are oriented so that you are looking straight at the symphyseal face. Red arrows have been added to show the location of the ischiopubic ramus. Compare the pinched, ridge-like appearance of the ramus on the female to the blunt and round appearance of the male ramus.

Directions: View the pubic bone as if you are looking directly at the pubic symphysis (from the medial aspect). Observe the bone directly inferior to the symphyseal face. Klales and colleagues (2012) have provided a revised, five-stage scoring system for this trait (**Table 6.2**). As with the ventral arc, lower scores are more frequent among females and higher scores are more frequent among males. While intermediate scores are observed among both males and females, they are more common among females for this trait.

Feature 3: Subpubic Contour

The subpubic contour describes the shape of the inferior, or medial, margin of the ischiopubic ramus. A lateral recurve in the ischiopubic ramus directly inferior to the lower margin of the pubic symphysis is referred to as a *subpubic concavity* (Phenice, 1969). This trait develops during puberty and reflects the different growth processes of the bone experienced by males and females (Klales et al., 2012). Typically, females possess a concavity and males possess a flat or convex surface right below the symphyseal face. In **Figure 6.7** (top), the curvature of the ischiopubic ramus is illustrated with a dotted red line.

Directions: View the pubic bone from the ventral/anterior aspect. Klales and colleagues (2012) have presented a revised, five-stage scoring system for this trait (**Table 6.2**). As with the ventral arc, lower scores are more frequently observed among females, higher scores are more frequent among males, and intermediate scores are observed among both males and females.

Feature 4: Greater Sciatic Notch

The greater sciatic notch is located on the posterior aspect of the ilium. This part of the os coxa is thick and preserves well in fragmentary cases (Walker, 2008). For this reason, the sciatic notch could be the only sex assessment feature observable in very fragmentary or burned remains. Unfortunately, it is also the least accurate of the five features discussed here. Females tend to have a broad notch (50–75 degrees) while males have a much narrower notch (50 degrees or less) as shown in **Figure 6.7** (middle). While it is possible to measure the width of the sciatic notch, there are few anatomical landmarks in this area of the os coxa that allow one to define replicable measurement locations. For this reason, a visual assessment is often used.

Directions: Hold the os coxa and view the sciatic notch from the medial aspect. Compare the notch (blue arrow) to the schematic drawings ranked from 1 to 5. Align the contour of the sciatic notch to those shown in **Figure 6.7** (middle). Males tend to have a narrow notch (5); females tend to have a wide notch (1). In this system, a 3 is indeterminate, a 2 is a probable female, and a 4 is a probable male.

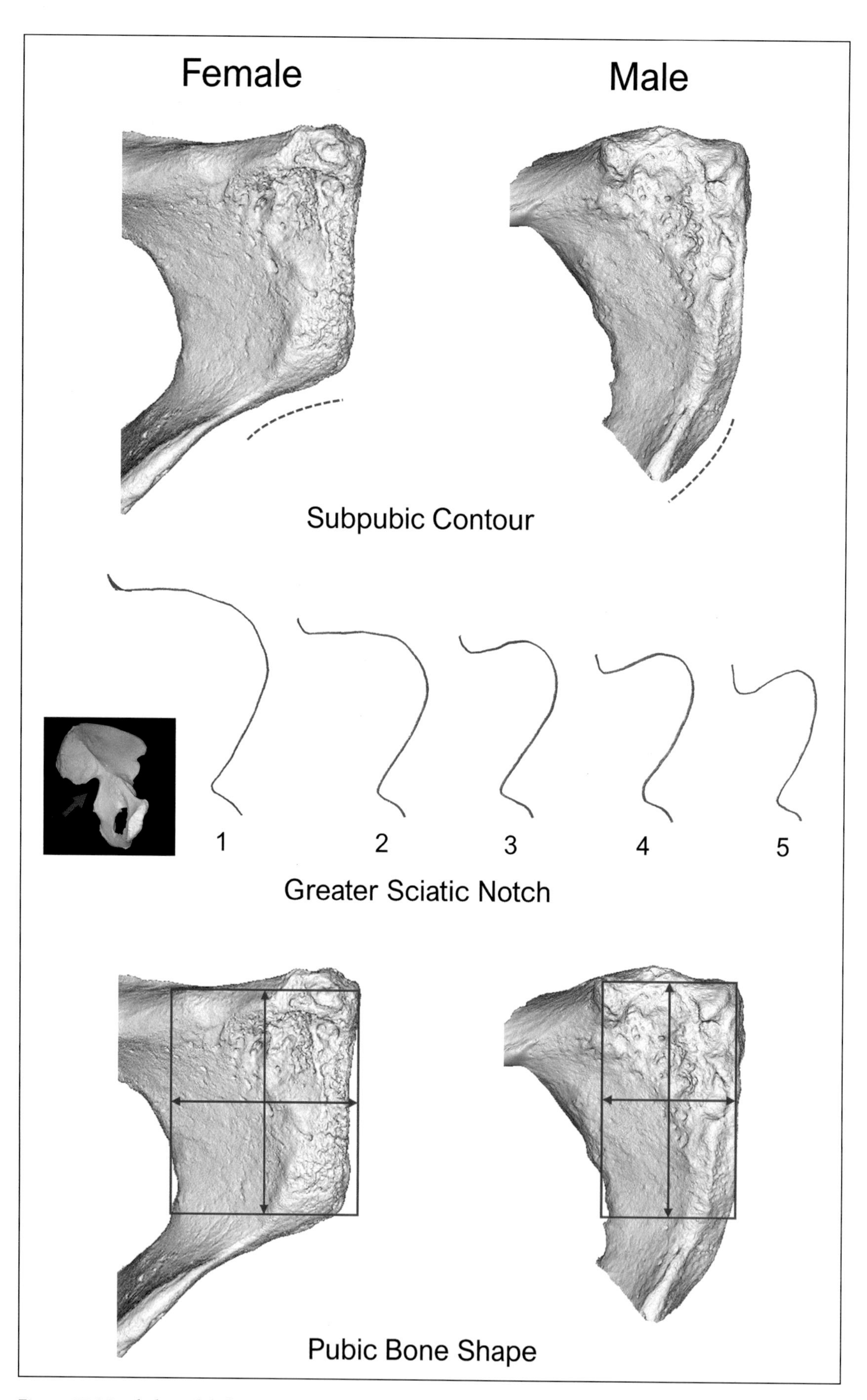

Figure 6.7. Morphological differences between male and female os coxa for the subpubic contour (*top*), greater sciatic notch (*middle*), and the shape of the pubic bone (*bottom*).

Feature 5: Pubic Bone Shape

The overall shape of the pubic bone also differs between the sexes. Females tend to have a broad pubic bone, giving the bone a more square-shaped appearance when viewed from the anterior. Males tend to have a narrower pubic bone that is also taller, giving it a more rectangular appearance when viewed from the anterior (**Figure 6.7,** bottom).

Directions: View the pubic bone from the posterior/dorsal surface. The presence of a square-shaped bone is scored a 1 (female), the presence of a rectangular bone is scored a 3 (male), and indeterminate cases are scored a 2.

LEARNING CHECK

Q9. Consider **Figure 6.8**. Based on the traits discussed, which picture shows the ventral surface of the pubic bone?
A) A
B) B
C) C

Q10. Is there a ventral arc present in **Figure 6.8**? Be sure to compare the correct image identified in Q9 to **Figure 6.6** (top).
A) Yes
B) No

Q11. Considering now Image B in **Figure 6.8,** is there an ischiopubic ramus ridge? Compare Image B to **Figure 6.6** (bottom).
A) Yes
B) No

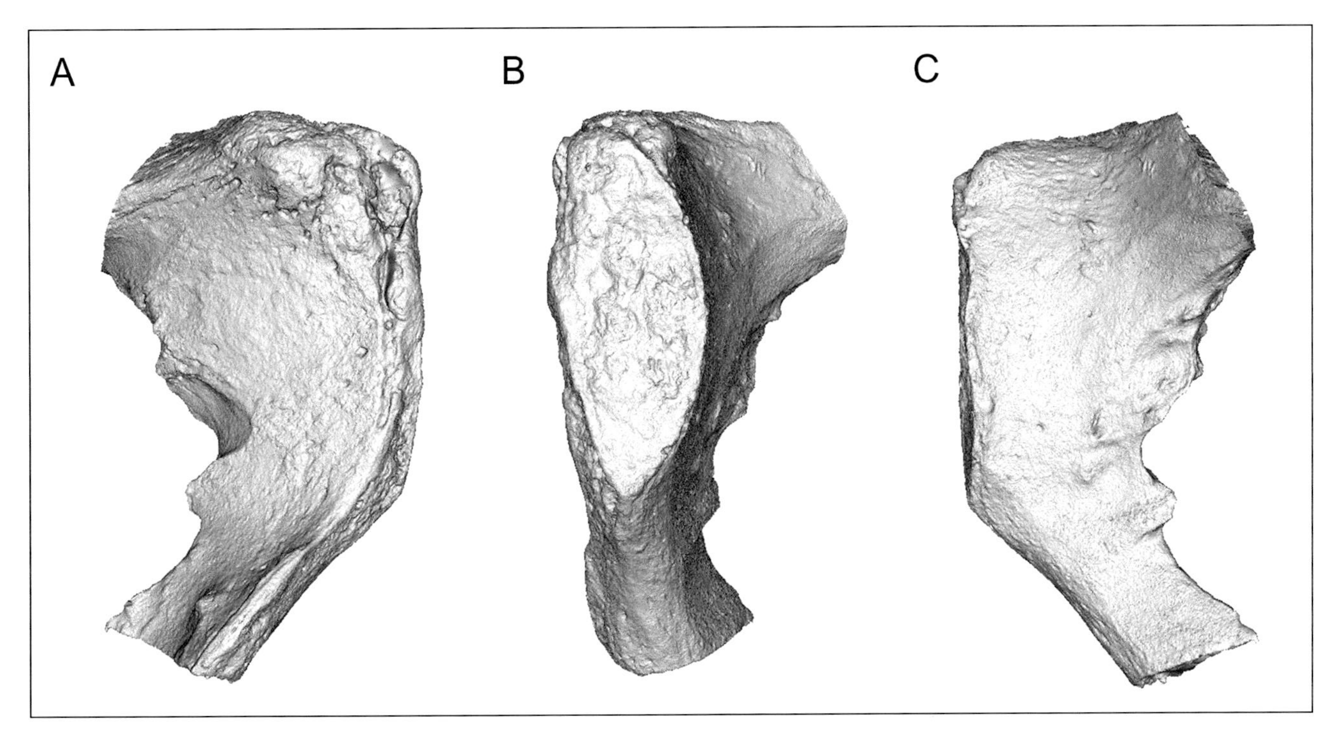

Figure 6.8. Three views of a pubic bone.

Q12. Based on your responses to the above, is the individual shown in **Figure 6.8** a male or female?
A) Male
B) Female

Q13. Which of these two bones in **Figure 6.9** exhibit a male form of the sciatic notch? Note the sciatic notch is located on the left side of each bone as depicted in this image. Compare to the scoring scale in **Figure 6.7**.
A) A
B) B

Q14. Considering the sciatic notch in **Figure 6.10,** does this appear to be a male or female individual? Compare the sciatic notch to **Figure 6.7**.
A) Female
B) Male

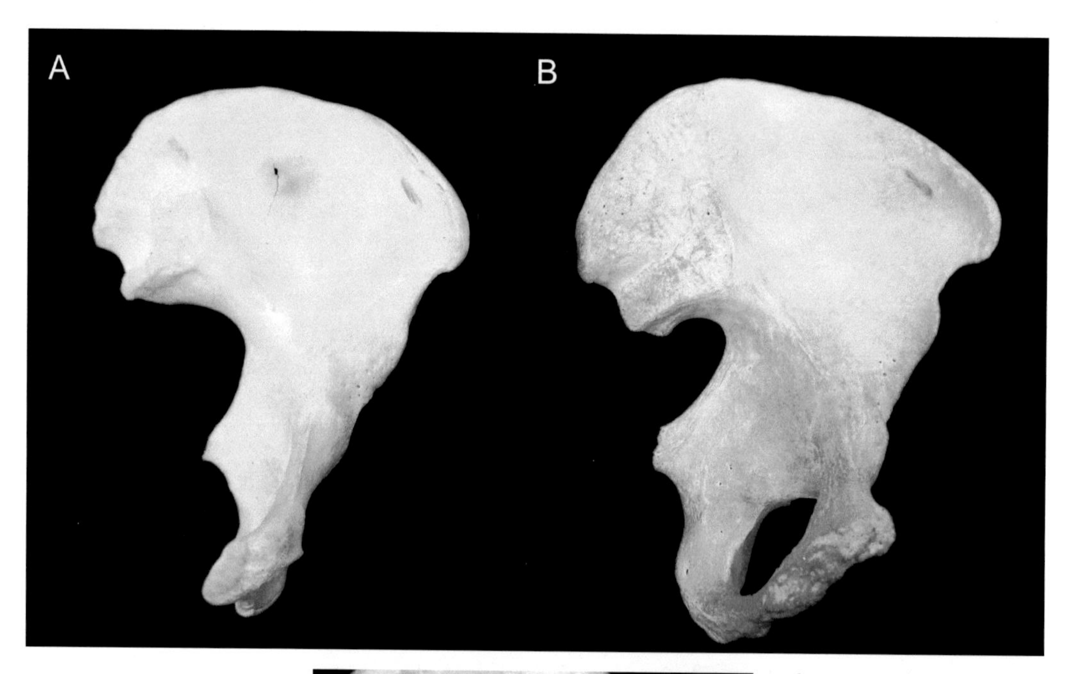

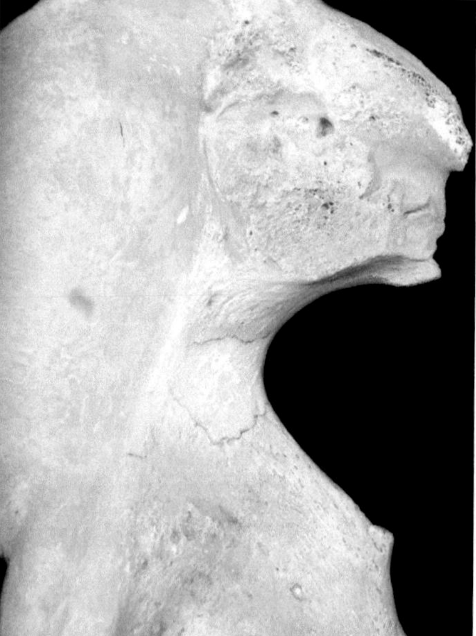

Above: Figure 6.9. Two ossa coxae, both left side, viewed from the medial aspect.

Left: Figure 6.10. Os coxa, right side, viewed from the medial aspect.

Sex Estimation Using the Skull

In the absence of pelvic remains, forensic anthropologists are still able to estimate the sex of a decedent based on visual skeletal indicators. The skull (the cranium and mandible) provides reliable information on sex estimation; however, there is a decline in accuracy in comparison to the pelvis and also in comparison to some long bone measurements (Spradley and Jantz, 2011). Pelvic features of sex estimation are more closely related to primary sexual characteristics between males and females, that is, differences that relate to the sex organs that are present at birth. Whereas differences in the skull reflect secondary sexual characteristics, changes that occur during growth and development.

Features of Skull Sex Estimation

The skulls of males are larger overall and tend to have better developed muscle attachment sites leading to thicker bones and more developed, raised areas of muscle attachment. This is most easily seen at the temporal lines for attachment of the temporalis muscle used in chewing. Male skulls tend to have sloping foreheads, less rounded, more angular cranial vaults, and a rougher (rugose) appearance overall. All muscle attachment sites are slightly more developed in males. The skulls of females are smaller overall and have a generally smoother appearance, with less well-developed muscle attachment sites.

Over the years dozens of features have been identified that differentiate male and female skulls. However, many of these have not been systematically tested and observer error rates are unknown. For this reason, forensic anthropologists have focused on five features of the skull that seem to estimate sex most reliably (Walker, 2008).

Feature 1: Nuchal Crest

The nuchal crest is a rugose (roughened) area on the posterior aspect of the occipital bone. The nuchal crest serves as an attachment site for the nuchal (neck) muscles. The crest is oriented transversely on the vertical portion of the occipital. Males tend to have a more developed nuchal crest than females. **Figure 6.11** shows the cranium from both a lateral view (*top*) and an inferior view (*bottom*). The white arrows indicate the location of a well-developed nuchal crest.

Directions: Examine the scoring system at the bottom of **Figure 6.11**. Be sure to view this area of the skull from the lateral aspect. Focus on the development of the ridge of bone that anchors the neck muscles, which appears as a disruption in the smooth contour of the occipital when viewed laterally. The external surface of the occipital bone is smooth with no bony projections in females (score = 1), while in males, the nuchal crest projects from the occipital bone and the entire area is rougher in appearance. In some males, a hook-like projection of bone may be present (score = 5). A score of 3 is considered ambiguous with regard to sex. In making this assessment, view the skull at approximately arm's length.

Feature 2: Mastoid Process

The mastoid process is a bony protrusion of the temporal bone that serves as another attachment for the neck muscles (**Figure 6.12**). The mastoid process can be palpated by feeling directly behind your ear lobe. Males tend to have larger, more inferiorly projecting mastoid processes than females. Although difficult to measure, the key difference between males and females relates to the overall volume of the mastoid process in comparison to surrounding structures like the auditory meatus and root of the cheek (zygomatic process).

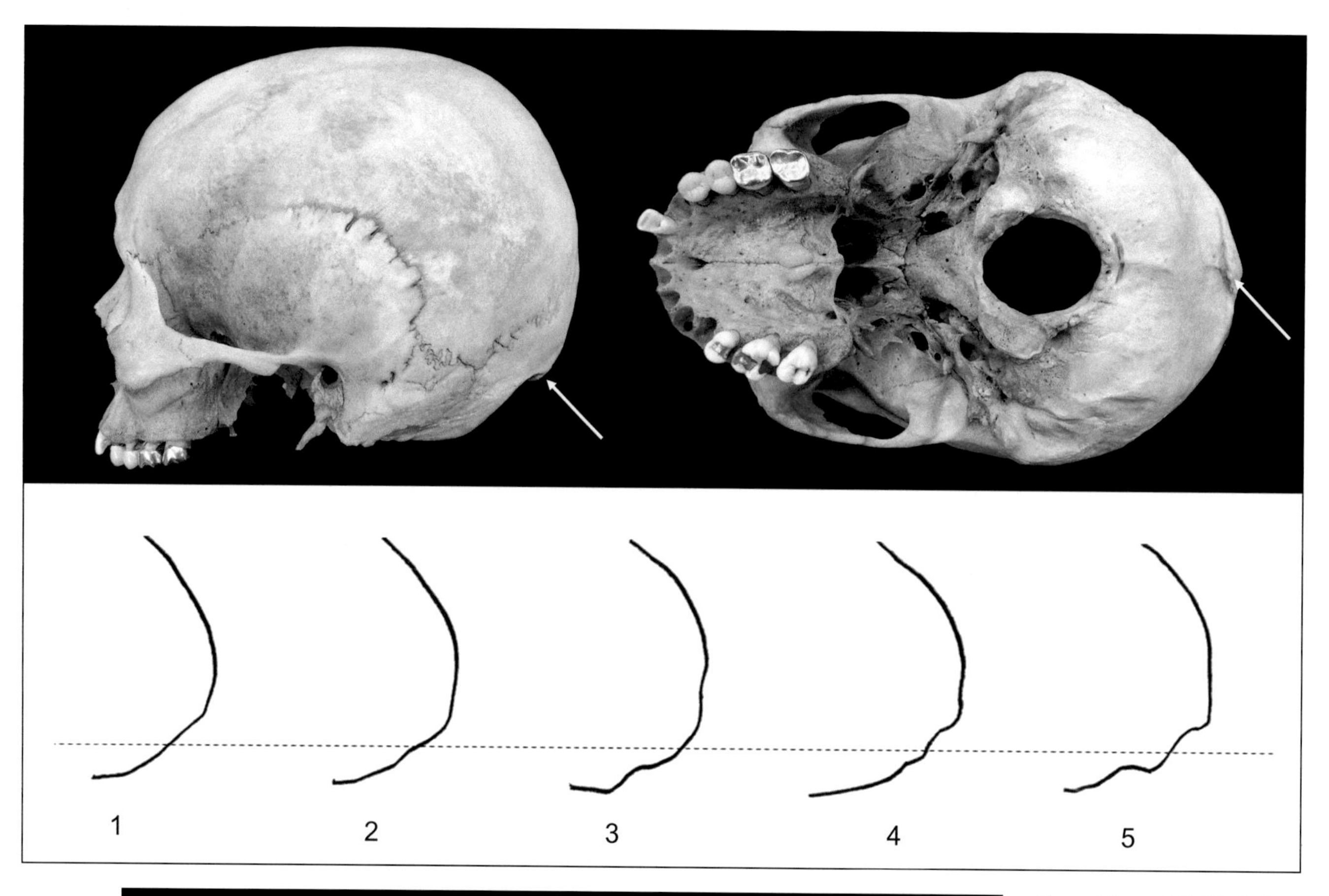

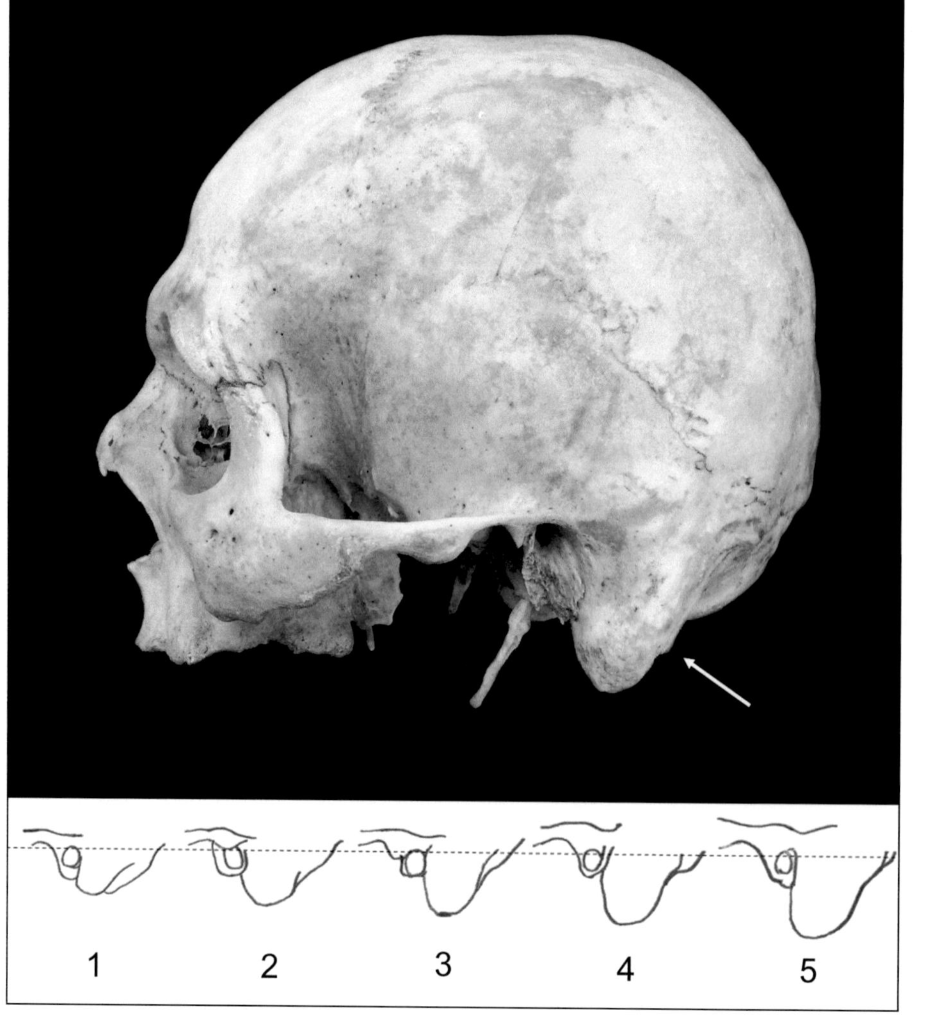

Above: Figure 6.11. Nuchal crest. The arrows indicate where to observe this feature on lateral and inferior views of the cranium (*top*); this trait is scored following the scale at the bottom of the figure.

Left: Figure 6.12. Mastoid process. The arrow indicates the location of the mastoid process on a lateral view of the cranium (*top*); this trait is scored following the scale at the bottom of the figure.

Directions: Examine the scoring system at the bottom of **Figure 6.12**. Be sure to view this area of the skull from the lateral aspect. Score this feature by comparing the size of the mastoid process to surrounding structures, such as the opening of the ear (external auditory meatus). The volume and mass of the process is more important than its length. Minimal expression (score = 1) is a very small mastoid process that extends only a small distance below the external auditory meatus. A very large mastoid process (score = 5) has a length and width that is several times the size of the opening for the ear. A score of 3 is considered ambiguous with regard to sex. In making this assessment, view the skull at approximately arm's length.

Feature 3: Supraorbital Margin

The supraorbital margins form the anterior upper roof of the orbits (**Figure 6.13**). Males tend to have rounder and thicker supraorbital margins, while females tend to have sharper and thinner supraorbital margins. The area of observation is the lateral half of the orbital roof, that is, lateral to any supraorbital notches or foramina that may be present.

Directions: Examine the scoring system at the bottom of **Figure 6.13,** which shows a lateral view of the orbital roof as if cut in half. Be sure to view this area of the skull in frontal or frontal-inferior view when scoring. The best strategy is to pinch the lateral half of the supraorbital margin and compare its thickness to the scale pictured in **Figure 6.13,** being careful not to damage the orbit. The margin is sharp in females (score = 1) and thick and rounded in males (score = 5). Walker (2008) states that a score of 5 approximates the feel of a pencil. A score of 3 is considered ambiguous with regard to sex. In making this assessment, view the skull at approximately arm's length.

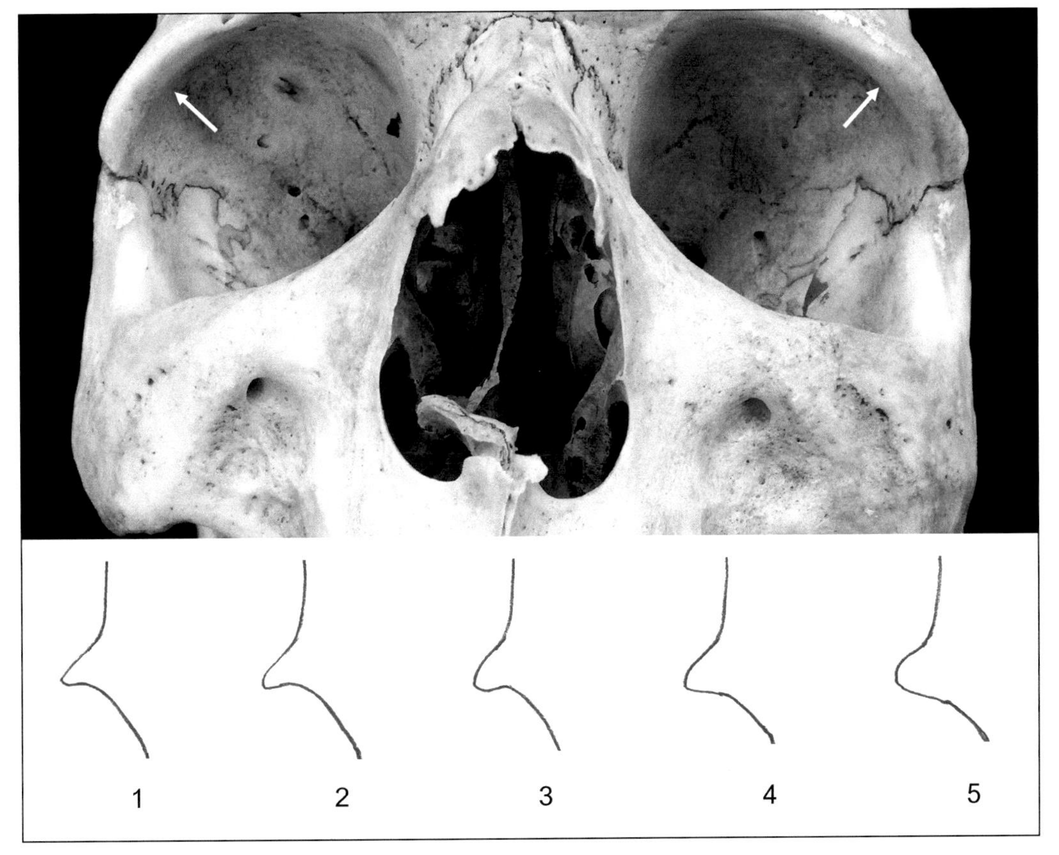

Figure 6.13. Supraorbital margin. The arrows indicate where this trait should be assessed—imagine squeezing this part of the cranium between your fingers to evaluate its thickness; this trait is scored following the scale at the bottom of the figure that shows the superior margin of the orbit from a lateral view.

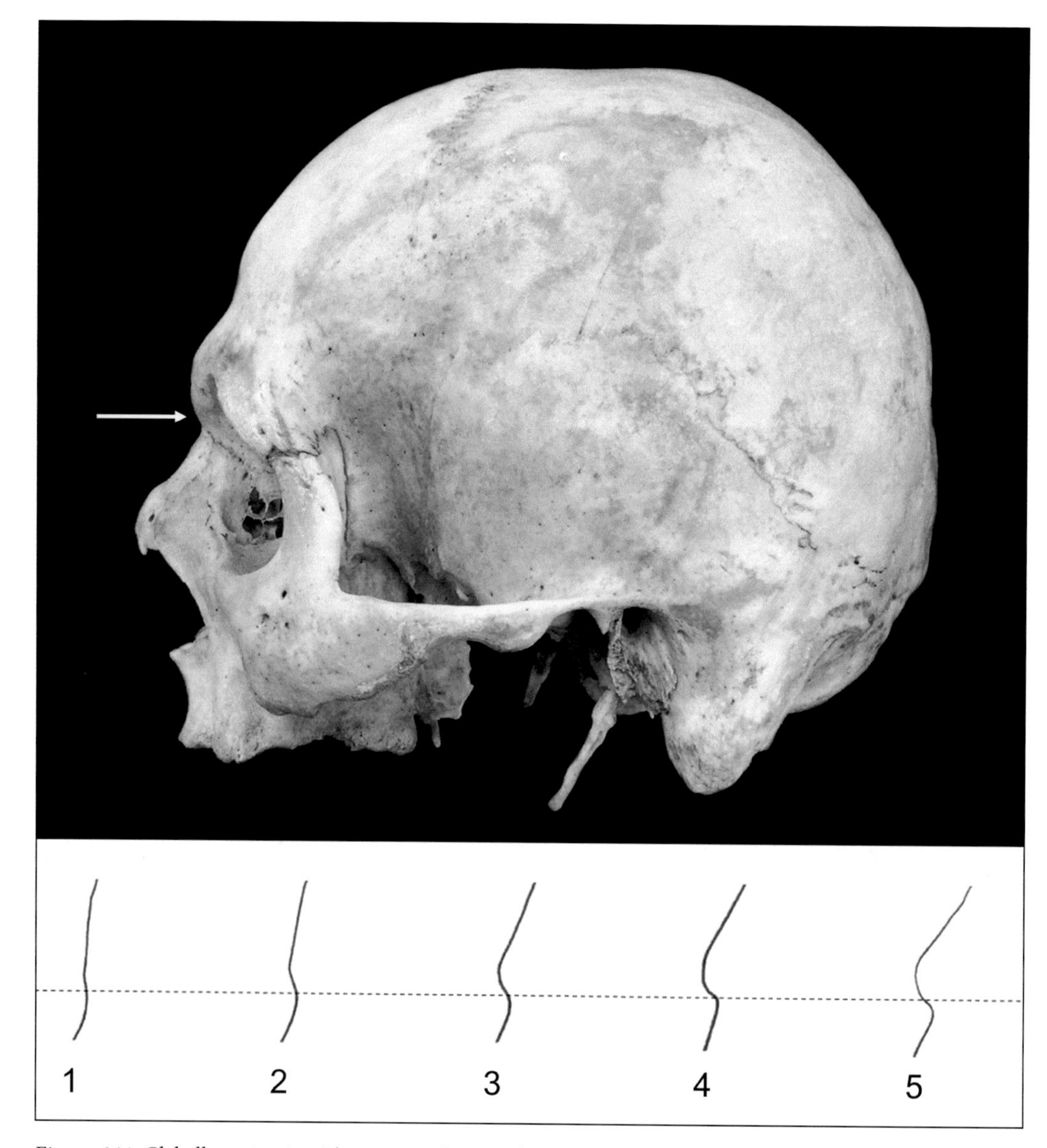

Figure 6.14. Glabella projection. The arrow indicates where this trait should be assessed on a lateral view of the cranium (*top*); this trait is scored following the scale at the bottom of the figure.

Feature 4: Glabella Projection

Glabella is an anatomical landmark in the midline of the frontal bone located slightly above and between the brow ridges (**Figure 6.14**). Males tend to have a more developed and forward-projecting glabella than females. In females, the contour of the bone is flat when traced in lateral view from the forehead to the bridge of the nose.

Directions: Examine the scoring system at the bottom of **Figure 6.14**. Be sure to view this area of the skull in lateral view. For a minimum score of 1, the contour of glabella is smooth, with little or no anterior projection at the midline. For maximum expression scores of 5, glabella forms a rounded, loaf-shaped projection that extends anteriorly beyond the plane of the orbits. A score of 3 is considered ambiguous with regard to sex. In making this assessment, view the skull at approximately arm's length.

Feature 5: Mental Eminence

The mental eminence refers to the area of the projecting chin on the front of the lower jaw in the midline that is the site of attachment for one of the muscles of the lower face (**Figure 6.15**). In females, the mental eminence tends to be weakly developed. The anterior midline tends to be smoother with a less well-defined mental eminence and more rounded inferior profile of the mandible. In males, the anterior midline often presents as an elevated area of bone that is triangular.

Directions: Examine the scoring system at the bottom of **Figure 6.15**. Be sure to view the mandible in anterior view, holding the mandible by the *ascending rami*. In minimal expressions (scores of 1), there is little or no projection of the mental eminence above the surrounding bone. In maximal expressions (scores of 5), the mental eminence occupies most of the anterior portion of the mandible. Pay attention to the lines drawn in the front of the mandible in **Figure 6.15**. Those lines begin small and do not touch superiorly at scores of 1. The lines get larger (scores 2 and 3) and eventually touch (scores of 4), forming an inverted T-shape projection of bone (scores of 5). Also note the change in the outline of the mandible's inferior profile. In scores of 1, the mandible is pointed and rounded with no inferior projections of bone visible on the inferior outline, but in scores of 5 the mandible's outline is more irregular and squared off with widely separated inferior projections of bone. It may help to view the mandible anteriorly but to rotate it up and down in your hands to better visualize the extent of bone development and projection development. Inferior views of the mandible can provide a better sense of the overall profile shape of the inferior margin.

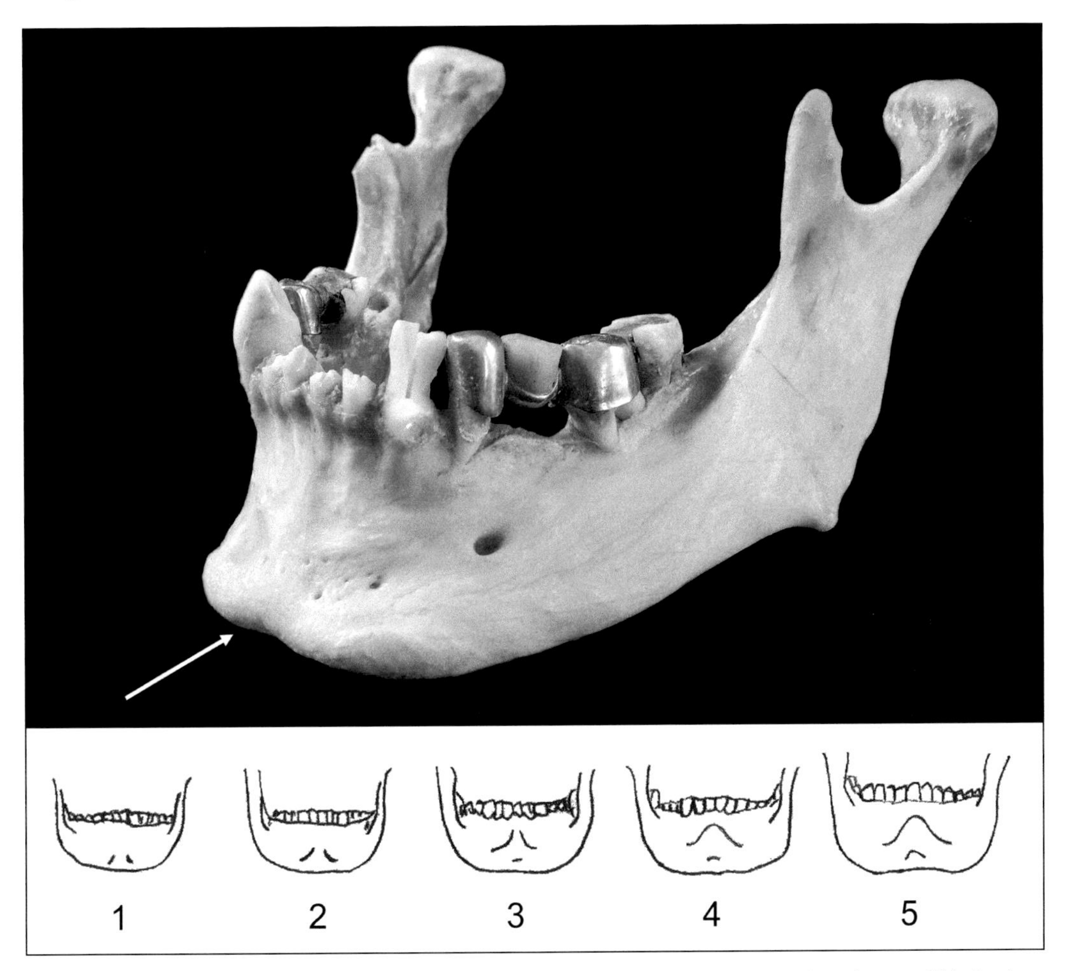

Figure 6.15. Mental eminence. The arrow indicates where this trait should be assessed on the mandible (*top*); this trait is scored following the scale at the bottom of the figure.

LEARNING CHECK

Q15. Using the features discussed in this chapter, estimate the sex of the individual in **Figure 6.16**.
A) Male
B) Female
C) Indeterminate

Q16. Using the features discussed in this chapter, estimate the sex of the individual in **Figure 6.17**.
A) Male
B) Female
C) Indeterminate

Q17. Using the features discussed in this chapter, estimate the sex of the individual in **Figure 6.18**.
A) Male
B) Female
C) Indeterminate

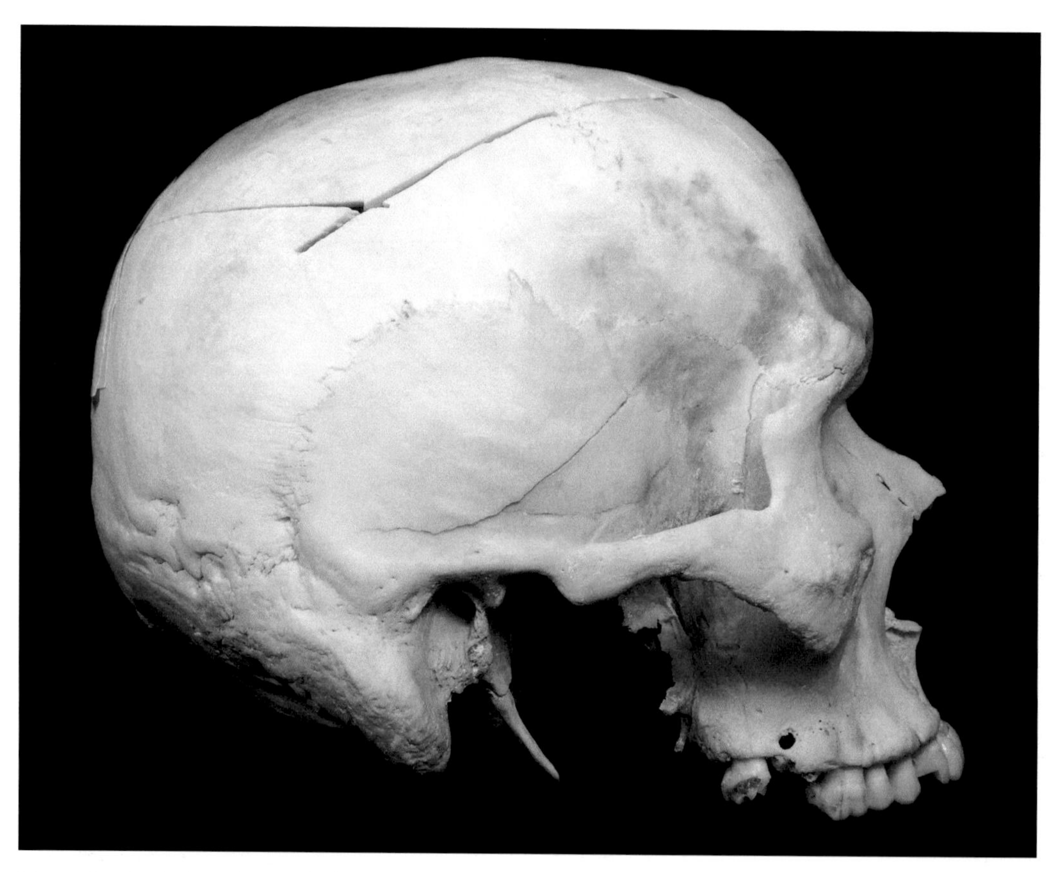

Figure 6.16. Right lateral view of a human cranium.

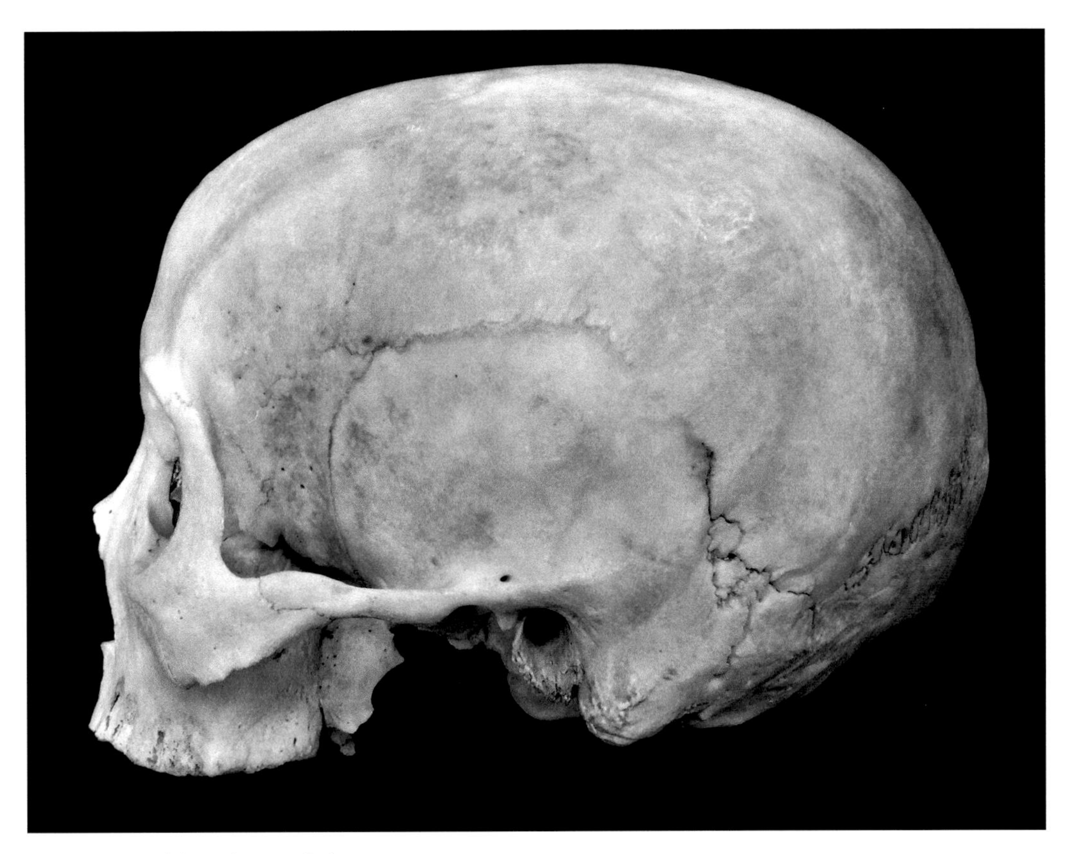

Figure 6.17. Left lateral view of a human cranium.

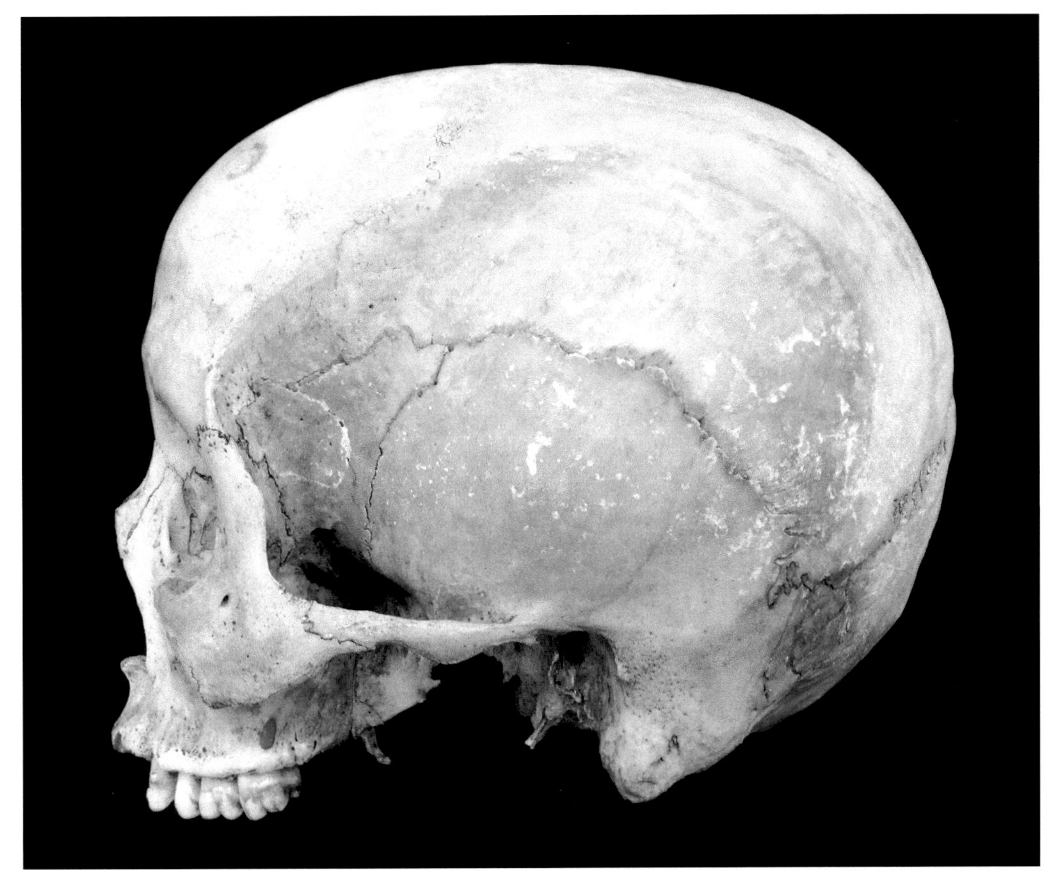

Figure 6.18. Left lateral view of a human cranium.

End-of-Chapter Summary

This chapter presented the methods used in sex estimation using the pelvis and skull based on morphoscopic, qualitative assessment of key features that are sexually dimorphic. Sexual dimorphism refers to size and shape differences between the sexes. Measurements that are most useful for estimating sex are those that show the greatest difference in the means of males and females and show the lowest within-sex spread or standard deviation. Currently, there are few reliable methods for identifying the sex of decedents less than 18 years of age. In such cases, DNA analyses may be the only technique available for identifying the victim's biological sex.

Five pelvic features useful for estimating the sex of a decedent were introduced: ventral arc, ischiopubic ramus ridge, subpubic concavity, sciatic notch, and pubic bone width. Orienting the os coxa is critical for using these features for sex estimation. With a complete os coxa and fully intact pubic bone, sex assessment accuracy is within the range of 95 percent. With respect to the skull, five features of sexual dimorphism were introduced: nuchal crest, mastoid process, glabella projection, supraorbital margin, and mental eminence. Individually, and as a group, these features perform worse than the pelvis for estimating sex.

End-of-Chapter Exercises

Exercise 1

Materials Required: **Figure 6.19**

Directions: Although not discussed in detail in this chapter, many metric methods for sex estimation exist and several of them have been shown to out-perform sex estimation using features of the cranium (Spradley and Jantz, 2011). In a large, American sample, Spradley and Jantz (2011) found males and females to be significantly different in many postcranial measurements. For example, the average diameter of the head of the humerus for females is 42.2 mm. For males, this number is 48.4 mm. Similarly, the maximum diameter of the femoral head averages 41.9 mm among females and 48.2 mm among males. **Figure 6.19** shows a completed data collection form for the purposes of sex estimation from a set of skeletal remains. The data in this form have been collected according to the protocols described in this chapter and also includes some metric assessments. Use the information in the form to answer the following questions.

Question 1: Which information included in this form can be considered as metadata (see chapter 2)?

Question 2: How many of Stevens' data types are represented by the data collected to estimate sex for this individual? Which types are represented?

Question 3: Based on the information in this form, what would your estimate of sex for this individual be?

Exercise 2

Materials Required: **Table 6.2, Figures 6.6,** and **6.7,** and a computer with internet access.

Directions: Let's practice scoring morphological traits of the pelvis for the purposes of sex estimation. On your computer, open your internet browser and go to www.morphosource.org. Each of the following questions will refer to a specific Media ID number that you will enter in the search bar at the top of this page. This will take you to a page where you can view a 3D structured light scan of a human os coxa. To orient each scan so that you can see the traits of interest, it may be helpful to switch the web viewer into "orbit" mode (click on

Case #: 20-11135 **Pathologist:** Dr. Smith-Johnson
Observer: CCS **Date:** 09/19/2020

Pelvis	L	R	Skull	L	R
Ventral Arc	2	2	Nuchal Crest	3	3
Ischiopubic Ramus Ridge	2	2	Mastoid Process	3	3
Subpubic Contour	3	3	Supraorbital Margin	2	2
Greater Sciatic Notch	3	3	Glabella	3	3
Pubic Bone Shape	Sq	Sq	Mental Eminence	3	3

Femoral Head Diameter: 44.0 mm
Humeral Head Diameter: 43.2 mm

Notes:

The remains are entirely skeletonized with no remaining adherent soft tissue and no smell of decomposition. The surface of the bones has been bleached as a result of sun exposure and is starting to exfoliate.

Figure 6.19. Sample data sheet with recorded data on estimated skeletal sex.

the "Tools" menu to the left of the view screen, then click on the button that says "Rotate"; it should switch to "Orbit"). Likewise, it may be helpful to view the scan in full-screen mode (click the open square on the lower right). Holding the left mouse button and moving the mouse will rotate the scan, holding the right mouse button and moving the mouse will shift the scan, and the mouse wheel will let you zoom in or out.

Question 1: Enter "000361332" in the search bar and then click on the thumbnail of the os coxa on the results page. Using the web viewer, observe this os coxa and score the following traits using **Table 6.2** and **Figures 6.6** and **6.7** as references:

Ventral Arc (1–5)
Medial Aspect of the Ischiopubic Ramus (1–5)
Subpubic Contour (1–5)
Greater Sciatic Notch (1–5)
Pubic Bone Shape (more square or more of a vertical rectangle)
Based on the scores you have recorded, what sex would you estimate this individual to be?

Question 2: Enter "000360971" in the search bar and then click on the thumbnail of the os coxa on the results page. Using the web viewer, observe this os coxa and score the following traits using **Table 6.2** and **Figures 6.6** and **6.7** as references:

Ventral Arc (1–5)
Medial Aspect of the Ischiopubic Ramus (1–5)
Subpubic Contour (1–5)
Greater Sciatic Notch (1–5)
Pubic Bone Shape (more square or more of a vertical rectangle)
Based on the scores you have recorded, what sex would you estimate this individual to be?

Question 3: Enter "000361019" in the search bar and then click on the thumbnail of the os coxa on the results page. Using the web viewer, observe this os coxa and score the following traits using **Table 6.2** and **Figures 6.6** and **6.7** as references:

Ventral Arc (1–5)
Medial Aspect of the Ischiopubic Ramus (1–5)
Subpubic Contour (1–5)
Greater Sciatic Notch (1–5)
Pubic Bone Shape (more square or more of a vertical rectangle)
Based on the scores you have recorded, what sex would you estimate this individual to be?

Question 4: One of the primary advantages of the system presented by Klales and colleagues (2012) is the inclusion of a logistic regression equation for the estimation of sex and the resulting ability to generate a probability that your estimation of sex is correct. Since the latter requires slightly more advanced mathematics, we'll focus on the logistic regression equation since it is more straightforward to apply. The equation presented by Klales and colleagues (2012:111) is as follows:

2.726(VA) + 1.214(MA) + 1.073(SPC)–16.312

In this equation, VA stands for the score assigned for the Ventral Arc, MA stands for the score assigned to the Medial Aspect of the ischiopubic ramus, and SPC stands for the score assigned for Sub-Pubic Contour. If the number resulting from this equation is less than zero, then the individual is classified as a female; if the result is greater than zero, then the individual is classified as a male. Similarly, the farther from 0 the score is (i.e., the more negative or the more positive), the more reliable the estimate. As an example, if an individual scored a 5 for the ventral arc, a 4 for the medial aspect of the ischiopubic ramus, and a 5 for the subpubic contour, then their resulting score would be

2.726*5 + 1.214*4 + 1.073*5–16.312 =

13.63 + 4.856 + 5.365–16.312 = 7.539

This number is well above 0, indicating that this individual is most likely a male.

Use the equation of Klales and colleagues (2012) shown above to estimate the sex of the three individuals whose traits you scored in Questions 1–3.

Exercise 3

Materials Required: **Figures 6.11–6.15,** and a computer with an internet connection.

Directions: Let's practice estimating sex using morphological traits of the skull. Each of the following questions will provide you with a link to a 3D model of a skull or cranium on Sketchfab. The links provided are long, but providing the link should ensure that you are working with the correct model. As with the viewer in MorphoSource, holding the left mouse button and moving the mouse will rotate the model, holding the right mouse button and moving the mouse will shift the model's position in the view space, and the mouse wheel can be used to zoom in or out. If a given model has annotations, you can turn them off by right clicking on the annotations bar (bottom center of the view screen) and clicking "Hide annotations." Doing this may make it easier to see the morphological traits of interest in these exercises. Use the specified model to answer the following questions.

Question 1: Observe the 3D model of a cranium that can be found at the following link: https://sketchfab.com/3d-models/skull-with-annotations-2793801cdc724347ab741dc9b-65ba2f8.

Using the scoring scales in **Figures 6.11–6.15,** score all of the following traits that can be observed on this model:

Nuchal Crest (1–5)
Mastoid Process (1–5)
Supraorbital Margin (1–5)
Glabella (1–5)
Mental Eminence (1–5)
Based on the scores you have recorded, what sex would you estimate this individual to be?

Question 2: Observe the 3D model of a cranium that can be found at the following link: https://sketchfab.com/3d-models/craneo-5519188ba2614f3abfed873f0ca793de.

Using the scoring scales in **Figures 6.11–6.15,** score of the following traits that can be observed on this model:

Nuchal Crest (1–5)
Mastoid Process (1–5)
Supraorbital Margin (1–5)
Glabella (1–5)
Mental Eminence (1–5)
Based on the scores you have recorded, what sex would you estimate this individual to be?

Question 3: Observe the 3D model of a cranium that can be found at the following link: https://sketchfab.com/3d-models/skull-bzn-2-d1b2649903e94266a91d7cef529799ec.

Using the scoring scales in **Figures 6.11–6.15,** scores of the following traits that can be observed on this model:

Nuchal Crest (1–5)
Mastoid Process (1–5)
Supraorbital Margin (1–5)
Glabella (1–5)
Mental Eminence (1–5)
Based on the scores you have recorded, what sex would you estimate this individual to be?

Question 4: Walker (2008) presents a series of logistic regression equations for use with different populations and different combinations of cranial traits. Two of these equations (based on a modern American population) are as follows (Walker 2008:47):

$Y = -1.375(\text{glabella}) - 1.185(\text{mastoid}) - 1.151(\text{mental}) + 9.128$

$Y = -1.568(\text{glabella}) - 1.459(\text{mastoid}) + 7.434$

Note that these two equations use different combinations of traits. The value of Y that results from substituting trait scores into these equations can be used to estimate the sex of the individual. Y-values less than zero are more likely to represent male individuals and Y-values greater than zero are more likely to represent female individuals. The further from 0 that a Y-value is, the more reliable the sex estimate. As an example, if an unknown decedent has a score of 3 for glabella and 3 for the mastoid process, then, using the second equation, the Y-value is:

$Y = -1.568*3 - 1.459*3 + 7.434 = -4.704 - 4.377 + 7.434 = -1.647$

This number is below 0 and this individual would therefore be classified as a male.

Use the two equations from Walker (2008) shown above to estimate the sex of the individuals whose traits you scored in Questions 1 through 3.

References

Holcomb SMC, Konigsberg LW. 1995. Statistical study of sexual dimorphism in the human fetal sciatic notch. *American Journal of Physical Anthropology* 97: 113–125.

Klales AR, Ousley SD, Vollner JM. 2012. A revised method of sexing the human innominate using Phenice's nonmetric traits and statistical methods. *American Journal of Physical Anthropology* 149: 104–114.

Loth SR, Henneberg M. 2001. Sexually dimorphic mandibular morphology in the first few years of life. *American Journal of Physical Anthropology* 115: 179–186.

Mittler DM, Sheridan SG. 1992. Sex determination in subadults using auricular surface morphology: a forensic science perspective. *Journal of Forensic Sciences* 37: 1068–1075.

Phenice T. 1969. A newly developed visual method of sexing in the os pubis. *American Journal of Physical Anthropology* 30: 297–301.

Rogers T, Saunders S. 1994. Accuracy of sex determination using morphological traits of the human pelvis. *Journal of Forensic Sciences* 39: 1047–1056.

Schutkowski H. 1993. Sex determination of infant and juvenile skeletons I: morphognostic features. *American Journal of Physical Anthropology* 90: 199–205.

Sobotta J. 1909. *Atlas and Text-Book of Human Anatomy.* Philadelphia, PA: W.B. Saunders Co.

Spradley MK, Jantz RL. 2011. Sex estimation in forensic anthropology: skull versus postcranial elements. *Journal of Forensic Sciences* 56: 289–296.

Stull KE, Cirillo LE, Cole SJ, Hulse CN. 2020. Subadult sex estimation and Kidstats. In: Klales A (Ed.), *Sex Estimation of the Human Skeleton. History, Methods, and Emerging Techniques.* London: Academic Press. p. 219–242.

Vlak D, Roksandic M, Schillaci MA. 2008. Greater sciatic notch as a sex indicator in juveniles. *American Journal of Physical Anthropology* 137: 309–315.

Walker PL. 2008. Sexing skulls using discriminant function analysis of visually assessed traits. *American Journal of Physical Anthropology* 136: 39–50.

Weaver DS. 1980. Sex differences in the ilia of a known sex and age sample of fetal and infant skeletons. *American Journal of Physical Anthropology* 52: 191–195.

7

Estimating Age of Subadults

Learning Goals

By the end of this chapter, the student will be able to:

- Describe tooth anatomy, development, and eruption sequence as relevant to forensic anthropology.
- Apply the methods of age estimation using subadult age estimation criteria.
- Interpret charts of standard measurements and growth curves.
- Describe the concept of seriation, or ordering of observations.

Introduction

Age-at-death is the second component of the biological profile of interest to forensic anthropologists. Estimating age-at-death is more difficult than sex estimation because the latter is primarily concerned with identifying decedents as belonging to one of two classes: male or female (with rare exceptions). Age-at-death, however, encompasses much more uncertainty because the variable being estimated is continuous, not dichotomous, and is subject to infinite amounts of potential precision. By this we mean that age-at-death could be stated in years, months, weeks, minutes, and even seconds. However, the forensic anthropologist is rarely interested in such a degree of specificity, and, in practice, age estimates are broader. As a continuous variable, age estimates should always have an error estimate associated with them and should *not* be stated as a single point estimate. For example, saying someone was 2 years old at the time of death is overly precise. Rather, it is more correct to say that they were 2 years old ± 2 months (or 22–26 months).

Although age estimates help narrow the pool of decedents in a missing persons case, there are other reasons why age-at-death is important. It may be critical to know whether someone was legally an adult for the purposes of prosecution. This issue has become controversial with respect to the human rights crisis at the U.S.-Mexico border, where determining an immigrant's legal status as an adult has a significant effect on that person's legal rights and immigration status (Anderson, 2008; Lewis and Senn, 2010). Knowing a decedent's age-at-death can also inform sex estimates. Walker (1995) showed that the skulls of females can develop more male characteristics as they age because the robusticity of the skull increases with age. Finally, estimates of stature (chapter 9) are also age dependent, initially because of growth processes prior to and during adolescence, but also because we lose stature as we age (Galloway, 1988). Therefore, knowing someone's age can help increase the accuracy of stature estimates. It should be apparent now that estimating the biological profile is complicated by the interdependence of the variables of interest. Aging processes are sex-specific yet knowing one's age also produces better sex estimates in some cases.

In non-anthropological settings we tend to think of age in a singular way, i.e., how old you are. But this is only one conceptualization of age. Anthropologists recognize, for example, **social age**, which is the way society defines expected behaviors based loosely on how old you are in years. In some cultures, children that have reached puberty may be expected to work outside the home, something we generally delay in the U.S. Social age is not relevant to forensic anthropologists except to the extent that social age and legal age are intertwined in our society. This could affect one's legal status in a criminal proceeding, such as "being tried as an adult." Forensic anthropologists are more interested in knowing the **chronological age** of a decedent, which is what would be reported on official documents (e.g., a birth certificate) and in missing persons reports. Chronological age refers to how old one is in years since birth, that is, how long one has been alive. This is the age one typically thinks of when asked how old they are. However, we cannot estimate chronological age from the skeleton. Rather, what we estimate is **biological age**. Biological age (or physiological age) is how old your body appears to be. This is related to chronological age, but biological age encompasses a wider variety of factors. For example, one's biological age is affected by one's rate of growth and development (largely determined by genes) as well as by degenerative changes in your body affected by diet, activity and exercise patterns, overall health, and other lifestyle choices. Smoking, sun exposure, drug use, and poor diet can all make someone "look old for their age." This is what biological age refers to.

Skeletal and dental traits that are useful age estimators generate biological age estimates that are highly correlated with chronological age, produce observable changes in the skeleton that are unidirectional (i.e., do not reverse appearance or heal), and reflect biological processes that are experienced by most people regardless of sex or where in the world they live (Christensen et al., 2014). However, few (if any) age estimators satisfy these criteria entirely. Females develop earlier than males, so it is useful to know the sex of a decedent prior to estimating age-at-death, which is not possible for subadult (juvenile) decedents (see chapter 6). In addition, lifestyle has such a significant impact on biological age that age estimation is population-specific to some degree. In general, age estimates for subadult decedents are more accurate and precise than age estimates for adults. Furthermore, the older one gets the less precise the age estimates will be, and as one enters the sixth decade of life, age-at-death estimates are often stated in decades rather than specific years. However, it is better to be accurate than precise when estimating age-at-death. That is, the estimate produced should include the *true* age of the individual even if the predicted range is overly broad.

The methods used to estimate age depend on the relative age of the individual with a coarse division based on whether one is an adult or a subadult (juvenile).

Age-at-death estimation in subadults is based on **developmental traits**, that is, the methods use the growth process to estimate age, which is easy to measure in living individuals. These methods include dental formation, dental eruption, epiphyseal appearance, epiphyseal closure, and long bone length. **Degenerative traits** are used to estimate the age of adults. These include changes in joint surfaces in the pelvis and to a lesser extent in the ribs and cranium, the subject of chapter 8. Developmental traits are generally seen as determined by genetic factors and are therefore more predictable and regular than biological changes associated with degenerative traits.

The basic division of developmental traits into dental methods and osteological methods is also relevant. Dental methods are more accurate than osteological methods of subadult age estimation. This is because tooth development is under even stronger genetic control than osteological development—the latter is more affected by poor diet and health during childhood and adolescence. And because teeth begin to form in utero, dental methods for age estimation can be used very early in fetal development and all throughout childhood and adolescence into a person's early 20s when third molar development finishes. This is another example of the difference between social age and biological age. We consider someone an adult when they reach the chronological age of 18. However, biologically an individual is not an adult until all teeth have erupted into occlusion, all root apices have closed (see below), and all epiphyses have fused, which can occur well into a person's mid-20s. Once these growth processes have completed, however, estimation of age must rely on degenerative changes as discussed in chapter 8.

Dental Methods of Subadult Age Estimation

Basic Dental Anatomy

Dentition is among the most useful of biological indicators about an individual's life. Teeth are the hardest substance in the human body and preserve well in many different environments. In addition, teeth provide an indelible record of one's life because they do not remodel after they have formed (unlike your skeleton, which replaces itself entirely—more or less—every 10 years). Even the day of your birth is recorded in your teeth in the form of a microscopic stress line called the neonatal line. Teeth are also sexually dimorphic, making them useful for sex estimation (chapter 6) and, for our purposes in chapter 7, providing information on growth and development that can be used to estimate age-at-death in young individuals. An understanding of the anatomy of human teeth is key to using these methods properly.

A schematic of a human molar is presented in **Figure 7.1** with key aspects of anatomy indicated. The crown of the tooth is that part visible in the mouth of living individuals. The root anchors the tooth into the alveolar bone and the blood supply to the tooth is located within the alveolar bone. This blood supply provides nourishment to the cells within the tooth. The outer, white part of the tooth is enamel, which does not maintain active, living cells after the tooth has formed. Enamel cannot heal itself, which is why cavities must be filled

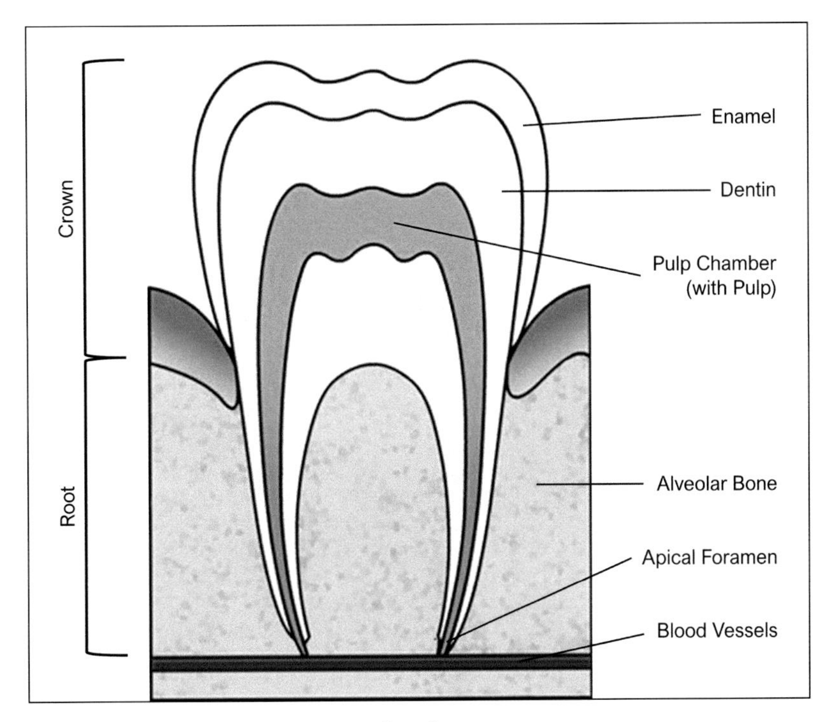

Figure 7.1. Schematic representation of tooth anatomy.

by a dentist. But enamel also cannot sense pain (because there are no living cells within the enamel matrix). Instead, when you have a cavity it is first detected in the dentin, which does contain living cells (dentinoblasts) that can sense and respond to pain and thus repair itself to a certain degree. Dentin is yellowish-brown in color and is visible once teeth begin to wear. The pulp chamber is where the blood and nerve supply of the tooth—the dental pulp—is maintained. These nerve and blood vessels connect to the body through the apical foramen, a small hole in the very end of the root. All teeth in the human mouth, regardless of shape, have this same basic anatomy.

Humans, as mammals, are *heterodonts*. This means they have different types of teeth in their mouths, each with different functions. Heterodonts have four different tooth types, called tooth classes: incisors, canines, premolars (your dentist incorrectly calls them bicuspids because they can have more than two cusps), and molars. **Figure 7.2** (*left*) shows the lower jaw of a human with the tooth types labeled. Each quadrant of your mouth has two incisors, one canine, two premolars, and three molars. Identifying the different classes of teeth is crucial for applying age estimation methods.

Incisors are the mesial-most teeth in the dental arcade with two per quadrant or eight total. They are spatula-shaped with a linear biting edge, with maxillary incisors being larger than mandibular incisors. Incisors only have one root.

Canines are located directly distal to the incisors. They are conical, single-cusped, pointed teeth that also only have one root. There is only one canine per quadrant, or four total. Maxillary canines are larger than mandibular canines.

Premolars are distal to the canines. Premolars are generally round or oval in outline and usually have two cusps in the maxilla but can have between one and four cusps in the mandible. Premolars usually have one root but can also have two roots. There are two premolars

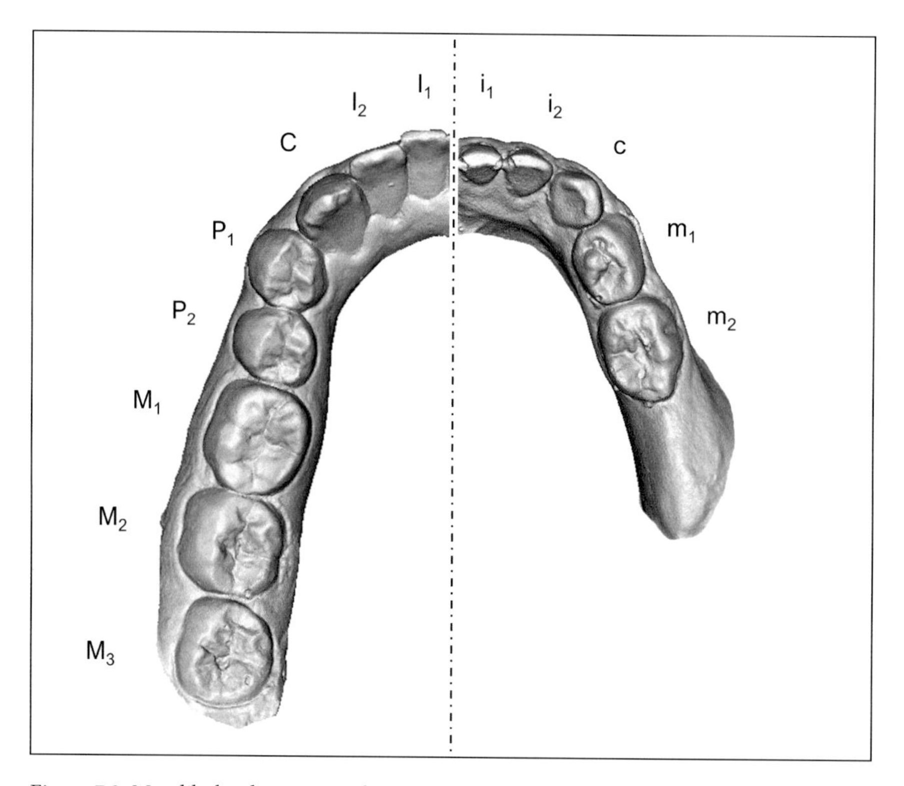

Figure 7.2. Mandibular dentition with permanent teeth on the left and deciduous teeth on the right. I indicates incisors; C indicates canines; P indicates premolars; M indicates molars. Capital letters indicate permanent teeth and lowercase indicates deciduous, or "baby" teeth.

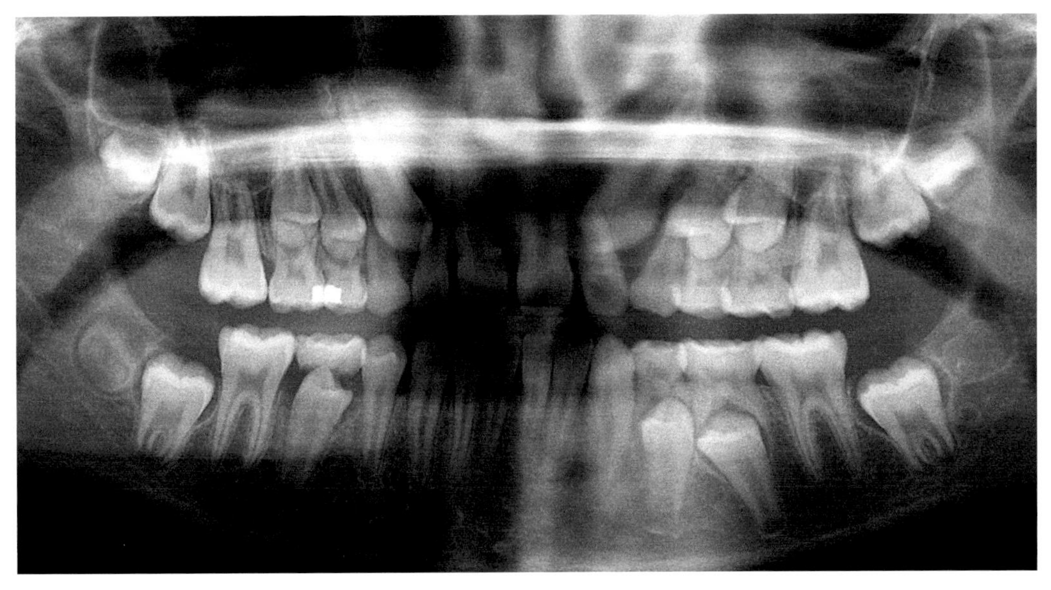

Figure 7.3. Panoramic radiograph showing an individual with mixed dentition. Note the developing permanent dentition in their crypts.

per quadrant for eight total in the dental arcade. Maxillary premolars have two nearly equally sized, large cusps. Mandibular premolars have one cusp that is always much larger than the other cusps and there could be three or four total cusps. Some mandibular premolars could be mistaken for canines.

Molars are distal to the premolars and generally square or rectangular in outline with between 3 and 5 cusps and between 2 and 3 roots. There are 3 molars per quadrant or 12 total in the dental arcade, although the third molars are often missing or have been extracted due to being impacted. Maxillary molars typically have 4 cusps and are rhomboid in shape. Mandibular molars have 4–5 cusps and are rectangular in shape.

In addition to being heterodont, humans are *diphyodont*—meaning we have two sets of teeth that erupt sequentially. The first set is called the *deciduous* (or baby, or milk) dentition. The second is called the *permanent* (or adult) dentition. Unlike the adult dentition, there are no deciduous premolars, so a complete set of deciduous teeth has *2 incisors, 1 canine, and 2 molars* per quadrant, or 20 total teeth (**Figure 7.2,** *right*). This means that when a child is growing, they have a *mixed dentition,* with many deciduous teeth still in place and many adult teeth developing in the crypts hidden within the bone or starting to replace their deciduous precursors (**Figure 7.3**).

Differentiating deciduous and permanent teeth is relatively straightforward. Deciduous teeth are smaller than their permanent replacements, have thinner, more translucent enamel, and shorter roots in comparison to permanent teeth. Specific conventions are used to denote each tooth type: incisor = I, canine = C, premolar = P, and molar = M. For deciduous teeth the abbreviation is always lower case (i, c, m). The specific position of the tooth is indicated by a number, which is superscripted for a maxillary tooth and subscripted for a mandibular tooth. This system allows us to identify the specific tooth in the mouth. For example, I^1 is a permanent maxillary first incisor, i^1 is a deciduous maxillary first incisor, M_1 is a permanent mandibular first molar, etc.

Dental Growth and Development

During growth, teeth develop in the crypts within the body of the mandible and maxilla and eventually erupt into the oral cavity where they function in food breakdown. The process of dental development is well studied, with standards based on longitudinal and cross-sectional studies of growth and development in children. Longitudinal studies are those that record information from the *same* children repeatedly as they age, which allows researchers to reconstruct the average age when specific developmental milestones are met. Cross-sectional studies combine data from multiple individuals observed only one time, each of which is observed at a different age but then grouped into samples based on these ages. Both types of research have contributed to our knowledge of dental development, and each has its benefits (see Smith, 1991; Stull et al., 2014).

The concept of dental growth and development describes two processes. **Dental formation** refers to the growth of individual teeth. Teeth begin forming at the occlusal surface, or the biting or chewing surface of the tooth. The crown forms first, then the root as new material is added to the growing tooth in a crown-to-root direction. **Dental eruption** refers to the emergence of the teeth through the gum tissue into the oral cavity where they enter functional occlusion (i.e., used for biting or chewing). There is a specific pattern in which this happens, allowing one to estimate age-at-death based on which teeth are present in the mouth at the time of death, which deciduous teeth have been lost and replaced, and which have yet to emerge into the oral cavity. Dental formation is considered more accurate than dental eruption for subadult age estimation (Ubelaker, 1987; WEA, 1980).

Dental Formation

The power of the dental formation method for forensic analysis is that a single tooth can provide accurate information on age-at-death. The method is particularly useful for younger children, less than 12 years of age, when many teeth are still developing or actively resorbing and preparing to be replaced. The application of the method is relatively straightforward.

First, identify the teeth available. For teeth still in the socket, this is fairly obvious—however, one may have to use X-rays to see the development stage of the roots. For loose teeth, the forensic anthropologist must be certain they have identified the specific tooth correctly, including whether it is a deciduous or permanent tooth, the tooth type (incisor, canine, premolar, molar), the tooth position (1st or 2nd or 3rd), and the arcade (maxilla or mandible).

Second, for each tooth, score the degree of dental development using stage-based standards, as shown in **Figure 7.4** for single rooted teeth (incisors, canines, premolars) and **Figure 7.5** for molars (see AlQahtani et al., 2010; Moorrees et al., 1963a, b). In addition, the process of tooth replacement in the diphyodont dentition requires that the deciduous teeth fall out. Once the tooth has completely formed, it will be used in occlusion and eventually the roots will begin to resorb as the body prepares to replace each tooth with its permanent successor. The process of tooth formation is reversed, and the roots will begin to decalcify beginning at the root apex and proceeding toward the crown. A small remnant of root will remain as the tooth is shed. The resorption process is also described with respect to stages, as shown in **Figure 7.6**. A possible point of confusion is whether a root is actively forming or resorbing; however, other data on age such as overall body size and whether the teeth are showing wear are helpful in making this determination.

Third, now that the teeth have been identified and their phase of development or resorption scored, these observations are compared to tables and charts summarizing the ages at which specific milestones are met. AlQahtani et al. (2010) provide a series of tables summarizing the minimum, median, and maximum stages of tooth development by age. When

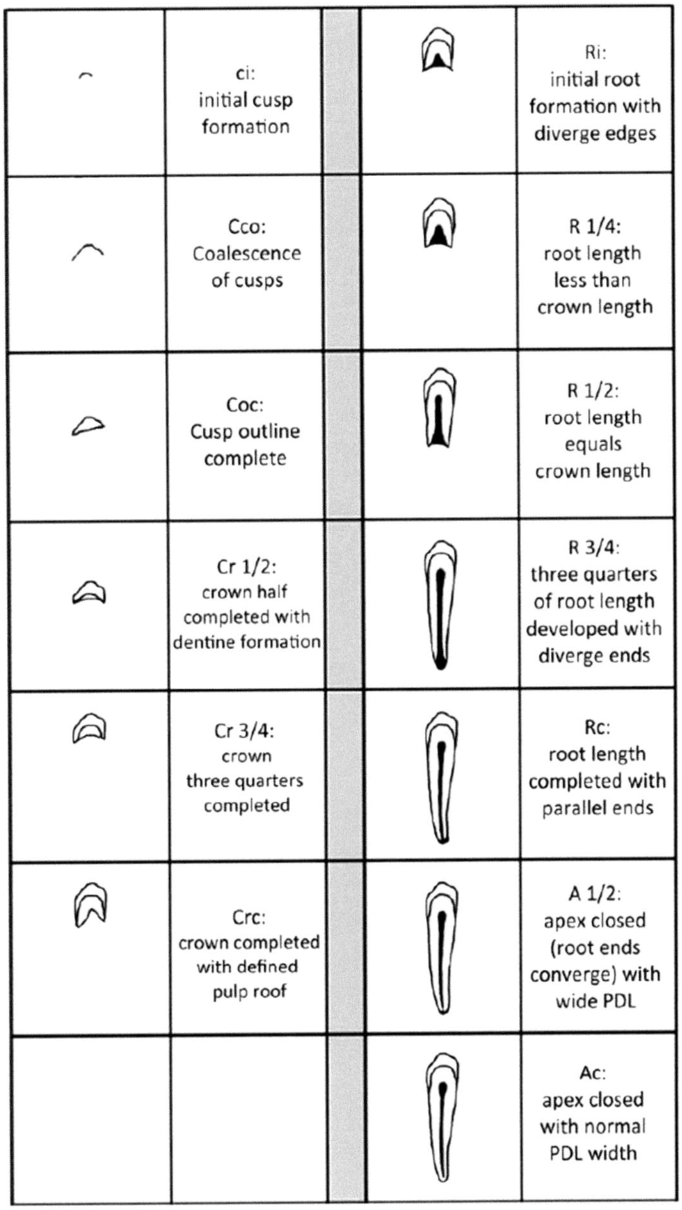

ci: initial cusp formation	Ri: initial root formation with diverge edges
Cco: Coalescence of cusps	R 1/4: root length less than crown length
Coc: Cusp outline complete	R 1/2: root length equals crown length
Cr 1/2: crown half completed with dentine formation	R 3/4: three quarters of root length developed with diverge ends
Cr 3/4: crown three quarters completed	Rc: root length completed with parallel ends
Crc: crown completed with defined pulp roof	A 1/2: apex closed (root ends converge) with wide PDL
	Ac: apex closed with normal PDL width

Figure 7.4. Scoring stages of dental development for single rooted teeth.

accessing these tables, it is important to consider the tooth formation stage abbreviations (**Figure 7.4** and **Figure 7.5**) as well as the tooth resorption abbreviations (**Figure 7.6**). These data have been excerpted in **Table 7.1** for two teeth—the maxillary first deciduous incisor and the permanent incisor that replaces it.

The left three columns give data for the deciduous incisor. Read through this description while considering **Table 7.1** to understand how to use these data to estimate an age-at-death for an isolated tooth. The first deciduous incisor begins forming (C^{oc}) prior to 30 weeks in utero, the crown will be half complete ($Cr^{1/2}$) between 30 and 38 weeks in utero, three-quarters complete ($Cr^{3/4}$) between 30 weeks in utero and 1.5 months after birth, and the crown will be completely formed (Cr^{c}) between birth and 4.5 months. The root initiates formation (R^{i}) between the age of 1.5 and 4.5 months and is complete (R^{c}) around 2.5 years. Between 2.5 and 6.5 years the tooth is used for chewing. Between 5.5 and 7.5 years the root begins to resorb (Res ¼ to ¾) and is lost at the age of 6.5–7.5 years. Therefore, if you found an isolated deciduous first incisor at the stage of (R^{i}) the median age would be 4.5 months with a range of 1.5 to 10.5 months. Its permanent replacement begins forming (C^{i}) at 4.5 months, with a

Crown stages	Root stages
Ci: initial cusp formation	
Cco: Coalescence of cusps	R 1/4: root length less than crown length with visible bifurcation area
Coc: Cusp outline complete	R 1/2: root length equals crown length
Cr 1/2: crown half completed with dentine formation	R 3/4: three quarters of root length developed with diverge ends
Cr 3/4: crown three quarters completed	Rc: root length completed with parallel ends
Crc: crown completed with defined pulp roof	A 1/2: apex closed (root ends converge) with wide PDL
Ri: initial root formation with diverge edges	Ac: apex closed with normal PDL width

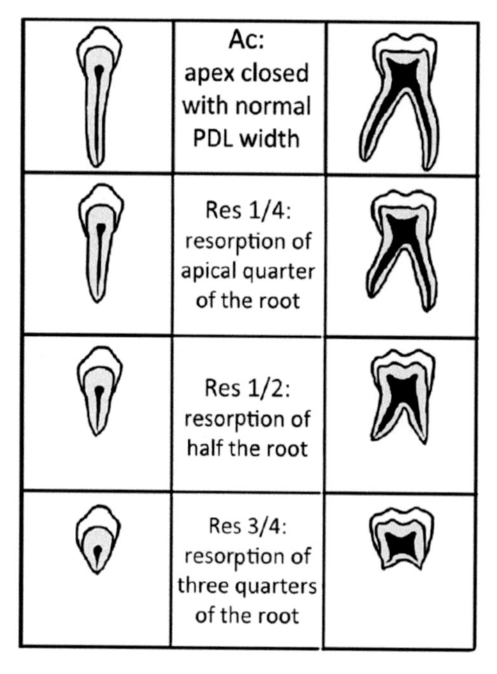

Above: Figure 7.5. Scoring stages of dental development for molars.

Left: Figure 7.6. Scoring stages of root resorption.

Table 7.1. Dental formation data for the maxillary deciduous and permanent first incisor, after AlQahtani et al. (2010)

	Tooth Formation Stage					
	i1			I1		
Age	Min	Med	Max	Min	Med	Max
30 weeks in utero	C^{oc}	**$Cr^{1/2}$**	$Cr^{3/4}$	-	-	-
34 weeks in utero	$Cr^{1/2}$	**$Cr^{3/4}$**	$Cr^{3/4}$	-	-	-
38 weeks in utero	$Cr^{1/2}$	**$Cr^{3/4}$**	Cr^{c}	-	-	-
Birth	$Cr^{3/4}$	**$Cr^{3/4}$**	Cr^{c}	-	-	-
1.5 months	$Cr^{3/4}$	**Crc**	R^{i}	-	-	-
4.5 months	Cr^{c}	**Ri**	$R^{1/4}$	C^{i}	**Ci**	C^{oc}
7.5 months	R^{i}	**$R^{1/4}$**	$R^{1/2}$	C^{oc}	**Coc**	$Cr^{1/2}$
10.5 months	R^{i}	**$R^{1/2}$**	$R^{1/2}$	C^{oc}	**$Cr^{1/2}$**	$Cr^{1/2}$
1.5 years	$R^{1/4}$	**$R^{3/4}$**	$R^{3/4}$	$Cr^{1/2}$	**$Cr^{1/2}$**	$Cr^{1/2}$
2.5 years	R^{c}	**Ac**	A^{c}	$Cr^{1/2}$	**$Cr^{3/4}$**	$Cr^{3/4}$
3.5 years	A^{c}	**Ac**	A^{c}	$Cr^{3/4}$	**$Cr^{3/4}$**	R^{i}
4.5 years	A^{c}	**Ac**	A^{c}	$Cr^{3/4}$	**Crc**	R^{i}
5.5 years	A^{c}	**Ac**	$Res^{1/4}$	Cr^{c}	**Ri**	$R^{1/4}$
6.5 years	A^{c}	**$Res^{3/4}$**	-	Cr^{c}	**$R^{1/4}$**	$R^{3/4}$
7.5 years	$Res^{3/4}$	-	-	$R^{1/4}$	**$R^{3/4}$**	R^{c}
8.5 years	-	-	-	$R^{1/2}$	**Rc**	$A^{1/2}$
9.5 years	-	-	-	$R^{3/4}$	**Rc**	$A^{1/2}$
10.5 years	-	-	-	R^{c}	**$A^{1/2}$**	A^{c}
11.5 years	-	-	-	R^{c}	**Ac**	A^{c}
12.5 years	-	-	-	R^{c}	**Ac**	A^{c}
13.5 years	-	-	-	A^{c}	**Ac**	A^{c}

complete crown (Cr^c) formed between 4.5 and 6.5 years, complete roots (R^c) between 7.5 and 10.5 years, and a completely formed tooth with closed roots (A^c) as early as 10.5 years. The bold columns in **Table 7.1** indicate the median ages at which a specific stage of development is attained.

Dental Eruption

After the tooth crown forms it will erupt from its crypt into occlusion, a process called dental eruption. Dental eruption has specific, well-known milestones: the upper incisors are replaced at around the age of 6 or 7, the first adult molar erupts around the age of 6, the second adult molar around the age of 12, and the third adult molar around the age of 18. However, these are general "rules of thumb" and more detailed patterns have been established based on studies of living individuals (AlQahtani et al., 2010; Schour and Massler, 1941a, b; Ubelaker, 1978).

When a tooth crown crosses the gum line, it has *erupted.* However, in dry bone specimens there is no gum tissue to observe so the focus is on *alveolar eruption,* when the tooth initially breaks through the alveolar bone and is visible within the crypt (**Figure 7.7**). When the tooth is fully erupted, such that the crown is level with other teeth in the mouth, then the tooth has reached *functional occlusion* and is said to be fully erupted. Small amounts of wear may be visible, evidenced by polished facets of enamel or in more severe cases exposed brown patches of dentine (see **Figure 7.7**), indicating the teeth are being used to chew. Tables of

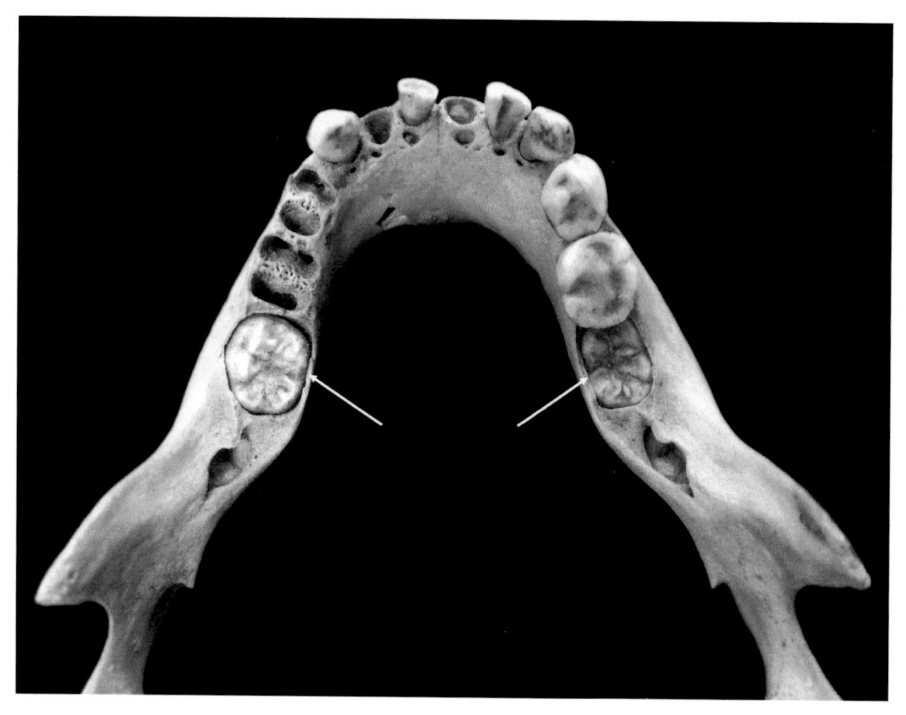

Figure 7.7. Mandible with arrows indicating permanent first molars at the stage of alveolar emergence but not fully erupted. Note the exposed dentine patches on the deciduous molars on the right side.

median ages at which the deciduous and permanent dentition reaches these stages are provided in **Table 7.2.**

A more detailed graphic can be seen in **Figure 7.8,** which is drawn from the London Atlas of Human Tooth Development and Eruption (AlQahtani et al., 2010). The London Atlas was developed using a comparatively large and modern sample and provides more continuous coverage for individuals older than 12 years of age than older systems. The single image contains information on both dental formation and dental eruption. In interpreting this figure,

Table 7.2. Dental eruption data for the deciduous and permanent dentition, sexes combined, after AlQahtani et al. (2010)

Tooth	Alveolar Eruption	Full Eruption	Tooth	Alveolar Eruption	Full Eruption
	DECIDUOUS MAXILLARY			DECIDUOUS MANDIBULAR	
i^1	4.5 mos	10.5 mos	i_1	4.5 mos	10.5 mos
i^2	7.5 mos	1.5 yrs	i_2	7.5 mos	1.5 yrs
c	10.5 mos	2.5 yrs	c	10.5 mos	2.5 yrs
m^1	10.5 mos	1.5 yrs	m_1	10.5 mos	1.5 yrs
m^2	1.5 yrs	2.5 yrs	m_2	1.5 yrs	2.5 yrs
	PERMANENT MAXILLARY (YEARS)			PERMANENT MANDIBULAR (YEARS)	
I^1	6.5	7.5	I_1	5.5	7.5
I^2	7.5	9.5	I_2	6.5	7.5
C	11.5	12.5	C	9.5	11.5
P^1	10.5	11.5	P_1	10.5	11.5
P^2	11.5	12.5	P_2	11.5	12.5
M^1	5.5	6.5	M_1	5.5	6.5
M^2	10.5	13.5	M_2	10.5	12.5
M^3	16.5	20.5	M_3	16.5	20.5

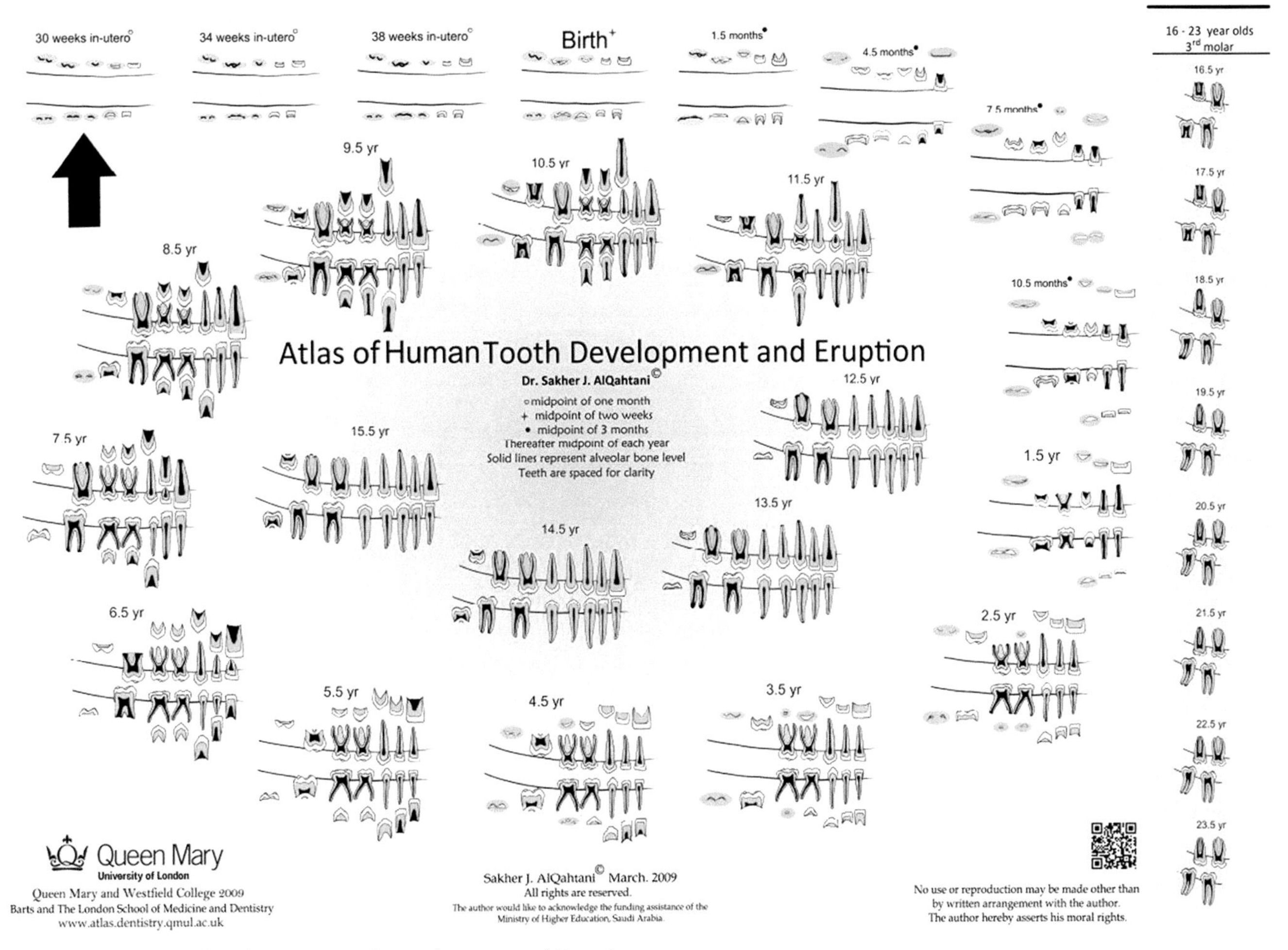

Figure 7.8. London Atlas of Human Tooth Development and Eruption.

note that deciduous teeth are white and permanent teeth are blue. Begin with the arrow in the top-left corner of the figure and proceed in a clockwise direction as age increases. Please note that this figure does not contain ranges, but rather point estimates for given degrees of dental eruption at a specific age. Adams and colleagues (2019) have criticized the London Atlas for producing age ranges that are too narrow to consistently capture the true age of a decedent—a good illustration of the tradeoff that exists between the accuracy of an age estimate and its precision. Despite this, the London Atlas appears to perform as well or better than earlier atlases (e.g., Schour and Massler, 1941a, b; Ubelaker, 1978) but, like them, tends to produce age estimates that are slightly younger than an individual's true age (Adams et al., 2019; AlQahtani et al., 2014). Due to variability in the development of the third molars, dental age estimation beyond the age of 19 is not recommended (AlQahtani et al., 2014).

LEARNING CHECK

Q1. Which of these abbreviations refers to a deciduous maxillary first molar?

A) M_1

B) m_1

C) M^1

D) m^1

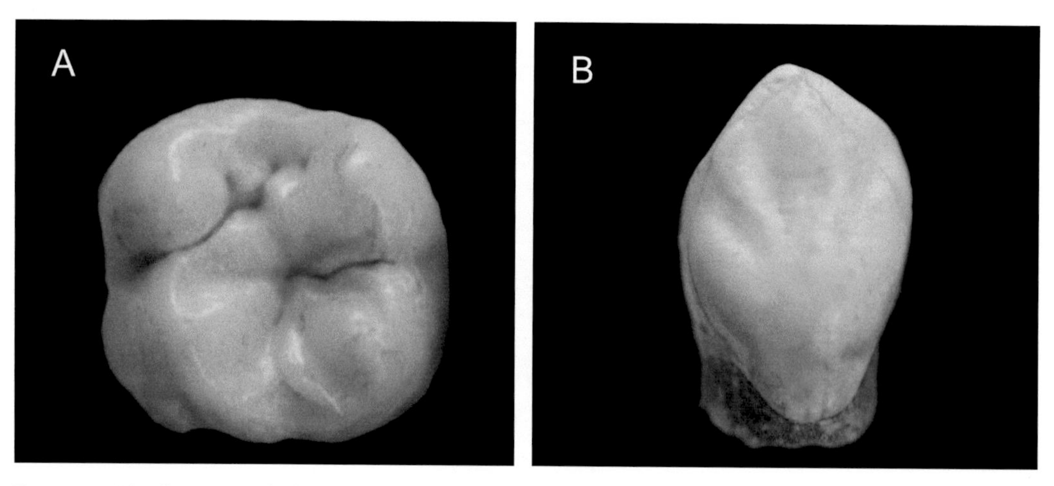

Figure 7.9. Two human teeth. (*A*) shown in occlusal view, (*B*) shown in lingual view.

Q2. What type of tooth is shown in **Figure 7.9A**?
A) Incisors
B) Canines
C) Premolars
D) Molars

Q3. What type of tooth is shown in **Figure 7.9B**?
A) Incisors
B) Canines
C) Premolars
D) Molars

Q4. Using **Table 7.1,** at what median degree of development would you expect the deciduous first maxillary incisor to be at the time of birth?
A) C^{oc}
B) R^{i}
C) A^{c}
D) $Cr^{3/4}$

Q5. Using **Table 7.1,** what median degree of development would you expect the permanent first maxillary incisor to be at the time of birth?
A) C^{oc}
B) unformed
C) A^{c}
D) $Cr^{3/4}$

Q6. Using **Table 7.1,** what is the absolute latest you would expect to find a deciduous first maxillary incisor still in a decedent's mouth?
A) 7.5 years
B) 8.5 years
C) 9.5 years
D) 10.5 years

Q7. Using **Table 7.2,** about how long does it take the permanent first molar to go from alveolar eruption to full eruption?
A) 1.5 years
B) 1 year
C) .5 years
D) 2 years

Q8. Using **Table 7.2,** a decedent with the following teeth fully erupted would be about how old: I^1, I^2, M^1?
A) 9.5 years
B) 6 years
C) 7.5 years
D) 10.5 years

Q9. Using **Figure 7.8,** at what age would you expect the second permanent molar to reach the stage of alveolar eruption?
A) 6.5–7.5 years
B) 12.5–14.5 years
C) 9.5–11.5 years
D) 15.5–16.5 years

Q10. Using **Figure 7.8,** at what age would you expect the first permanent tooth to reach the stage of alveolar eruption?
A) 1.5–2.5 years
B) 2.5–3.5 years
C) 6.5–7.5 years
D) 4.5–5.5 years

Skeletal Growth and Development

As discussed in chapter 3, many bones that appear as a single element in the adult skeleton form from multiple growth centers that fuse during adolescence. For example, the long bones develop from one primary growth center for the shaft (*diaphysis*) and one or more growth centers for each of the proximal and distal *epiphyses,* which form the joints (**Figure 7.10**). Some bones, such as the femur, have four distinct epiphyses, one for the proximal joint, one for the distal joint, and one each for the structures called the greater and lesser trochanters. As you develop, the diaphysis of the bone increases in length and circumference, and the epiphyses increase in size until the body directs the long bone to stop growing. At this point the epiphyses fuse to the diaphysis. Most other bones in the body also form from multiple growth centers that fuse together in adolescence, including the sacrum, scapula, vertebrae, and as discussed in chapter 6, the pelvis, which in addition to being formed by three primary growth centers also has multiple secondary epiphyses that fuse to the os coxa. The skull is also a highly composite anatomical structure that forms from many distinct growth centers that fuse as we age.

In forensic anthropology, there are four methods of age-at-death estimation based on osteological evidence. The first method uses *fusion of the growth centers of the skull.* Because human brain development occurs rapidly after birth, our skull bones are formed from distinct embryological growth centers that fuse early in life, making them very useful for age estimation for individuals under the age of 5. Scheuer and Black (2000) list many of these

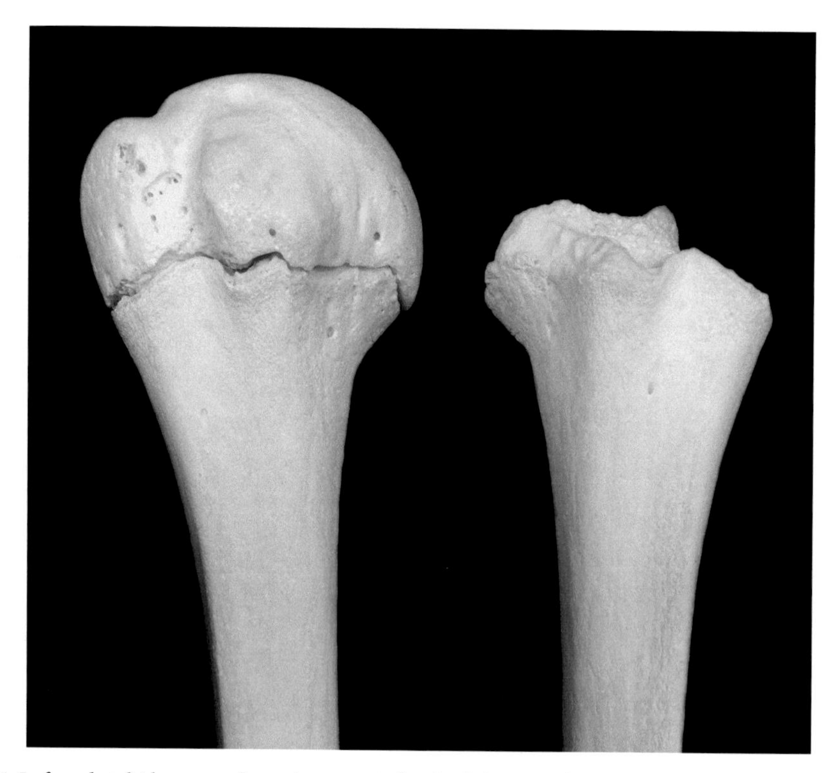

Figure 7.10. Left and right humerus from the same individual showing the diaphysis without the epiphysis (*right*) and with the proximal epiphysis in place (*left*). Note the epiphysis is not fused to the shaft as indicated by the clear line of separation between the diaphysis and epiphysis.

developmental milestones, the most relevant of which are the closure of the fontanelles (i.e., the soft spots; the anterior fontanelle is the most well known and fuses between the ages of 1–2 years), the fusion of the two halves of the frontal bone (2–4 years—see **Figure 7.11**) and the fusion of the two halves of the mandible (during the first year of life).

The second method uses the *appearance of the secondary growth centers,* which occur at known ages and are useful for age estimation for pre-adolescents (< 12 years). Scheuer and Black (2000) also provide a comprehensive review of this topic and can be consulted on a case-by-case basis. For the femur, the distal epiphysis appears between 36 and 40 weeks in utero, the femoral head epiphysis appears between .5 and 1 years of age, the greater trochanter between 2 and 5 years, and the lesser trochanter between 7 and 12 years (Scheuer and Black, 2000). Similar data can be found for all major elements of the skeleton in that volume. The issue with epiphyseal appearance is that the epiphyses are very small fragments of bone, and it is often difficult to identify small epiphyses that have not yet taken on the appearance of the fully formed joint. In addition, the absence of an epiphysis associated with a decedent's remains could reflect the individual's young age or it could reflect recovery bias because small epiphyses are easily missed during recovery and may not be preserved at all at crime scenes with poor preservation. Therefore, the forensic utility of this method is limited (Stewart, 1979).

The third method uses *long bone lengths* to estimate age-at-death. Long bones increase in length and diameter during development until epiphyseal union occurs. Measurements of the diaphysis length (usually not including the epiphyses) can be compared to tables of long bone lengths based on known-age individuals to estimate how old the individual was at the time of death. This method is useful for estimating the age of fetal remains (Fazekas and Kósa, 1978) all the way through late adolescence (Scheuer and Black, 2000). Stull et al. (2014) provide analysis options using certain breadths of the epiphyseal ends, though these newer

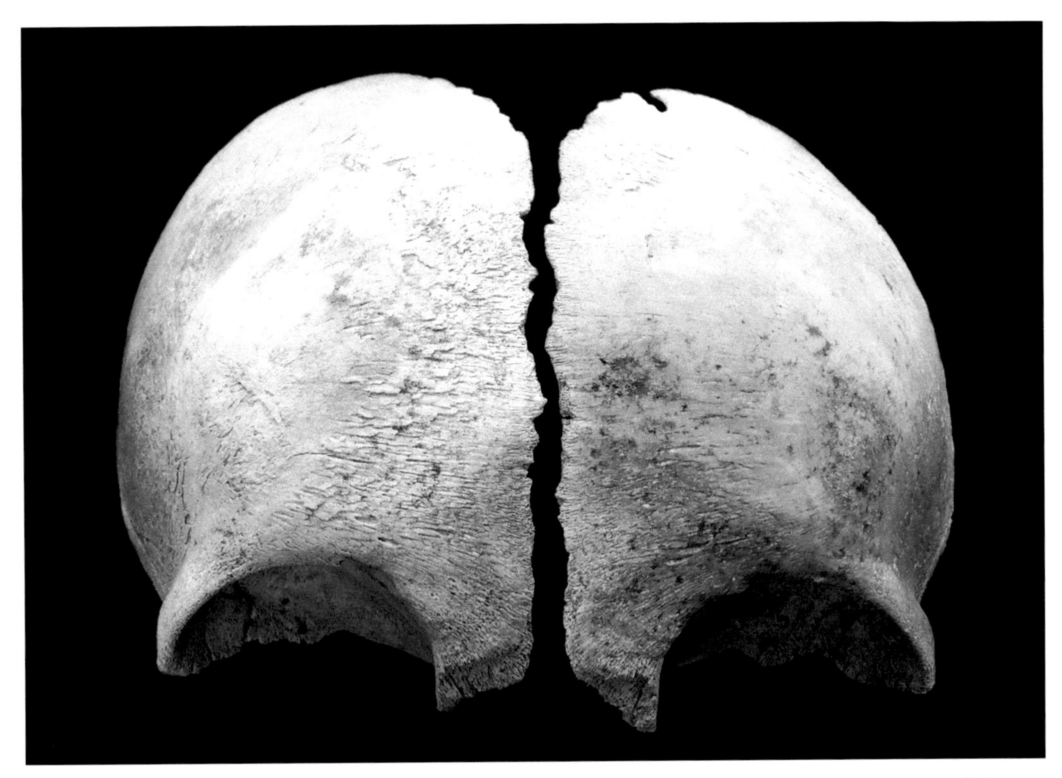

Figure 7.11. Infant frontal bone showing the two unfused halves. This midline suture can persist into adulthood where it is referred to as a metopic suture. It normally fuses between the ages of 2 and 4 years.

methods are still in development. Relevant comparative data compiled by Scheuer and Black (2000) are presented in **Table 7.3** for the humerus and femur. Application of the method is relatively straightforward. First, the immature long bone is identified and measured, being sure not to include the epiphyses. Second, the resulting measurement is compared against known-age individuals to estimate the age-at-death based on the measured bone length. This method is complicated by sex-specific growth patterns during adolescence and the difficulty of estimating sex for subadult remains. In addition, juvenile long bones are fragile and must be complete for the observation to be accurate.

The final method uses the timing of the *fusion of the epiphyses* to estimate age-at-death. Table 7.4 provides ranges of age (in years) that major epiphyses begin to fuse to the diaphysis. A more comprehensive list of epiphyseal union data is presented in Scheuer and Black (2000). To use these data an epiphysis must be assigned to one of three categories:

1) Unfused, the epiphysis is completely separated from the diaphysis,
2) Fusing, the epiphysis is attached to the diaphysis but a clear line of separation is still present (**Figure 7.12A**), and
3) Fused, the diaphysis and epiphysis are completely fused and no line of fusion is evident (**Figure 7.12B**).

As with dental development, there is normal intra-populational variation (from person to person), variation related to sex (females mature earlier than males), and variation by population (some parts of the world exhibit earlier development than others) that contribute error to the age estimates. In addition, several years can pass between when an epiphysis begins to fuse and its final completion, thus adding to the range of error associated with this method (see Ubelaker, 1987).

Table 7.3. Humerus and femur long bone lengths by age (without epiphyses) for males and females, after Scheuer and Black (2000) (all data in mm)

	Humerus				Femur			
	Male		Female		Male		Female	
Age	Mean	SD	Mean	SD	Mean	SD	Mean	SD
0.125	72.4	4.5	71.8	3.6	86.0	5.4	87.2	4.3
0.25	80.6	4.8	80.2	3.8	100.0	4.8	100.8	3.6
0.5	88.4	5.0	86.8	4.6	112.2	5.0	111.1	4.6
1	105.5	5.2	103.6	4.8	136.6	5.8	134.6	4.9
1.5	118.8	5.4	117.0	5.1	155.4	6.8	153.9	6.4
2	130.0	5.5	127.7	5.8	172.4	7.3	170.8	7.1
2.5	139.0	5.9	136.9	6.1	187.2	7.8	185.2	7.7
3	147.5	6.7	145.3	6.7	200.3	8.5	198.4	8.7
3.5	155.0	7.8	153.4	7.1	212.1	11.4	211.1	10.0
4	162.7	6.9	160.9	7.7	224.1	9.9	223.2	10.1
4.5	169.8	7.4	169.1	8.3	235.7	10.5	235.5	11.4
5	177.4	8.2	176.3	8.7	247.5	11.1	247.0	11.5
5.5	184.6	8.1	182.6	9.0	258.2	11.7	257.0	12.2
6	190.9	7.6	190.0	9.6	269.7	12.0	268.9	13.5
6.5	197.3	8.1	196.7	9.7	280.3	12.6	279.0	13.8
7	203.6	8.7	202.6	10.0	291.1	13.3	288.8	13.6
7.5	210.4	8.9	209.3	10.5	301.2	13.5	299.8	15.2
8	217.3	9.8	216.3	10.4	312.1	14.6	309.8	15.6
8.5	222.5	9.2	221.3	11.2	321.0	14.6	318.9	15.8
9	228.7	9.6	228.0	11.8	330.4	14.6	328.7	16.8
9.5	235.1	10.7	234.2	12.9	340.0	15.8	338.8	18.6
10	241.0	10.3	239.8	13.2	349.3	15.7	347.9	19.1
10.5	245.8	11.0	245.9	14.6	357.4	16.2	356.5	21.4
11	251.7	10.7	251.9	14.7	367.0	16.5	367.0	22.4
11.5	257.4	11.9	259.1	15.3	375.8	18.1	378.0	23.4
12	263.0	12.8	265.6	15.6	386.1	19.0	387.6	22.9

Table 7.4. Timing of union for long bone epiphyses (in years), after Schaefer et al. (2009)

	Males			Females		
Bone	Unfused	Fusing	Fused	Unfused	Fusing	Fused
HUMERUS						
Proximal	<20	16–21	>18	<17	14–19	>16
Distal	<15	14–18	>15	<15	13–15	>13
RADIUS						
Proximal	<18	14–18	>16	<15	12–13	>13
Distal	<19	16–20	>17	<18	14–19	>15
ULNA						
Proximal	<16	14–18	>15	<15	12–15	>12
Distal	<20	17–20	>17	<18	15–19	>15
FEMUR						
Proximal	<18	16–19	>16	<15	14–17	>14
Distal	<19	16–20	>17	<16	14–19	>17
TIBIA						
Proximal	<18	16–20	>17	<17	14–18	>18
Distal	<18	16–18	>16	<17	14–17	>15
FIBULA						
Proximal	<19	16–20	>17	<17	14–17	>15
Distal	<18	15–20	>17	<17	14–17	>15

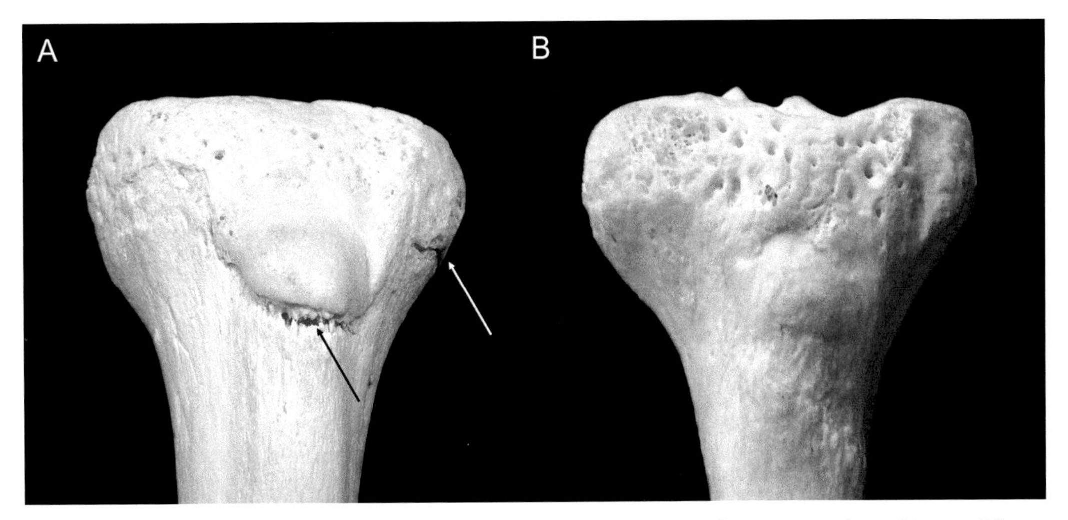

Figure 7.12. (*A*) Proximal tibia with actively fusing epiphysis and arrows indicating open line of fusion; (*B*) proximal tibia with completely fused epiphysis.

LEARNING CHECK

Q11. Using the data from **Table 7.3,** what age range is the most appropriate for a humerus (either sex) that is 200 mm in length?

A) 1–1.5 years
B) 3–3.5 years
C) 4.5–5.5 years
D) 6–7 years

Q12. Using the data from **Table 7.3,** what age range is the most appropriate for a femur (either sex) that is 200 mm in length?

A) 1–1.5 years
B) 3–3.5 years
C) 4.5–5.5 years
D) 6–7 years

Q13. Using the data from **Table 7.4,** what age range is the most appropriate for a female with a fused proximal tibial epiphysis?

A) <17 years
B) 14–18 years
C) >18 years
D) <15 years

Q14. Using the data from **Table 7.4,** what age range is the most appropriate for a male with an unfused proximal humeral epiphysis?

A) >17 years
B) 14–18 years
C) >18 years
D) <20 years

Q15. Using the data from **Table 7.4** and assuming a male sex estimate, how old is the individual in **Figure 7.12A**?
A) <16 years
B) 16–20 years
C) >21 years
D) <12 years

KidStats and Probabilistic Models of Age Estimation

The methods discussed above reflect standard practices in forensic anthropology dating back several decades. However, their use in this format is complicated by the *Daubert* criteria (see chapter 2) because they result in point estimates or broad ranges-of-age estimates that are not strongly based in statistical probability theory. Stull et al. (2014) developed a solution to this (through software called KidStats) that can be used to generate age estimates and *prediction intervals* from long bone lengths and certain breadths of the long bone ends. The use of breadth data provides a solution to the common problem that lengths are not possible to measure due to damage to the proximal or distal end. KidStats is useful because it uses a database of over 1,300 known-age subadults between the ages of birth and 12 years to generate tailored age estimates based on which specific measurements are available for a case. KidStats can use a single long bone measurement or a combination of long bone measurements to generate an age estimate with a 95 percent prediction interval. For example, a femur diaphyseal length of 350 mm returns the following: estimated age 11.03 years with a 95 percent prediction interval from 8.55 to 13.51 years. This means that 95 percent of subadults with a femur length of 350 mm will be between the ages of 8.55 and 13.51. Because this method is based in probability theory, we know the error rate is 5 percent, meaning 5 percent of individuals with a femur length of 350 mm will fall outside of this age range. This approach satisfies the *Daubert* criteria.

End-of-Chapter Summary

Subadult age estimation is based on well-documented processes of human growth and development. Because of this, the age estimates provided by these methods are generally more accurate than those based on adult age estimation methods and have smaller ranges of error. Forensic anthropologists can use several methods for estimating age-at-death. These include dental formation, dental eruption, epiphyseal appearance, long bone length, and epiphyseal union. Because dental development is considered more regimented than skeletal growth, dental methods are more accurate than long bone or epiphyseal methods to estimate age-at-death in subadult remains. Furthermore, teeth are durable, and an accurate age estimate can be generated from a single tooth that was still forming or resorbing at the time of death. To the contrary, epiphyses are difficult to identify when first forming and are subject to breakdown in the soil. Long bone length methods require that the bone is complete enough to measure, which is less assured with younger individuals that have thinner bones.

End-of-Chapter Exercises

Exercise 1

Materials Required: **Table 7.2, Figure 7.8,** and **Figure 7.13**

Scenario: The burned remains of two adults and two children were recovered from a vehicle that had caught on fire following a traffic accident. The identities of the adults were confirmed by matching their dental remains to the dental records of the individuals whose driver's licenses were recovered from the vehicle. They were known to have two children, ages 8 and 14. Radiographs of the mandibular dentition of the two children found in the car are shown in **Figure 7.13**. Using this information as well as the information included in **Figure 7.8** and **Table 7.2,** answer the following questions.

Question 1: The dental development chart shown in **Figure 7.8** is an example of which of the Stevens' data types?

Question 2: Based on dental development, which one of the individuals shown in **Figure 7.13** is older, Individual A or Individual B?

Question 3: Is the age of Individual A consistent with either of the family's two children? Remember that **Figure 7.8** provides point estimates of age and that age-at-death should be reported as a range.

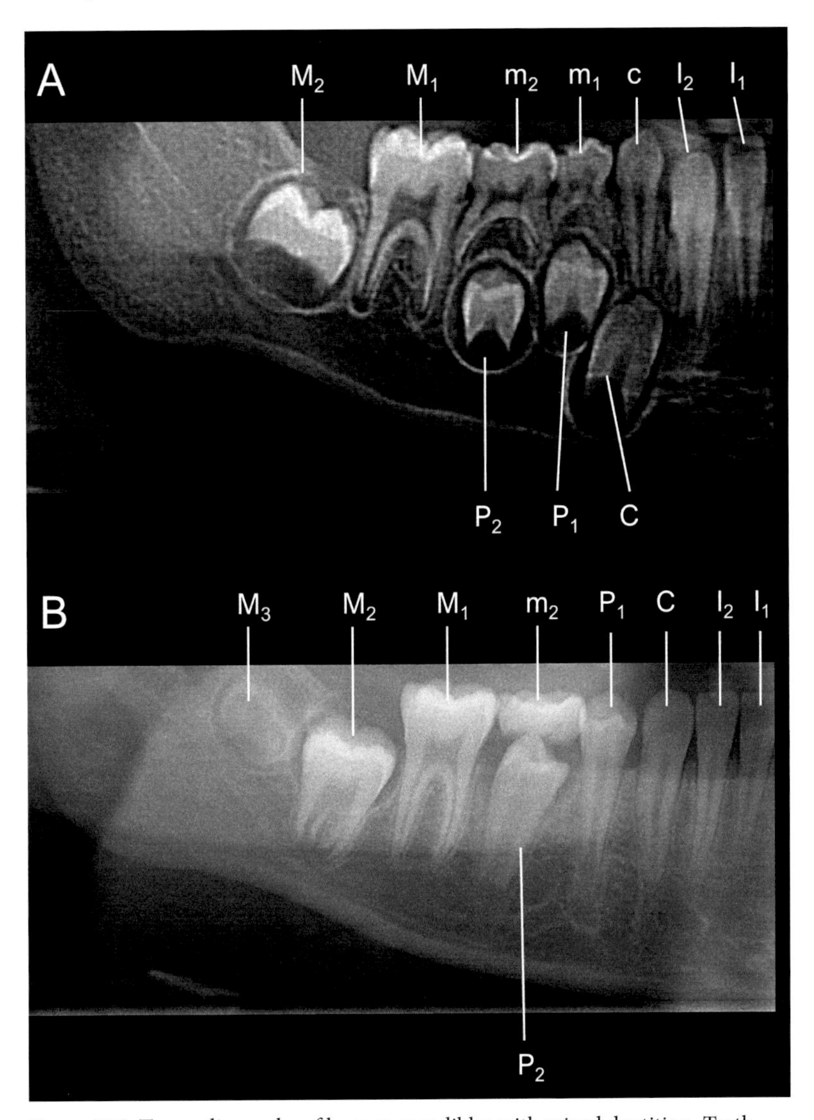

Figure 7.13. Two radiographs of human mandibles with mixed dentition. Teeth have been labeled according to type and position.

Question 4: Is the age of Individual B consistent with either of the family's two children? Remember that **Figure 7.8** provides point estimates of age and that age-at-death should be reported as a range.

Exercise 2

Materials Required: **Table 7.3**

Directions: In any skeletal analysis, it is important to consider the normal range of variation in whatever traits are being observed. For metric data, this is often presented as a mean (average) and a standard deviation, which represents the spread, or dispersion, of the data around the mean. When data are normally distributed (i.e., a frequency distribution appears like a classic bell curve), approximately 67 percent of the data are within one standard deviation of the mean and approximately 95 percent of the data are within two standard deviations of the mean. For example, if the average height of males in a population is 5 feet, 9 inches (or 69 inches) with a standard deviation of 2.5 inches, then 67 percent of males in that population are expected to be between 66.5 and 71.5 inches tall. Assume that the data presented in **Table 7.3** are normally distributed and use this information to answer the following questions.

Question 1: 95 percent of females who are 3.5 years of age are expected to have a femur length (without epiphyses) between what values?

Question 2: What percentage of 9-year-old males is expected to have a humeral length (without epiphyses) of 219.1 mm and 238.3 mm?

Question 3: Skeletonized remains of a child have been found in a wooded area. You measure the length of the femur (without epiphyses) and find it to be 310 mm long. Based on this measurement, what age range would you estimate if this individual were a male? What age range would you estimate if they were female? Remember to include all ages for which this measurement falls into the 95 percent range (i.e., falls within two standard deviations of the mean).

Exercise 3

Materials Required: **Tables 7.2 and 7.4, Figure 7.14.**

Scenario: After hikers in the desert reported finding human remains in a dry wash, law enforcement searched the area and recovered a left humerus, a left femur, and a mandible. These elements are shown (not to scale) in **Figure 7.14**. Use what you have learned in this and preceding chapters to answer the following questions.

Question 1: Which bone and which portion (i.e., proximal or distal) of that bone is shown in **Figure 7.14A**?

Question 2: Which bone and which portion (i.e., proximal or distal) of that bone is shown in **Figure 7.14B**?

Question 3: A colleague of yours has reported that, since there are no repeated elements, the minimum number of individuals represented by these bones is one. Do you agree with their assessment? Why or why not?

References

Adams DM, Ralston CE, Sussman RA, Heim K, Bethard JD. 2019. Impact of population-specific dental development on age estimation using dental atlases. *American Journal of Physical Anthropology* 168: 190–199.

AlQahtani SJ, Hector MP, Liversidge HM. 2010. Brief communication. The London atlas of human tooth development and eruption. *American Journal of Physical Anthropology* 142: 481–490.

AlQahtani SJ, Hector MP, Liversidge HM. 2014. Accuracy of dental age estimation charts: Schour and Massler, Ubelaker, and the London Atlas. *American Journal of Physical Anthropology* 154: 70–78.

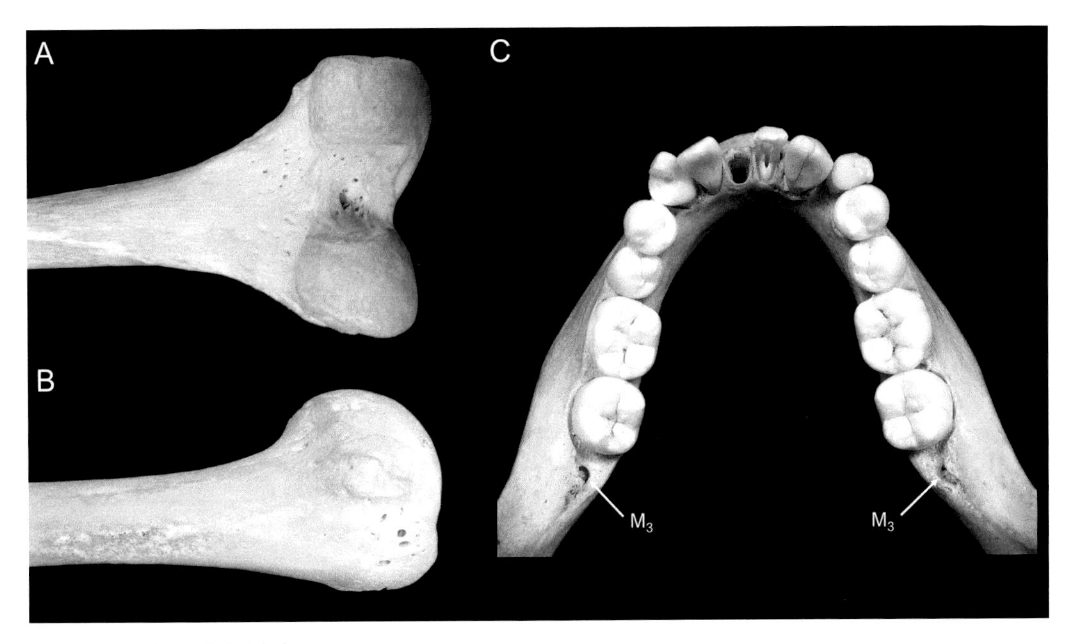

Figure 7.14. Three skeletal elements.

Anderson BE. 2008. Identifying the dead: methods utilized by the Pima County (Arizona) Office of the Medical Examiner for undocumented border crossers: 2001–2006. *Journal of Forensic Sciences* 53: 8–15.

Christensen AM, Passalacqua NV, Bartelink EJ. 2014. *Forensic Anthropology. Current Methods and Practice.* Oxford: Elsevier.

Fazekas GI, Kósa F. 1978. *Forensic Fetal Osteology.* Budapest: Akadémiai Kiadó.

Galloway A. 1988. Estimating actual height in the older individual. *Journal of Forensic Sciences* 33: 126–136.

Lewis JM, Senn DR. 2010. Dental age estimation utilizing third molar development: a review of principles, methods, and population studies in the United States. *Forensic Science International* 201: 79–83.

Moorrees CFA, Fanning EA, Hunt EE Jr. 1963a. Formation and resorption of three deciduous teeth in children. *American Journal of Physical Anthropology* 21: 205–213.

Moorrees CFA, Fanning EA, Hunt EE Jr. 1963b. Age formation stages for ten permanent teeth. *Journal of Dental Research* 42: 1490–1502.

Schaefer M, Black S, Scheuer L. 2009. *Juvenile Osteology: A Laboratory and Field Manual.* San Diego: Academic Press.

Scheuer L, Black S. 2000. *Developmental Juvenile Osteology.* San Diego, CA: Academic Press.

Schour I, Massler M. 1941a. The development of the human dentition. *Journal of the American Dental Association* 28: 1153–1160.

Schour I, Massler M. 1941b. *Development of Human Dentition Chart,* Second Edition. Chicago: American Dental Association.

Smith BH. 1991. Standards of human tooth formation and dental age assessment. In: Kelley MA, Larsen CS (Eds.), *Advances in Dental Anthropology.* New York, NY: Wiley-Liss. p. 143–168.

Stewart TD. 1979. *Essentials of Forensic Anthropology. Especially as Developed in the United States.* Springfield, IL: Charles C. Thomas.

Stull KE, L'Abbé EN, Ousley SD. 2014. Using multivariate adaptive regression splines to estimate subadult age from diaphyseal dimensions. *American Journal of Physical Anthropology* 154: 376–386.

Ubelaker DH. 1978. *Human Skeletal Remains: Excavations, Analysis, Interpretation.* Washington, DC: Taraxacum.

Ubelaker DH. 1987. Estimating age at death from immature human skeletons: an overview. *Journal of Forensic Sciences* 32: 1254–1263.

Walker PL. 1995. Problems of preservation and sexism in sexing: some lessons from historical collections for palaeodemographers. In: Saunders SR, Herring A (Eds.), *Grave Reflections. Portraying the Past through Cemetery Studies.* Toronto: Canadian Scholars' Press. p. 31–47.

Workshop of European Anthropologists. 1980. Recommendations for age and sex diagnoses of skeletons. *Journal of Human Evolution* 9: 517–549.

8

Estimating Age of Adults

Learning Goals

By the end of this chapter, the student will be able to:

- Describe the biological basis for adult age estimation using skeletal indicators.
- Describe the biological and anatomical basis for age-related changes in the pubic symphysis, auricular surface, and sternal rib ends.
- Evaluate cranial suture closure as an indicator of age and apply a standard approach for scoring cranial suture closure and generating an age estimate.

Introduction

Adult age estimation methods are based primarily on processes of senescence and generally involve specific joint surfaces. While joint breakdown is part of the normal course of aging, not all joints do so at a rate that is relatively constant and follows a progression that is easily observable. Here we discuss four regions of relevance to forensic anthropology: the pubic symphysis, the auricular surface, the sternal ends of the ribs (the fourth rib in particular), and the cranial sutures. In this chapter, we emphasize the *pubic symphysis* because it is one of the oldest and most widely used methods (Garvin and Passalacqua, 2012) that is accessible to forensic anthropologists through standard autopsy and demonstrates the general principles of joint-related age deterioration very well. It is also one of the easiest methods to apply. The other methods are discussed in their order of utility: sternal rib ends, auricular surface, and cranial suture closure (Garvin and Passalacqua, 2012), with some discussion of newly emerging methods and more statistically rigorous approaches.

Pubic Symphysis Age Estimation

The pubic symphysis is the joint that connects the ossa coxae ventrally (**Figure 8.1**). Although mostly immobile, the joint experiences some movement during life, which results in cumulative changes as one ages. This region of the body has been recognized for over a century as relevant to age-at-death estimation. Multiple scoring systems have been developed over the years such as those by Todd (1920, 1921), McKern and Stewart (1957), and Gilbert and McKern (1973). Until recently, the most commonly used method was the Suchey-Brooks method (Brooks and Suchey, 1990). Recent modifications to this method have been proposed by Berg (2008) and Hartnett and Fulginiti (Hartnett, 2010a). Here, we emphasize the Hartnett-Fulginiti method due to its growing popularity among forensic anthropologists, its foundation upon a more recent sample, and its enhanced ability to estimate age-at-death for older individuals.

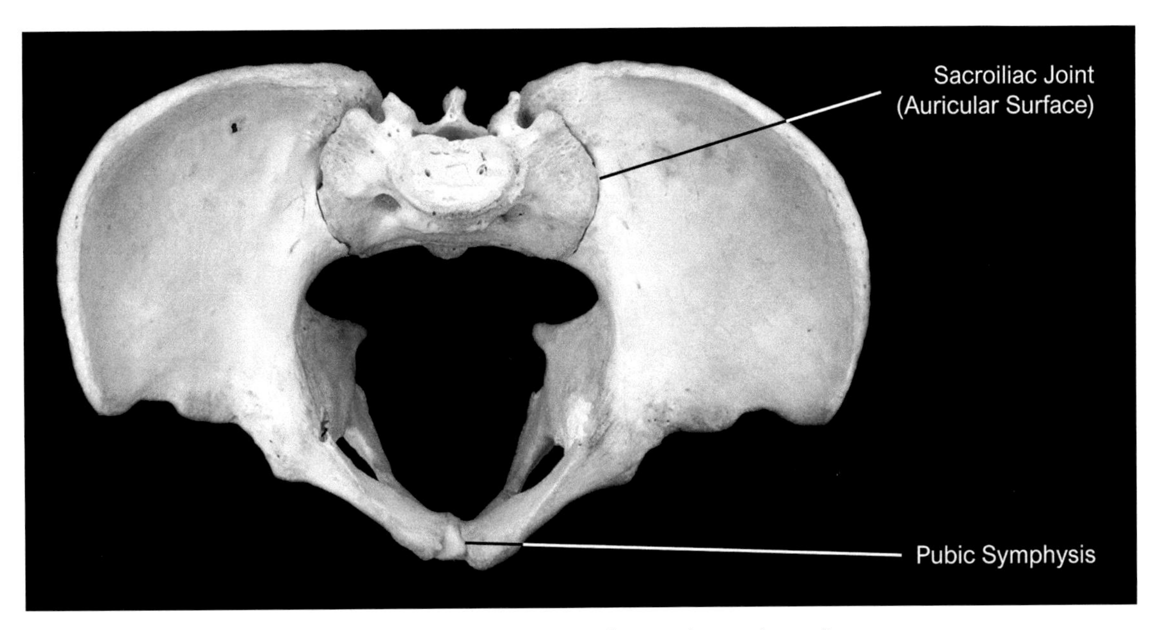

Figure 8.1. Pelvic girdle showing the location of the pubic symphysis and auricular surface.

Orienting the Pubic Symphysis

Imagine separating the pelvis in **Figure 8.1** at the pubic symphysis and looking straight at the exposed joint surface, as shown in a real bone example in **Figure 8.2**. The joint surface itself is roughly oval in appearance. There is an upper and lower end (or extremity), and the face is divided into a dorsal (posterior) and ventral (anterior) rim/margin. For the rest of this chapter, the pubic symphyses will only be seen from the left side where dorsal is always to the left and ventral is always to the right.

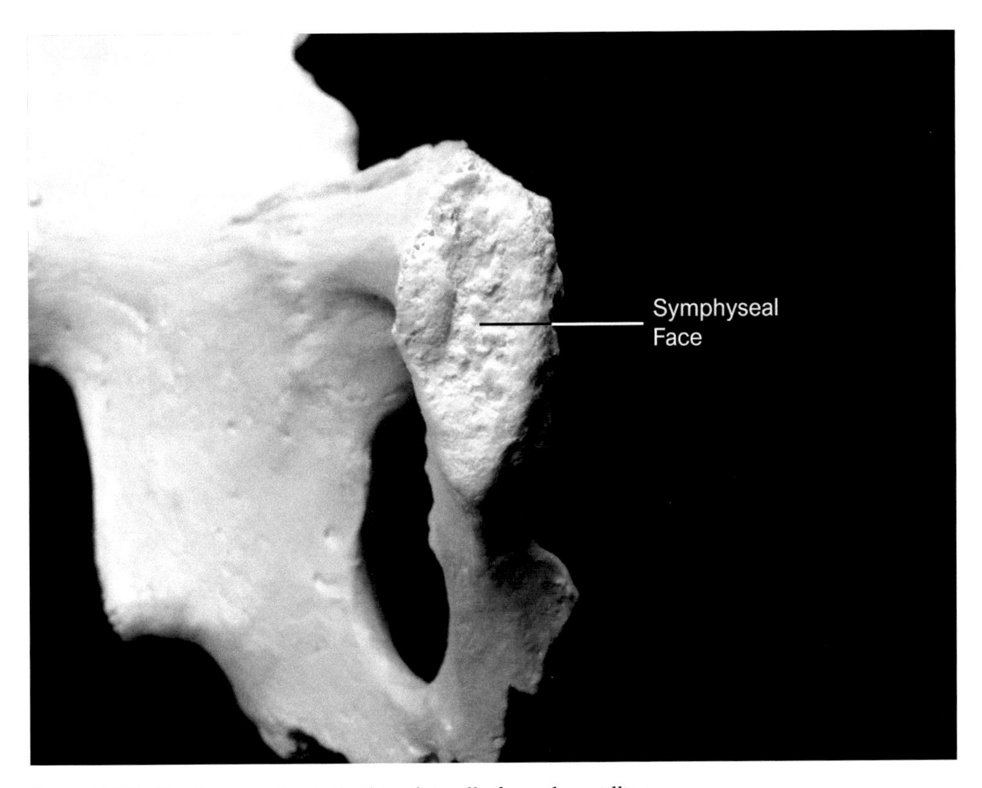

Figure 8.2. Left pubic symphysis, looking laterally from the midline.

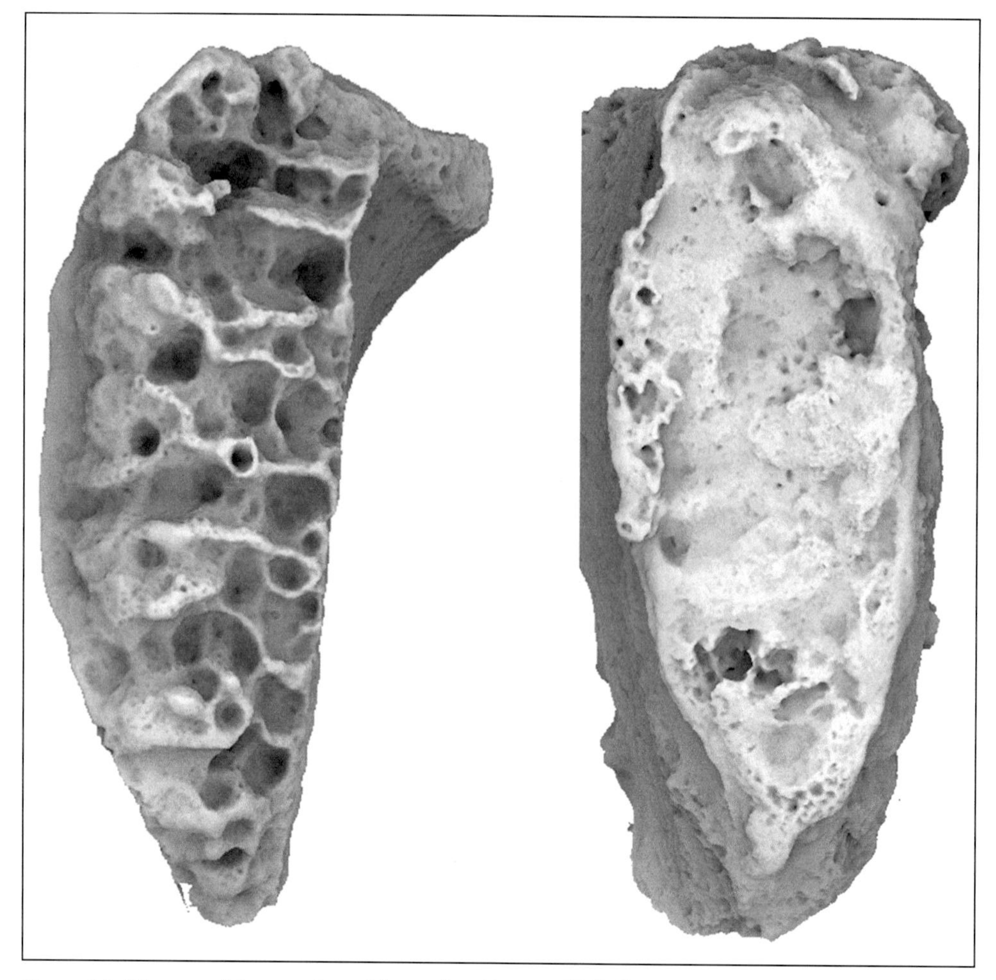

Figure 8.3. 3D scans of the symphyseal faces of an 18-year-old (*left*) and a 90-year-old (*right*).

Young vs. Old: General Considerations

Consider the two pubic bones shown in **Figure 8.3**. The left is a very young individual. The right is an elderly individual that represents the opposite end of age-related variation. In comparing these models, you should notice the following:

The young individual shows clear ridges that are organized horizontally across the symphyseal face; this is called *billowing*. These billows are absent in the older individual.

The young individual has a relatively smooth outline to the symphyseal face. The older individual has a more irregular outline.

The young individual shows no clear bottom limit of the symphyseal face. The older individual has a clearly demarcated rim surrounding the entire symphyseal face.

Whereas the younger individual shows undulating ridges, the older individual has a symphyseal face that is irregular: it is pitted, depressed, and generally disorganized.

Pubic Symphysis Scoring Systems

The Hartnett-Fulginiti scoring system (Hartnett 2010a) uses seven phases to describe changes in the pubic symphysis with advancing age. Analysts use written descriptions and photographic examples to identify the phase to which an unknown pubic symphysis is most similar. These phases are then compared with tables of means and age ranges from known-age individuals to determine age-at-death.

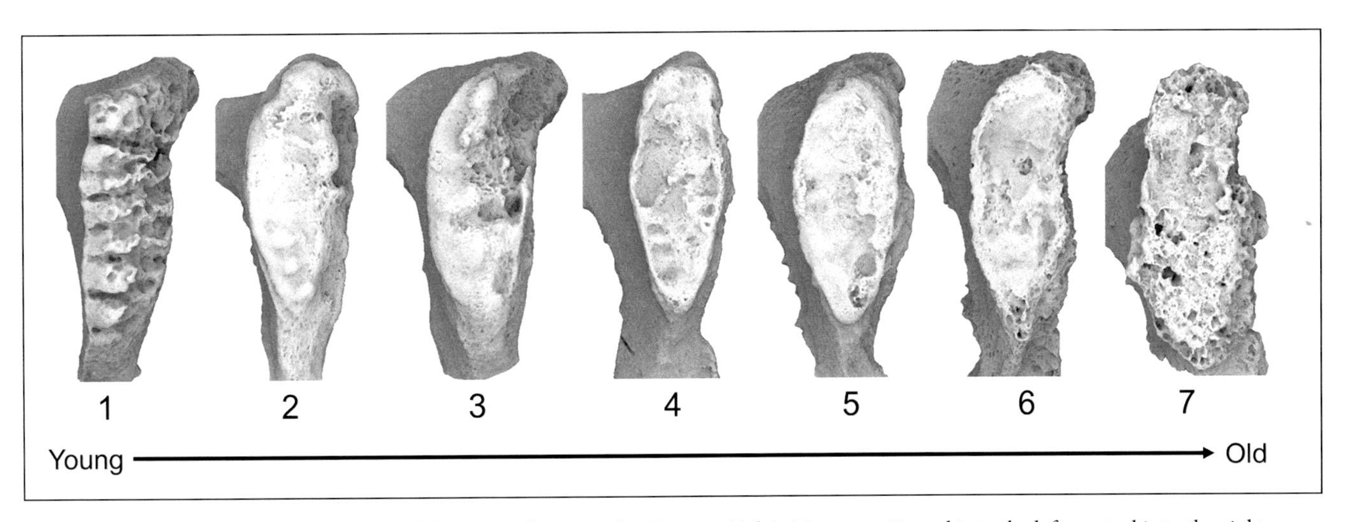

Figure 8.4. 3D scans of examples of each of the seven phases in the Hartnett-Fulginiti system. Dorsal is to the left, ventral is to the right.

The critical first step in applying this method is to understand what each phase represents in terms of changes to the appearance of the pubic bone. Scans of examples of the seven phases—with Phase 1 being the youngest and Phase 7 the oldest—are shown in **Figure 8.4.** Complete phase descriptions are provided in **Table 8.1,** with associated sex-specific age ranges shown in **Table 8.2**. For all images that follow, dorsal (posterior) is to the left, and ventral (anterior) is to the right.

To understand the anatomy of the pubic symphysis and the changes the joint undergoes during aging, it is useful to consider three interrelated aspects of the pubic bone: the *symphyseal face,* the buildup and breakdown of the *symphyseal rim,* and the *quality of the bone* (**Figure 8.5**).

Symphyseal Face: In Phase 1, the symphyseal face is characterized by a pronounced and transversely oriented *ridge and furrow system* (tinted blue in **Figure 8.5**). In Phase 2, this ridge and furrow system begins to be filled in, with the result that ridges start to become flattened and furrows become shallower through the addition of new bone onto the surface. This process begins along the dorsal part of the symphyseal face. In addition, the ventral half of the face is covered by an epiphysis (called the ventral rampart) that fuses much later than most other epiphyses in the skeleton. The ridge and furrow system becomes progressively

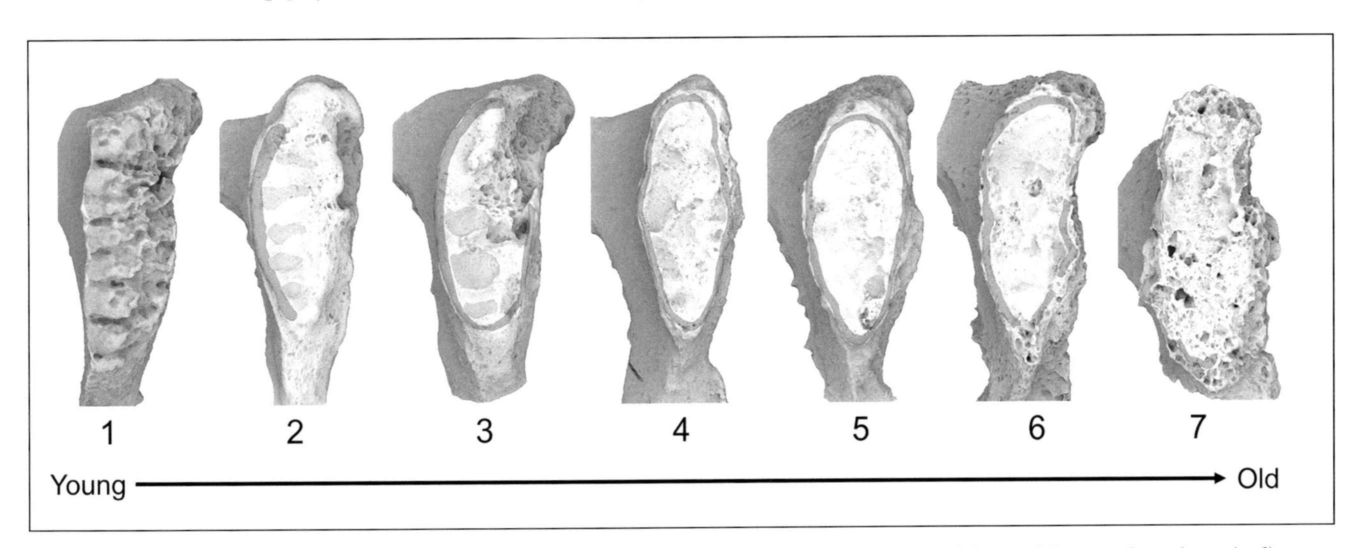

Figure 8.5. Age-related changes in the ridge and furrow system (blue) and the development and breakdown of the symphyseal rim (red). Note, too, the increasing porosity and irregularity of the symphyseal face with age.

Table 8.1. Hartnett (2007, 2010a) pubic symphysis phase descriptions

Phase 1	A clear system of ridges and furrows that are deep, well-defined, and unworn extending from the pubic tubercle onto the inferior ramus. No dorsal lipping. Bone is of excellent quality: firm, heavy, dense, and smooth on the ventral and dorsal body. No rim formation. Dorsal plateau is not formed; ridges and furrows extend to the dorsal edge.
Phase 2	The rim is in the process of forming, but mainly consists of a flattening of the ridges on the dorsal aspect of the face and ossific nodules present along the ventral border. Ridges and furrows are still present but are becoming shallow; they may appear worn down or flattened, especially on the dorsal aspect of the face. The upper and lower rim edges are not formed. There is no dorsal lipping. The bone quality is very good; the bone is firm, heavy, dense, and smooth on the ventral and dorsal body, with little porosity. The pubic tubercle may appear separate from the face.
Phase 3	The lower rim is complete on the dorsal side of the face and extends until it ends approximately halfway up the ventral face, leaving a medium to fairly large gap between the lower and upper extremities on the ventral face. This enlarged "V" is longer on the dorsal side than the ventral side. Some ridges and shallow furrows are still visible but appear worn down. In some cases, the face is becoming slightly porous. The rim is forming both on the dorsal aspect of the face and the upper and lower extremities. In some cases, there is a rounded buildup of bone in the gap between the upper and lower extremities above the enlarged "V." Bone quality is good; the bone is firm, heavy, dense, and has little porosity. The dorsal surface of the body is smooth, and there are small bony projections near the medial aspect of the obturator foramen. The ventral aspect of the body is not elaborate. Very slight to no dorsal lipping. Quality of bone and rim completion are important deciding factors. Variant: In some cases, a deep line or epiphysis is visible on the ventral aspect parallel to and adjacent to the face (males only).
Phase 4	In most cases, the rim is complete at this stage, but may have a small ventral hiatus on the superior and ventral aspect of the rim. The face is flattened and not depressed. Remnants of ridges and furrows may be visible on the face, especially on the lower half. The quality of bone is good, but the face is beginning to appear more porous. The dorsal and ventral surfaces of the body are roughened and becoming coarse. There is slight dorsal lipping. In females with large parturition pits, there tends to be more dorsal lipping. The ventral arc may be large and elaborate in females.
Phase 5	The face is becoming more porous and is depressed but maintains an oval shape. The face is not irregularly shaped or erratic. The rim is complete at this stage. In general, the rim is not irregular. Ridges and furrows are absent on the face. There may be some breakdown of the rim on the ventral border, which appears as irregular bone (not rounded/solid). The ventral surface of the body is roughened and irregular, with some bony excrescences. The dorsal surface of the body is coarse and irregular. Projections are present on the medial aspect of the obturator foramen. Bone quality is good to fair; it is losing density and is not smooth. The bone is becoming lighter. In females the ventral arc is prominent.
Phase 6	The face is losing its oval shape and is becoming irregular. The rim is complete, but breaking down, especially on the ventral border. The rim and face are irregular, porous, and macroporous. Bone quality is fair, and the bone is lighter and more porous, even with bony buildup on the ventral body surface. The rim is eroding. The dorsal surface of the bone is rough and coarse. There are no ridges and furrows. Dorsal lipping is present. Projections are present at the medial aspect of the obturator foramen. Bone weight is a major deciding factor between phases 6 and 7.
Phase 7	The face and rim are very irregular in shape. The rim is complete but is eroding and breaking down. There are no ridges and furrows. The face is porous and macroporous. Dorsal lipping is pronounced. Bone quality is poor, and the bone is very light and brittle. The dorsal surface of the bone is roughened. The ventral surface of the body is very rough and porous. Projections are present at the medial aspect of the obturator foramen. Bone weight is a major deciding factor between phases 6 and 7.
VARIANT:	The rim is complete except for a lytic/sclerotic appearing hiatus at the superior ventral margin that extends toward the pubic tubercle and sometimes underneath the ventral rim, which should not be confused with a hiatus.

Table 8.2. Age ranges for Hartnett-Fulginiti pubic symphysis phases, after Hartnett (2010a)

	Males				Females			
Phase	*n*	*Mean*	*SD*	*Range*	*n*	*Mean*	*SD*	*Range*
1	14	19.29	1.93	18–22	5	19.80	1.33	18–22
2	14	22.14	1.86	20–26	5	23.20	2.38	20–25
3	36	29.53	6.63	21–44	25	31.44	5.12	24–44
4	69	42.54	8.80	27–61	35	43.26	6.12	33–58
5	90	53.87	8.42	37–72	32	51.47	3.94	44–60
6	34	63.76	8.06	51–83	35	72.34	7.36	56–86
7	96	77.00	9.33	58–97	56	82.54	7.41	62–99

flatter through Phase 3 and generally disappears by Phase 5. Although there may be some residual ridges on the lower half of the symphyseal face in Phase 4, the surface is generally flat. During Phase 5, the symphyseal face takes on a slightly depressed, or concave, appearance before becoming progressively more irregular in later phases. *Porosity* can become apparent on the symphyseal face as early as Phase 3 but is usually evident by Phase 4. The number and size of these holes then increases with advancing age and decreasing bone quality.

Symphyseal Rim: The formation and breakdown of the symphyseal rim is illustrated by red tinting in **Figure 8.5**. In Phase 1, the symphyseal rim has yet to begin forming. This process starts during Phase 2 and is related to the flattening of the dorsal aspect of the ridge and furrow system as well as the development of small nodules of bone—*ossific nodules*—along the ventral border of the symphyseal face. In Phase 3, the rim is largely complete along the dorsal margin of the symphyseal face, has a defined lower extremity, and extends approximately halfway up the ventral margin, thereby creating a "V" shape. By Phase 4, the rim is generally complete although there may be a small gap on the superior portion of the ventral margin—this is referred to as a *ventral hiatus.* It is often difficult to tell the difference between a ventral hiatus and the breakdown of the symphyseal rim that may begin as early as Phase 5. In Phase 6, the symphyseal rim is becoming irregular and losing definition along the ventral margin. This process continues into Phase 7 as declining bone quality compromises the integrity of the symphyseal rim.

Bone Quality: The Hartnett-Fulginiti method differs from earlier methods in its greater emphasis on bone quality—a trait that primarily refers to the surface texture and density of the bone (**Figure 8.6**). For the first three phases, bone quality is good—surface textures are

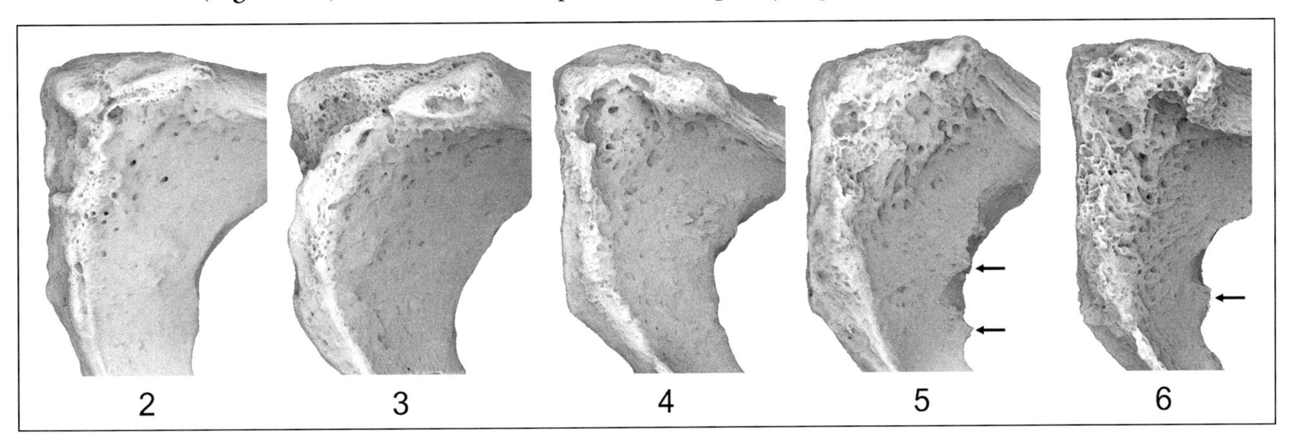

Figure 8.6. Ventral surface of the pubic bone for Hartnett-Fulginiti Phases 2 through 6 illustrating age-related changes to the surface texture of the bone. These changes are accompanied by a progressive loss of bone density. Black arrows indicate spicules on the medial margin of the obturator foramen.

smooth, and the bone feels relatively heavy and dense. Beginning in Phase 4, the surface starts to become roughened—especially on the dorsal and ventral surfaces. The bone surface becomes increasingly irregular with advancing age, and this is sometimes accompanied by the development of bony spicules on the medial margin of the obturator foramen (black arrows in **Figure 8.6**). Beginning in Phase 5, the bone starts to become less dense, with the result that it begins to feel lighter than you would expect it to. By Phase 7, these changes lead to the bone feeling very light and brittle. Bone weight (density) is a major modifying factor in the Hartnett-Fulginiti method and can be used to assign an individual to an older or a younger phase than that which would be assigned based on morphology alone.

LEARNING CHECK

Q1. Based on your understanding of age-related changes in the pubic symphysis, which of the individuals in **Figure 8.7A** was older?
A) The individual on the left
B) The individual on the right

Q2. Based on your understanding of age-related changes in the pubic symphysis, which of the individuals in **Figure 8.7B** was older?
A) The individual on the left
B) The individual on the right

Q3. Based on your understanding of age-related changes in the pubic symphysis, which of the individuals in **Figure 8.7C** was older?
A) The individual on the left
B) The individual on the right

Sternal Rib Age Estimation

The articulation between the sternal (anterior) end of the ribs and the costal cartilage is another joint that undergoes age-related changes. The methodology for age estimation from the sternal rib ends was developed by İşcan and colleagues (İşcan et al., 1984a, b, 1985, 1987) for use on the fourth rib, which is easily accessible through autopsy. The approach identified eight phases to which an unknown decedent could be ascribed, each phase having an associated range of ages. Ribs 3–9 (Yoder et al., 2001) and rib 1 (DiGangi et al., 2009) undergo similar changes, though the specific process of age progression may vary from rib to rib. Based on a more recent sample, Hartnett (Hartnett 2010b) has proposed revisions to the earlier phase systems of İşcan and colleagues. Her revised phase descriptions, also referred to as the Hartnett-Fulginiti method (for ribs), are provided in **Table 8.3**. The sex-specific age ranges associated with each phase are given in **Table 8.4**. Examples of each of the phases illustrating the relevant morphology are shown in **Figure 8.8**.

As with the pubic symphysis, age-related changes in the sternal fourth ribs can be discussed through consideration of three interrelated aspects of sternal rib morphology: the *sternal pit*, the *rim and walls* that surround the pit, and *overall bone quality*.

Sternal pit: In reading these descriptions, please compare to the top row in **Figure 8.8** that shows the sternal pit from an anterior view. Beginning in Phase 1, the sternal pit of the rib is either flat or a very shallow U-shape in cross-section and often exhibits billowing. In Phase 2, the sternal pit develops a distinct indentation and takes on a V-shaped cross section

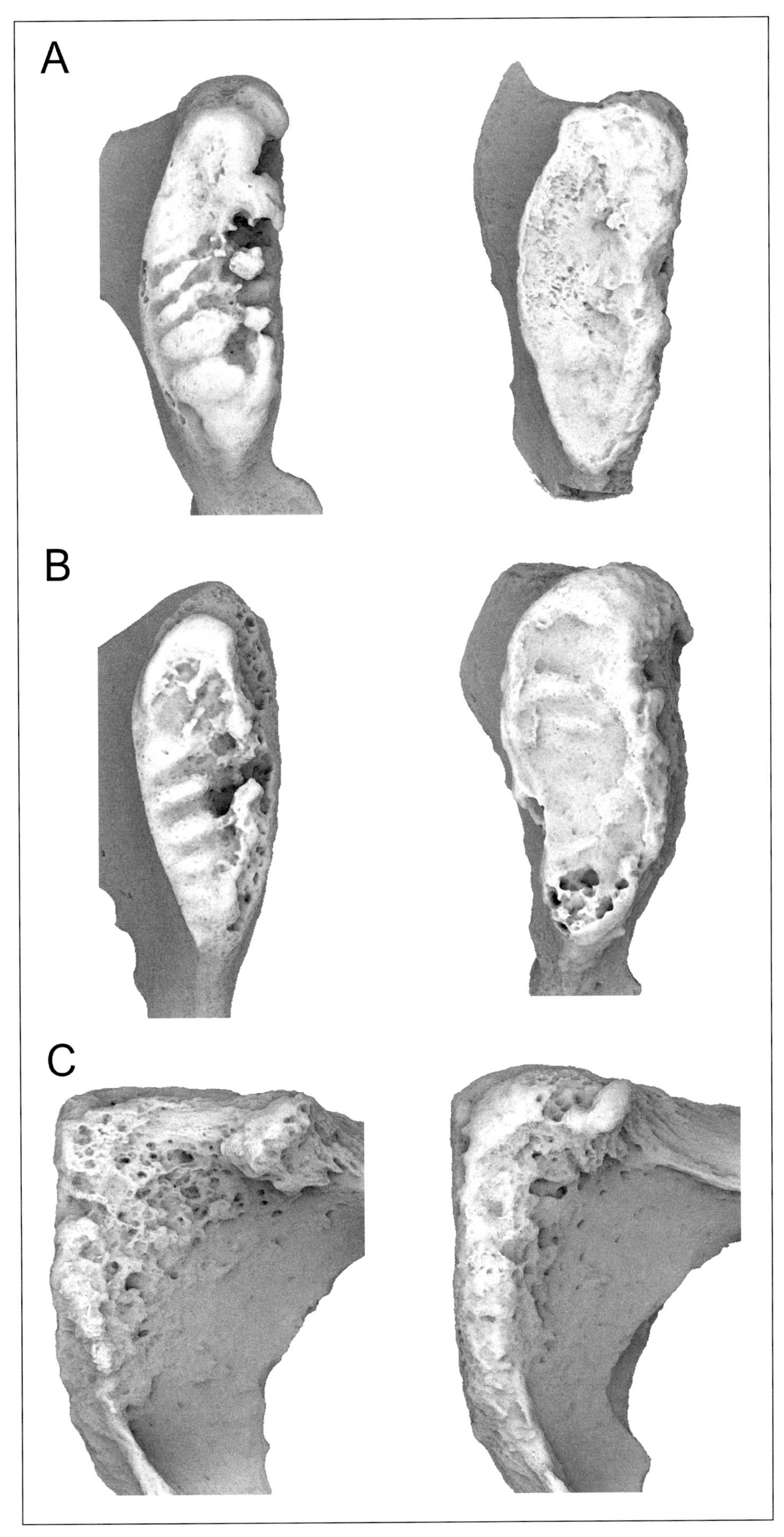

Figure 8.7. Three sets of pubic bones of different ages. (*A and B*) shown from the medial view, (*C*) shown from the ventral view.

Table 8.3. Hartnett (2007, 2010b) sternal rib phase descriptions

Phase 1	The pit is shallow and flat, and there are billows in the pit. The pit is shallow and U-shaped in cross-section. The bone is very firm and solid, smooth to the touch, dense, and of good quality. The walls of the rim are thick. The rim may show the beginnings of scalloping.
Phase 2	There is an indentation to the pit. The pit is V-shaped in cross-section, and the rim is well-defined with round edges. The rim is regular with some scalloping. The bone is firm and solid, smooth to the touch, dense, and of good quality. There is no flare to the rim edges; they are parallel to each other. The pit is still smooth inside, with little to no porosity. In females, the central arc, which manifests on the anterior and posterior walls as a semicircular curve, is visible.
Phase 3	The pit is V-shaped, and there is a slight flare to the rim edges. The rim edges are becoming undulating and slightly irregular, and there may be remnants of scallops, but they look worn down. There are no bony projections from the rim. There is porosity inside the pit. The bone quality is good; it is firm, solid, and smooth to the touch. The rim edges are rounded, but sharp. In many females, there is a build-up of bony plaque, either in the bottom of the pit or lining the interior of the pit, creating the appearance of a two-layer rim. An irregular central arc is apparent in many cases.
Phase 4	The pit is deep and U-shaped. The edges of the pit flare outwards, expanding the oval area inside the pit. The rim edges are not undulating or scalloped but are irregular. There are no long bony projections from the rim, and the rim edges are thin, but firm. The bone quality is good and does not feel dense or heavy. There is porosity inside the pit. In some males, two distinct impressions are visible in the pit. In females, the central arc may be present and irregular; however, the superior and inferior edges of the rim have developed, decreasing the prominence of the central arc.
Phase 5	There are frequently small bony projections along the rim edges, especially at the superior and inferior edges of the rim. The pit is deep and U-shaped. The rim edges are irregular, flared, sharp, and thin. There is porosity inside the pit. The bone quality is fair; the bone is coarse to the touch and feels lighter than it looks.
Phase 6	The bone quality is fair to poor, the surfaces of the bone feel coarse and brittle, and the bone is lightweight. There are bony projections along the rim edges, especially at the superior and inferior edges, some of which may be over 1 cm long. The pit is deep and U-shaped. The rim is very irregular, thin, and fragile. There is porosity inside the pit. In some cases, there may be small bony extrusions inside the pit.
Phase 7	The bone is very poor quality, and in many cases, translucent. The bone is very light, sometimes feeling like paper, and feels coarse and brittle to the touch. The pit is deep and U-shaped. There may be long, bony growths inside the pit. The rim is very irregular with long bony projections. In some cases, much of the cartilage has ossified and window formation occurs. In some females, much of the cartilage in the interior of the pit has ossified into a bony projection extending more than 1 cm in length.
VARIANT:	In some males, the cartilage has completely or almost completely ossified. The ossification tends to be a solid extension of bone, rather than a thin projection. All of the bone is of very good quality, including the ossification. It is dense, heavy, and smooth. In these instances, bone quality should be the determining factor. There are probably other factors, such as disease, trauma, or substance abuse that caused premature ossification of the cartilage. When the individual is truly very old, the bone quality will be very poor. Be aware of these instances where a rib end may appear very old because of ossification of the cartilage but is actually a young individual, which can be ascertained by bone quality. In these cases, consult other age indicators in conjunction with the rib end.

Table 8.4. Age ranges for Hartnett-Fulginiti sternal rib end phases, after Hartnett (2010b)

	Males				Females			
Phase	*n*	*Mean*	*SD*	*Range*	*n*	*Mean*	*SD*	*Range*
1	20	20.00	1.45	18–22	7	19.57	1.67	18–22
2	27	24.63	2.00	21–28	7	25.14	1.17	24–27
3	27	32.27	3.69	27–37	22	32.95	3.17	27–38
4	47	42.43	2.98	36–48	21	43.52	3.08	39–49
5	76	52.05	3.50	45–59	32	51.69	3.31	47–58
6	61	63.13	3.53	57–70	18	67.17	3.41	60–73
7	75	80.91	6.60	70–97	71	81.20	6.95	65–99

that persists through Phase 3. Older individuals (Phases 4 and above) are characterized by a sternal pit that is deep and U-shaped.

Rim and walls: The rim surrounding the sternal pit is initially thick, rounded, and flat (see Phase 1 in **Figure 8.8**). By Phase 2, the rim takes on a scalloped appearance, although these may appear as early as Phase 1 and may persist through Phase 3. Beginning in Phase 3, the walls of the sternal pit flare slightly outward, widening the area of the sternal pit. The rim edges during this phase are becoming irregular and undulating and the rim itself, while still rounded, is less thick than in younger phases. The rim becomes increasingly irregular, thin, and sharp throughout the older phases. Beginning in Phase 5, bony projections can frequently be observed at the superior and inferior margins of the rim, and these will typically increase in length with advancing age.

Bone quality: In younger individuals, the bone of the sternal rib ends is smooth, firm, and dense. Beginning in Phase 3, porosity is observable within the sternal pit and this persists through later phases. In Phase 4, the bone begins to feel less dense, although it remains smooth in texture and firm to the touch. Throughout the later phases, the texture of the bone becomes increasingly coarse to the touch and brittle. As bone quality decreases, the bone also becomes increasingly light.

Figure 8.8. Photographic examples of sternal rib ends consistent with each of the seven phases described in the Hartnett-Fulginiti method.

LEARNING CHECK

Q4. Based on your understanding of age-related changes in the sternal fourth rib, which of the individuals in **Figure 8.9A** was older?
- A) The individual on the left
- B) The individual on the right

Q5. Based on your understanding of age-related changes in the sternal fourth rib, which of the individuals in **Figure 8.9B** was older?
- A) The individual on the left
- B) The individual on the right

Q6. Based on your understanding of age-related changes in the sternal fourth rib, which of the individuals in **Figure 8.9C** was older?
- A) The individual on the left
- B) The individual on the right

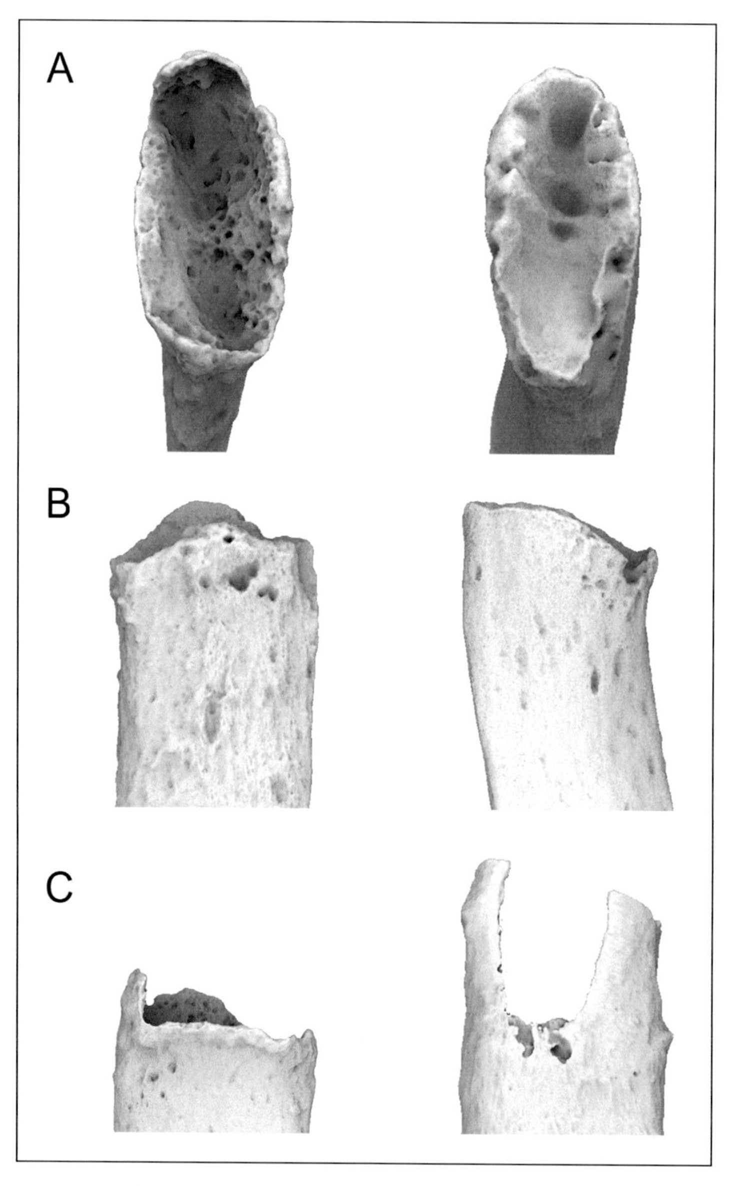

Figure 8.9. Three rib ends of different ages. (*A*) shown from the sternal view, (*B and C*) shown from the anterior or posterior view.

Auricular Surface Age Estimation

The auricular surface is located on the ossa coxae. It articulates with the sacrum to form the sacroiliac joint on the posterior aspect of the pelvis (**Figure 8.1**). The age-related morphological changes in the auricular surface roughly parallel those seen in the pubic symphysis, and both joints are likely subject to similar kinds of biomechanical forces. The original auricular surface aging method was developed by Lovejoy et al. (1985), with multiple scoring systems developed over the years (e.g., Buckberry and Chamberlain, 2002; Igarashi et al., 2005; Osborne et al., 2004). The method most commonly used is that of Lovejoy et al. (1985) (Garvin and Passalacqua, 2012), which uses an eight-phase scoring system similar to the Suchey-Brooks method for the pubic symphysis. More recently, Osborne et al. (2004) evaluated the original method and proposed a simplified six phase system, which we present here.

The auricular surface is less commonly used for age estimation than the pubic symphysis, which reflects the fact that the method is more difficult to apply and less easily accessible in partially or fully fleshed decedents. This latter fact also means it tends to preserve better in fragmentary cases. The difficulty applying the method results from the subtler changes seen in the auricular surface in comparison to the pubic symphysis. In fact, the description of the auricular surface includes textural changes that cannot be seen in images and can only be felt in real bone specimens, making this an especially challenging exercise for students.

The subtler changes in the auricular surface result from key differences in the aging process. For example, although delayed epiphyseal union is important for pubic symphysis age estimation, it is *not* part of the auricular surface aging process. While the joint surfaces in both methods show evidence of billowing and horizontal organization in younger ages, this is much subtler in the auricular surface. Both joints show evidence of increasing porosity and overall surface irregularity in older ages and increasingly irregular additions of bone in the areas surrounding the joint surface in older ages. We break down the auricular surface into components for the purposes of describing age-related changes in the bone with a focus on three areas outlined in **Figure 8.10**: the *surface,* the *margin/rim* including the *apex,* and the *retroauricular area.*

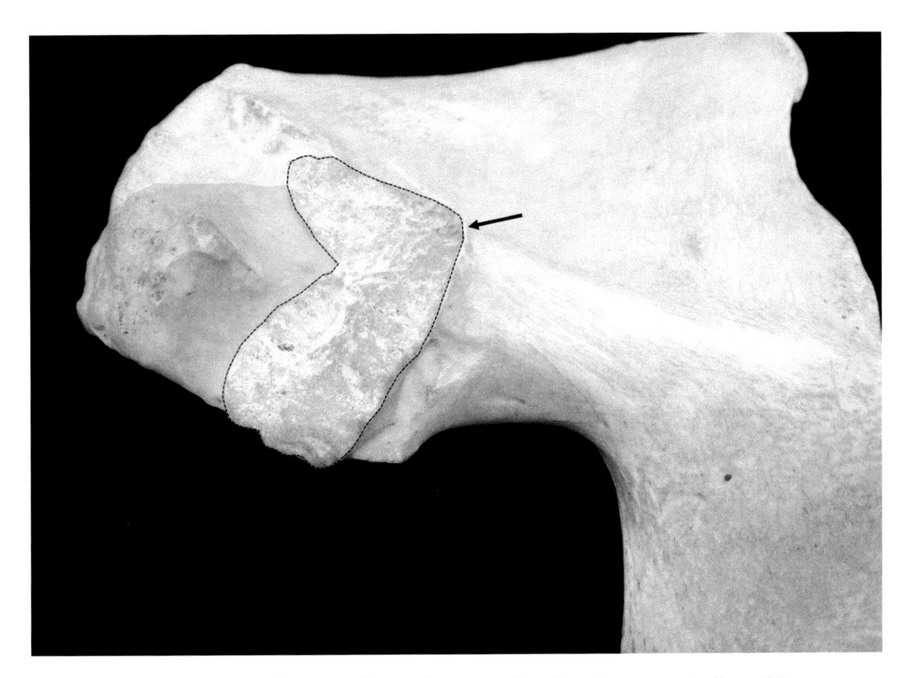

Figure 8.10. Left os coxa. The auricular surface is outlined with its apex indicated by an arrow; the retroauricular area is tinted blue.

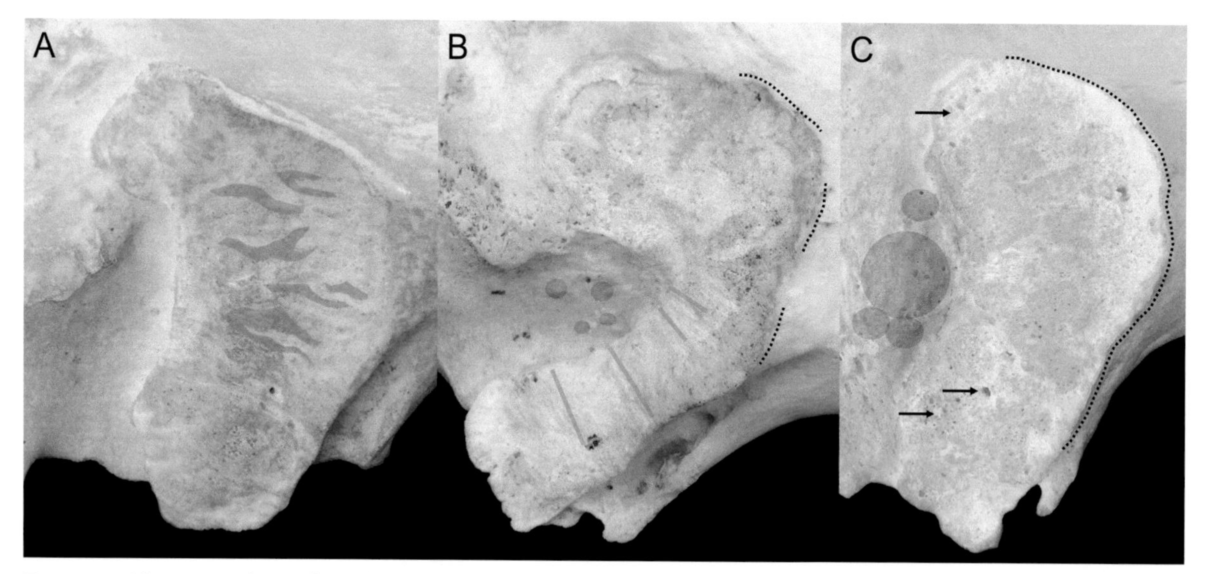

Figure 8.11. Three auricular surfaces showing three different ages. Individual A is the youngest with billowing (blue) and small patches of microporosity (green). The retroauricular area is smooth and there is no lipping of the rim. Individual B is more advanced in age: there are no billows, but some striae remain (blue) and there is more microporosity (green) across the surface. The rim shows lipping (dotted line) and there are some bony spicules in the retroauricular area (red). Individual C is the oldest. Billowing and striae are absent and the face is extensively covered with microporosity (green) with some macroporosity (black arrows). There is clear lipping (dotted line) along the rim and the retroauricular area shows extensive bony spicules and porosity (red).

The surface: The ear-shaped joint surface shows age-related changes in three ways—*organization, texture,* and *porosity*. **Figure 8.11** illustrates these changes with features of organization tinted in blue and increasing porosity tinted in green (macroporosity is indicated by black arrows). The progression of age-related changes in **Figure 8.11** demonstrates how the *organization* of the face changes from horizontal billowing (thicker blue lines, **Figure 8.11A**) to horizontal striations (thin blue lines) to a smooth surface with no horizontal organization (**Figure 8.11C**). The *texture* of the surface changes from finely grained sandpaper to coarsely grained sandpaper, finally becoming dense and smooth. These changes in texture are hard to see in images. Textural changes are complicated by *porosity*. There is limited porosity in younger individuals (**Figure 8.11A**), but this increases with age as indicated by the increasing area of green microporosity (small holes) shown in **Figure 8.11A** and **Figure 8.11B**. Macroporosity (large holes indicated by black arrows) appears on the surface at more advanced stages of surface breakdown.

The margin/rim: This refers to the outline of the joint surface. The margin starts with a smooth rim that is round to the touch and not sharp and irregular. The rim gradually changes to a sharp rim with raised and irregular spicules of bone present. This lipping is most often present on the inferior margin of the auricular surface. The formation of the rim around the apex (see **Figure 8.10**) is a key indicator in age estimation. Progressive rim formation is illustrated by the dotted lines in **Figure 8.11B**.

The retroauricular area: This refers to the area posterior and superior to the joint surface that is the attachment site for ligaments (**Figure 8.10**). This area is smooth in younger individuals with few bone spicules present and increasingly develops a roughened appearance with many bone spicules and/or porosity present with advancing age. These changes are illustrated with red tinting in **Figure 8.11**.

Once you understand the various ways in which the auricular surface changes with age, you can apply these to the phase-based system of Osborne et al. (2004). Phase descriptions and age ranges associated with each phase are presented in **Table 8.5**.

Table 8.5. Age ranges for auricular surface phases, after Osborne et al. (2004)

Phase 1	Billowing with possible striae; mostly fine granularity with some coarse granularity possible *Mean*: 21.1; *Standard Deviation*: 2.98; *Suggested age range*: < 27
Phase 2	Striae; coarse granularity with residual fine granularity; retroauricular activity may be present *Mean*: 29.5; *Standard Deviation*: 8.20; *Suggested age range*: < 46
Phase 3	Decreased striae with transverse organization; coarse granularity; retroauricular activity present, beginnings of apical changes *Mean*: 42.0; *Standard Deviation*: 13.74; *Suggested age range*: < 69
Phase 4	Remnants of transverse organization; coarse granularity becoming replaced by densification; retroauricular activity present; apical changes; macroporosity is present *Mean*: 47.8; *Standard Deviation*: 13.95; *Suggested age range*: 20–75
Phase 5	Surface becomes irregular; surface texture is largely dense; moderate retroauricular activity; moderate apical changes; macroporosity *Mean*: 53.1; *Standard Deviation*: 14.14; *Suggested age range*: 24–82
Phase 6	Irregular surface; densification accompanied by subchondral destruction; severe retroauricular activity; severe apical changes; macroporosity *Mean*: 59.9; *Standard Deviation*: 15.24; *Suggested age range*: 29–89

LEARNING CHECK

Q7. Based on your understanding of age-related changes in the auricular surface, which of the individuals in **Figure 8.12** had more billowing?

A) A
B) B

Q8. Based on your understanding of age-related changes in the auricular surface, which of the individuals in **Figure 8.12** had more porosity? Consider the entire auricular surface joint but focus especially on the lower half.

A) A
B) B

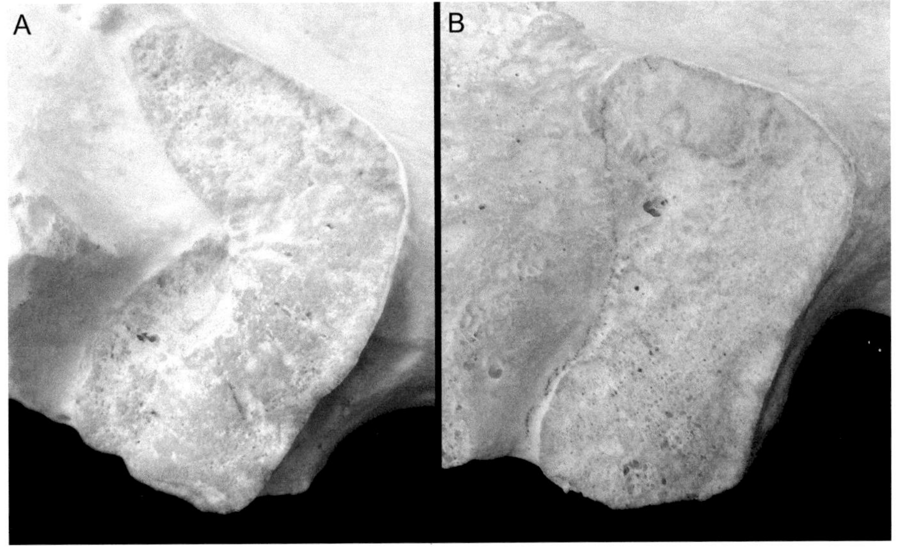

Figure 8.12. Two auricular surfaces of different ages, inferior is down, posterior is to the left.

Q9. Based on your understanding of age-related changes in the auricular surface, which of the individuals in **Figure 8.12** had more retroauricular activity?
A) A
B) B

Q10. Based on your responses above, which of the individuals in **Figure 8.12** was younger?
A) A
B) B

Cranial Suture Closure Age Estimation

The sutures between the cranial bones fuse together in a progressive manner as individuals age. Sutures gradually "close" through the addition of new bone across the suture line. Thus, what appears to be a dark line separating the two cranial bones slowly fills in with new bone and becomes more difficult to see. Cranial suture closure is one of the earliest developed skeletal techniques for age estimation; however, the method most often used is that published by Meindl and Lovejoy (1985), which was expanded on by Nawrocki (1998).

Unfortunately, cranial suture closure is less accurate than other age assessment methods. It only provides a crude approximation of age-at-death and should only be used when other methods are not available, such as when an isolated skull has been recovered. Whereas pubic symphysis, auricular surface, and sternal rib end methods rely on matching a decedent to a series of phases, the cranial suture method works by scoring multiple suture locations with regard to degree of closure, summing those scores, and comparing this composite score to known-age ranges. The use of a composite scoring system like this acknowledges the fact that

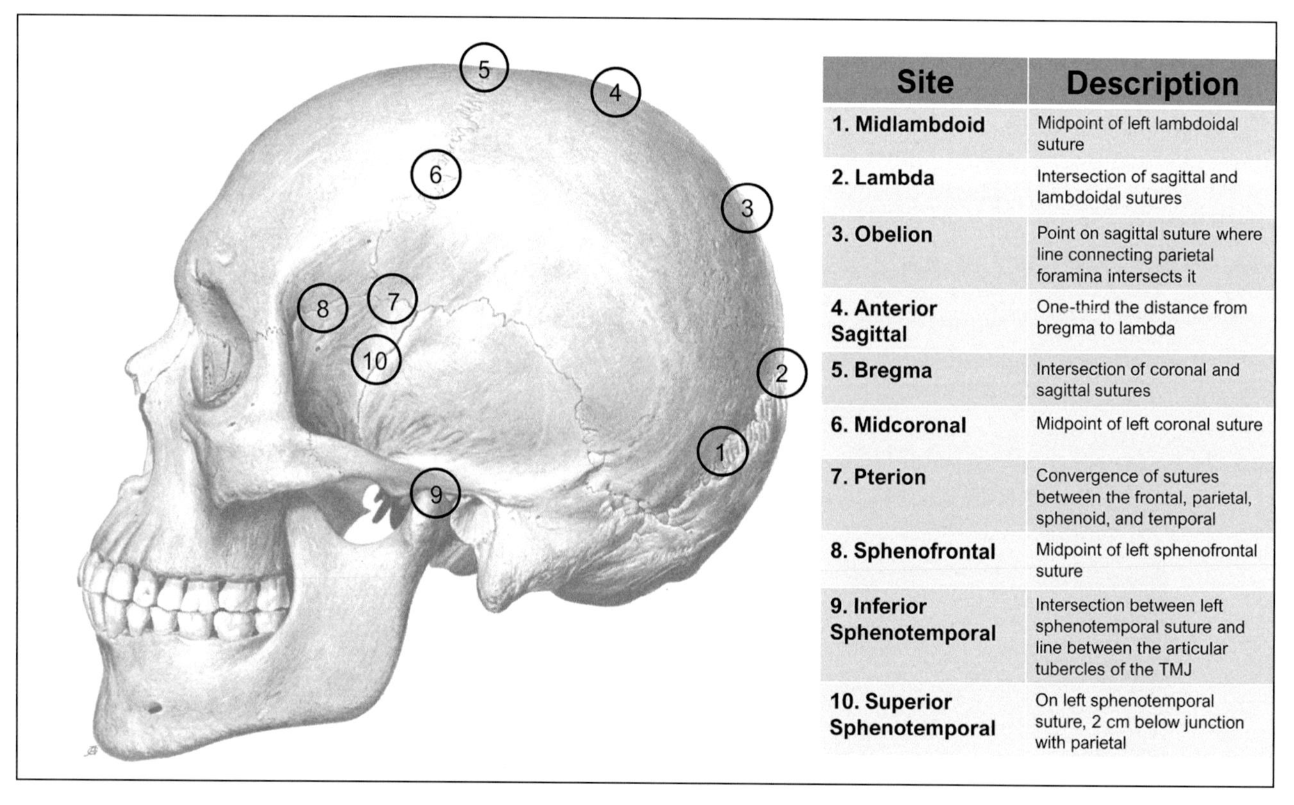

Site	Description
1. Midlambdoid	Midpoint of left lambdoidal suture
2. Lambda	Intersection of sagittal and lambdoidal sutures
3. Obelion	Point on sagittal suture where line connecting parietal foramina intersects it
4. Anterior Sagittal	One-third the distance from bregma to lambda
5. Bregma	Intersection of coronal and sagittal sutures
6. Midcoronal	Midpoint of left coronal suture
7. Pterion	Convergence of sutures between the frontal, parietal, sphenoid, and temporal
8. Sphenofrontal	Midpoint of left sphenofrontal suture
9. Inferior Sphenotemporal	Intersection between left sphenotemporal suture and line between the articular tubercles of the TMJ
10. Superior Sphenotemporal	On left sphenotemporal suture, 2 cm below junction with parietal

Figure 8.13. Meindl and Lovejoy (1985) suture observation locations with a description of where they are located. Technique summarized from Buikstra and Ubelaker (1994). Sites 1–5 are part of the vault scoring system and sites 6–10 are part of the lateral-anterior scoring system.

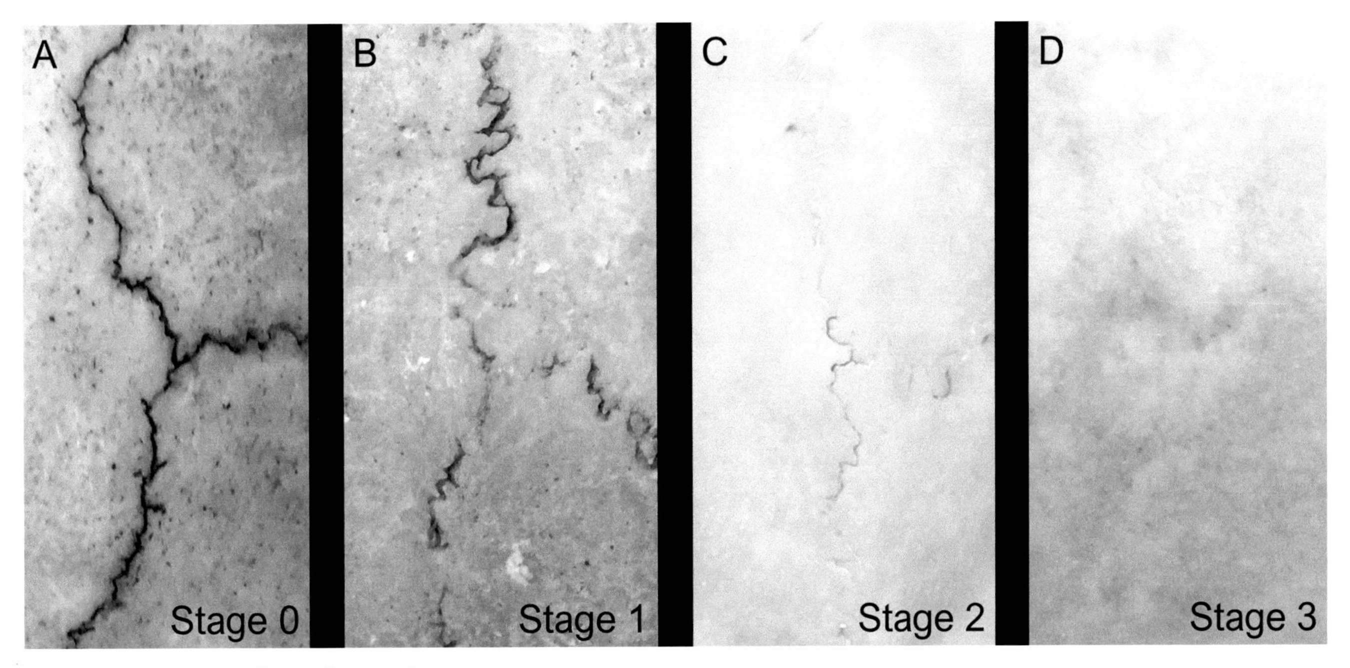

Figure 8.14. Four stages of cranial suture closure.

individual sutures contain very little useful information about age. For this reason, multiple suture sites are observed throughout the cranium (**Figure 8.13**), each being scored between 0 and 3. These are added together to generate a composite score for age estimation. Each suture is scored along a 1 cm section using the following stages.

0—open: the two bones do not touch and there is no bridging across the bones. The bones are separated by a dark and distinct line.
1—minimal closure: the bones are pushed together tightly but are still clearly defined, there are some bony bridges that have formed across the two bones.
2—significant closure: the suture is partially obliterated.
3—obliterated: little trace of the suture remains.

Figure 8.14 illustrates the four stages of suture closure. This is the view of the top of the skull near bregma, where the frontal bone meets the parietal bones (point #5 in **Figure 8.13**).

LEARNING CHECK

Q11. Evaluate the degree of cranial suture closure, paying attention to how open the suture is and any evidence for bone bridges forming. Are there gaps in the suture that you can see in this image? Please use the four stages of suture closure from **Figure 8.14** to score the individual shown in **Figure 8.15A**.
A) 0—open
B) 1—minimal closure
C) 2—significant closure
D) 3—obliterated

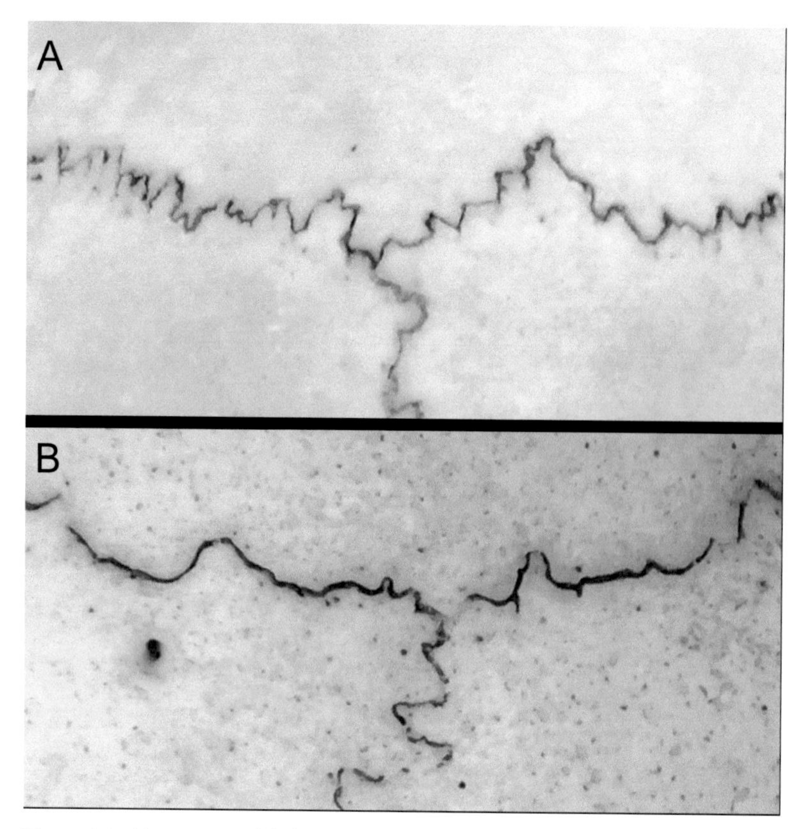

Figure 8.15. Two crania of different ages showing a close-up of the cranial sutures. The intersection of the two sutures is the osteometric point bregma.

Q12. Evaluate the degree of cranial suture closure, paying attention to how open the suture is and any evidence for bone bridges forming. Are there gaps in the suture that you can see in this image? Please use the four stages of suture closure from **Figure 8.14** to score the individual shown in **Figure 8.15B**.

A) 0—open
B) 1—minimal closure
C) 2—significant closure
D) 3—obliterated

Combining Multiple Indicators

Thus far we have focused on applying individual methods. However, in well-preserved remains the forensic anthropologist may have multiple age estimates from different methods. How does one combine these different age estimates into a single estimate, and how does one do so in a statistically valid manner? It is generally recognized that **multifactorial aging methods** are more accurate than single indicator ages; however, this is not always the case. Combining age estimates from different methods is statistically complex and beyond the scope of this book. Suffice it to say one should not just take the average of the different age estimates. **Transition analysis** is one of the more promising methods worth considering (Boldsen et al., 2002). Transition analysis uses complex statistical methods to combine characteristics of each of the scoring systems discussed above as well as information from a suite of other generalized age indicators, such as observation of the late fusing epiphysis of the medial clavicle or the presence and distribution of osteoarthritis. This method has been implemented in freely available software called Transition Analysis 3 (or TA3) that forensic anthropologists can use to generate a multifactorial age-at-death estimate, even when the remains are fragmentary and certain observations are missing. Importantly, the TA3 software is still being revised and expanded as a result of ongoing research.

End-of-Chapter Summary

In this chapter, four methods of adult age assessment were presented: 1) pubic symphysis morphology, 2) sternal fourth rib morphology, 3) auricular surface morphology, and 4) cranial suture closure. These methods share an anatomical and biological basis related to joint deterioration and breakdown. None of these methods are as accurate as age estimation techniques for subadult decedents, which are based on growth and development and not joint deterioration. Although not discussed in detail here, more advanced methodological training addresses how to handle error ranges for these techniques, which typically cover a decade or more. The older the individual was when he or she died, the larger the error range generally will be. For example, an age estimate for a 5-year-old could have an error range measured in months (e.g., ± 18–24 months), whereas the range of error for an 80-year-old is more likely to be measured in decades (e.g., ± 10 years).

End-of-Chapter Exercises

Exercise 1

Materials Required: **Figure 8.16, Tables 8.1** and **8.2.**

Directions: The left pubic bones of three different individuals are shown in **Figure 8.16,** with the symphyseal face shown on the left and the ventral surface shown on the right. Use what you have learned in chapter 6 and the Hartnett-Fulginiti method (**Tables 8.1** and **8.2**) to answer the following questions.

Question 1: What is the sex of Individual A? Based on the characteristics of their pubic bone, which Hartnett-Fulginiti phase should they be assigned to? What is the age range associated with this phase?

Question 2: What is the sex of Individual B? Based on the characteristics of their pubic bone, which Hartnett-Fulginiti phase should they be assigned to? What is the age range associated with this phase?

Question 3: What is the sex of Individual C? Based on the characteristics of their pubic bone, which Hartnett-Fulginiti phase should they be assigned to? What is the age range associated with this phase?

Exercise 2

Materials Required: **Figure 8.17, Tables 8.1–8.4.**

Scenario: After receiving a tip concerning the location of the body of a missing 37-year-old male, law enforcement discovered a box containing a pile of human bones within a storage unit. They have provided you with a pubic bone and the sternal end of a fourth rib as shown in **Figure 8.17**.

Question 1: Based on what you learned in chapter 6 and using the Hartnett-Fulginiti method (**Tables 8.1** and **8.2**), is the pubic bone shown in **Figure 8.17** consistent with the profile of the missing person?

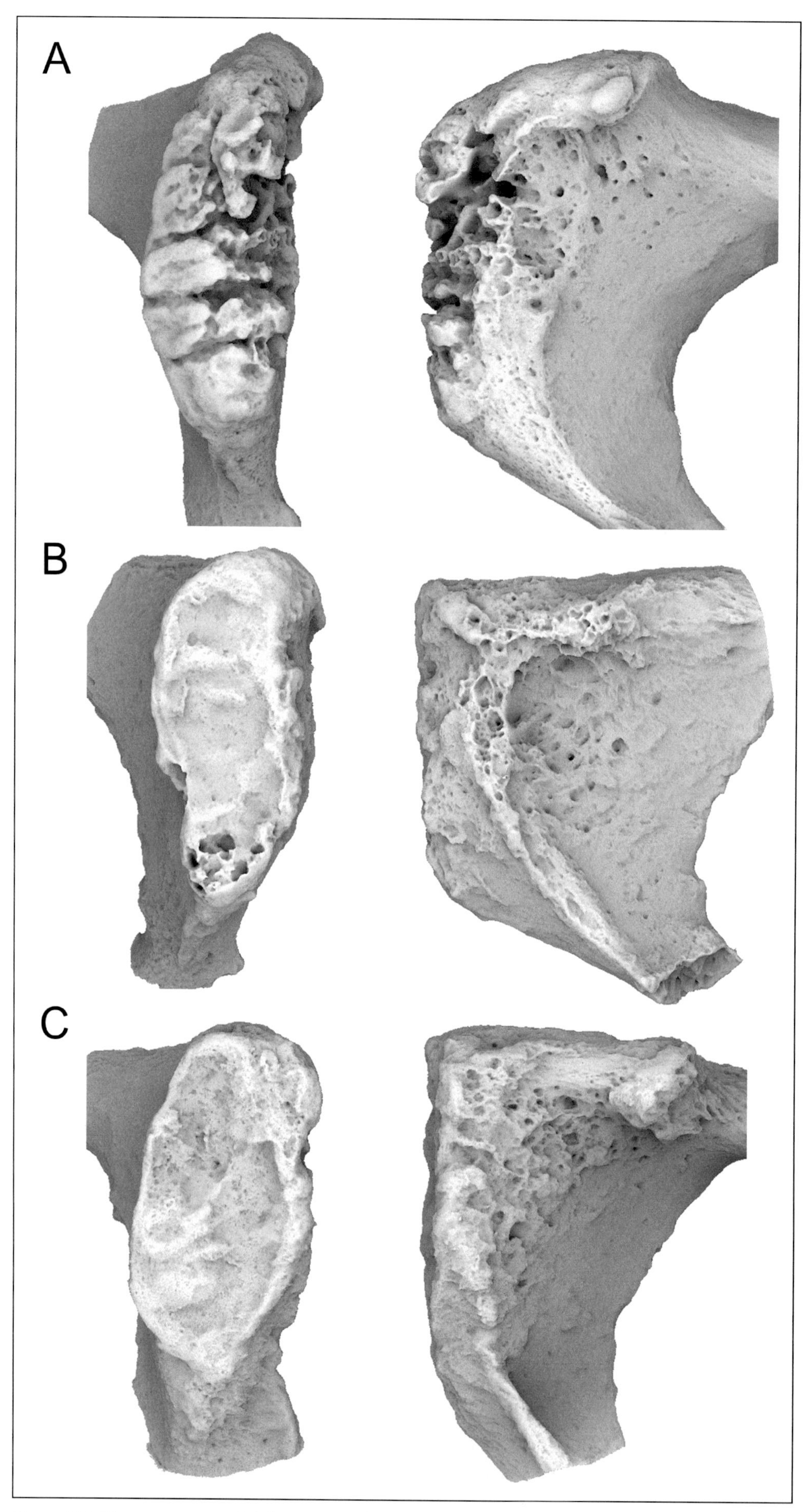

Figure 8.16. Three ossa coxae of different ages. The medial view of the pubic symphysis is on the left and the ventral view is on the right.

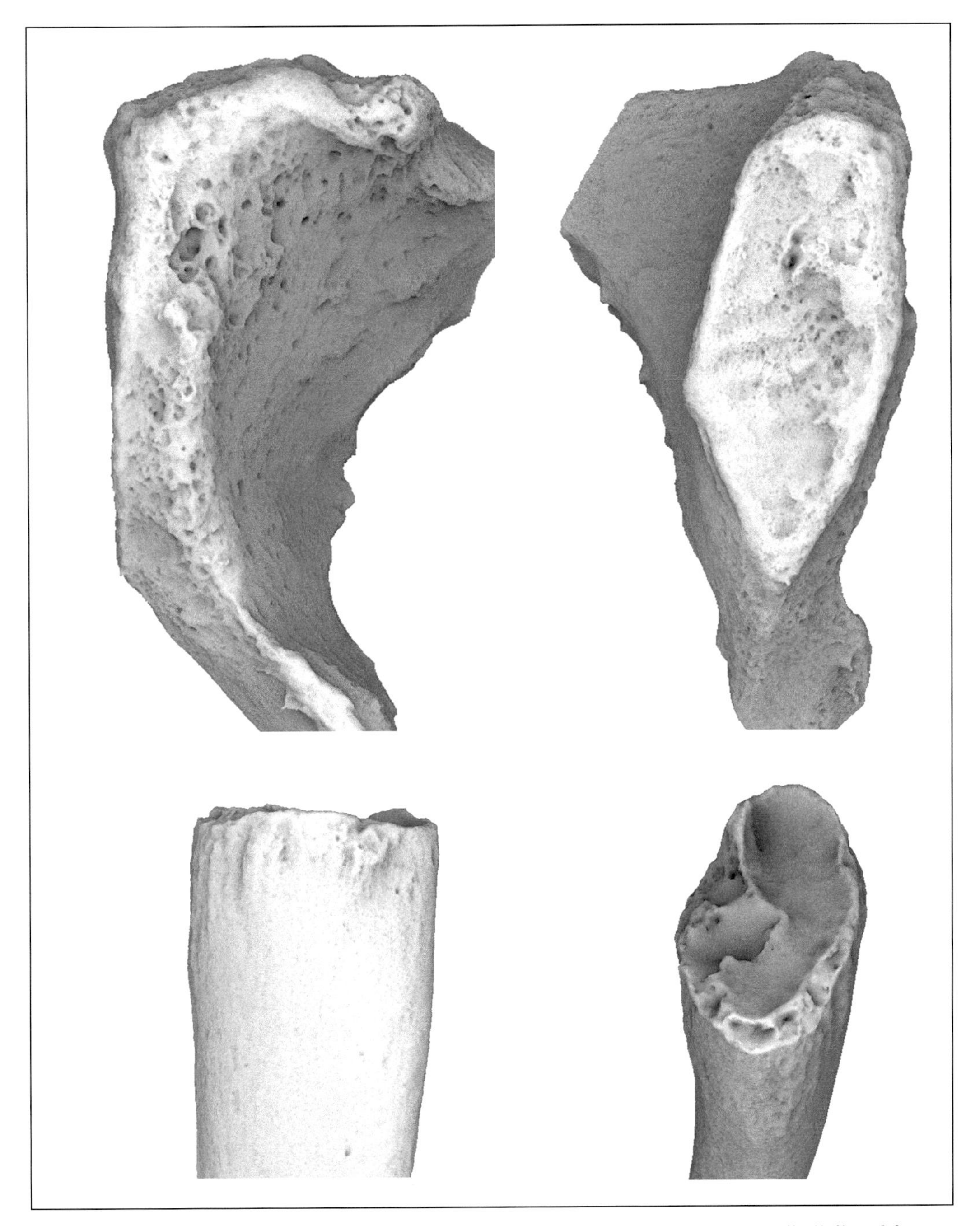

Figure 8.17. Pubic bone (*top*) and sternal rib end (*bottom*). The pubic bone is shown ventrally (*left*) and from the medial (*right*). The rib end is shown anteriorly (*left*) and from the sternal end (*right*).

Question 2: Using the Hartnett-Fulginiti method (**Tables 8.3** and **8.4**), is the sternal rib shown in Figure 8.19 consistent with the profile of the missing person?

Question 3: Based on your answers to the previous two questions, what is the MNI of the bones sent to you for analysis?

Exercise 3

Materials Required: **Figure 8.13** and **Figure 8.18**

Scenario: While searching the same storage unit as in the previous exercise, law enforcement discovered a second box containing a human cranium. Morphologically, the cranium is consistent with a male, but is its age-at-death consistent with that of the missing person?

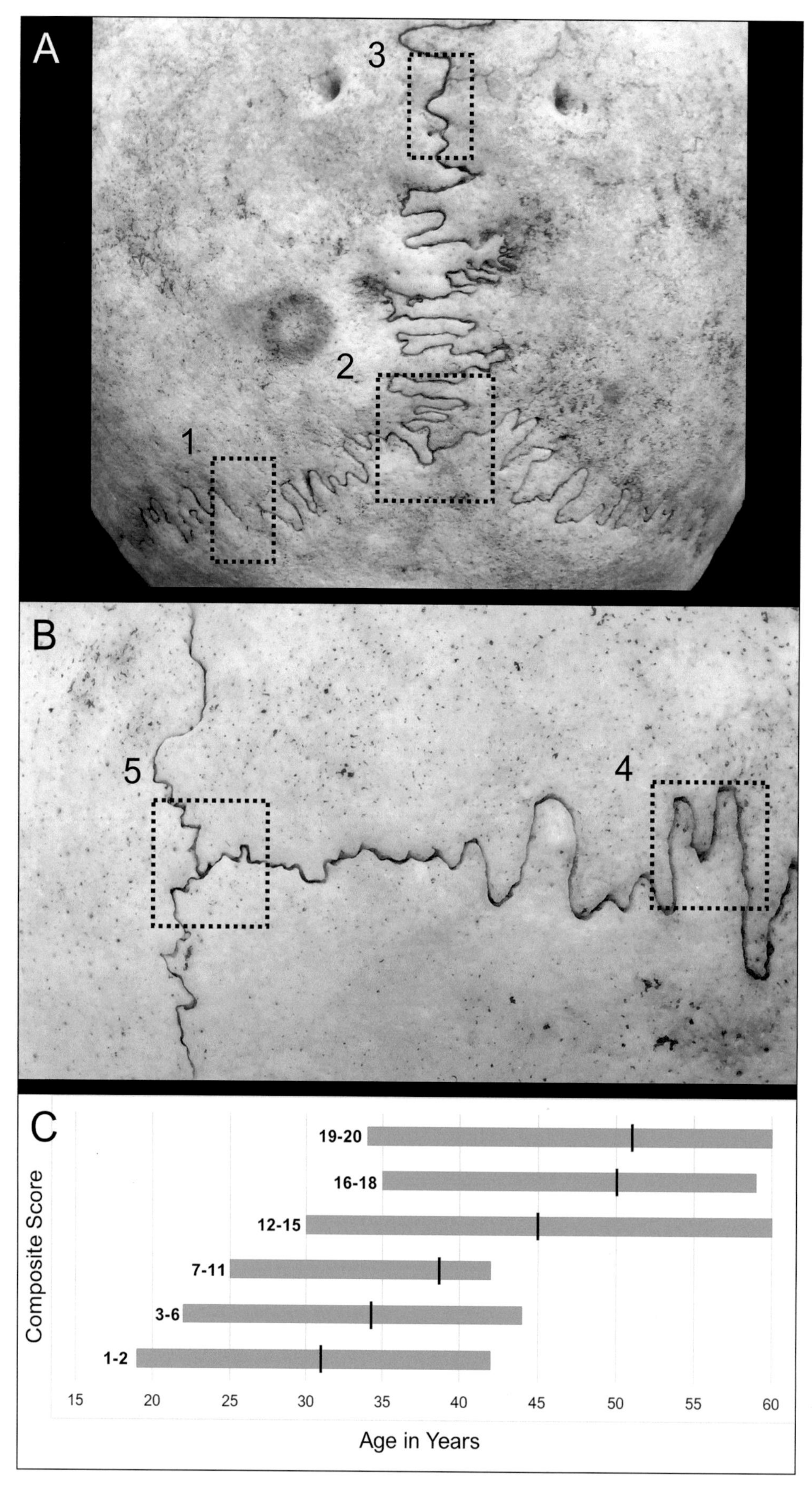

Figure 8.18. Close-up view of five cranial suture scoring locations. (*A*) posterior view of skull, (*B*) superior view of skull with bregma (5) and sagittal suture (4) indicated. (*C*) presents the scoring methodology based on summed composite scores.

Use **Figure 8.18** and the cranial suture scoring system presented in this chapter to answer the following questions.

Question 1: **Figure 8.18A** and **8.18B** show cranial suture sites 1 through 5 as defined in **Figure 8.13**. For each site, score the section of suture enclosed in the black box. Add your scores from the five sites together to get a composite score. What is the composite score for this skull (hint: it will be a number between 1 and 20)?

Question 2: **Figure 8.18C** shows the mean ages (black lines) and age ranges (blue bars) associated with each cranial suture composite score. Based on this figure, what is the age range associated with the composite score that you obtained in Question 1?

Question 3: Based on its pattern of cranial suture closure, is the age-at-death of this cranium consistent with that of the missing person? Is the age of the missing person *in*consistent with any of the possible composite scores for cranial suture closure?

References

Berg GE. 2008. Pubic bone age estimation in adult women. *Journal of Forensic Sciences* 53: 569–577.

Boldsen JL, Milner GR, Konigsberg LW, Wood JW. 2002. Transition analysis: a new method for estimating age from skeletons. In: Hoppa RD, Vaupel J (Eds.), *Palaeodemography: Age Distributions from Skeletal Samples.* Cambridge: Cambridge University Press. p. 73–106.

Brooks ST, Suchey JM. 1990. Skeletal age determination based on the os pubis: a comparison of the Acsádi-Nemeskéri and Suchey-Brooks methods. *Journal of Human Evolution* 5: 227–238.

Buckberry JL, Chamberlain AT. 2002. Age estimation from the auricular surface of the ilium: a revised method. *American Journal of Physical Anthropology* 119: 231–239.

Buikstra JE, Ubelaker DH (Eds.). 1994. *Standards for Data Collection from Human Skeletal Remains.* Fayetteville: Arkansas Archeological Survey.

DiGangi EA, Bethard JD, Kimmerle EH, Konigsberg LW. 2009. A new method for estimating age-at-death from the first rib. *American Journal of Physical Anthropology* 138: 164–176.

Garvin HM, Passalacqua NV. 2012. Current practices by forensic anthropologists in adult skeletal age estimation. *Journal of Forensic Sciences* 57: 427–433.

Gilbert BM, McKern TW. 1973. A method for aging the female *Os pubis. American Journal of Physical Anthropology* 38: 31–38.

Hartnett KM. 2007. *A Re-evaluation and Revision of Pubic Symphysis and Fourth Rib Aging Techniques.* PhD dissertation, School of Human Evolution and Social Change, Arizona State University, Tempe.

Hartnett KM. 2010a. Analysis of age-at-death estimation using data from a new, modern autopsy sample—part I: pubic bone. *Journal of Forensic Sciences* 55: 1145–1151.

Hartnett KM. 2010b. Analysis of age-at-death estimation using data from a new, modern autopsy sample—part II: sternal end of the fourth rib. *Journal of Forensic Sciences* 55: 1152–1156.

Igarashi Y, Uesu K, Wakebe T, Kanazawa F. 2005. New method for estimation of adult skeletal age at death from the morphology of the auricular surface of the ilium. *American Journal of Physical Anthropology* 128: 324–339.

İşcan MY, Loth SR, Wright RK. 1984a. Metamorphosis at the sternal rib end: a new method to estimate age at death in white males. *American Journal of Physical Anthropology* 65: 147–156.

İşcan MY, Loth SR, Wright RK. 1984b. Age estimation from the rib by phase analysis: white males. *Journal of Forensic Sciences* 29: 1094–1104.

İşcan MY, Loth SR, Wright RK. 1985. Age estimation from the rib by phase analysis: white females. *Journal of Forensic Sciences* 30: 853–863.

İşcan MY, Loth SR, Wright RK. 1987. Racial variation in the sternal extremity of the rib and its effect on age determination. *Journal of Forensic Sciences* 32: 452–466.

Lovejoy CO, Meindl RS, Pryzbeck TR, Mensforth RP. 1985. Chronological metamorphosis of the auricular surface of the ilium: a new method for the determination of adult skeletal age at death. *American Journal of Physical Anthropology* 68: 15–28.

McKern TW, Stewart TD. 1957. Skeletal age changes in young American males. In *Headquarters Quartermaster Research and Development and Command Technical Report RP-45.* Natick, MA.

Meindl RS, Lovejoy CO. 1985. Ectocranial suture closure: a revised method for the determination of skeletal age at death based on the lateral-anterior sutures. *American Journal of Physical Anthropology* 68: 57–66.

Nawrocki SP. 1998. Regression formulae for the estimation of age from cranial suture closure. In: Reichs KJ (Ed.), *Forensic Osteology: Advances in the Identification of Human Remains.* Second edition. Springfield, IL: Charles C. Thomas. p. 276–292.

Osborne DL, Simmons TL, Nawrocki SP. 2004. Reconsidering the auricular surface as an indicator of age at death. *Journal of Forensic Sciences* 49: 905–911.

Sobotta J. 1909. *Atlas and Text-Book of Human Anatomy.* Philadelphia, PA: W.B. Saunders Co.

Todd TW. 1920. Age changes in the pubic bone: I. The male white pubis. *American Journal of Physical Anthropology* 3: 285–334.

Todd TW. 1921. Age changes in the pubic bone. *American Journal of Physical Anthropology* 4: 1–70.

Yoder C, Ubelaker DH, Powell JF. 2001. Examination of variation in sternal rib end morphology relevant to age assessment. *Journal of Forensic Sciences* 46: 223–227.

9

Ancestry Estimation

Learning Goals

By the end of this chapter, the student will be able to:

Describe patterns of global human biological variation.
Explain why skin color does not reflect biological relationships.
Explain why human races are social rather than biological categories.
Differentiate between morphoscopic and metric methods of ancestry estimation.
Describe the basic mechanics of discriminant analysis.
Evaluate the reliability of an ancestry estimate.
Identify the limitations of ancestry estimation methods.

Introduction

This chapter is intended to convey the complexities of ancestry estimation within the practice of forensic anthropology. Ancestry, one of the primary components of the biological profile, is a topic that must be contextualized within broader discussions of human variation and the concept of race. For this reason, this chapter differs from others in this book in that it attempts to provide some necessary components of this background before presenting an overview of some of the more widely used methods of ancestry estimation. Rather than focusing on the description of morphological traits or anatomical landmarks incorporated into standardized measurements, this chapter places an emphasis on how to interpret ancestry estimates and understand their limitations.

Due to its entanglement with the concept of race, ancestry estimation has been a controversial topic for decades (e.g., Armelagos and Goodman, 1998; Bethard and DiGangi, 2020; Kennedy, 1995; Sauer, 1992; Smay and Armelagos, 2000; Stull et al., 2021) and some forensic anthropologists have recently called for its abandonment within forensic casework (DiGangi and Bethard, 2021). As of this writing, however, proficiency in the methods used to estimate ancestry is required for certification from the American Board of Forensic Anthropology (ABFA) and, therefore, is implicitly involved in the accreditation process of the National Association of Medical Examiners (NAME). As such, ancestry estimation continues to play a role in the contemporary practice of forensic anthropology.

As used in forensic anthropology, **ancestry** refers to the geographic origins of an individual's ancestors (Christensen et al., 2019; Sauer et al., 2016). The estimation of ancestry from skeletal remains is possible because human variation exhibits geographic patterning when viewed over long distances (Li et al., 2008; Ousley et al., 2009; Relethford, 2009; Templeton, 2013). Importantly, this same patterning provides some of the strongest evidence *against* the existence of biological races among humans. Within biology, the word *race* is generally used as a synonym for *subspecies* (Templeton, 2013). It is worth clarifying, here, that the prefix *sub-* does not imply something that is inferior, but simply indicates a meaningful division *within* a species. Current biological definitions of subspecies are based on the structure of a species' genetic variation (Templeton, 2013).

Human Biological Variation

Genetic variation refers to the totality of the different genes and the different versions (**alleles**) of those genes that are found among the members of a species. Portions of genetic variation exist within populations (i.e., among individuals in the same population) as well as between them. When the proportion of a species' genetic variation that exists between populations reaches a certain threshold, usually around 25 percent, those populations are considered to represent distinct subspecies, or races. Further, the genetic differentiation between populations must occur as a relatively abrupt change across a boundary (usually geographic) rather than a gradual change with increasing distance (Templeton, 2013). In humans, the proportion of genetic variation that exists between populations is generally reported to be around 15 percent (Barbujani and Colonna, 2010), but several studies suggest that it may be even lower (e.g., Elhaik, 2012; Li et al., 2008; Weir et al., 2005). In comparison, 38 percent of the genetic variation found among gorillas exists between populations. In fact, human populations are more closely related to each other than the populations of nearly every other primate species (Barbujani and Colonna, 2010).

Further, the 15 percent of genetic variation that exists between human populations is patterned in a particular way. Humans tend to mate with members of either their own population or a neighboring one. As the distance between populations increases, it becomes less likely that mate exchange will occur between them. Instead, the exchange of genes between distant populations tends to occur in a down-the-line fashion and takes many generations. The pattern that results from this is one in which the amount of genetic differentiation between two populations increases gradually with the geographic distance between them. In other words, all else being equal, neighboring populations tend to be more genetically similar than geographically distant ones. Human variation is, therefore, **clinal** and does not show abrupt transitions.

Figure 9.1 provides a simple illustration of this concept. In the top bar, populations B and C are more similar to one another than populations A and D and the differences between populations increase gradually with their distance from each other. Contrast this with the lower bar, where populations A and B are similar to one another but there is an abrupt change between populations B and C, despite their relative proximity in space. Where the patterning illustrated by the lower bar is what we would expect to see if human populations could be grouped into races, human biological variation is better approximated by the patterning shown in the top bar. This is a pattern that is referred to as isolation-by-distance, and it appears repeatedly in studies of human biological variation (e.g., Brace, 2005; Manica et al., 2007; Ousley et al., 2009; Relethford, 2004; Templeton, 2013). The frequency with which this patterning occurs indicates that human populations have never experienced the kind of extended reproductive isolation required for the development of distinct biological races.

There are, of course, exceptions to this trend. Skin color, for example, does not follow an isolation-by-distance pattern because it is an adaptive trait. It is also the quintessential "racial" trait in the United States. Different skin colors result primarily from differing amounts and kinds of melanin as well as the size and distribution of melanosomes (Jablonski, 2012). Eumelanin, a pigment that varies through several shades of brown, is particularly good at

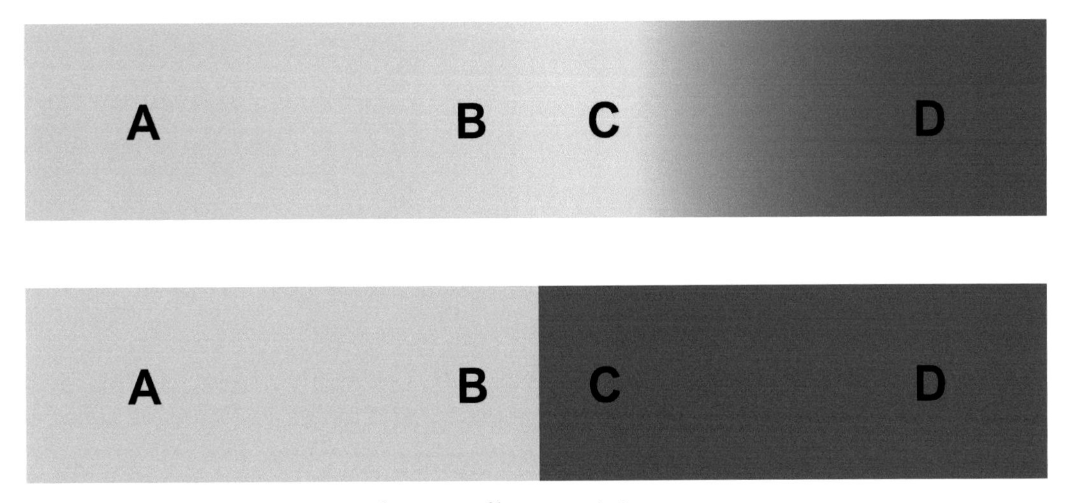

Figure 9.1. Schematic representation of patterns of human variation.

absorbing the free radicals produced by ultraviolet radiation (UVR) and, as such, plays a role in the prevention of skin cancer (Brace, 2005; Jablonski, 2012). More importantly, eumelanin helps to prevent the breakdown of folate by UVR. Folate is not only required for the maintenance of eggs and the production of sperm, but it is also necessary for healthy fetal growth and development (Jablonski, 2012; Jablonski and Chaplin, 2010). Darker skin pigmentation is therefore advantageous in regions characterized by high levels of UVR, such as those near the equator.

While ultraviolet radiation can cause skin cancer and deplete folate reserves in the body, it is also required for the synthesis of vitamin D. Vitamin D enables the absorption of calcium from the diet and is essential for the mineralization of the skeleton. Vitamin D also helps to maintain normal cellular function in the brain, the heart, and the immune system as well as other organs of the body. For human populations living at higher latitudes with lower overall UVR exposure and/or in regions with dramatic seasonal fluctuations in the amount of UVR that they receive, darker pigmentation may lead to vitamin D deficiency and related health problems. A loss of skin pigmentation, however, enables vitamin D production under such conditions (Jablonski, 2012).

The global distribution of skin color is therefore strongly correlated to both regional and seasonal levels of exposure to ultraviolet radiation, with darker pigmentation favored nearer the equator and gradual depigmentation with increasing latitude (see **Figure 9.2**) (Brace, 2005; Jablonski and Chaplin, 2010). This correlation is far from perfect, however, in that skin color is also mediated by social and behavioral factors. The degree of depigmentation exhibited by populations living in higher latitudes, for example, is affected by the amount of vitamin D that is in their diet. Populations whose diets are rich in vitamin D tend to exhibit less depigmentation (Jablonski, 2012). In India, cultural norms that dictate marriage patterns between different social classes have produced a pattern in which geographically close but socially distinct populations exhibit significantly different levels of skin pigmentation (Iliescu et al., 2018). This situation is further complicated by the complexity of the genetic architecture of skin pigmentation.

There are approximately 3 billion base pairs in the human genome. Between 25 percent and 38 percent of the difference in skin color between African and European populations is associated with a change in just one of these DNA bases (a *single nucleotide polymorphism*, or *SNP*) (Jablonski, 2012). While that may make it seem like the genetics of skin color are relatively straightforward, the polymorphisms that are associated with lighter skin tone in

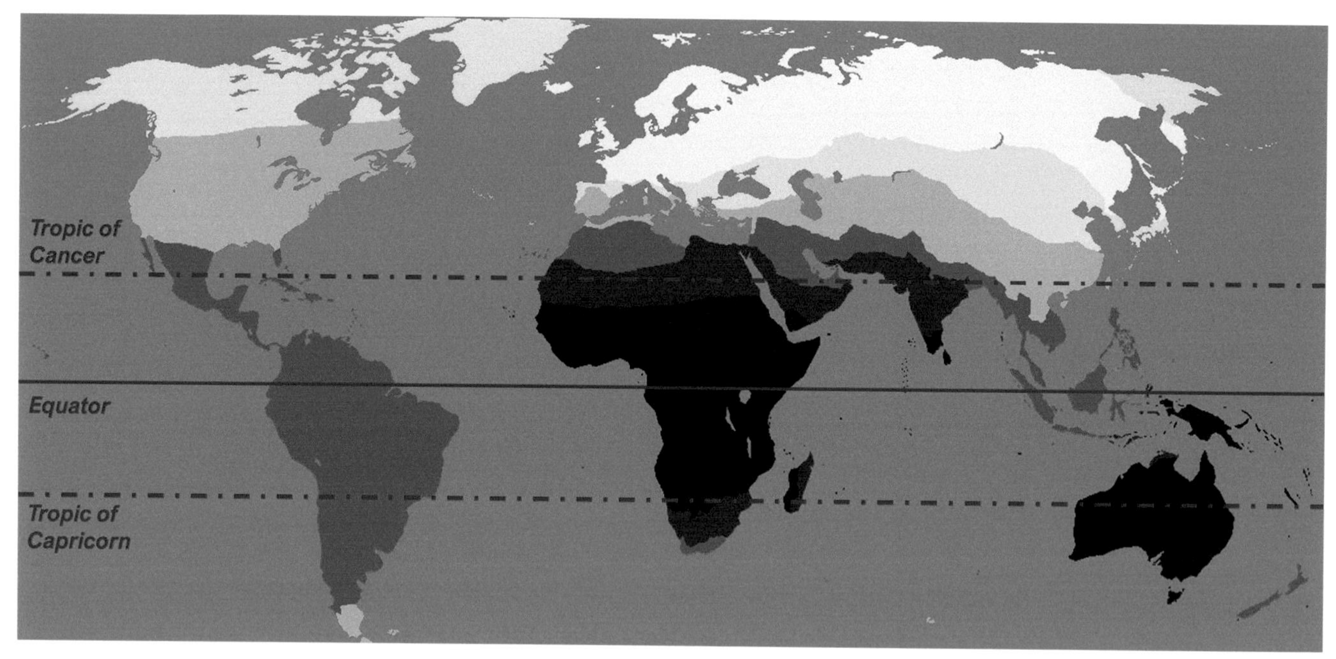

Figure 9.2. Global distribution of skin color.

Europe differ from those responsible for the lighter skin of populations in Asia (Jablonski, 2012; Quillen et al., 2019). Further, recent research has documented that the genes associated with skin pigmentation are often subject to *epistasis*—where the effects of one gene change depending on the presence or absence of other genes. Thus, a gene that increases pigmentation in one population may not have the same effect in a different population with different alleles (Quillen et al., 2019). Similar skin colors can therefore result from different combinations of genes.

Due to the adaptive nature of skin pigmentation and the complexity of the genes that contribute to it, skin color carries very little information about overall genetic similarity. Populations in Melanesia and sub-Saharan Africa, for example, exhibit very dark skin due to the high levels of UVR in those regions. In contrast, European populations tend to have lighter skin as the result of lower UVR exposure. Despite the differences in skin color, populations in sub-Saharan Africa are genetically more similar to European populations than they are to Melanesians (Relethford, 2010; Templeton, 2013). In general, characteristics that have been subject to natural selection exhibit geographic distributions that vary clinally in response to some selective pressure. As such, they tend to obscure rather than reflect underlying biological relationships (Brace, 2005).

Ancestry estimation in forensic anthropology is possible because of the geographic patterning of human biological variation. Yet it is this same patterning that, as discussed above, demonstrates that human populations are too interrelated to be or to ever have been considered as distinct biological races. So, are human races real? Not as a meaningful biological category. As social categories and lived experiences, however, races are *very* real.

LEARNING CHECK

Q1. Sickle cell trait is when an individual has inherited one normal version of the gene that makes hemoglobin and one version with a mutation that creates crescent-shaped red blood cells. While having two copies of the gene with the mutation results in sickle cell disease, sickle cell trait provides some protection against malarial infection. Since sickle cell trait is adaptive, which of the following would you expect to be true?

A) The geographic distribution of sickle cell trait among populations will only be affected by the biological relationships between populations.
B) Sickle cell trait should be more common among populations that are generally not exposed to malaria.
C) The geographic distribution of sickle cell trait among populations will not accurately reflect the biological relationships between them.
D) Sickle cell trait will be clinally distributed in response to variations in UVR.

Q2. Which of the following is an accurate statement about skin pigmentation?
A) Skin pigmentation does not follow a clinal distribution.
B) The genetic underpinning of skin pigmentation is simple.
C) Skin pigmentation varies more longitudinally than latitudinally.
D) Skin pigmentation is problematic as an indicator of overall genetic similarity and difference.

Race as a Social Category

Hubbard (2017) provides a simple and effective definition of race: a **race** (or a racial group) is "a culturally variable term describing a group of people who are *perceived* as sharing *biological features*." Similarly, she defines an **ethnicity** (or an ethnic group) as "a culturally variable term describing a group of people who share or are perceived to share *cultural features* (e.g., language, dress, cuisine, etc.)" (Hubbard, 2017: 519, emphases in original). Both race and ethnicity are examples of **folk taxonomies**, or the ways in which people categorize each other in their day-to-day interactions (Edgar, 2009). While residents of the United States tend to think of race and ethnicity in terms of the census categories that they are familiar with (see **Table 9.1**), other societies categorize race in different ways (e.g., Dikötter, 1992; Telles and Paschel, 2014).

Racial and ethnic categories also change over time. **Table 9.2** illustrates changes in the categories used on the U.S. census since its inception in 1790. The first census was primarily interested in how many free white people and how many slaves resided in each state; everyone else was lumped into a single "other free persons" category. Native Americans were not even included on the census until 1870. In the same year, "Chinese" was added as a census category, primarily in response to the influx of migrants working in railroad construction. "Mexican" first appeared as a census category in 1930, only to disappear and emerge as an ethnic category 40 years later. In 1997, "Native Hawaiian or Pacific Islander" was adopted as a new racial category as a compromise between political pressure from Hawaiians to be included in the American Indian category and concerns that this change would negatively impact the administration of federal programs for Native Americans (Office of Management and Budget, 1997). Currently, there is an ongoing discussion about the addition of a Middle Eastern and North African ethnic category (Federal Interagency Working Group, 2017). The historical changes of the United States census categories clearly demonstrate that racial and ethnic categories are fluid and largely contingent on both public perceptions and political interests.

While racial categories change depending on cultural and historical contexts, the concept of race used in the United States maintains the illusion of stability because of the way in which it groups people based on perceived biological similarities. For example, in the United States, skin color is the racial characteristic par excellence. This is evident in the fact that census questions used *race* and *color* interchangeably for nearly 200 years (see **Table**

Table 9.1. Standards for the classification of federal data on race and ethnicity (Office of Management and Budget, 1997)

Race	Description
American Indian or Alaska Native	Individuals who identify with any of the original peoples of North and South America (including Central America) and who maintain Tribal affiliation or community attachment. It includes people who identify as "American Indian" or "Alaska Native" and includes groups such as Navajo Nation, Blackfeet Tribe, Mayan, Aztec, Native Village of Barrow Inupiat Traditional Government, and Nome Eskimo Community.
Asian	Individuals who identify with one or more nationalities or ethnic groups originating in the Far East, Southeast Asia, or the Indian subcontinent. Examples of these groups include, but are not limited to, Chinese, Filipino, Asian Indian, Vietnamese, Korean, and Japanese. The category also includes groups such as Pakistani, Cambodian, Hmong, Thai, Bengali, Mien.
Black or African American	Individuals who identify with one or more nationalities or ethnic groups originating in any of the black racial groups of Africa. Examples of these groups include, but are not limited to, African American, Jamaican, Haitian, Nigerian, Ethiopian, and Somali. The category also includes groups such as Ghanaian, South African, Barbadian, Kenyan, Liberian, and Bahamian.
Native Hawaiian and Pacific Islander	Individuals who identify with one or more nationalities or ethnic groups originating in Hawaii, Guam, Samoa, or other Pacific islands. Examples of these groups include, but are not limited to, Native Hawaiian, Samoan, Chamorro, Tongan, Fijian, and Marshallese. The category also includes groups such as Palauan, Tahitian, Chuukese, Pohnpeian, Saipanese, Yapese
White	Individuals who identify with one or more nationalities or ethnic groups originating in Europe, the Middle East, or North Africa. Examples of these groups include, but are not limited to, German, Irish, English, Italian, Lebanese, Egyptian, Polish, French, Iranian, Slavic, Cajun, and Chaldean.

Ethnicity	Description
Hispanic or Latino	A person of Cuban, Mexican, Puerto Rican, South or Central American, or other Spanish culture or origin, regardless of race

Table 9.2. United States census categories through the years, after Pratt et al. (n.d.)

Year	Question	Response Categories
1790	N/A	Record the number of the following: free White males, free White females, other free persons, slaves
1850	Color	White, Black, or Mulatto
1870	Color	White, Black, Mulatto, Chinese, or Indian
1890	Race	White, Black, Mulatto, Quadroon, Octoroon, Chinese, Japanese, or Indian
1930	Color or Race	White, Negro, Mexican, Indian, Chinese, Japanese, Filipino, Hindu, Korean, or Other
1960	Race or Color	White, Negro, American Indian, Japanese, Chinese, Filipino, Hawaiian, Part Hawaiian, Aleut, Eskimo
1970	Color or Race	White, Negro or Black, Indian (Amer.), Japanese, Chinese, Filipino, Hawaiian, Korean, or Other
	Origin or Descent	Mexican, Puerto Rican, Cuban, Central or South American, other Spanish, or None of these
2000	Spanish/Hispanic/ Latino	Mexican, Mexican American, Chicano; Puerto Rican; Cuban; other Spanish/Hispanic/Latino
	Race	White; Black, African American, or Negro; American Indian or Alaska Native; Asian (Asian Indian, Chinese, Filipino, Japanese, Korean, Vietnamese, or Other Asian); Native Hawaiian or Pacific Islander (Native Hawaiian, Guamanian or Chamorro, Samoan, Other Pacific Islander); or Other
2020	Hispanic, Latino, or Spanish Origin	Mexican, Mexican American Chicano; Puerto Rican; Cuban; Another Hispanic, Latino, or Spanish origin
	Race	White; Black or African American; American Indian or Alaska Native; Asian (Chinese, Filipino, Asian Indian, Vietnamese, Korean, Japanese, Other Asian); Native Hawaiian or Pacific Islander (Native Hawaiian, Samoan, Chamorro, Other Pacific Islander); or Other

9.2) and that White and Black continue to be used as racial labels. Even though skin color is a poor proxy for genetic similarity, individuals with similar skin tones are perceived as being more closely related than individuals with differently pigmented skin. Thus, social races are misunderstood as having biological validity and differences between racial groups are mistakenly ascribed to differences in group biology (Caspari, 2009, 2010). This characterization is intimately related to the establishment of racial hierarchies and, consequently, racism (Blakey, 1999).

There is a substantial body of research that documents health disparities between racially defined groups in the United States. African Americans exhibit higher age-adjusted mortality rates and are at substantially greater risk than their white counterparts for a number of health concerns, including diabetes and hypertension. In addition, African Americans exhibit higher rates of infant mortality, lower average birth weights, and increased rates of prematurity (Earnshaw et al., 2013; Gravlee, 2009; Kuzawa and Sweet, 2009; Lukachko et al., 2014). While such disparities have often been attributed to genetic differences between racial groups, it has been demonstrated that racial discrimination can increase maternal stress and result in lower birth weight (Collins et al., 2004; Earnshaw et al., 2013). Lower birth weight, in turn, is associated with poorer health outcomes later in life, including an increased risk of developing cardiovascular disease (Kuzawa and Sweet, 2009). Such differences in health outcomes are exacerbated by structural inequalities that restrict access to education, income, credit, and employment for communities of color (Lukachko et al., 2014) (**Figure 9.3**). Racism, not genetic differences, may be the primary cause of health disparities among racial groups in the United States. Although race is a social category, it has biological consequences.

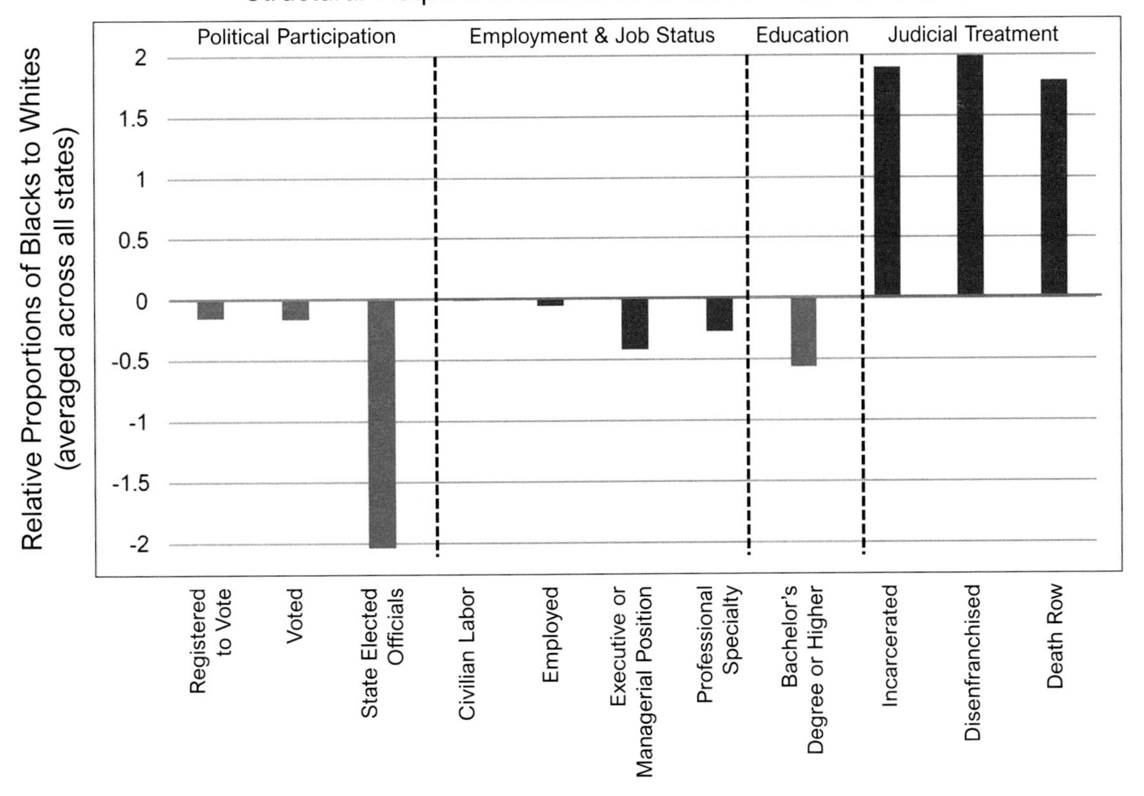

Figure 9.3. Structural inequalities in the United States after data from Lukachko and colleagues (2014). Values below zero indicate the status applies to proportionately more people who identify as white; values above zero indicate the status applies to proportionately more people who identify as Black.

While race is often assumed to reflect ancestry, the two are not the same. Yet, it is not uncommon for forensic anthropologists to conclude in their case reports that unidentified skeletal remains represent an individual who would likely have identified as Black (or white, or Native American, or some other racial category) in life. If race is only a social category, how is this possible? The major populations present throughout much of the history of the United States represent groups whose ancestral populations were separated by large geographic distances: Indigenous Native Americans, colonists from Western Europe, people sold into slavery from Western Africa, and immigrants from Eastern Asia (Edgar, 2009). Moreover, the racial folk taxonomy developed in the United States and the racist policies and practices that it generated have acted to keep gene flow between these populations relatively low.

For example, in 1967, interracial marriage was still illegal in 16 states (Ousley et al., 2009). These historical circumstances have produced a certain amount of agreement between social race and skeletal ancestry in the United States (SWGANTH, 2013). Forensic anthropologists make observations pertaining to morphological characteristics of the skeleton and estimate an individual's ancestry based on documented population differences that arise from the geographic patterning of human biological variation. They then translate the estimated ancestry to a racial or ethnic category (Edgar, 2014). While such translation inevitably incorporates a certain amount of error, it is aided by the fact that, in the United States, both racial and ethnic categories are currently defined in terms of ancestral geographic origins (see **Table 9.1**).

LEARNING CHECK

Q3. According to **Table 9.2,** the word "Race" first appeared in the United States Census in which year?

A) 1790
B) 1890
C) 1870
D) 1930

Q4. According to **Figure 9.3,** which of the following exhibits a uniformly high level of inequality between Black and white Americans?

A) Political participation
B) Employment and job status
C) Education
D) Judicial treatment

Ancestry Estimation

Contemporary methods for ancestry estimation involve the classification of an unknown decedent by comparing observed skeletal features to those of documented skeletal samples derived from various populations. As such, techniques are limited in their application by the reference samples on which they are based. For example, if a method is based on African and European samples, it can only classify unidentified remains into one of these two groups, regardless of their actual ancestry. For this reason, forensic anthropologists must be aware of the limitations of the methods that they use in their analyses and take these constraints into consideration when interpreting results. Further, current standards for best practices strongly suggest that ancestry estimations be accompanied by a statistical statement of the probability that an assessment is correct (SWGANTH, 2013). While several researchers have developed methods for ancestry estimation based on the postcranial skeleton (e.g., Baker et al., 1990; Gilbert and Gill, 1990; Holliday and Falsetti, 1999; Meeusen et al., 2015), these techniques are generally based on small sample sizes and have performed poorly in validation studies (e.g., Berg et al., 2007). Several researchers have likewise devised methods to estimate ancestry using features of the dentition (e.g., Edgar, 2005, 2013; Irish, 2015; Lease and Sciulli, 2005; Pilloud et al., 2014; Scott et al., 2018). While such methods are promising in that they have the potential to be applied to younger individuals, a lack of validation studies and the fact that few forensic anthropologists are currently trained in dental morphology have prevented their widespread adoption (Pilloud et al., 2019). The more commonly used methods for ancestry estimation are based on features of the skull.

Morphoscopic Methods

Morphoscopic traits are composite morphological characteristics of the skull that take on different forms or grades of expression, and can be scored as categorical, ordinal, or binary (i.e., presence/absence) variables. Until relatively recently, morphoscopic traits were embedded within a trait list approach to ancestry estimation that was uncritically derived from racially motivated typological studies where certain traits were assumed to be effectively diagnostic of specific ancestries (e.g., Rhine, 1990). Recent reappraisals of morphoscopic traits have demonstrated that they are far more variable than previously thought and exhibit some degree of trait intercorrelation (e.g., Hefner, 2009). Current morphoscopic methods for ancestry estimation are based on a reduced list of morphological characteristics that exhibit documented population differences in their frequency of occurrence as well as their mode or degree of expression. Standardized trait definitions and scoring procedures for morphoscopic traits are provided by Hefner (2009) and discussions of their heritability, gross anatomy, and functional morphology are presented by Hefner and Linde (2018).

The **Optimized Summed Scored Attributes** (**OSSA**) is a morphoscopic method that is frequently used in forensic anthropology. This method uses six morphoscopic traits for ancestry assessment: anterior nasal spine (ANS), inferior nasal aperture (INA), interorbital breadth (IOB), nasal aperture width (NAW), nasal bone contour (NBC), and post-bregmatic depression (PBD) (see **Figures 9.4** and **9.5**). All six traits must be observable for the method to be applied. OSSA proceeds by dichotomizing the recorded scores for each trait in such a way as to maximize the differences between its two reference samples—American Blacks and American whites. Scores for trait expressions that are more common among American Blacks are set to "0" and scores more frequently observed among American whites are set to "1." The scores for each of the six traits are then added together, and the sum is used to classify the decedent as either Black or white (Hefner and Ousley, 2014). While the original method description suggests that summed scores of 3 or less should be classified as Black, a validation study has suggested that classifying scores of 4 or less as Black yields a higher overall correct classification rate (79.2 percent) (Kenyhercz et al., 2017). Although advantageous in its simplicity, OSSA is limited by its reference samples to American Blacks and whites. Application of this method outside of these groups can be problematic (Pilloud et al., 2018).

Several other techniques have been developed for morphoscopic ancestry estimation using advanced computational techniques including decision tree modeling (Hefner and Ousley, 2014), naïve Bayesian classifiers (e.g., Herrmann et al., 2016; d'Oliveira Coelho and Navega, 2019), and artificial neural networks (Spiros and Hefner, 2020). For ease of use, many of these approaches have been operationalized as downloadable analytical worksheets or

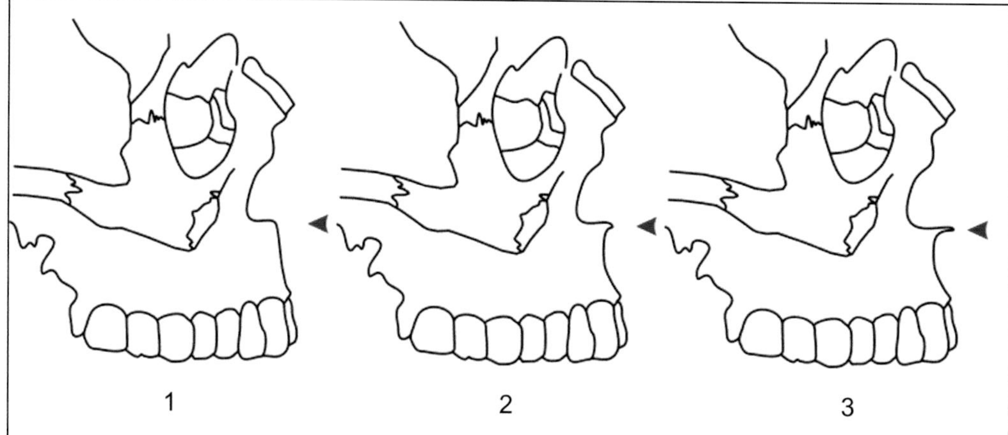

Anterior Nasal Spine (ANS): 1) Slight – minimal or no projection of the ANS beyond the inferior nasal aperture; 2) Intermediate – a moderate projection of the ANS beyond the inferior nasal aperture; 3) Marked – a pronounced projection of the ANS beyond the inferior nasal aperture.

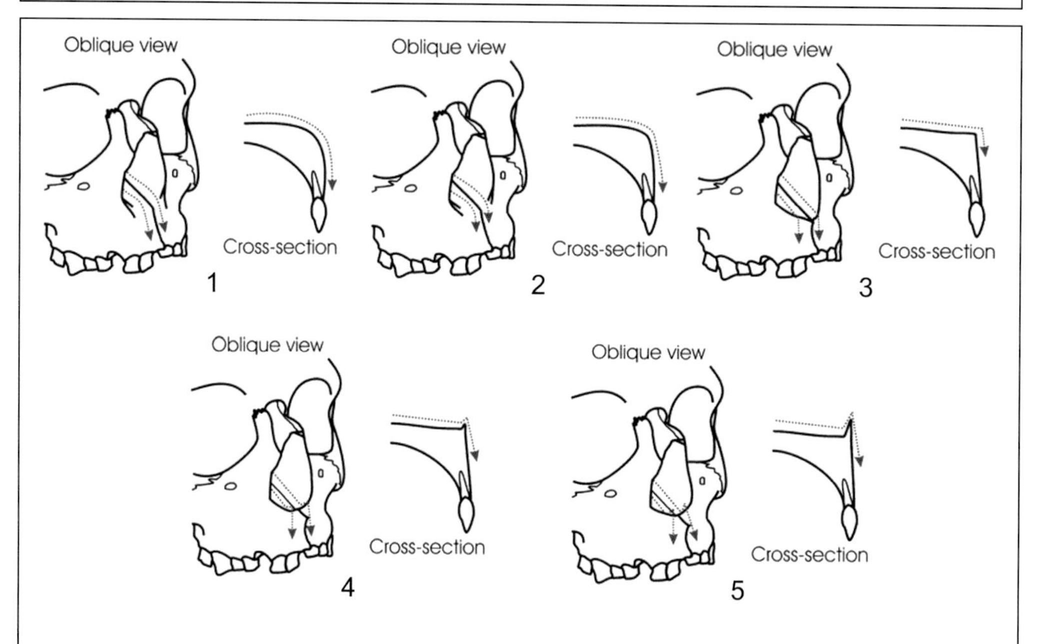

Inferior Nasal Aperture (INA): 1) An inferior sloping of the nasal floor which begins within the nasal cavity and terminates on the vertical surface of the maxilla, producing a smooth transition; 2) sloping of the nasal aperture beginning more anteriorly than in INA 1, and with more angulation at the exit of the nasal opening; 3) the transition from nasal floor to the vertical maxilla is not sloping, nor is there an intervening projection, or sill; 4) any superior incline of the anterior nasal floor, creating a weak (but present) vertical ridge of bone that traverses the inferior nasal border (partial nasal sill); 5) a pronounced ridge (nasal sill) obstructing the floor-to-maxilla transition.

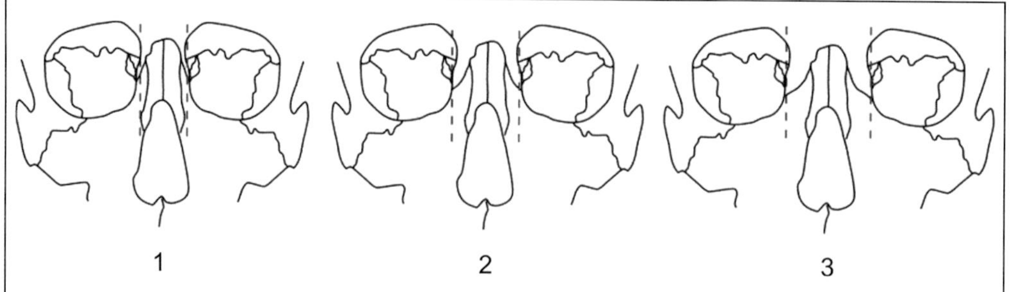

Interorbital Breadth (IOB): This assessment is made relative to the facial skeleton. 1) the IOB is narrow; 2) the IOB is intermediate, and 3) the IOB is broad.

Figure 9.4. Morphoscopic traits used in OSSA, Part I.

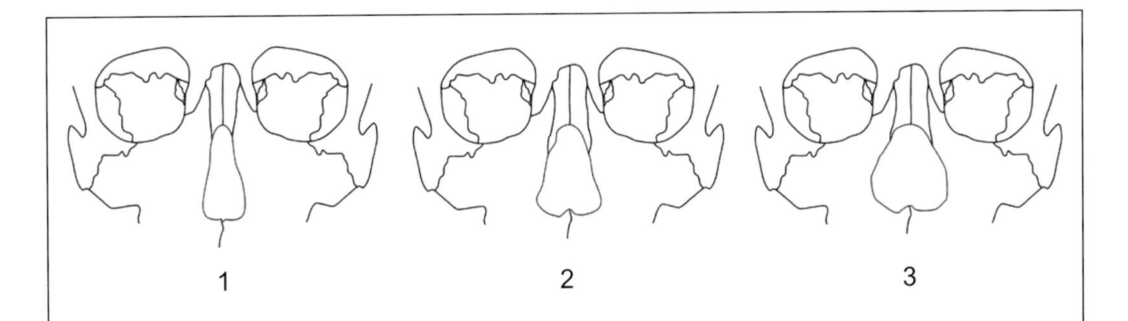

Nasal Aperture Width (NAW): This assessment is made relative to the facial skeleton. 1) nasal aperture width is narrow; 2) nasal aperture width is medium; 3) nasal aperture width is broad.

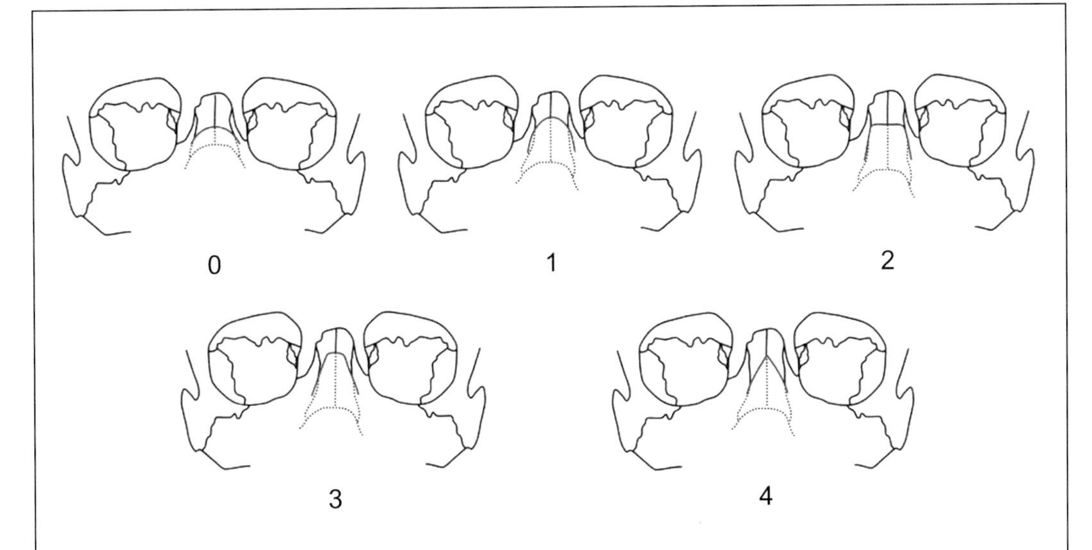

Nasal Bone Contour (NBC): The contour of the nasal bones and the frontal processes of the maxillae about 1cm below the anatomical landmark nasion. Most reliably assessed by using a contour gauge oriented perpendicular to the palate and parallel to the orbits. 0) the NBC is low and rounded; 1) an oval contour, with elongated, high, and rounded lateral walls; 2) steep lateral walls and a broad (roughly 7 mm or more), flat superior surface "plateau"; 3) steep-sided lateral walls and a narrow superior surface "plateau"; 4) triangular cross-section, lacking a superior surface "plateau."

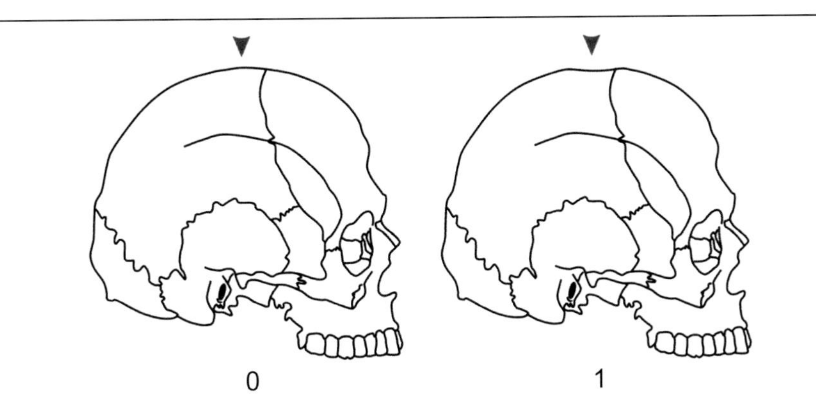

Postbregmatic Depression (PBD): A slight to broad depression along the sagittal suture, posterior to the anatomical landmark bregma, that is not pathological in origin. Observed in lateral profile. 0) a postbregmatic depression is absent; 1) a postbregmatic depression is present.

Figure 9.5. Morphoscopic traits used in OSSA, Part II.

web-based applications. Most of these methods, however, are currently not appropriate for forensic casework as they have yet to be validated (Dunn et al., 2020).

Craniometric Methods

Craniometric methods of ancestry estimation are based on measurements or indices (i.e., combinations of measurements) of the skull. The more widely used craniometric methods employ **linear discriminant analysis (LDA)** to classify an unknown decedent into one of several groups. In the context of forensic anthropology, LDA assumes that it is possible to separate different populations of people based on some suite of measurements. The procedure works by finding the linear combination of the original measurements that maximizes the differences between the groups included in the analysis. This combination of variables is called the **discriminant function** (Shennan, 1997). A discriminant function is used to calculate discriminant function scores for each member of the predefined groups and each group is subsequently characterized by its **centroid**, or the mean value of its discriminant function scores. After calculating the discriminant function score of an unknown case, it is classified into the group whose centroid is closest to its own position in multivariate space (Klecka, 1980) (see **Figure 9.6**).

Early applications of LDA for estimating ancestry (e.g., Giles and Elliot, 1962) were based on unrepresentative samples and it was soon recognized that their application to forensic cases was limited (e.g., Ayers et al., 1990; Birkby, 1966; Fisher and Gill, 1990). A desire to be able to generate custom discriminant functions based on contemporary forensic casework led to the development of the computer program Fordisc (Jantz and Ousley, 2005; Ousley

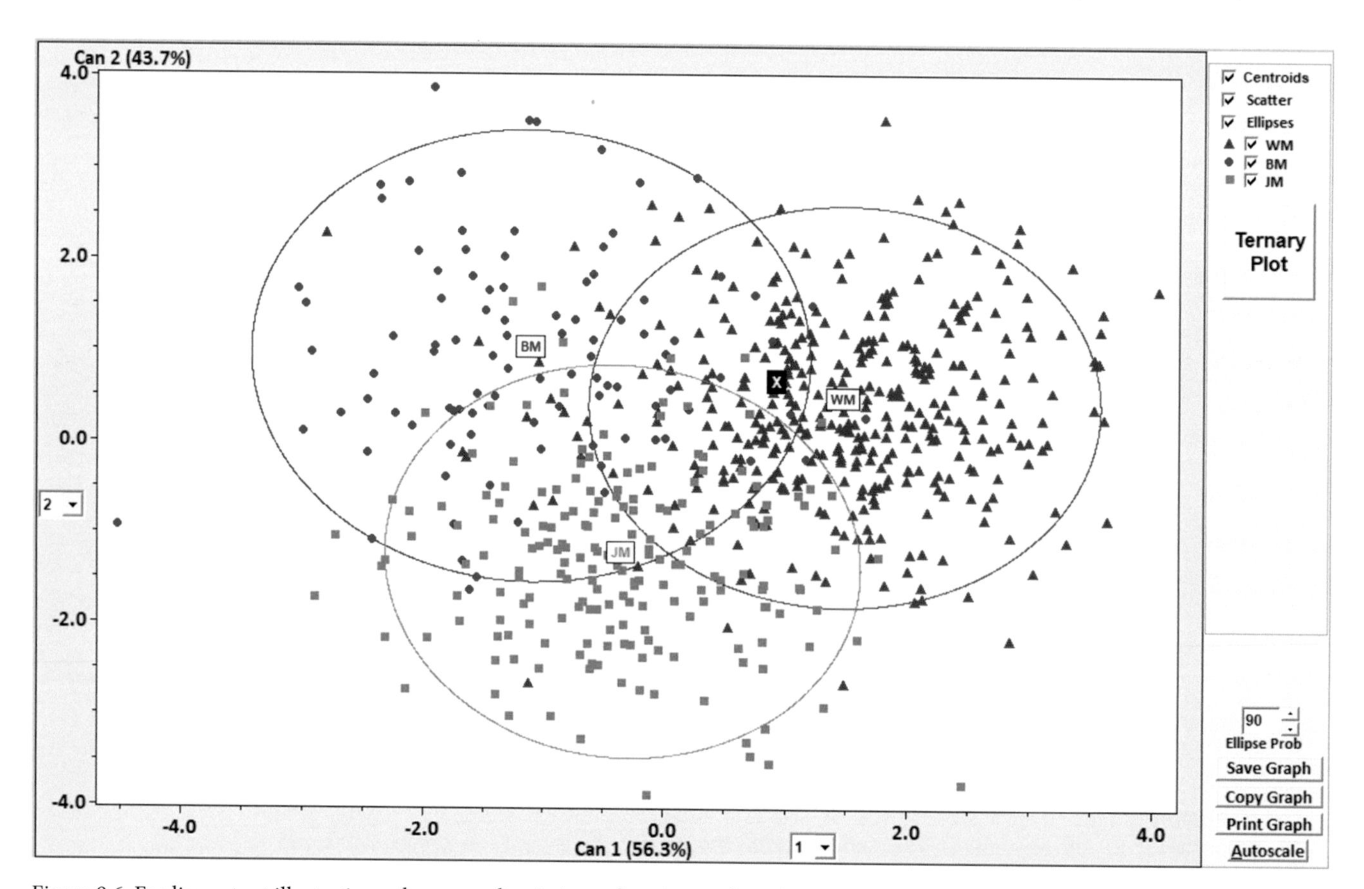

Figure 9.6. Fordisc output illustrating a three-way discriminant function analysis for craniometric ancestry estimation. The labels "BM," "JM," and "WM" represent the group centroids for Black males, Japanese males, and white males, respectively. The position of the unknown case is indicated by the white X in a black rectangle. In this example, the unknown case would be classified as a white male.

and Jantz, 2012). The current version, Fordisc 3.1, uses the **Forensic Anthropology Data Bank (FDB)** as its primary reference database. At its inception, the FDB contained information pertaining to 715 individuals whose remains had been analyzed by forensic anthropologists in the United States (Moore-Jansen and Jantz, 1986). Today, the FDB includes information derived from over 3,400 individuals, more than 2,400 of whom are positively identified and therefore of known sex and ancestry. Forensic anthropologists are encouraged to submit data from their positively identified cases to the FDB to enable its continued growth and refinement (SWGANTH, 2013). The sheer amount of information contained in the FDB gives Fordisc a competitive edge when compared to other methods of ancestry estimation.

Ancestry assessment using Fordisc generally follows the same basic procedure (Ousley and Jantz, 2012):

Record a standardized set of cranial measurements.
Check the accuracy of the measurements obtained.
Check the sample sizes of the populations being compared.
Run and refine analyses.
Evaluate the results.

Definitions of the standard measurements involved in ancestry estimation and descriptions of the cranial landmarks upon which they are based are provided by Langley and colleagues (2016) and freely available online (see Langley et al., 2016 in the references for a weblink to this resource; see also **Figures 9.7** and **9.8**). The process of collecting the measurements is relatively straightforward but does require specialized equipment in the form of highly precise and accurate calipers. There are minimally two calipers needed (spreading and sliding—see

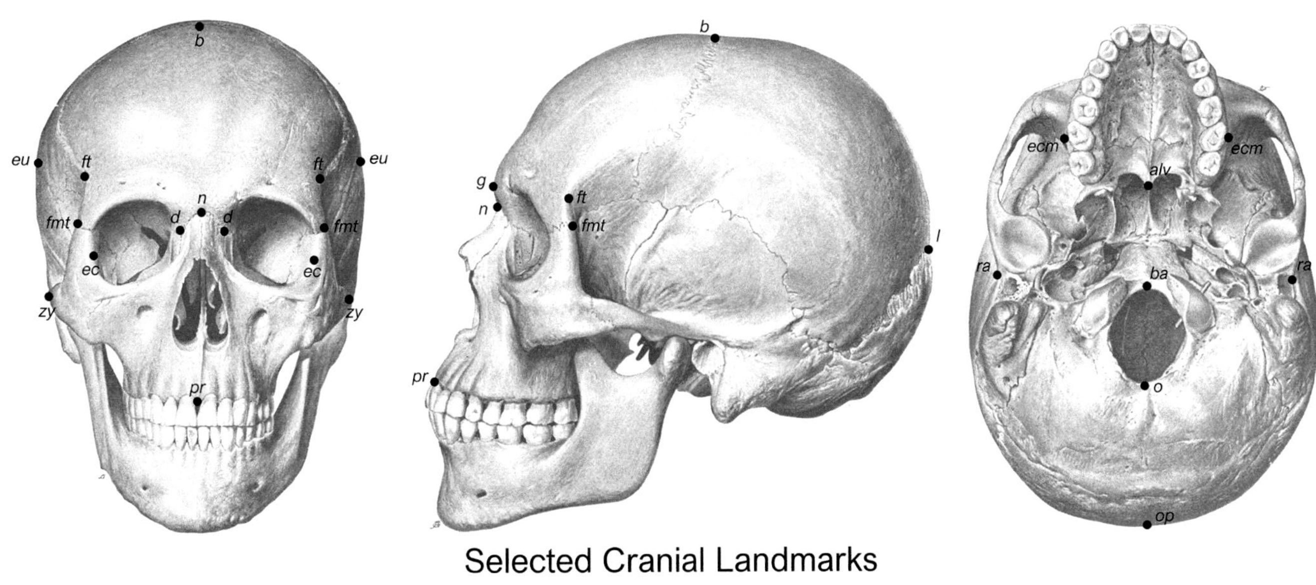

Figure 9.7. Selected cranial landmarks.

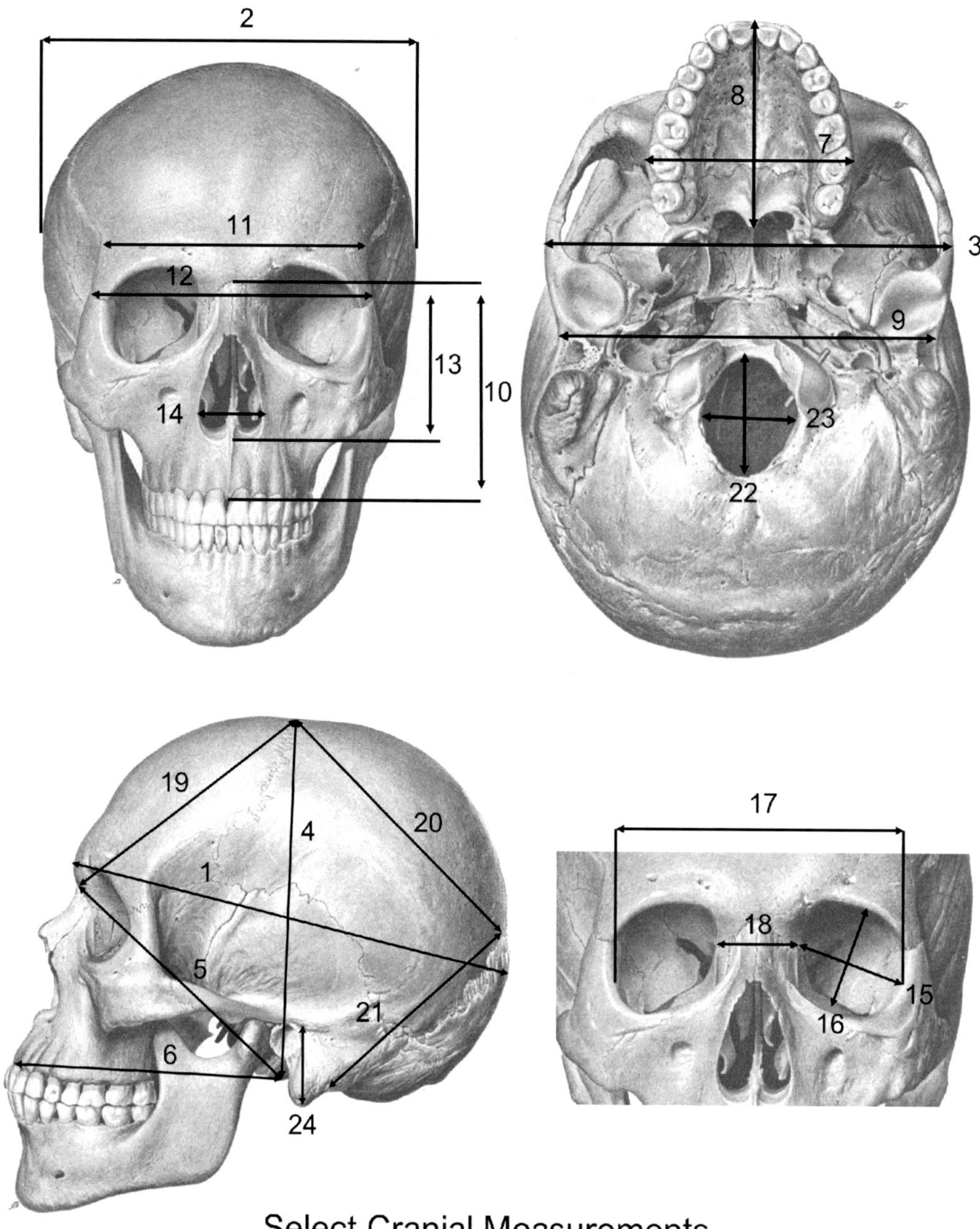

Select Cranial Measurements

1. Maximum Length (GOL)
2. Maximum Breadth (XCB)
3. Bizygomatic Breadth (ZYB)
4. Basion-Bregma Height (BBH)
5. Basion-Nasion Length (BNL)
6. Basion-Prosthion Length (BPL)
7. Maximum Alveolar Breadth (MAB)
8. Maximum Alveolar Length (MAL)
9. Biauricular Breadth (AUB)
10. Upper Facial Height (UFHT)
11. Minimum Frontal Breadth (WFB)
12. Upper Facial Breadth (UFBR)
13. Nasal Height (NLH)
14. Nasal Breadth (NLB)
15. Orbital Breadth (OBB)
16. Orbital Height (OBH)
17. Biorbital Breadth (EKB)
18. Interorbital Breadth (DKB)
19. Frontal Chord (FRC)
20. Parietal Chord (PRC)
21. Occipital Chord (OCC)
22. Foramen Magnum Length (FOL)
23. Foramen Magnum Breadth (FOB)
24. Mastoid Length (MDH)

Figure 9.8. Selected cranial measurements typically used in ancestry estimation. Standardized measurement abbreviations are included in parentheses for each measurement.

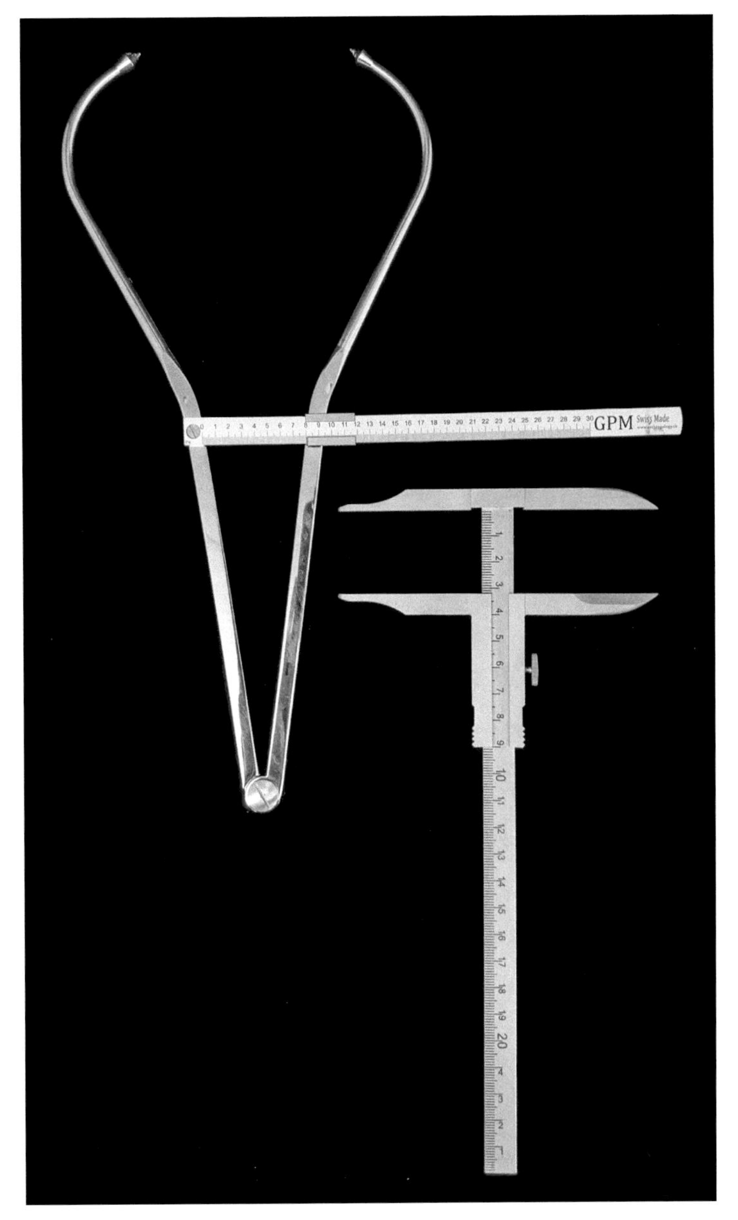

Figure 9.9. Spreading (*top*) and sliding (*bottom*) calipers.

Figure 9.9), though more specialized calipers exist for some measurements. Collecting these measurements requires training—even minor errors with placing the caliper tips on the landmarks can result in problematic ancestry assessments.

Once the measurements are entered, an initial analysis should be run using all groups considered appropriate for comparison. For example, if sex has been confidently estimated from the pelvis as male, then only male groups should be included in the analysis (SWGANTH, 2013). The initial run will double-check the values of the measurements that were entered by comparing them to the range of values included in the FDB. **Figures 9.10** and **9.11** illustrate the output from an analysis using a series of cranial measurements taken from the skull of a bear. The blue shaded regions flag measurements that seem problematic and these should be either checked against the specimen and corrected or removed from the analysis. Caution should be taken when removing measurements, however, as using too few measurements will compromise the analysis. If, for example, we were to retain only the non-problematic measurements from **Figure 9.10** (i.e., AUB, FRC, MAB, MDH, and WFB), then Fordisc will

DFA results using 18 measurements:

AUB BBH BNL BPL DKB FOL FRC GOL MAB MDH
NLB OBH OCC PAC UFHT WFB XCB ZYB

Measurements removed: UFBR FOB

Measurement Checks — Group Means

Current Case		Chk	AM 51	BM 105	CHM 73	GTM 69	HM 178	JM 183	VM 48	WM 291	GS Imp %	CC Imp %
AUB	126		132.1	120.7	123.6	123.9	124.1	123.2	122.8	123.2	7.5	5.6
BBH	**117**	----	133.4	137.8	139.5	133.2	136.7	138.2	137.8	141.7	7.1	8.5
BNL	**189**	++++	103.0	104.5	100.6	98.4	101.0	101.6	97.6	106.2	10.2	11.4
BPL	**272**	++++	100.1	104.2	96.0	97.7	98.7	97.4	95.4	98.3	4.9	4.2
DKB	**70**	++++	22.5	23.7	22.1	21.6	21.2	21.4	21.3	21.2	2.6	0.5
FOL	**25**	----	36.6	36.5	35.7	35.5	36.5	35.9	34.5	37.5	3.8	1.5
FRC	109		110.8	113.0	112.8	106.7	111.1	110.8	112.1	114.8	5.7	4.1
GOL	**23**	----	180.1	186.9	179.4	173.2	178.5	179.2	172.4	188.1	12.9	29.7
MAB	67	+	66.2	66.2	64.2	64.5	65.4	63.9	66.4	61.5	4.8	4.3
MDH	29		29.5	32.3	29.5	31.1	28.6	30.1	26.5	32.4	5.5	2.5
NLB	**36**	++++	26.1	26.2	26.2	25.5	24.9	25.5	26.2	23.8	5.4	2.6
OBH	**41**	+++	35.2	35.1	33.7	36.1	35.1	34.1	33.8	33.9	3.7	1.0
OCC	**66**	----	93.9	98.7	98.3	95.7	97.2	100.1	98.4	100.9	3.9	3.2
PAC	**65**	----	110.1	117.1	115.3	112.2	112.1	113.3	110.4	118.4	5.6	7.3
UFHT	**123**	++++	73.4	72.7	73.1	71.6	73.0	72.1	71.5	71.9	0.6	0.1
WFB	98	+	97.1	96.2	91.7	92.9	94.3	93.0	94.7	96.9	4.2	3.4
XCB	**105**	----	143.0	135.6	139.5	136.7	138.4	138.3	140.5	140.2	3.0	3.3
ZYB	**179**	++++	141.2	130.5	132.9	131.6	131.2	133.2	130.0	129.6	8.5	6.8

+/- measurement deviates higher/lower than all group means; ++/-- deviates 1 to 2 STDEVs
+++/--- deviates two to three STDEVs; ++++/---- deviates at least 3 STDEVs

Outliers detected in reference groups: 8

Natural Log of VCVM Determinant = 44.4546

Classification Table

From Group	Total Number	AM	BM	CHM	GTM	HM	JM	VM	WM	Correct
AM	51	33	2	2	4	2	4	2	2	64.7 %
BM	105	0	74	3	7	7	5	0	9	70.5 %
CHM	73	3	2	39	0	6	17	6	0	53.4 %
GTM	69	2	4	1	50	5	3	3	1	72.5 %
HM	178	9	20	9	29	64	17	15	15	36.0 %
JM	183	11	5	36	15	20	74	12	10	40.4 %
VM	48	0	0	6	4	2	1	35	0	72.9 %
WM	291	5	12	8	3	9	15	2	237	81.4 %

(Into Group (counts): AM through WM)

Total Correct: 606 out of 998 (60.7 %) *** CROSS-VALIDATED ***

Multigroup Classification of Current Case

Group	Classified into	Distance from	Posterior	Typ F	Typ Chi	Typ R
GTM		6349.9	1.000	0.000	0.000	**0.014 (69/70)**
VM		6398.0	0.000	0.000	0.000	**0.020 (48/49)**
JM		6403.6	0.000	0.000	0.000	**0.005 (183/184)**
BM		6410.0	0.000	0.000	0.000	**0.009 (105/106)**
AM		6423.4	0.000	0.000	0.000	**0.019 (51/52)**
HM		6428.0	0.000	0.000	0.000	**0.006 (178/179)**
CHM		6434.2	0.000	0.000	0.000	**0.014 (73/74)**
WM		6506.6	0.000	0.000	0.000	**0.003 (291/292)**

(Probabilities: Posterior, Typ F, Typ Chi, Typ R)

Current Case is too dissimilar to all groups; all TPs < 0.01

Figure 9.10. Fordisc output from an attempt to classify a bear skull. Blue shaded region illustrates measurement checks; green shaded region illustrates sample sizes for groups under consideration; red shaded region illustrates classification results.

```
DFA results using 5 measurements:
 AUB   FRC   MAB   MDH   WFB
 ---------------------------------------------------------------
Measurement Checks                Group Means     GS Imp    CC Imp
                               GTM      HM      VM       %         %
 Current Case     Chk           76     208      48
 ---------------------------------------------------------------
  AUB      126      +        123.8  124.2  122.8       2.9       3.2
  FRC      109               106.5  111.0  112.1      50.1      38.5
  MAB       67      +         64.5   65.0   66.4       9.1       9.8
  MDH       29                31.0   28.6   26.5      30.7      37.5
  WFB       98      +         92.8   94.1   94.7       7.3      11.0
 ---------------------------------------------------------------
 +/- measurement deviates higher/lower than all group means; ++/-- deviates 1 to 2 STDEVs
 +++/--- deviates two to three STDEVs; ++++/---- deviates at least 3 STDEVs
 ---------------------------------------------------------------
 Outliers detected in reference groups: 5

 Natural Log of VCVM Determinant =  14.6252
 ---------------------------------------------------------------
Classification Table
 --------------------------------------------------------------------
 From     Total             Into Group (counts)
 Group    Number    GTM     HM      VM     Correct
 --------------------------------------------------------------------
    GTM     76        56     16       4      73.7 %
     HM    208        58     88      62      42.3 %
     VM     48         5     12      31      64.6 %
 --------------------------------------------------------------------
 Total Correct:  175 out of 332 (52.7 %) *** CROSS-VALIDATED ***
 --------------------------------------------------------------------

Multigroup Classification of Current Case

 ------------------------------------------------------------------------------
   Group      Classified      Distance          Probabilities
                 into           from      Posterior   Typ F   Typ Chi   Typ R
 ------------------------------------------------------------------------------

     HM        **HM**            1.4         0.404     0.930    0.928    0.919 (17/209)
     VM                          1.9         0.308     0.870    0.862    0.816 (9/49)
    GTM                          2.0         0.287     0.849    0.842    0.727 (21/77)
 ------------------------------------------------------------------------------
  Current Case is closest to HMs
 ------------------------------------------------------------------------------
```

Figure 9.11. Fordisc output from an attempt to classify a bear skull while using too few measurements. Blue shaded region illustrates measurement checks; green shaded region illustrates sample sizes for groups under consideration; red shaded region illustrates classification results.

assign even a bear skull to a human group (see **Figure 9.11**), albeit with some indications that the results are unreliable (see below).

Just as too few measurements can be problematic, too many measurements can also cause issues. The green shaded areas in **Figures 9.10** and **9.11** show the sample sizes for each of the groups being compared in an ancestry assessment. If we let n be the smallest sample size for the groups under consideration, then it has been suggested that the number of measurements used in an analysis (m) should not exceed $n/3$ (Ousley and Jantz, 2012). From this standpoint, the 18 measurements used in the analysis shown in **Figure 9.10** are 2 too many since the smallest sample size is 48 (for Vietnamese males). While the statistical rationale for this suggestion is beyond the scope of this chapter (interested students are referred to Ousley and Jantz, 2012), the number of measurements used in an analysis should generally be adjusted such that it is as large as possible without violating this rule of thumb.

After checking measurements and sample sizes, comparisons are made iteratively, removing the group that is the least similar to the unknown decedent after each run and, if necessary, adjusting the number of measurements used until only a small number of groups remain and a classification is made. Fordisc, as with any LDA, will *always* result in a classification, even if it is incorrect (such as with the bear skull in **Figure 9.11**). This is because, in order to calculate the posterior probabilities of group membership, classification using LDA assumes that the unknown case belongs to one of the groups included in the analysis (Klecka, 1980).

Fortunately, Fordisc provides multiple ways to evaluate the strength of both the models it generates and the classifications that they produce. After each run, the success of the discriminant analysis in classifying the constituents of its reference groups is evaluated and a **classification accuracy** rate is provided (this can be seen in **Figures 9.10** and **9.11,** located just below the green shaded regions). This gives an estimate of how well the model can be expected to classify an unknown individual. For a discriminant analysis to be useful in a forensic context, the classification accuracy should be appreciably better than would be expected by chance (Ousley and Jantz, 2012). For example, in a two-group classification, the chances of randomly assigning an unknown individual to the correct group are one in two, or 50 percent. A discriminant function that yields a classification accuracy of 54 percent is not performing much better than chance in this case. In a three-group comparison, however, where the rate expected by chance is one in three, or 33 percent, a classification accuracy of 54 percent is a marked improvement.

As a means of assessing whether an unknown case *actually* belongs to one or more of the groups under consideration, Fordisc provides three different kinds of **typicality probabilities** (Typ F, Typ Chi, and Typ R). Typicality probabilities provide an index of how far an unknown individual's measurements are from a given group's centroid relative to known members of that group. For example, a typicality probability of 0.65 indicates that 65 percent of the reference sample for that group is as far or farther from the group centroid than the unknown individual. In general, typicality probabilities should exceed 0.05. Values lower than this threshold may indicate that the case is anomalous or derived from a population that is not well represented in the FDB (Jantz and Ousley, 2005; Ousley and Jantz, 2012). An example of this can be found in the red-shaded region of **Figure 9.10,** clearly illustrating that the measurements from the bear skull do not fit any of the human groups under consideration.

If there are no measurement errors and if the classification accuracy and typicality probabilities are all acceptable, then an unknown individual can usually be classified into the group that it is most similar to (i.e., the group whose centroid is closest to the case in question). The likelihood of this being a correct classification is then provided by the associated **posterior probability**. Posterior probabilities in a discriminant analysis are constrained to sum to 1 (or 100 percent). Therefore, the higher the posterior probability of belonging to one group, the lower the posterior probability of belonging to the other groups included in the analysis. Look again at the red-shaded regions of **Figures 9.10** and **9.11,** both of which provide posterior probabilities for the different analyses of the bear skull. In **Figure 9.10,** the high posterior probability (1.0) can be dismissed due to the extremely low typicality probabilities. In **Figure 9.11,** the posterior probabilities for all three groups are very similar, suggesting that the LDA in this case is not producing a good separation of the groups under consideration because of using too few measurements. Given these considerations, the reported results of an ancestry estimation conducted using Fordisc should include the group that the unknown was classified into, the associated posterior and typicality probabilities, and the classification accuracy rate (Ousley and Jantz, 2012). The combination of these parameters allows for an assessment of the estimate's reliability.

Drawing Conclusions

While presented under separate headings above, morphoscopic, craniometric, dental, and postcranial methods for estimating ancestry can be (and often are) used in concert. Several techniques have been developed that use combined datasets (e.g., morphoscopic and craniometric) to produce ancestry estimates, and initial assessments suggest that they result in improved classification accuracy over methods using only one type of data (e.g., Berg and Kenyhercz, 2017; Hefner et al., 2014; Spiros and Hefner, 2020). When drawing conclusions based on the results of multiple methods, it is important for forensic anthropologists to keep in mind the limitations of each method used. For example, if a craniometric assessment of an unknown decedent indicates a high posterior probability of Asian ancestry, then it is unlikely that using OSSA will contribute any meaningful information to the ancestry estimate since it is limited to American Black and white populations. If, to the contrary, an unknown decedent has nearly equal posterior probabilities of being Black or white based on craniometrics, then OSSA may provide clarifying information.

Ultimately, ancestry estimates are based not only on the results of methods like those described above, but also on a forensic anthropologist's familiarity with the range of skeletal morphology that characterizes the populations living in the region in which they work. Hispanic individuals in the Southwest, for example, are frequently classified as Japanese when performing craniometric assessments using Fordisc 3.1 (Dudzik and Jantz, 2016). Forensic anthropologists should be aware of regional idiosyncrasies like this and not adhere too strictly to the classifications produced by ancestry estimation methods when drawing their conclusions (Ousley and Jantz, 2012; Stock and Rubin, 2020). Lastly, an ancestry estimate can be reported as ambiguous or not possible given the data but should never be presented as a certainty (SWGANTH, 2013). Not only does this misrepresent the mathematics underlying the methods used, but it also ignores the error introduced by the imposition of categorical classifications on continuous, clinal patterns of human biological variation.

LEARNING CHECK

Q5. **Figure 9.12** shows a lateral view of three crania. Based on the trait description in **Figure 9.4,** how would these crania be ordered from least expression of the anterior nasal spine (ANS) to greatest?

A) A < B < C
B) C < B < A
C) A < C < B
D) C < A < B

Q6. **Figure 9.13** shows a lateral view of two crania with the location of the landmark bregma indicated by a white arrow. Based on the trait description in **Figure 9.5,** which of these crania exhibits a post-bregmatic depression (PBD), or a depression immediately posterior to bregma?

A) Cranium A
B) Cranium B
C) Neither of the crania exhibit a post-bregmatic depression.
D) Both crania exhibit a post-bregmatic depression.

Q7. You have been asked to peer review a case report written by another forensic anthropologist for a complete set of unidentified skeletal remains that were recovered from a Native American reservation. You notice that the ancestry estimation included in the report is based on an analysis using OSSA. Is this an appropriate choice of method?

A) Given the context, OSSA is an appropriate choice of method.
B) Given the context, OSSA is an inappropriate choice of method.

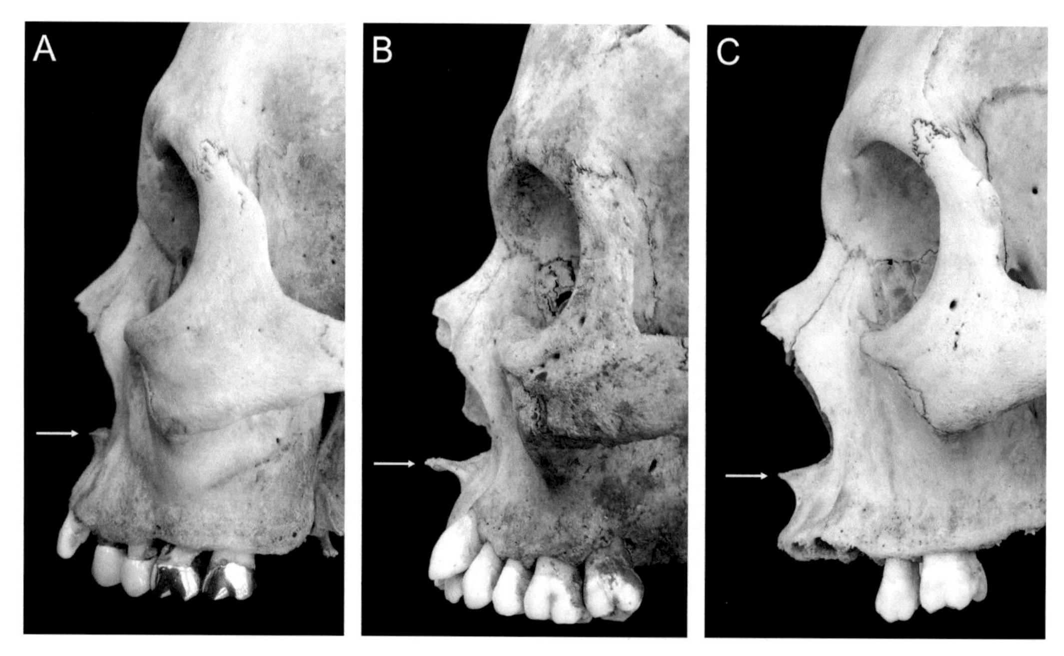

Figure 9.12. Three views of the human cranium, left lateral view.

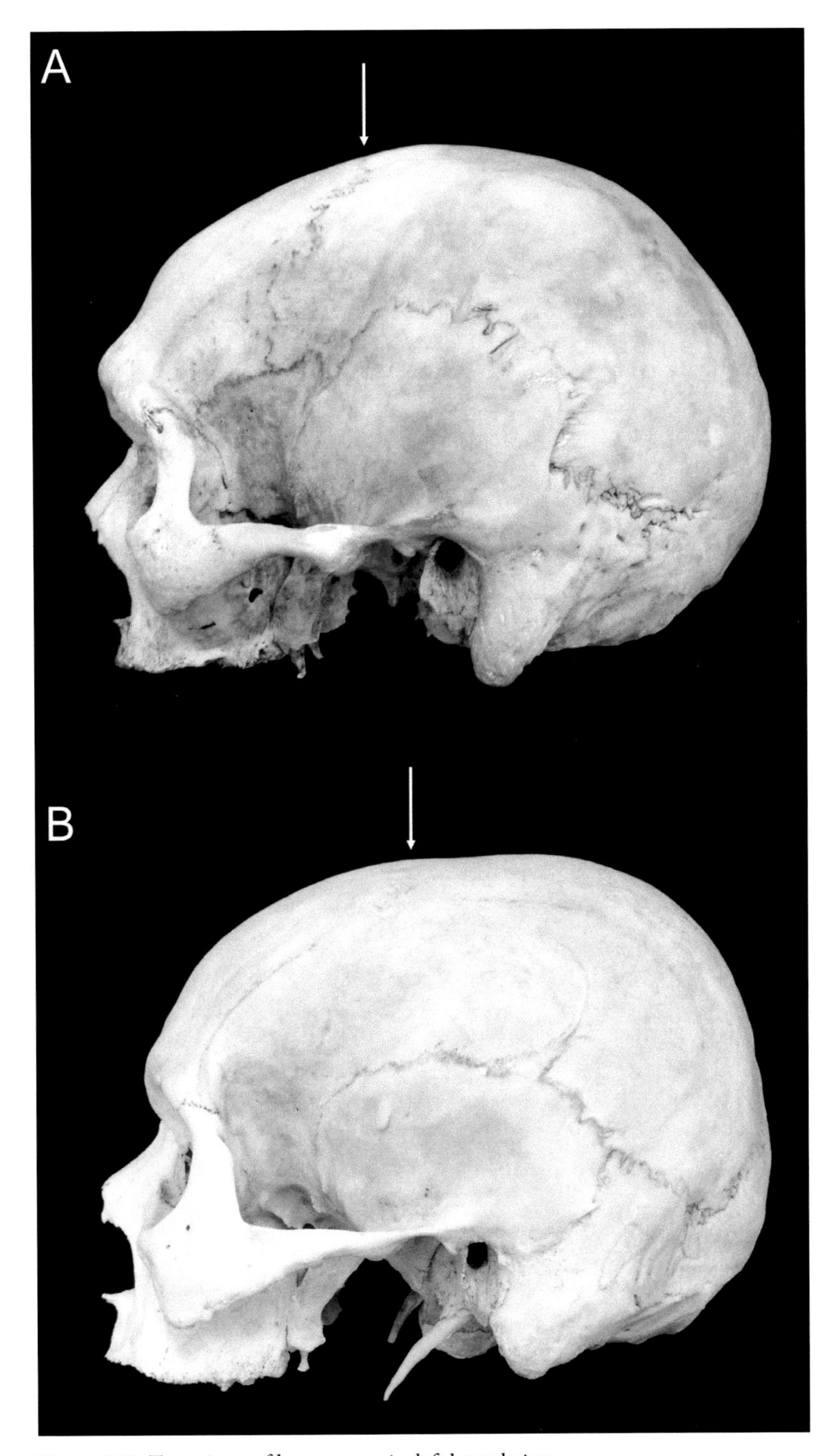

Figure 9.13. Two views of human crania, left lateral view.

Q8. While conducting a craniometric ancestry estimation using Fordisc, you notice that the smallest sample size for the groups under consideration is 51. Based on the guidelines discussed above, what is the largest number of measurements that it would be appropriate to use in this analysis?

A) 24
B) 21
C) 17
D) 15

Q9. **Figure 9.14** shows the Fordisc output from a craniometric ancestry estimation for an unidentified decedent. Based on this output, is the classification accuracy rate of the model used better than what can be expected by chance for a three-group comparison?
A) Yes
B) No

Q10. Based on the output shown in **Figure 9.14,** what is the posterior probability that the unidentified individual can be classified as a Japanese male (JM)?
A) 0.113
B) 0.073
C) 0.215
D) 0.814

Current Debates and Future Directions

For decades, forensic anthropologists have been criticized for appearing to lend support to the biological concept of race through ancestry estimation (e.g., Albanese and Saunders, 2006; Armelagos and Goodman, 1998; Smay and Armelagos, 2000). The use of administrative racial categories within ancestry estimation has generally been explained as a practical consideration necessitated by forensic anthropology's dual articulation with both medicolegal and law enforcement agencies (e.g., Brace, 1995; Kennedy, 1995; Sauer, 1992; St. Hoyme and İşcan, 1989). Recently, however, some forensic anthropologists have argued that, by appearing to validate the existence of biological races, the practice of ancestry estimation tacitly supports racist ideologies and should therefore be abandoned (Bethard and DiGangi, 2020; DiGangi and Bethard, 2021).

Does ancestry estimation reify race? Maybe. Within the methodological descriptions above, the reference sample descriptors used within the cited research were maintained. The reader will have noticed labels like "white" and "New Mexico Hispanic" being used as categorizations within methods purporting to estimate ancestry. This is representative of a larger problem within forensic anthropology—the conflation of terms used to designate racial and/or ethnic categories, ancestry, geography, and aspects of identity resulting from a near absence of stated, much less standardized, definitions of these terms (Maier et al., 2021). If forensic anthropologists are not explicit about what is meant by *ancestry, race,* and *ethnicity,* how can we expect to convey the differences between these concepts to the public?

While such terminological confusion can potentially be resolved in a way that brings the vocabulary of ancestry estimation in line with current understandings of the ecogeographic patterning of human biological variability, a more insidious problem is that the reference samples upon which our methods are built are themselves racially defined. The ancestry estimation methods described above rely on statistical procedures, usually linear discriminant analysis, that are designed to maximize the separation between predefined, mutually exclusive groups (Klecka, 1980). When reference groups are racially defined, these techniques create the illusion that the differences between racial groups exceed their similarities. Although this illusion is dispelled by an appreciation of human biological variation and some understanding of multivariate statistics, contemporary methods of ancestry estimation can indeed appear to validate the existence of biological races.

Does the fact that discriminant analyses can effectively separate racially defined groups suggest that human races have a biological basis? No. Discriminant analyses can also reliably

```
DFA results using 18 measurements:
 BBH   BNL   BPL   DKB   FOB   FOL   FRC   GOL   MAL   MDH
 NLB   NLH   OBB   OBH   OCC   PAC   UFHT  WFB
 ----------------------------------------------------------------
Measurement Checks              Group Means     GS Imp   CC Imp
                             BM      JM      WM     %        %
 Current Case     Chk       101     193     377
 ----------------------------------------------------------------
  BBH       146     +      137.4   138.3   141.2     4.1      3.4
  BNL       112     ++     104.6   101.7   105.7     6.5      7.8
  BPL       110     ++     104.1    97.5    97.5     8.4      9.7
  DKB        17     --      23.5    21.4    21.3     5.3      4.0
  FOB        31             29.9    29.8    32.0     9.7      0.6
  FOL        36             36.7    35.9    37.4     3.7      1.3
  FRC       109     -      112.8   110.9   114.8     5.2      5.2
  GOL       188     +      187.3   179.3   187.7    12.1     34.5
  MAL        60     +       57.9    52.7    53.9     9.4     10.5
  MDH        32             32.6    30.2    32.3     4.2      2.1
  NLB        22     --      26.4    25.5    24.1     8.4      4.6
  NLH        50     -       52.7    51.4    52.7     2.0      0.5
  OBB        41             40.8    39.3    41.1     7.7      4.1
  OBH        33     -       35.2    34.1    34.0     2.0      0.8
  OCC        99     -       99.0   100.1   100.8     0.7      0.3
  PAC       124     +      116.9   113.5   118.0     4.2      5.7
 UFHT        67     --      72.7    72.0    71.7     0.4      0.1
  WFB        89     -       96.0    93.1    96.9     6.0      4.9
 ----------------------------------------------------------------
 +/- measurement deviates higher/lower than all group means; ++/-- deviates 1 to 2 STDEVs
 +++/--- deviates two to three STDEVs; ++++/---- deviates at least 3 STDEVs
 ----------------------------------------------------------------
 Outliers detected in reference groups: 5

 Natural Log of VCVM Determinant =  39.1133
 ----------------------------------------------------------------
Classification Table
 ---------------------------------------------------------------------
 From     Total             Into Group (counts)
 Group    Number     BM      JM      WM    Correct
 ---------------------------------------------------------------------
     BM    101       77      13      11     76.2 %
     JM    193       17     160      16     82.9 %
     WM    377       30      28     319     84.6 %
 ---------------------------------------------------------------------
 Total Correct:  556 out of 671 (82.9 %) *** CROSS-VALIDATED ***
 ---------------------------------------------------------------------

Multigroup Classification of Current Case

 --------------------------------------------------------------------------------------
   Group      Classified      Distance         Probabilities
                 into           from      Posterior   Typ F   Typ Chi   Typ R
 --------------------------------------------------------------------------------------

    WM        **WM**            18.4        0.814     0.466    0.430    0.503 (188/378)
    BM                          22.3        0.113     0.256    0.217    0.412 (60/102)
    JM                          23.2        0.073     0.215    0.183    0.144 (166/194)
 --------------------------------------------------------------------------------------
  Current Case is closest to WMs
 --------------------------------------------------------------------------------------
```

Figure 9.14. Sample Fordisc output.

separate white males born between 1840 and 1890 from those born between 1930 and 1980 (Ousley et al., 2009). Accurate classification using LDA can result from reduced gene flow as a consequence of temporal, geographic, or socially prescribed distance between the populations under consideration. Racial classifications are not the only and probably not the best way to define human populations. As long as missing persons in our society are categorized using racial labels, however, such categories will remain practical for use in forensic ancestry estimation (Brace, 1995; Hochman, 2013; St. Hoyme and İşcan, 1989).

Are forensic ancestry estimates accurate? Available research indicates that ancestry estimates are correct in more than 90 percent of resolved cases (Parsons, 2017; Thomas et al., 2017). These numbers may not reflect the true accuracy of ancestry estimates, however, in that they are derived from case reports generated by forensic anthropologists affiliated with large medical examiners' offices that employ standardized protocols and peer-review processes. As such, they may differ from the accuracy rates of forensic anthropologists not embedded within such systems. Further, reported rates are derived from resolved cases, or cases in which the decedent has been positively identified. There is no way of evaluating the accuracy of ancestry estimates in unresolved casework and it is possible that incorrect ancestry assessments are actively hindering their resolution (DiGangi and Bethard, 2021). In at least one study, however, the majority of case resolutions were achieved independently of their associated ancestry estimates (e.g., through DNA database comparisons). This suggests that the reported accuracy rates of ancestry estimates are not overly inflated by being based on resolved casework and are likely to be reliable (Thomas et al., 2017).

Does ancestry estimation do more harm than good? It has been asserted that, in appearing to validate the existence of biological races for both law enforcement and the public, ancestry estimation inadvertently serves to maintain racist ideologies (DiGangi and Bethard, 2021). To the extent that the public understanding of race is derived from or supported by forensic anthropology, this may be true. However, ancestry estimation also contributes to the identification of unknown decedents. The sample size of missing persons consistent with unidentified remains has been reported to decrease by as much as 86 percent when ancestry is added to a biological profile (Walsh-Haney and Boys, 2015). Although not conclusive, this suggests that ancestry estimates enable investigators to prioritize their searches, thereby promoting an efficient use of resources and decreasing the time required to establish a positive identification. Further, ancestry estimation can aid in predicting the geographic origins of migrants, thereby making meaningful contributions to their identification and repatriation (e.g., Algee-Hewitt et al., 2018; Hughes et al., 2013; Spradley, 2014). In New Zealand, the Maori treat their dead in a culturally prescribed way and thus have a vested interest in determining whether unidentified remains are of Maori origin (Cox et al., 2006). In this scenario and others like it, a forensic anthropologist's refusal to provide an ancestry estimate can undermine the autonomy of Indigenous communities. In the United States, ancestry estimation may benefit Indigenous communities through its role in the repatriation of human remains mandated under the Native American Graves Protection and Repatriation Act (Ousley et al., 2005). Thus, while the assertion that ancestry estimation causes harm to communities of color by lending credence to the misconception that race is biological likely has some validity, the abandonment of ancestry estimates within forensic casework may carry consequences that are equally, if not more, harmful.

Ultimately, the decision as to whether to retain or abolish ancestry estimation within forensic casework must be made in consultation with the public and with a firm understanding of the costs and benefits involved (Stull et al., 2021). The hypothesis that ancestry estimates may adversely affect the identification rate of Black, Indigenous, and People of Color

(BIPOC) should be thoroughly investigated (DiGangi and Bethard, 2021). If identification biases are observed, research should be undertaken to discern whether they result from structural inequalities or discrepancies in investigative effort (e.g., Algee-Hewitt et al., 2018). The difference in the amount of time and resources needed to establish a positive identification for cases with associated ancestry estimates, without them, and with erroneous estimates should be assessed through robust simulation studies. The results of research like this will allow forensic anthropologists and the communities that they serve to make an informed decision as to whether ancestry should continue to be a component of the biological profile.

If ancestry estimation remains a prominent aspect of the practice of forensic anthropology, then it will be necessary to standardize the vocabulary used within ancestry research and establish definitions of terms like *ancestry, race,* and *ethnicity* that reflect contemporary understandings of human biological variability (Maier et al., 2021). Further, the use of racially defined reference samples within ancestry estimation methods introduces an unknown amount of error in that racial labels homogenize and obscure individual variability in terms of ancestry (e.g., Caspari, 2010; Parra et al., 1998). An increased emphasis should therefore be placed on the development of methods that are firmly rooted in evolutionary theory and can adequately handle the complexities of population history (Ross and Pilloud, 2021). Such steps will help to distance ancestry estimation from biological misconceptions of race.

End-of-Chapter Summary

This chapter began with an overview of why patterns of human biological diversity do not support the existence of biological races. Following this, the existence of human races as social categories was briefly discussed as well as the probability that many "racial" health disparities are likely rooted in structural racism rather than biology. In sum, race is not biology, but racism has biological consequences. As practiced in forensic anthropology, ancestry estimation capitalizes on geographic patterns of human skeletal variation to estimate the probability that an unknown decedent is affiliated with a particular ancestral population. The more widely used methods for ancestry estimation involve either morphological or metric assessment of the skull, although methods have also been developed for dental and postcranial remains. No matter the method employed, a forensic anthropologist should always be aware of its limitations and take them into consideration when drawing conclusions. Although critiques of ancestry estimation have existed for decades, there have been recent calls for its abandonment within forensic casework. This chapter therefore concluded with a brief discussion of both the potential and the problems inherent in contemporary approaches to ancestry estimation.

End-of-Chapter Exercises

Exercise 1

Materials Required: **Figure 9.15**

Scenario: You have been asked to conduct an ancestry estimation for skeletal remains recovered from a desert context. Since the skull is relatively intact, you decide to begin with a craniometric assessment using Fordisc. The output from the final run of your analysis is presented in **Figure 9.15**.

Question 1: Based on this output, are you concerned about any of the measurements that you entered?

DFA results using 21 measurements:

AUB BBH BNL BPL DKB FOB FOL FRC GOL MAL
MDH NLB NLH OBB OBH OCC PAC UFBR UFHT WFB
XCB

Measurement Checks

Current Case		Chk	Group Means BM 89	HM 160	WM 323	GS Imp %	CC Imp %
AUB	123		121.1	123.8	123.4	1.8	0.9
BBH	139		136.9	136.4	141.0	7.1	1.4
BNL	107	+	104.2	100.7	105.6	9.6	15.3
BPL	105	+	104.0	98.5	97.7	7.4	10.1
DKB	18	--	23.3	21.1	21.4	4.2	3.3
FOB	29	-	30.0	30.7	32.0	5.4	3.4
FOL	36	-	36.8	36.4	37.4	1.7	0.6
FRC	116	+	112.7	111.1	114.7	4.2	4.1
GOL	186		186.8	178.0	187.7	12.7	32.2
MAL	62	+	58.0	55.4	54.1	6.3	6.3
MDH	29		32.6	28.6	32.3	3.5	2.3
NLB	25		26.3	25.0	24.2	6.9	2.3
NLH	52	-	52.8	52.1	52.7	0.5	0.1
OBB	43	+	40.7	39.8	41.2	4.1	1.8
OBH	35		35.3	35.4	34.0	4.3	1.0
OCC	101	+	98.8	97.3	100.8	3.2	3.0
PAC	114		116.7	112.0	118.0	6.2	3.8
UFBR	105		106.8	103.7	104.9	2.7	1.1
UFHT	68	-	72.8	73.1	71.7	1.3	0.4
WFB	96		95.9	94.0	96.8	2.9	1.6
XCB	140		135.6	138.2	140.3	4.0	5.1

+/- measurement deviates higher/lower than all group means; ++/-- deviates 1 to 2 STDEVs
+++/--- deviates two to three STDEVs; ++++/---- deviates at least 3 STDEVs

Outliers detected in reference groups: 9

Natural Log of VCVM Determinant = 48.2435

Classification Table

From Group	Total Number	Into Group (counts) BM	HM	WM	Correct
BM	89	64	16	9	71.9 %
HM	160	23	119	18	74.4 %
WM	323	25	19	279	86.4 %

Total Correct: 462 out of 572 (80.8 %) *** CROSS-VALIDATED ***

Multigroup Classification of Current Case

Group	Classified into	Distance from	Probabilities Posterior	Typ F	Typ Chi	Typ R
WM	**WM**	16.6	0.538	0.767	0.732	0.685 (102/324)
BM		18.1	0.265	0.696	0.645	0.622 (34/90)
HM		18.7	0.197	0.655	0.607	0.646 (57/161)

Current Case is closest to WMs

Figure 9.15. Sample Fordisc output.

Question 2: Based on this output, does the model have a reasonable classification accuracy rate?

Question 3: Based on this output, is the unknown individual closest to the centroid of the sample of white males (WM), Black males (BM), or Hispanic males (HM)?

Question 4: Based on the typicality probabilities shown in this output, does your answer to Question 3 mean that this individual is unlikely to belong to one of the other two groups under consideration?

Question 5: How would you report the ancestry estimate for these remains and are you certain of your results?

Question 6: Would using the OSSA method potentially provide clarifying information?

Exercise 2

Materials Required: **Figure 9.16** and **Table 9.3**

Scenario: A colleague working in Mississippi has sent you photographs of a case that they are working on (see **Figure 9.16**) and asked you to give them an ancestry estimate. Since you are unable to take measurements from the photographs, you decide to use OSSA.

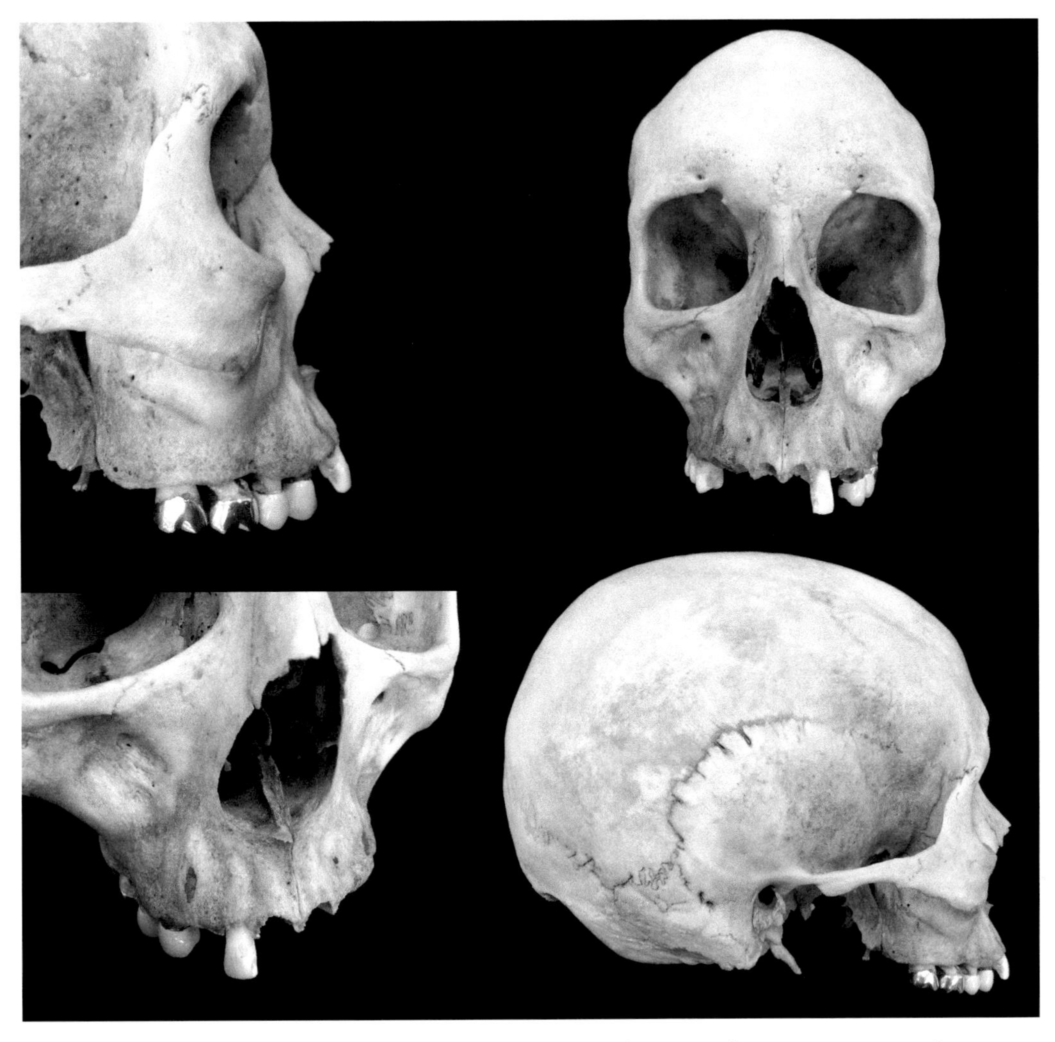

Figure 9.16. Four views of the same human cranium showing key locations of ancestry assessment observations.

Table 9.3. Converting character state scores to OSSA scores

Character State	Score	OSSA Score
Anterior Nasal Spine (ANS)	1	0
	2	1
	3	1
Inferior Nasal Aperture (INA)	1	0
	2	0
	3	0
	4	1
	5	1
Interorbital Breadth (IOB)	1	1
	2	1
	3	0
Nasal Aperture Width (NAW)	1	1
	2	1
	3	0
Nasal Bone Contour (NBC)	0	0
	1	0
	2	1
	3	1
	4	1
Postbregmatic Depression (PBD)	0	1
	1	0

Modified after Hefner and Ousley (2014).

Directions: Using the trait diagrams and descriptions provided in **Figures 9.4** and **9.5,** score ANS, INA, IOB, NAW, NBC, and PBD for the individual shown in **Figure 9.16.** Then, convert the character state scores you recorded to OSSA scores using **Table 9.3** and add together the OSSA scores to produce a summed score.

Question 1: Were there any character states that were particularly difficult to score from using images alone?

Question 2: As originally presented, summed scores of 0–3 are associated with American Black individuals while summed scores greater than 3 are associated with American white individuals. Based on your summed score, which group is this individual classified into?

Question 3: Kenyhercz and colleagues (2017) suggest that OSSA performs better when summed scores of 0–4 are associated with American Blacks and summed scores of 5 or 6 are associated with American whites. Does your assessment of this individual's ancestry change with this adjustment?

Exercise 3

Materials Required: **Figures 9.17–9.19**

Directions: **Figures 9.17, 9.18,** and **9.19** provide the results of a craniometric ancestry estimation using Fordisc for the same individual but using different numbers of measurements. Use these results to answer the following questions.

Question 1: Based on the sample sizes of the groups under consideration, which figure represents the analysis that uses the most appropriate number of measurements?

Question 2: Does the group to which the unknown individual is assigned change based on the number of measurements used in the analysis?

```
DFA results using 16 Forward Wilks selected (min: 1 max: 16, out of 21) measurements:
 GOL   NLB   ZYB   WFB   MDH   AUB   OBB   NLH   UFHT  FRC
 DKB   OBH   BPL   BNL   MAL   XCB
 Measurements removed: UFBR FOB
 -----------------------------------------------------------------
Measurement Checks                Group Means      GS Imp    CC Imp
                               JM      VM      WM      %         %
 Current Case     Chk         193      48     383
 -----------------------------------------------------------------
  GOL       181              179.3  172.4  187.8      23.5      36.4
  NLB        25               25.5   26.2   24.0       9.1       1.3
  ZYB       131              133.2  130.0  129.8       4.1       1.6
  WFB        97      +        93.1   94.7   96.9       7.0       6.7
  MDH        32               30.2   26.5   32.3      13.6      21.1
  AUB       122      -       123.1  122.8  123.3       0.1       0.0
  OBB        40               39.3   38.4   41.1      13.4       3.2
  NLH        54      +        51.4   53.1   52.8       2.3       1.3
 UFHT        69      -        72.0   71.5   71.7       0.1       0.0
  FRC       118      +       110.9  112.1  114.8       6.8       6.1
  DKB        21      -        21.4   21.3   21.3       0.0       0.0
  OBH        33      -        34.1   33.8   34.0       0.1       0.0
  BPL        91      --       97.5   95.4   97.4       0.5       0.2
  BNL       100              101.7   97.6  105.7      16.1      20.5
  MAL        49      --       52.7   52.2   53.8       1.8       0.9
  XCB       140              138.1  140.5  140.4       1.6       0.8
 -----------------------------------------------------------------
 +/- measurement deviates higher/lower than all group means; ++/-- deviates 1 to 2 STDEVs
 +++/--- deviates two to three STDEVs; ++++/---- deviates at least 3 STDEVs
 -----------------------------------------------------------------
 Outliers detected in reference groups: 9

 Natural Log of VCVM Determinant =  34.8424
 -----------------------------------------------------------------
Classification Table
 ---------------------------------------------------------------------
 From     Total              Into Group (counts)
 Group    Number      JM      VM      WM     Correct
 ---------------------------------------------------------------------
     JM    193      170      17       6      88.1 %
     VM     48        3      43       2      89.6 %
     WM    383       29       6     348      90.9 %
 ---------------------------------------------------------------------
 Total Correct:  561 out of 624 (89.9 %) *** CROSS-VALIDATED ***
 ---------------------------------------------------------------------

Multigroup Classification of Current Case

 ---------------------------------------------------------------------------------
  Group      Classified      Distance          Probabilities
                into            from      Posterior   Typ F   Typ Chi    Typ R
 ---------------------------------------------------------------------------------

    WM        **WM**            9.6         0.717     0.899    0.889     0.878 (47/384)
    VM                         12.2         0.190     0.765    0.729     0.694 (15/49)
    JM                         13.6         0.093     0.655    0.626     0.557 (86/194)
 ---------------------------------------------------------------------------------
  Current Case is closest to WMs
 ---------------------------------------------------------------------------------
```

Figure 9.17. Sample Fordisc output.

Question 3: How does the number of measurements used affect the posterior probabilities associated with each group assignment?

Question 4: Using too many measurements can result in what is called overfitting—where a predictive model is so finely tuned to the data that it is based on that its ability to make predictions from new data are compromised. Based on the results shown in these figures, how would you expect overfitting to affect the classification accuracy of the model?

Question 5: How would you report the results of this ancestry assessment?

```
DFA results using 4 Forward Wilks selected (min: 1 max: 4, out of 21) measurements:
 GOL   NLB   ZYB   XCB
 Measurements removed: UFBR FOB
 --------------------------------------------------------------
Measurement Checks                 Group Means      GS Imp    CC Imp
                              JM      VM      WM       %         %
 Current Case      Chk        196     48      533
 --------------------------------------------------------------
  GOL       181              179.4  172.4  187.7      64.1      92.1
  NLB        25               25.6   26.2   24.0      23.3       3.0
  ZYB       131              133.2  130.0  129.9       9.2       3.5
  XCB       140              138.2  140.5  140.5       3.4       1.4
 --------------------------------------------------------------
 +/- measurement deviates higher/lower than all group means; ++/-- deviates 1 to 2 STDEVs
 +++/--- deviates two to three STDEVs; ++++/---- deviates at least 3 STDEVs
 --------------------------------------------------------------
 Outliers detected in reference groups: 4

 Natural Log of VCVM Determinant =  11.5898
 --------------------------------------------------------------
Classification Table
 -------------------------------------------------------------------
 From      Total          Into Group (counts)
 Group     Number      JM     VM      WM     Correct
 -------------------------------------------------------------------
     JM     196       139     44      13      70.9 %
     VM      48        12     34       2      70.8 %
     WM     533        75     30     428      80.3 %
 -------------------------------------------------------------------
 Total Correct:  601 out of 777 (77.3 %) *** CROSS-VALIDATED ***
 -------------------------------------------------------------------

Multigroup Classification of Current Case

 ------------------------------------------------------------------------------
  Group       Classified      Distance          Probabilities
                 into           from      Posterior   Typ F   Typ Chi    Typ R
 ------------------------------------------------------------------------------

     JM         **JM**          0.7         0.468     0.950    0.949    0.898 (20/197)
     WM                         1.5         0.311     0.821    0.820    0.824 (94/534)
     VM                         2.2         0.221     0.704    0.694    0.694 (15/49)
 ------------------------------------------------------------------------------
  Current Case is closest to JMs
 ------------------------------------------------------------------------------
```

Figure 9.18. Sample Fordisc output.

```
DFA results using 21 measurements:
 AUB   BBH   BNL   BPL   DKB   EKB   FOL   FRC   GOL   MAL
 MDH   NLB   NLH   OBB   OBH   OCC   PAC   UFHT  WFB   XCB
 ZYB
 Measurements removed: UFBR FOB
 ----------------------------------------------------------------
Measurement Checks              Group Means      GS Imp    CC Imp
                              JM      VM      WM      %         %
 Current Case      Chk       193      48     373
 ----------------------------------------------------------------
  AUB       122     -       123.1  122.8  123.3      0.1       0.0
  BBH       139             138.3  137.8  141.1      3.4       1.5
  BNL       100             101.7   97.6  105.7     12.8      17.0
  BPL        91    --        97.5   95.4   97.4      0.4       0.2
  DKB        21     -        21.4   21.3   21.3      0.0       0.0
  EKB        97              97.2   95.3   97.8      1.6       0.7
  FOL        36              35.9   34.5   37.4      7.1       2.0
  FRC       118     +       110.9  112.1  114.7      5.4       5.0
  GOL       181             179.3  172.4  187.8     18.8      30.4
  MAL        49    --        52.7   52.2   53.8      1.5       0.8
  MDH        32              30.2   26.5   32.3     10.8      17.7
  NLB        25              25.5   26.2   24.0      7.2       1.1
  NLH        54     +        51.4   53.1   52.8      1.8       1.0
  OBB        40              39.3   38.4   41.1     10.7       2.7
  OBH        33     -        34.1   33.8   34.0      0.0       0.0
  OCC        95     -       100.1   98.4  100.8      0.7       0.4
  PAC       117             113.5  110.4  118.0      7.3      11.6
 UFHT        69     -        72.0   71.5   71.7      0.0       0.0
  WFB        97              93.1   94.7   96.9      5.8       5.8
  XCB       140             138.1  140.5  140.4      1.3       0.7
  ZYB       131             133.2  130.0  129.8      3.3       1.3
 ----------------------------------------------------------------
 +/- measurement deviates higher/lower than all group means; ++/-- deviates 1 to 2 STDEVs
 +++/--- deviates two to three STDEVs; ++++/---- deviates at least 3 STDEVs
 ----------------------------------------------------------------
 Outliers detected in reference groups: 8

 Natural Log of VCVM Determinant =  45.5582
 ----------------------------------------------------------------
Classification Table
 ---------------------------------------------------------------------
 From     Total          Into Group (counts)
 Group    Number      JM     VM     WM    Correct
 ---------------------------------------------------------------------
     JM    193       166     19      8     86.0 %
     VM     48         3     43      2     89.6 %
     WM    373        29      7    337     90.3 %
 ---------------------------------------------------------------------
 Total Correct:  546 out of 614 (88.9 %) *** CROSS-VALIDATED ***
 ---------------------------------------------------------------------

Multigroup Classification of Current Case

 ------------------------------------------------------------------------------------
  Group      Classified      Distance          Probabilities
               into            from      Posterior   Typ F   Typ Chi    Typ R
 ------------------------------------------------------------------------------------

    WM        **WM**          10.8         0.851     0.972    0.967     0.963 (14/374)
    VM                        15.2         0.093     0.850    0.813     0.755 (12/49)
    JM                        16.2         0.056     0.789    0.757     0.675 (63/194)
 ------------------------------------------------------------------------------------
  Current Case is closest to WMs
 ------------------------------------------------------------------------------------
```

Figure 9.19. Sample Fordisc output.

References

Albanese J, Saunders SR. 2006. Is it possible to escape racial typology in forensic identification? In: Schmitt A, Cunha E, Pinheiro J (Eds.), *Forensic Anthropology and Medicine: Complementary Sciences from Recovery to Cause of Death.* Totowa, NJ: Humana Press. p. 281–316.

Algee-Hewitt BFB, Hughes CE, Anderson BE. 2018. Temporal, geographic and identification trends in craniometric estimates of ancestry for persons of Latin American origin. *Forensic Anthropology* 1: 4–17.

Armelagos GJ, Goodman AH. 1998. Race, racism, and anthropology. In: Goodman AH, Leatherman TL (Eds.), *Building a New Biocultural System: Political-Economic Perspectives on Human Biology.* Ann Arbor: University of Michigan Press. p. 359–377.

Ayers HG, Jantz RL, Moore-Jansen PH. 1990. Giles & Elliot race discriminant functions revisited: a test using recent forensic cases. In: Gill GW, Rhine S (Eds.), *Skeletal Attribution of Race: Methods for Forensic Anthropology.* Albuquerque, NM: Maxwell Museum of Anthropology Anthropological Papers No. 4. p. 65–71.

Baker SJ, Gill GW, Kieffer DA. 1990. Race & sex determination from the intercondylar notch of the distal femur. In: Gill GW, Rhine S (Eds.), *Skeletal Attribution of Race: Methods for Forensic Anthropology.* Albuquerque, NM: Maxwell Museum of Anthropology Anthropological Papers No. 4. p. 91–96.

Barbujani G, Colonna V. 2010. Human genome diversity: frequently asked questions. *Trends in Genetics* 26: 285–295.

Berg GE, Kenyhercz MW. 2017. Introducing human mandible identification [(huMANid)]: a free, web-based GUI to classify human mandibles. *Journal of Forensic Sciences* 62: 1592–1598.

Berg GE, Ta'Ala SC, Kontanis EJ, Leney SS. 2007. Measuring the intercondylar shelf angle using radiographs: intra- and inter-observer error tests of reliability. *Journal of Forensic Sciences* 52: 1020–1024.

Bethard JD, DiGangi EA. 2020. Letter to the editor—moving beyond a lost cause: forensic anthropology and ancestry estimates in the United States. *Journal of Forensic Sciences* 65: 1791–1792.

Birkby WH. 1966. An evaluation of race and sex identification from cranial measurements. *American Journal of Physical Anthropology* 24: 21–27.

Blakey ML. 1999. Scientific racism and the biological concept of race. *Literature and Psychology* 45: 29–43.

Brace, CL. 1995. Region does not mean "race"—reality versus convention in forensic anthropology. *Journal of Forensic Sciences* 40: 171–175.

Brace CL. 2005. *"Race" Is a Four-Letter Word: The Genesis of the Concept.* Oxford: Oxford University Press.

Caspari R. 2009. 1918: Three perspectives on race and human variation. *American Journal of Physical Anthropology* 139: 5–15.

Caspari R. 2010. Deconstructing race: racial thinking, geographic variation, and implications for biological anthropology. In: Larsen CS (Ed.), *A Companion to Biological Anthropology.* Malden, MA: Blackwell. p. 104–123.

Christensen AM, Passalacqua NV, Bartelink EJ. 2019. *Forensic Anthropology: Current Methods and Practice,* Second Edition. London: Academic Press.

Collins JW Jr, David RJ, Handler A, Wall S, Andes S. 2004. Very low birthweight in African American infants: the role of maternal exposure to interpersonal racial discrimination. *American Journal of Public Health* 94: 2132–2138.

Cox K, Tayles NG, Buckley HR. 2006. Forensic identification of "race": the issues in New Zealand. *Current Anthropology* 47: 869–874.

DiGangi EA, Bethard JD. 2021. Uncloaking a lost cause: decolonizing ancestry estimation in the United States. *American Journal of Physical Anthropology* 175: 422–436.

Dikötter F. 1992. *The Discourse of Race in Modern China.* Hong Kong: Hong Kong University Press.

d'Oliveira Coelho J, Navega D. 2019. *hefneR: Cranial Nonmetric Traits Ancestry Estimation.* Available from: http://osteomics.com/hefneR.

Dudzik B, Jantz RL. 2016. Misclassification of Hispanics using Fordisc 3.1: comparing cranial morphology in Asian and Hispanic populations. *Journal of Forensic Sciences* 61: 1311–1318.

Dunn RR, Spiros MC, Kamnikar KR, Plemons AM, Hefner JT. 2020. Ancestry estimation in forensic anthropology: a review. *WIREs Forensic Science* 2: e1369. Available from: https://doi.org/10.1002/wfs2.1369.

Earnshaw VA, Rosenthal L, Lewis JB, Stasko EC, Tobin JN, Lewis TT, Reid AE, Ickovics JR. 2013. Maternal experiences with everyday discrimination and infant birth weight: a test of mediators and moderators among young, urban women of color. *Annals of Behavioral Medicine* 45: 13–23.

Edgar HJH. 2005. Prediction of race using characteristics of dental morphology. *Journal of Forensic Sciences* 50: 269–273.

Edgar HJH. 2009. Biohistorical approaches to "race" in the United States: biological distances among African Americans, European Americans, and their ancestors. *American Journal of Physical Anthropology* 139: 58–67.

Edgar HJH. 2013. Estimation of ancestry using dental morphological characteristics. *Journal of Forensic Sciences* 58: S3-S8.
Edgar HJH. 2014. Dental morphological estimation of ancestry in forensic contexts. In: Berg G, Ta'ala SC (Eds.), *Biological Affinity in Forensic Identification of Human Skeletal Remains: Beyond Black and White.* Boca Raton, FL: CRC Press. p. 191–207.
Elhaik E. 2012. Empirical distributions of F_{ST} from large-scale human polymorphism data. *PLoS ONE* 7(11): e49837. doi: 10.1371/journal.pone.0049837.
Federal Interagency Working Group for Research on Race and Ethnicity. *Interim Report to the Office of Management and Budget: Review of Standards for Maintaining, Collecting, and Presenting Federal Data on Race and Ethnicity* (March 3, 2017). Available from: https://www.whitehouse.gov/wp-content/uploads/legacy_drupal_files/briefing-room/presidential-actions/related-omb-material/r_e_iwg_interim_report_022417.pdf.
Fisher TD, Gill GW. 1990. Application of the Giles & Elliot discriminant function formulae to a cranial sample of Northwestern Plains Indians. In: Gill GW, Rhine S (Eds.), *Skeletal Attribution of Race: Methods for Forensic Anthropology.* Albuquerque, NM: Maxwell Museum of Anthropology Anthropological Papers No. 4. p. 59–63.
Gilbert R, Gill GW. 1990. A metric technique for identifying American Indian femora. In: Gill GW, Rhine S (Eds.), *Skeletal Attribution of Race: Methods for Forensic Anthropology.* Albuquerque, NM: Maxwell Museum of Anthropology Anthropological Papers No. 4. p. 97–99.
Giles E, Elliot O. 1962. Race identification from cranial measurements. *Journal of Forensic Sciences* 7: 147–157.
Gravlee CC. 2009. How race becomes biology: embodiment of social inequality. *American Journal of Physical Anthropology* 139: 47–57.
Hefner JT. 2007. *The Statistical Determination of Ancestry Using Nonmetric Traits.* PhD Dissertation, University of Florida.
Hefner JT. 2009. Cranial nonmetric variation and estimating ancestry. *Journal of Forensic Sciences* 54: 985–995.
Hefner JT, Linde KC. 2018. *Atlas of Human Cranial Macromorphoscopic Traits.* Cambridge, MA: Elsevier.
Hefner JT, Ousley SD. 2014. Statistical classification methods for estimating ancestry using morphoscopic traits. *Journal of Forensic Sciences* 59: 883–890.
Hefner JT, Spradley MK, Anderson B. 2014. Ancestry assessment using random forest modeling. *Journal of Forensic Sciences* 59: 583–589.
Herrmann NP, Plemons A, Harris EF. 2016. Estimating ancestry of fragmentary remains via multiple classifier systems: a study of the Mississippi State Asylum skeletal assemblage. In: Pilloud MA, Hefner JT (Eds.), *Biological Distance Analysis: Forensic and Bioarchaeological Perspectives.* London: Elsevier. p. 285–299.
Hochman A. 2013. Racial discrimination: how not to do it. *Studies in History and Philosophy of Biological and Biomedical Sciences* 44: 278–286.
Holliday TW, Falsetti AB. 1999. A new method for discriminating African-American from European-American skeletons using postcranial osteometrics reflective of body shape. *Journal of Forensic Sciences* 44: 926–930.
Hubbard AR. 2017. Teaching race (bioculturally) matters: a visual approach for college biology courses. *The American Biology Teacher* 79: 516–524.
Hughes CE, Tise ML, Trammel LH, Anderson BE. 2013. Cranial morphological variation among contemporary Mexicans: regional trends, ancestral affinities, and genetic comparisons. *American Journal of Physical Anthropology* 151: 506–517.
Iliescu FM, Chaplin G, Rai N, Jacobs GS, Mallick CB, Mishra A, Thangaraj K, Jablonski NG. 2018. The influences of genes, the environment, and social factors in the evolution of skin color diversity in India. *American Journal of Human Biology* 30: e23170. https://doi.org/10.1002/ajhb.23170.
Irish JD. 2015. Dental nonmetric variation around the world: using key traits in populations to estimate ancestry in individuals. In: Berg GE, Ta'ala SC (Eds.), *Biological Affinity in Forensic Identification of Human Skeletal Remains: Beyond Black and White.* Boca Raton, FL: CRC Press. p. 165–190.
Jablonski NG. 2012. *Living Color: The Biological and Social Meaning of Skin Color.* Berkeley, CA: University of California Press.
Jablonski NG, Chaplin G. 2010. Human skin pigmentation as an adaptation to UV radiation. *Proceedings of the National Academy of Sciences of the United States of America* 107: 8962–8968.
Jantz RL, Ousley SD. 2005. *FORDISC 3.0: Personal Computer Forensic Discriminant Functions.* Knoxville: University of Tennessee.
Kennedy KAR. 1995. But professor, why teach race identification if races don't exist? *Journal of Forensic Sciences* 40: 797–800.
Kenyhercz MW, Klales AR, Rainwater CW, Fredette SM. 2017. The optimized summed scored attributes method for the classification of U.S. Blacks and Whites: a validation study. *Journal of Forensic Sciences* 62: 174–180.

Klecka WR. 1980. *Discriminant Analysis.* Newbury Park, CA: SAGE Publications, Inc.
Kuzawa CW, Sweet E. 2009. Epigenetics and the embodiment of race: developmental origins of US racial disparities in cardiovascular health. *American Journal of Human Biology* 21: 2–15.
Langley NR, Meadows Jantz L, Ousley SD, Jantz RL, Milner G. 2016. *Data Collection Procedures for Forensic Skeletal Material 2.0.* Knoxville: University of Tennessee. Available from: http://fac.utk.edu/wp-content/uploads/2016/03/DCP20_webversion.pdf.
Lease LR, Sciulli PW. 2005. Brief communication: discrimination between European-American and African-American children based on deciduous dental metrics and morphology. *American Journal of Physical Anthropology* 126: 56–60.
Li JZ, Absher DM, Tang H, Southwick AM, Casto AM, Ramachandran S, Cann HM, Barsh GS, Feldman M, Cavalli-Sforza LL, Myers RM. 2008. Worldwide human relationships inferred from genome-wide patterns of variation. *Science* 319: 1100–1104.
Lukachko A, Hatzenbuehler ML, Keyes KM. 2014. Structural racism and myocardial infarction in the United States. *Social Science & Medicine* 103: 42–50.
Maier C, Craig A, Adams DM. 2021. Language use in ancestry research and estimation. *Journal of Forensic Sciences* 66: 11–24.
Manica A, Amos W, Balloux F, Hanihara T. 2007. The effect of ancient population bottlenecks on human phenotypic variation. *Nature* 448: 346–348.
Meeusen RA, Christensen AM, Hefner JT. 2015. The use of femoral neck axis length to estimate sex and ancestry. *Journal of Forensic Sciences* 60: 1300–1304.
Moore-Jansen PH, Jantz RL. 1986. *A Computerized Skeletal Data Bank for Forensic Anthropology.* Knoxville: University of Tennessee.
Office of Management and Budget. *Revisions to the Standards for the Classification of Federal Data on Race and Ethnicity,* 62 Fed. Reg. 58782 (October 30, 1997).
Ousley SD, Jantz RL. 2012. Fordisc 3 and statistical methods for estimating sex and ancestry. In: Dirkmaat DC (Ed.), *A Companion to Forensic Anthropology,* First Edition. London: Blackwell. p. 311–329.
Ousley SD, Billeck WT, Hollinger RE. 2005. Federal repatriation legislation and the role of physical anthropology in repatriation. *Yearbook of Physical Anthropology* 48: 2–32.
Ousley S, Jantz R, Fried D. 2009. Understanding race and human variation: why forensic anthropologists are good at identifying race. *American Journal of Physical Anthropology* 139: 68–76.
Parra EJ, Marcini A, Akey J, Martinson J, Batzer MA, Cooper R, Forrester T, Allison DB, Deka R, Ferrell RE, Shriver MD. 1998. Estimating African American admixture proportions by use of population-specific alleles. *American Journal of Human Genetics* 63: 1839–1851.
Parsons HR. 2017. *The Accuracy of the Biological Profile in Casework: An Analysis of Forensic Anthropology Reports in Three Medical Examiners' Offices.* PhD Dissertation, University of Tennessee, Knoxville.
Pilloud MA, Hefner JT, Hanihara T, Hayashi A. 2014. The use of tooth crown measurements in the assessment of ancestry. *Journal of Forensic Sciences* 59: 1493–1501.
Pilloud MA, Maier C, Scott GR, Hefner JT. 2018. Advances in cranial macromorphoscopic trait and dental morphology analysis for ancestry estimation. In: Latham K, Bartelink E, Finnegan M (Eds.), *New Perspectives in Forensic Human Skeletal Identification.* London: Academic Press. p. 23–34.
Pilloud MA, Adams DM, Hefner JT. 2019. Observer error and its impact on ancestry estimation using dental morphology. *International Journal of Legal Medicine* 133: 949–962.
Pratt BM, Hixson L, Jones NA. n.d. Infographic: measuring race and ethnicity across the decades, 1790–2010. United States Census Bureau. Available from: https://www.census.gov/data-tools/demo/race/MREAD_1790_2010.html.
Quillen EE, Norton HL, Parra EJ, Lona-Durazo F, Ang KC, Iliescu FM, Pearson LN, Shriver MD, Lasisi T, Gokcumen O, Starr I, Lin Y-L, Martin AR, Jablonski NG. 2019. Shades of complexity: new perspectives on the evolution and genetic architecture of human skin. *American Journal of Physical Anthropology* 168 (S67): 4–26.
Relethford JH. 2004. Boas and beyond: migration and craniometric variation. *American Journal of Human Biology* 16: 379–386.
Relethford JH. 2009. Race and global patterns of phenotypic variation. *American Journal of Physical Anthropology* 139: 16–22.
Relethford JH. 2010. The study of human population genetics. In: Larsen CS (Ed.), *A Companion to Biological Anthropology.* Malden, MA: Blackwell Publishing. p. 74–87.
Rhine S. 1990. Non-metric skull racing. In: Gill GW, Rhine S (Eds.), *Skeletal Attribution of Race: Methods for Forensic Anthropology.* Albuquerque, NM: Maxwell Museum of Anthropology Anthropological Papers No. 4. p. 9–20.

Ross AH, Pilloud M. 2021. The need to incorporate human variation and evolutionary theory in forensic anthropology: a call for reform. *American Journal of Physical Anthropology* 176: 672–683.
Sauer NJ. 1992. Forensic anthropology and the concept of race: if races don't exist, why are forensic anthropologists so good at identifying them? *Social Science & Medicine* 34: 107–111.
Sauer NJ, Wankmiller JC, Hefner JT. 2016. The assessment of ancestry and the concept of race. In: Blau S, Ubelaker DH (Eds.), *Handbook of Forensic Anthropology and Archaeology,* Second Edition. New York, NY: Routledge. p. 243–260.
Scientific Working Group for Forensic Anthropology (SWGANTH). 2013. Ancestry assessment, June 12, 2013. Available from: https://www.nist.gov/system/files/documents/2018/03/13/swganth_ancestry_assessment.pdf.
Scott GR, Pilloud MA, Navega D, d'Oliveira Coelho J, Cunha E, Irish JD. 2018. rASUDAS: a new web-based application for estimating ancestry from tooth morphology. *Forensic Anthropology* 1: 18–31.
Shennan S. 1997. *Quantifying Archaeology,* Second Edition. Iowa City: University of Iowa Press.
Smay D, Armelagos G. 2000. Galileo wept: a critical assessment of the use of race in forensic anthropology. *Transforming Anthropology* 9: 19–29.
Sobotta J. 1909. *Atlas and Text-Book of Human Anatomy.* Philadelphia, PA: W.B. Saunders Co.
Spiros MC, Hefner JT. 2020. Ancestry estimation using cranial and postcranial macromorphoscopic traits. *Journal of Forensic Sciences* 65: 921–929.
Spradley MK. 2014. Toward estimating geographic origin of migrant remains along the United States–Mexico border. *Annals of Anthropological Practice* 38: 101–110.
St. Hoyme LE, İşcan MY. 1989. Determination of sex and race: accuracy and assumptions. In: İşcan MY, Kennedy KAR (Eds.), *Reconstruction of Life from the Skeleton.* New York, NY: Alan R. Liss, Inc. p. 53–94.
Stock MK, Rubin KM. 2020. Race and the role of sociocultural context in forensic anthropological ancestry assessment. In: Garvin HM, Langley NR (Eds.), *Case Studies in Forensic Anthropology: Bonified Skeletons.* Boca Raton, FL: CRC Press. p. 39–50.
Stull KE, Bartelink EJ, Klales AR, Berg GE, Kenyhercz MW, L'Abbé EN, Go MC, McCormick K, Mariscal C. 2021. Commentary on: Bethard JD, DiGangi EA. Letter to the editor—moving beyond a lost cause: forensic anthropology and ancestry estimates in the United States. *Journal of Forensic Sciences* 66: 417–420.
Telles E, Paschel T. 2014. Who is Black, white, or mixed race? How skin color, status, and nation shape racial classification in Latin America. *American Journal of Sociology* 120: 864–907.
Templeton AR. 2013. Biological races in humans. *Studies in History and Philosophy of Biological and Biomedical Sciences* 44: 262–271.
Thomas RM, Parks CL, Richard AH. 2017. Accuracy rates of ancestry estimation by forensic anthropologists using identified forensic cases. *Journal of Forensic Sciences* 62: 971–974.
Walsh-Haney H, Boys S. 2015. Creating the biological profile: the question of race and ancestry. In: Crossland Z, Joyce RA (Eds.), *Disturbing Bodies: Perspectives on Forensic Anthropology.* Santa Fe, NM: SAR Press. p. 121–135.
Weir BS, Cardon LR, Anderson AD, Nielsen DM, Hill WG. 2005. Measures of human population structure show heterogeneity among genomic regions. *Genome Research* 15: 1468–1476.

10

Stature Estimation

Learning Goals

By the end of this chapter, the student will be able to:

Describe and apply the anatomical method of stature estimation.
Describe and apply the mathematical method of stature estimation.
Define regression and correlation with respect to their role in stature estimation.
Identify challenges with stature estimation, including possible adjustments to estimated stature.
Define forensic stature.

Introduction

Stature, one's adult standing height, is the fourth attribute of the biological profile of interest to forensic anthropologists. While one might think of one's height as something that is constant once adulthood is reached, this isn't actually the case. Height varies throughout the day, for example. You are taller in the morning before gravity has had time to compress the soft tissue connections in your joints, primarily in your spine (Kobayashi and Togo, 1993). In addition, your stature changes as you age. While we all remember growing taller as a child, less obvious is the loss of stature as we get older, which is why many of us fail to correct official records such as a driver's license with current and correct stature data. Researchers have debated when this age-related decline begins in life, but it appears to be sometime in your 40s (Ousley, 1995; Willey, 2016). Finally, people misreport their stature, both intentionally and unintentionally. For example, males (especially college age males), older individuals, and short-to-medium height individuals tend to overreport their stature when their measured height is compared against driver's license data (Giles and Hutchinson, 1991; Willey and Falsetti, 1991).

These complexities around the reporting of stature are why the concept of **forensic stature** was proposed (Ousley, 1995). Similar to social race discussed in chapter 9, forensic stature refers to the height one would expect to be reported in official documents such as a driver's license, police report or other public records, or could be derived from interviews with family members. Since our goal is positive identification by linking the biological profile to missing persons reports, forensic stature is usually the target of our analysis. However, some individuals also have data available for **measured stature**, which refers to one's stature based on direct measurement of your body either while living (such as at an annual physical exam by a doctor) or as a cadaver (which are on average 2.5 cm larger than living measured statures (Trotter and Gleser, 1952)). **Biological stature** is the physical height of an individual, which is variable throughout the day and during the course of one's life. Ousley (1995) more fully discusses the challenges and nuances associated with estimating stature in medicolegal contexts because of the differences between biological, measured, and forensic stature.

Approaches to Stature Estimation

There are two broad approaches to estimating stature that are principally used in forensic anthropology. The first is based on measuring the skeletal elements that directly contribute to one's standing height, referred to as the **anatomical method** of stature estimation. The second uses statistical techniques such as linear regression (and less commonly, bone ratios) to estimate stature from single or multiple long bones based on the known, positive correlation between skeletal element size (primarily length) and overall height. This is referred to as the **mathematical method** of stature estimation.

Anatomical Method

Methods using the complete skeleton are referred to as the Fully or anatomical method (Fully, 1956; Fully and Pineau, 1960; Raxter et al., 2006). The basis of this method is direct and straightforward. First, measure the skeletal elements that directly contribute to stature. These are shown in **Figure 10.1**. Second, add these measurements together to produce a *skeletal element sum*. Third, apply a correction factor to account for spinal curvature (which reduces stature) and soft tissue contributions to stature (which increase it). The latter include the spacing of the joints, especially the inter-vertebral joints occupied by the inter-vertebral discs, and the thickness of the heel and scalp.

The anatomical method of Fully (1956) has been the subject of considerable research and revision. Here we follow the approach of Raxter et al. (2006), which redefines and clarifies the Fully measurements that define the skeletal element sum as follows (all measurements are in cm):

1) *Cranial height,* measured from basion to bregma. Basion is the anterior most point of the foramen magnum in the base of the skull. Bregma is the point at the intersection of the coronal and sagittal sutures in the midline. This measurement is taken with a pair of spreading calipers (**Figure 10.2**).
2) *Maximum heights of the vertebral bodies,* measuring the second cervical through fifth lumbar vertebrae. Be sure to include the dens or odontoid process when measuring the second cervical vertebra (**Figure 10.3**). Note that Raxter et al. (2006) use the *maximum height of the body* anterior to the pedicles instead of the anterior midline height. Difficulties arise because of the need to have a completely preserved spine, although methods have been proposed for estimating the heights of missing segments (Auerbach, 2011). This measurement is taken with a pair of sliding calipers.
3) *Anterior height of the first sacral segment* (**Figure 10.4**). This measurement is taken with a pair of sliding calipers.
4) *Bicondylar femur length,* measured by placing both the medial and lateral distal condyles firmly against one end of an osteometric board. This measurement will always be shorter than maximum femur length for which only the medial condyle touches the end of the measuring device (**Figure 10.5**). Ideally the average of the left and right side should be used.
5) *Physiological tibia length,* defined as the length of the tibia *excluding* the spines of the proximal surface and *including* the medial malleolus (**Figure 10.6**). This measurement is taken with an osteometric board, preferably one without a fixed measuring

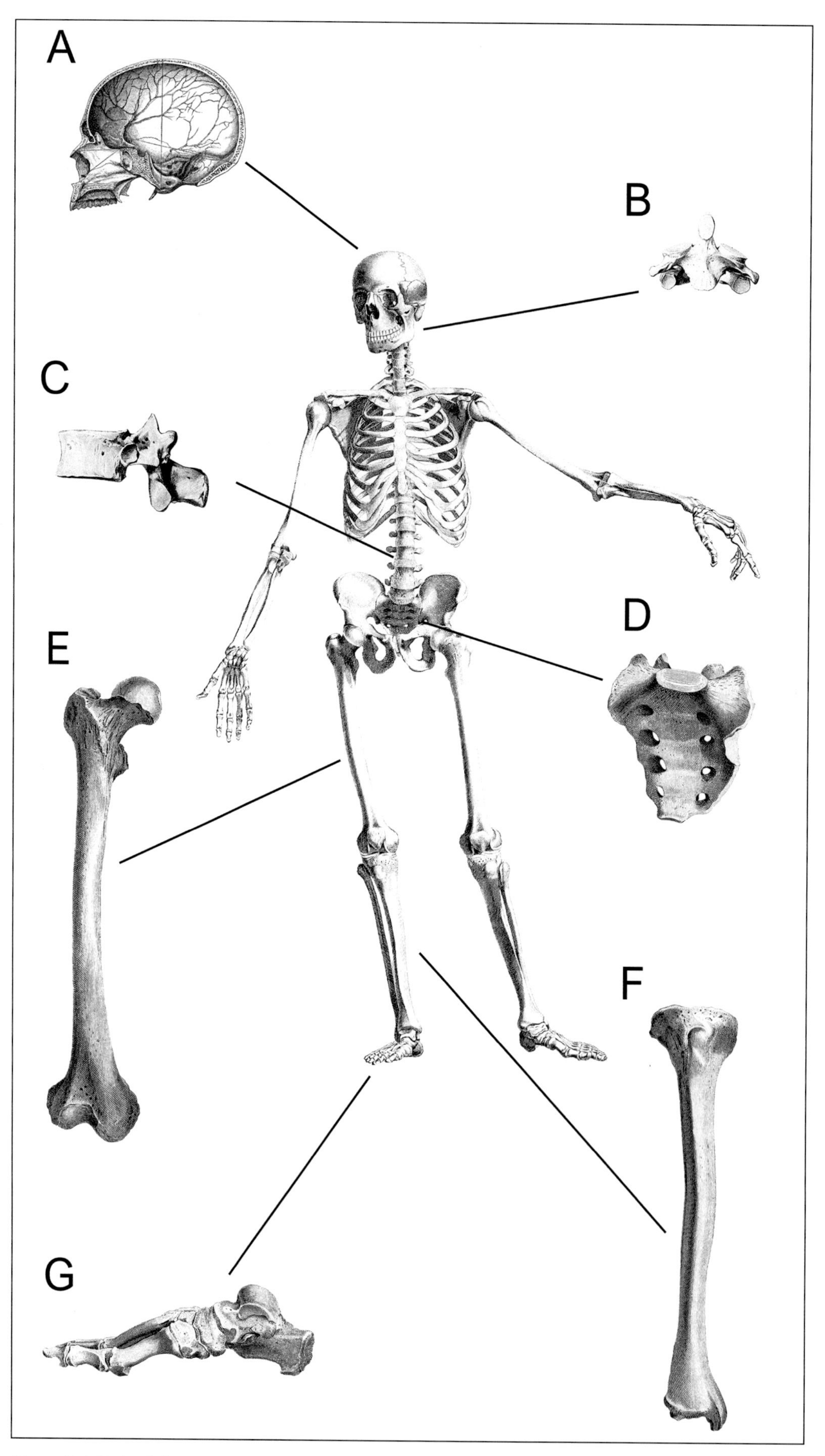

Figure 10.1. Skeletal elements measured in the anatomical method of stature estimation: (*A*) the cranium; (*B*) the axis, including the odontoid process; (*C*) vertebral elements C3 through L5; (*D*) the first sacral element; (*E*) the femur; (*F*) the tibia; (*G*) the height of the articulated foot.

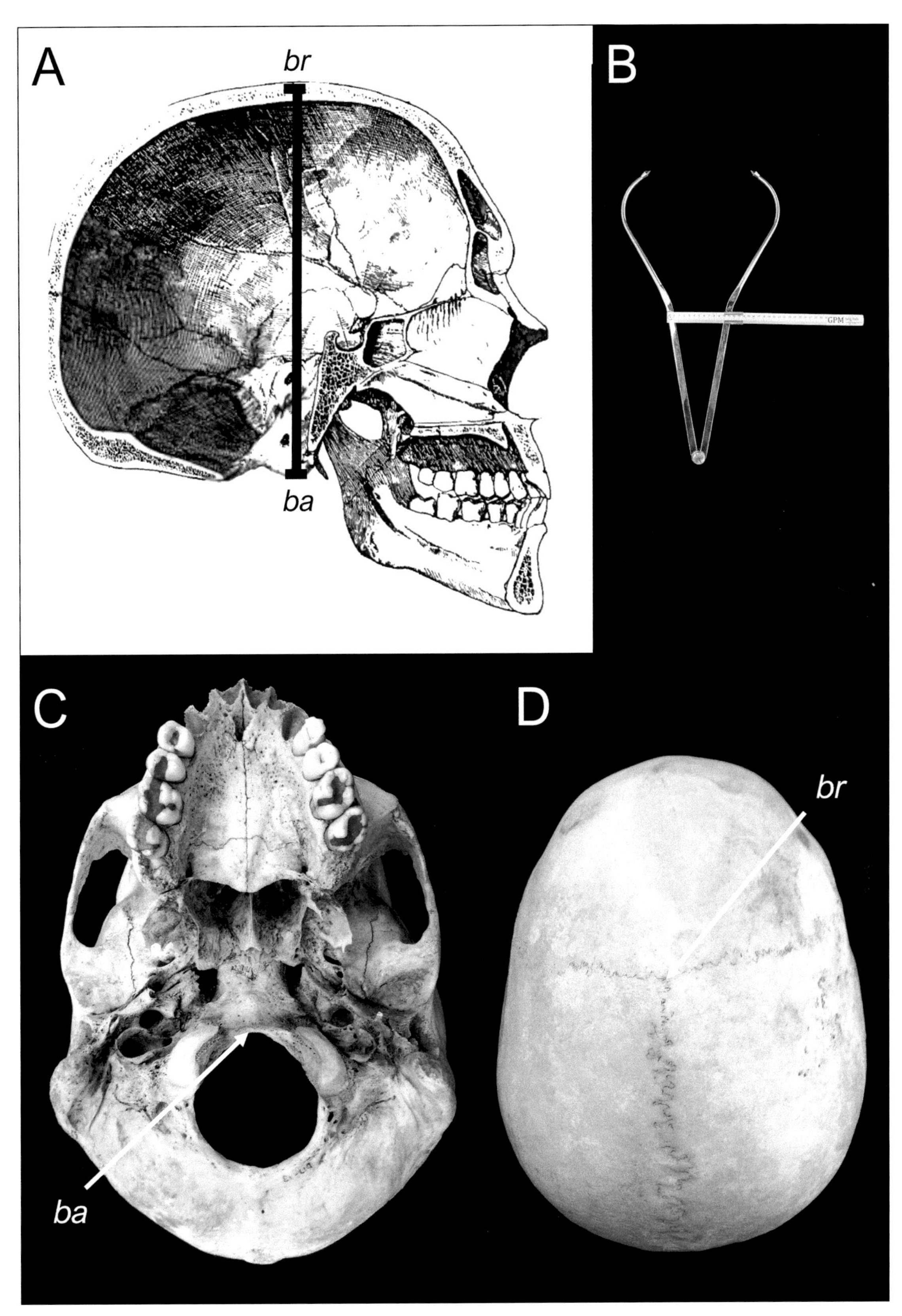

Figure 10.2. Cranial height. (*A*) hemisected cranium with the chord from basion (ba) to bregma (br); (*B*) spreading calipers—the instrument used to take this measurement; (*C*) cranium, inferior aspect, showing the location of basion; (*D*) cranium, superior aspect, showing location of bregma.

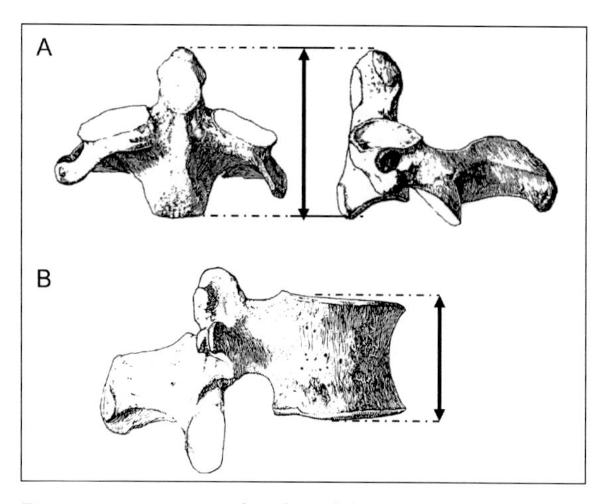

Figure 10.3. Maximum heights of the vertebral bodies as shown for (*A*) the axis (C2) and (*B*) a lumbar vertebra.

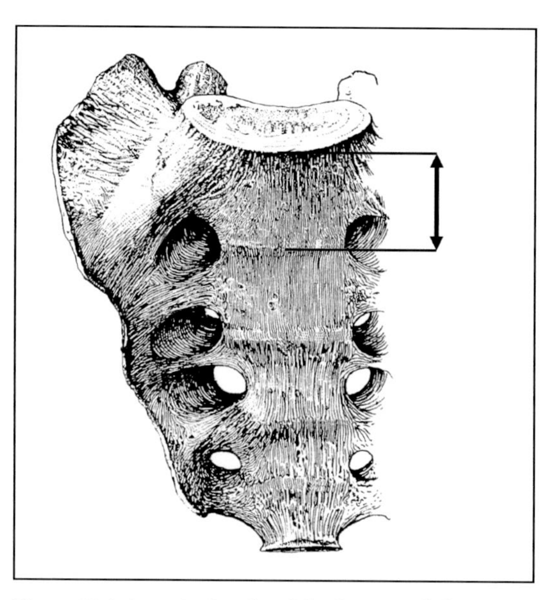

Figure 10.4. Anterior height of the first sacral element.

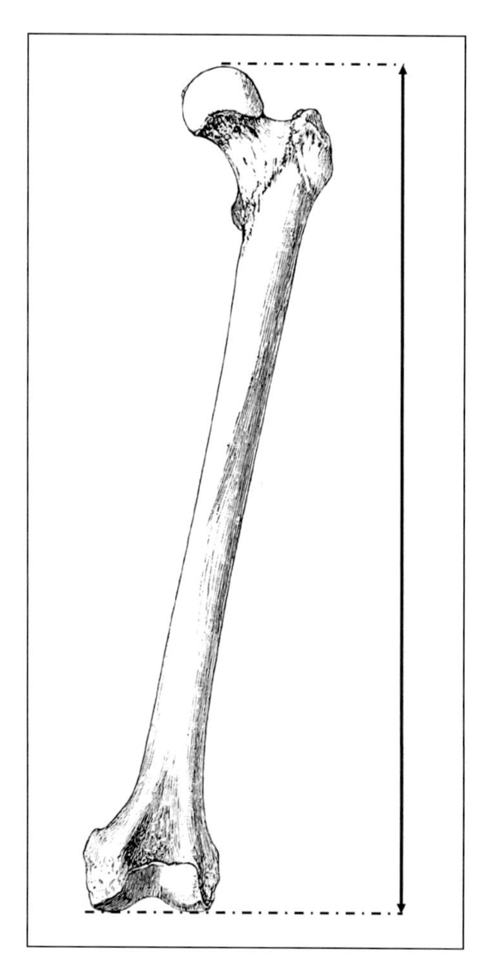

Figure 10.5. Bicondylar femur length.

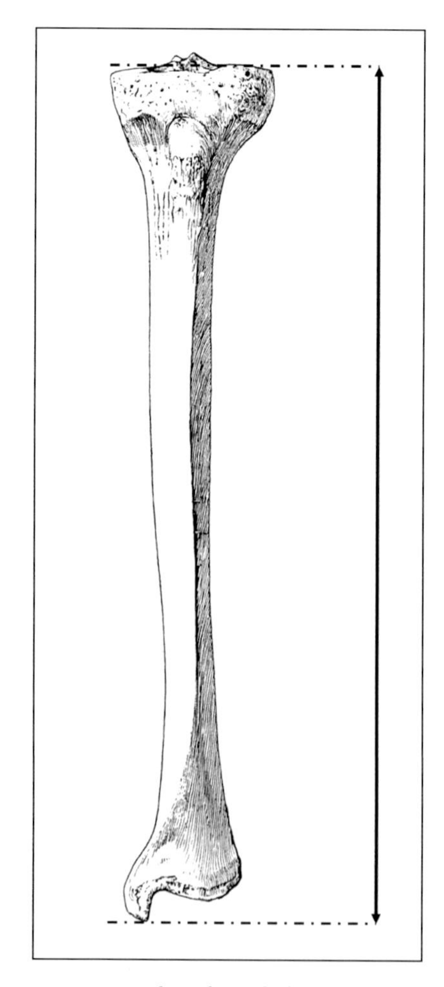

Figure 10.6. Physiological tibia length.

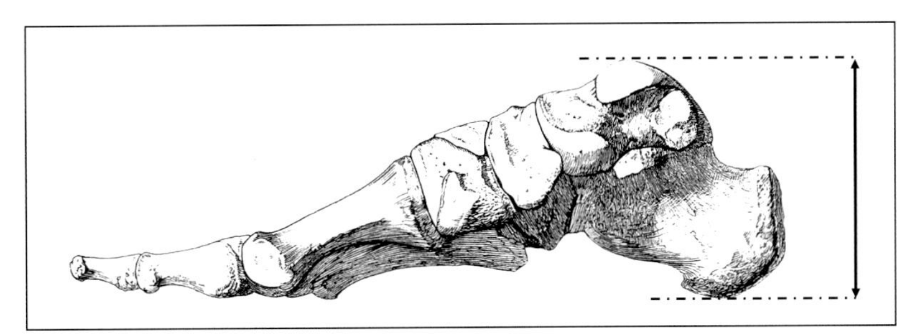

Figure 10.7. Height of the articulated talus and calcaneus.

block (see Raxter et al., 2006 for discussion). Ideally, the average of the left and right side should be used.

6) *Height of the articulated talus and calcaneus,* measured from the most superior point on the talus to the most inferior point on the calcaneus with the anterior end of the calcaneus elevated above the board in a horizontal position (i.e., in anatomical position) (**Figure 10.7**). This measurement is taken with a pair of extended jaw sliding calipers. Ideally, the average of the left and right side should be used.

The specific soft tissue correction factor added to the *skeletal element sum* is a source of continued revision (e.g., Bidmos and Manger, 2012). Fully and Pineau (1960) proposed the following:

1) If the sum of skeletal measurements is < 153.5 cm, add 10 cm.
2) If the sum of skeletal measurements is between 153.6 and 165.5 cm, add 10.5 cm.
3) If the sum of skeletal measurements is >165.5 cm, add 11.5 cm.

Subsequent research found that the Fully method underestimated stature in adults, which is likely caused by incorrect correction factors. Raxter et al. (2006) proposed an equation that better approximates known stature:

$$\text{(1) Living stature (age unknown)} = [0.996 \times (\text{summed skeletal elements in cm})] + 11.7\ \text{cm}$$
$$\text{(standard error is 2.31 cm)}$$

Because stature is known to decline with age, Raxter et al. (2006) also provided an equation for known-age individuals, which accounts for this trend in age-related stature decline.

$$\text{(2) Living stature (age known)} = [1.009 \times (\text{summed skeletal elements in cm})] - [.0426 \times \text{age}] + 12.1\ \text{cm}$$
$$\text{(standard error is 2.22 cm)}$$

These estimates in centimeters can be converted to imperial units (feet and inches) to calculate **forensic stature** as follows:

1) Divide the total stature in cm by 2.54 to convert centimeters to inches. If your stature estimate is 170 cm this converts to 66.93 in.
2) Divide this number (66.93 in) by 12 to get the number of feet (5.58 ft).
3) Then multiply the decimals (.58) by 12 to get the fractional inches (7 in).
4) 170 cm equals a forensic stature of 5 ft 7 in.

The benefits of the anatomical method are that it can be applied regardless of the sex of the individual and is relatively free of population-specific effects. This is because you are estimating stature directly from the skeleton of the individual and not making an inference based on a number of unknown factors. Because it is a more direct assessment of stature, it is also the most accurate method of estimation. The principal limitation of the anatomical method is that its application requires many skeletal elements to be present and well preserved. Recording the measurements, especially for the vertebrae, is also time consuming. For this reason, the anatomical method is less often used than the mathematical method.

LEARNING CHECK

Q1. Using the data in **Table 10.1,** compute the estimated total *skeletal element height.*
A) 168 cm
B) 171 cm
C) 175 cm
D) 180 cm

Q2. Using the data in **Table 10.1,** compute the estimated stature using the Fully and Pineau (1960) correction factor.
A) 178.5 cm
B) 184.5 cm
C) 185 cm
D) 182.5 cm

Q3. Using the data in **Table 10.1,** compute the estimated stature using the Raxter et al. (2006) correction factor as shown in equation (1) above.
A) 178 cm
B) 184 cm
C) 185 cm
D) 182 cm

Q4. Using the data in **Table 10.1,** compute the estimated stature using the Raxter et al. (2006) correction factor as shown in equation (2) above for someone who was estimated to be 68 years old at the time of death.
A) 181.7 cm
B) 184.3 cm
C) 185.2 cm
D) 178.2 cm

Mathematical Methods

Mathematical methods of stature reconstruction use statistical techniques to generate stature estimates based on a single or several skeletal elements, many of which do not directly contribute to one's living stature. These techniques are based on the statistical concept of *correlation,* in this case correlation between a measured skeletal element and stature as viewed on a bivariate (two-variable) plot (**Figure 10.8**). Correlation assumes there is a linear relationship between two variables—as one variable increases (a bone length) so does the other (living stature) in a predictable fashion (**Figure 10.8A, B**). That is, correlation assumes that individuals with longer femora tend to be taller than those with shorter femora, and vice versa. Correlations can vary from -1 to +1 (compare **Figure 10.8A, B to 10.8D, E**), where a correlation of 0 means there is no relationship between the two variables (**Figure 10.8C, F**). This means the variables cannot be used to predict each other. Correlations close to +1 or -1

Table 10.1. Sample skeletal measurements for stature estimation

Measurement	Data
Cranial height	14 cm
Heights of C2-L5 vertebrae	54 cm
Sacral segment 1 height	4 cm
Femur bicondylar length	49 cm
Tibia physiological length	42 cm
Talus and calcaneus height	8 cm

indicate a strong relationship between the variables. Compare how close the points are to the black line in **Figure 10.8A, D** (higher correlation) versus **Figure 10.8C, F** (lower correlation). Most correlations involving the human skeleton are *positive correlations*, and our goal is to identify those measurements that produce the highest positive correlation between that measurement and stature.

The statistical technique used to explore the correlation between a bone measurement and stature is called *linear regression*. Linear regression uses information on individuals for whom stature is known and for which different skeletal elements can be measured to mathematically characterize the correlation between the two variables (say femur length and stature). The rate at which one variable increases relative to another defines the slope of the *regression line*, where a slope of 1 means that for every 1 unit increase in femur length there is a 1 unit increase in stature. Statistical regression is the process of defining the line that best fits a plot of bone measurement data against stature data for a sample of individuals for whom both

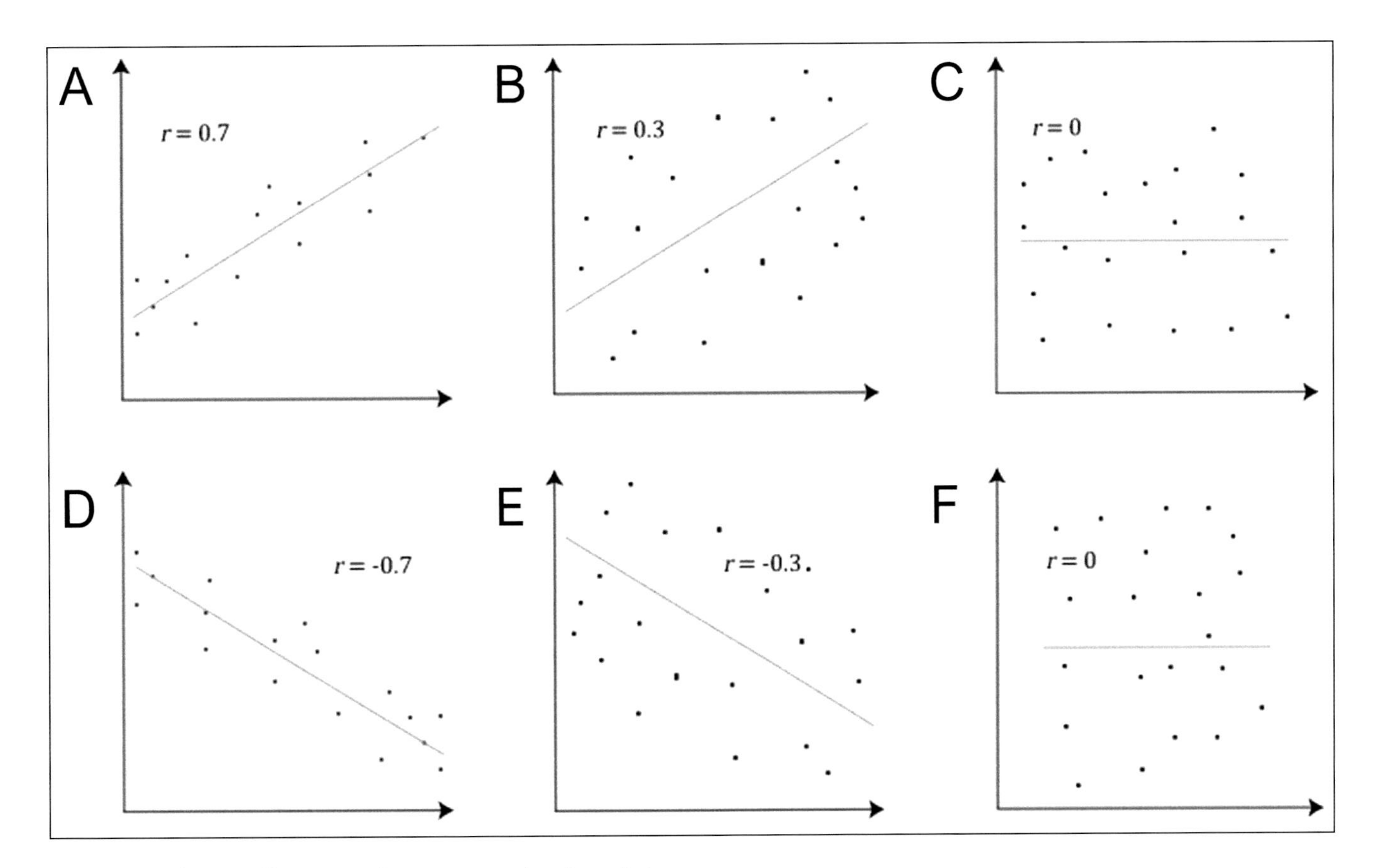

Figure 10.8. Schematic illustration of various strengths of correlation.

measurements are known (see **Figure 10.8**). The goal is to define an equation that can be used for predicting stature when *only* the bone measurement is available.

The general formula for a regression line is:

$$(3)\ y = mx + b$$

Where m is the slope of the line, x is the variable we can measure, and b is the y-intercept.

Consider again that the Raxter et al. equation (1) above demonstrates a regression line where the slope is .996 and the y-intercept is 11.7 cm. The variable y in equation (3) is the unknown variable (stature) that we want to predict from the known variable x (the bone measurement in our forensics case). Regression lines that better fit the data result when the correlation between stature and the bone measurement is high. Visually this appears when the points for each individual included in the study sample plot close to the regression line, such as **Figures 10.8A** and **10.8D**. The more spread out the points are from the regression line the worse the regression will perform at predicting the stature of an unknown individual.

One statistical measure of this performance is the *standard error,* which we report for various published stature regression formulae below. However, the standard error cannot easily be used to estimate the precision of stature estimates along the full range of bone lengths (Giles and Klepinger, 1988). For this reason, another measure of precision, the *prediction interval,* is now preferred (Ousley, 1995). The statistical details of standard errors, confidence intervals, and prediction intervals are beyond the scope of this chapter. However, prediction intervals, when calculated correctly such as in Fordisc, provide the best stature estimate ranges with known probabilities that satisfy the *Daubert* criteria. For example, a 90 percent prediction interval tells the investigator that 10 percent of actual statures for that bone length will be outside the range of the prediction interval. In other words, the forensic anthropologist will "miss" the true stature of the individuals 10 percent of the time. This can be reduced to 5 percent or even 1 percent, but doing so increases the stature range, which makes it less useful as part of the biological profile. For example, using a femur length of 45 cm we can generate prediction intervals as follows:

90 percent PI—62.8 to 69.8 in; 7 in range
95 percent PI—62.1 to 70.4 in; 8.3 in range
99 percent PI—60.8 to 71.8 in; 11 in range

The process of stature estimation using the mathematical method is quick, efficient, and relatively straightforward. Record a measurement using the appropriate definition; insert that measurement into the appropriate stature estimation formula, taking into consideration the sex, age, and population affinity of the decedent (if known); estimate stature and record the prediction interval; and apply appropriate corrections or adjustments to the final estimate.

Using documented collections of skeletons, researchers have defined numerous regression equations that describe linear correlations between skeletal elements and stature. The most commonly used methods, and the most accurate, are those based on the lengths of long bones. For long bones, those of the lower limb, in particular the femur, perform better than the bones of the upper limb because the former directly contribute to one's stature. Regression equations that use multiple long bones also tend to perform better than single bone equations. Nonetheless, when deciding between which equation to use, the one with the *lowest standard error* is always preferred. Averaging the estimates from multiple long bone equations (e.g., using the humerus equation and then the femur equation) is not statistically valid and should be avoided. Measurement definitions are presented for the long bones in **Table 10.2**.

Table 10.2. Long bone measurement definitions used to calculate stature

Measurement	Definition
Humerus Length	Maximum length from most proximal to most distal point as measured on an osteometric board. Hold the proximal end firmly against the non-moveable end of the board and slightly shift the distal end until the maximum measurement is obtained.
Radius Length	Maximum length from most proximal to most distal point as measured on an osteometric board. Hold the proximal end firmly against the non-moveable end of the board and slightly shift the distal end until the maximum measurement is obtained. Be sure to include the distal styloid process.
Ulna Length	Maximum length from most proximal to most distal point as measured on an osteometric board. Hold the proximal end firmly against the non-moveable end of the board and slightly shift the distal end until the maximum measurement is obtained. Be sure to include the distal styloid process.
Femur Length	Maximum length from most proximal portion of the head to the most distal point on the medial condyle as measured on an osteometric board. Hold the proximal end firmly against the non-moveable end of the board and slightly shift the medial condyle until the maximum measurement is obtained.
Tibia Length	Length from the most proximal point on the lateral condyle to the most distal point on the malleolus. This measurement is taken with an osteometric board but the distal end should be held against the non-moveable end. This is the most difficult long bone length measurement to record accurately.
Fibula Length	Maximum length from most proximal to most distal point as measured on an osteometric board. Hold the proximal end firmly against the non-moveable end of the board and slightly shift the distal end until the maximum measurement is obtained.

Trotter and Gleser (1952, 1958) and Trotter (1970) provided some of the most well-known examples of long bone equations that are sex and population-specific. These formulae are among the most commonly used by forensic anthropologists, providing stature estimation equations for the humerus, radius, ulna, femur, tibia, and fibula as well as various combinations of these individual long bone lengths. Unfortunately, the specific method used to measure the tibia in these studies was unclear and apparently misreported (Jantz et al., 1994, 1995). Tibia length can be defined as including or excluding the spines of the proximal epiphysis and including or excluding the medial malleolus, and it seems the Trotter equations were actually derived from tibia lengths *excluding* the medial malleolus, despite how those publications define tibia length. Wilson et al. (2010) provided updated equations based on a modern American sample (post-1944 births), which is more appropriate for contemporary U.S. forensic cases due to secular trends in stature during the twentieth century (Meadows Jantz and Jantz, 1995; Ousley and Jantz, 1998). We report selected stature estimation equations from Wilson et al. (2010) in **Table 10.3**.

Table 10.3. Forensic stature estimation equations

Population	Equation (Bone Length in cm)[a]	95% PI at mean[a]
Black Males	3.371 × Humerus + 62.046	11.306
	2.410 × Femur + 58.483	10.179
	2.628 × Tibia + 68.205	11.385
	2.916 × Fibula + 60.030	9.991
	1.323 × (Femur + Tibia) + 57.345	10.377
Black Females	5.01 × Humerus + 9.777	9.206
	2.802 × Femur + 37.852	7.618
	3.217 × Tibia + 43.66	7.568
	3.569 × Fibula + 33.128	8.344
	1.576 × (Femur + Tibia) + 33.78	5.111
White Males	3.541 × Humerus + 58.389	11.465
	2.835 × Femur + 41.967	9.724
	2.962 × Tibia + 68.205	9.342
	2.916 × Fibula + 64.052	10.140
	1.603 × (Femur + Tibia) + 37.933	8.796
White Females	2.527 × Humerus + 86.587	10.656
	2.637 × Femur + 48.549	6.978
	2.311 × Tibia + 81.485	9.070
	2.559 × Fibula + 73.747	8.531
	1.336 × (Femur + Tibia) + 57.754	8.065

[a] Excerpted from Wilson et al. (2010), Table 1. Students are referred to this source for further information and equations using additional skeletal elements.

Using the "White Male" equations we can estimate stature using maximum femur length. Assuming a measured maximum femur length of 45 cm, the equation is as follows:

$$(4)\ \text{forensic stature} = [2.835 \times (\text{femur length in cm})] + 41.967\ \text{cm}$$

—This equals 169.542 cm.

—The reported 95 percent prediction interval given in **Table 10.3** is 9.724.

—The 95 percent predicted range of stature for this individual is 159.818 to 179.266 cm, or we can round this to 160–179 cm.

—Converting these to imperial units yields a forensic stature of 63–70 in, or 5 ft 3 in to 5 ft 10 in.

—Because this is a 95 percent prediction interval, we expect only 5 percent of white males with a femur length of 45 cm to fall outside this range of variation.

Fordisc (Jantz and Ousley, 2005) provides another tool for estimating forensic stature that offers a number of benefits over published equations. In particular, Fordisc provides an option for when the sex and population affinity of the individual is unknown, provides options for which population you use to estimate stature including birth year data to account for secular trends, allows you to estimate 90 percent, 95 percent, and 99 percent prediction intervals, and can produce stature estimates for single long bones or multiple bones simultaneously. An example of Fordisc output for a femur length of 450 mm and tibia length of 350 mm is presented in **Figure 10.9**. The table at the top shows three stature estimates (tibia and femur, femur only, and tibia only). The "tibia and femur" estimate is preferred because it has the lowest prediction interval (PI) of 8.8 cm. Note that this estimate also has the highest value for R^2. Referred to as the coefficient of determination, R^2 can range in value from 0 to 1. For regression equations, like stature estimation, the closer the R^2 value is to 1, the better the regression fits the data. The other output gives the estimate of stature (165.4 cm), the lower and upper bounds for the prediction interval (156.7 to 174.2 cm) shown as red dotted lines on the graphic output, the sample size of the population used to estimate the equation (N = 545 individuals represented by blue triangles), and the slope (0.138) and y-intercept (54.64) of the regression line (shown in black). The stature estimation equation is:

$$\text{(5) forensic stature} = [0.13848 \times (\text{femur length} + \text{tibia length})] + 54.64 \text{ cm}$$

90 percent prediction interval = ± 8.8 cm
The same information is given for the “femur only” and “tibia only” equations.

LEARNING CHECK

Q5. Using the data in **Figure 10.9,** which of the following is the correct equation for estimating stature from the femur?
A) .26366 × (femur length) + 49.69
B) .27059 × (femur length) + 49.69
C) .26366 × (femur length) + 68.11
D) .27059 × (femur length) + 68.11

Q6. Using the data in **Figure 10.9,** which of the following is the correct equation for estimating stature from the tibia?
A) .26366 × (tibia length) + 49.69
B) .27059 × (tibia length) + 49.69
C) .26366 × (tibia length) + 68.11
D) .27059 × (tibia length) + 68.11

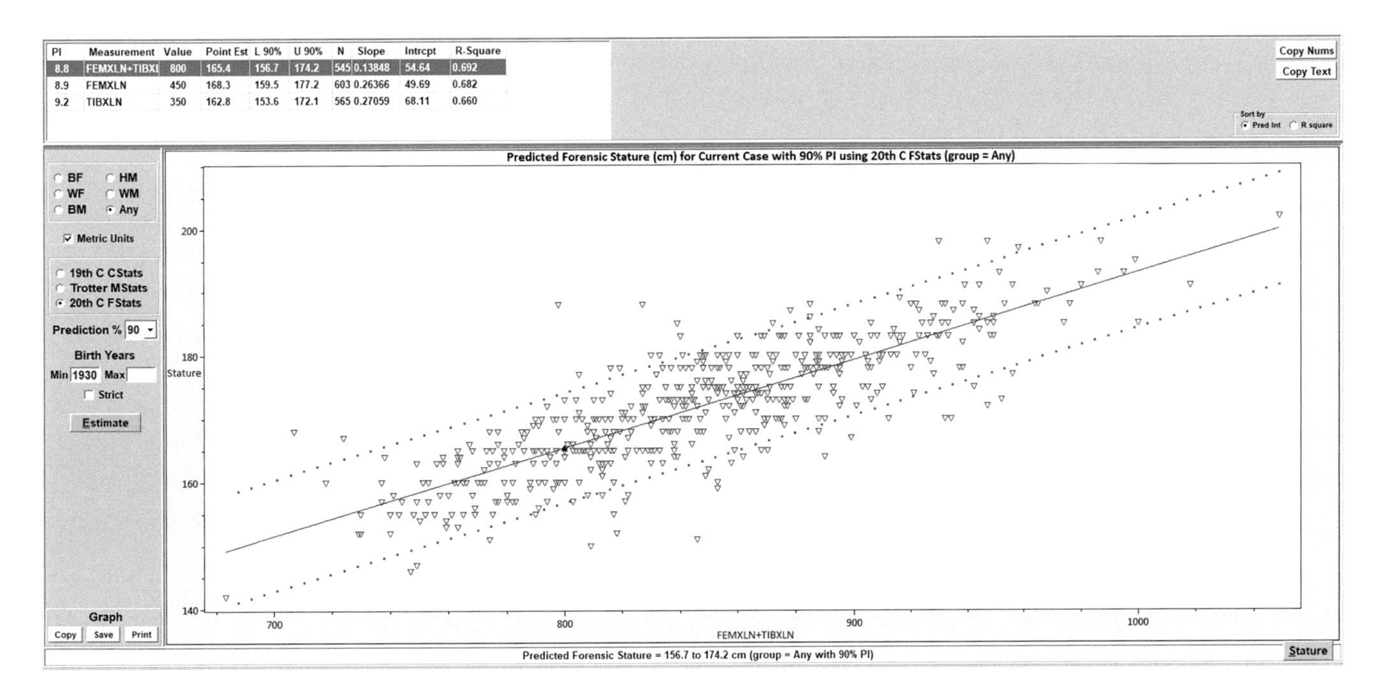

Figure 10.9. Sample Fordisc output for a stature estimate based on a femur length of 450 mm and a tibia length of 350 mm.

Q7. Using the data in **Figure 10.9,** which long bone by itself produces the better stature estimate based on the prediction intervals?
A) Femur
B) Tibia

Q8. Using the data in **Figure 10.9,** which of the following represents the prediction interval *in imperial units* for the femur + tibia stature equation?
A) 5 ft 6 in to 5 ft 8 in
B) 5 ft 2 in to 6 ft 1 in
C) 5 ft 2 in to 5 ft 9 in
D) 5 ft 4 in to 5 ft 10 in

Q9. Using the data in **Figure 10.9,** what is the point estimate of stature for a tibia length of 350 mm, rounded to the nearest centimeter?
A) 155 cm
B) 158 cm
C) 163 cm
D) 166 cm

Q10. Using the data in **Figure 10.9,** what is the 90 percent prediction interval for a tibia length of 350 mm, rounded to the nearest centimeter?
A) 157 to 174 cm
B) 160 to 177 cm
C) 154 to 174 cm
D) 154 to 172 cm

Other Methods

While long bones provide the most accurate stature estimates, many other measurements have been considered including those of the cranium, clavicle, femoral head diameter, foramen magnum, metacarpals, metatarsals, os coxa, sacrum, scapula, sternum, tarsals (calcaneus and talus), and vertebrae (see Willey, 2016 for a review). None of these methods work as well as long bone lengths, however, because they have lower correlations with stature. Stature can also be estimated from partial long bones, which could be the only valid approach for highly fragmentary cases (Steele, 1970). However, this process involves two estimation steps (estimating full long bone length from the partial length and then estimating stature from the reconstructed long bone length) that increase error significantly.

Stature Estimation Considerations

Because mathematical methods use statistical inference to estimate stature, they are more subject to population specificity, which includes differences between the sexes, country of origin, and even time period. Similarly, despite the debate about forensic anthropologists' use of ancestry and race (see chapter 9), it is well known that body proportions vary geographically with latitude and therefore population-specific equations are always used. Males and females also have different body proportions, so it is critical that sex be estimated prior to stature. This is why published equations are sex-specific. Finally, stature is highly affected by diet and nutrition and shows strong secular trends (Jantz and Jantz, 1999; Meadows and Jantz, 1995). This means that stature can change very rapidly over short periods of time in a population due to changes in health and diet, which could quickly make published stature equations invalid or error prone. For these reasons, highly specific stature estimation equations have been published (e.g., for Thai or Nigerian populations—Didia et al., 2009; Mahakkanukrauh et al., 2011) that may have limited use in the United States. Secular trends also explain issues with using the Trotter stature equations, which is why they have been replaced with data using modern Americans, such as in the Forensic Data Bank utilized by Fordisc (Ousley, 1995; Wilson et al., 2010).

Adjustments to Stature Estimates

The age of the decedent is another factor that should be considered when estimating stature. The cause is suggested to be the loss of height of the inter-vertebral discs as one ages as well as a reduction in height due to accumulating injuries of the spine and general loss of bone density. Early research suggested stature begins to decline around the age of 30, while other researchers found the decline in stature began later in life, around the age of 45 (Galloway, 1998; Giles, 1991; Trotter and Gleser, 1951). The rate of stature decline varies from around .6 to 1.6 mm per year; however, the rate of decrease may not be linear and is likely sex-specific (Giles, 1991). Age dependent decreases in stature are reflected in equation (2) above (after Raxter et al., 2006), which models a slight reduction in predicted stature for each year of increasing age. Nonetheless, the forensic anthropological community is still debating how the effects of age should be included in estimations of stature. Adjustments to stature for expected over or underreporting are also problematic and should be used carefully.

End-of-Chapter Summary

This chapter presented two methods of estimating adult stature based on the skeleton. The first method is the anatomical or Fully method, which uses measurements of the skeletal elements that directly contribute to stature and applies a tissue correction factor to generate a height estimate. Because this method uses information from the specific decedent, it is less prone to error from sex, age, or population effects. It is time consuming, however, and requires excellent preservation. The second method is the mathematical method, which uses information on bone size and stature for known individuals to generate equations that can be used to estimate stature in unknown individuals based on a single bone or multiple elements, primarily long bones. This method is based on the statistical concepts of correlation and regression. Because it uses statistical methods, it is important to match, as best as possible, the decedent's biological profile to the population used to generate the stature regression equation. This can be challenging, and there is more potential for bias using mathematical methods. Stature should be reported with prediction intervals and not with single point estimates. The effects of age on stature could be significant, but the forensics community is still determining how best to use such data to adjust stature estimates. Finally, this chapter introduced the concepts of biological vs. measured vs. forensic stature.

End-of-Chapter Exercises

Exercise 1

Materials Required: **Figure 10.10.**

Scenario: Four separate sets of skeletal remains have been submitted to your office for analysis. The skeletal inventories for each case are shown in **Figure 10.10.**

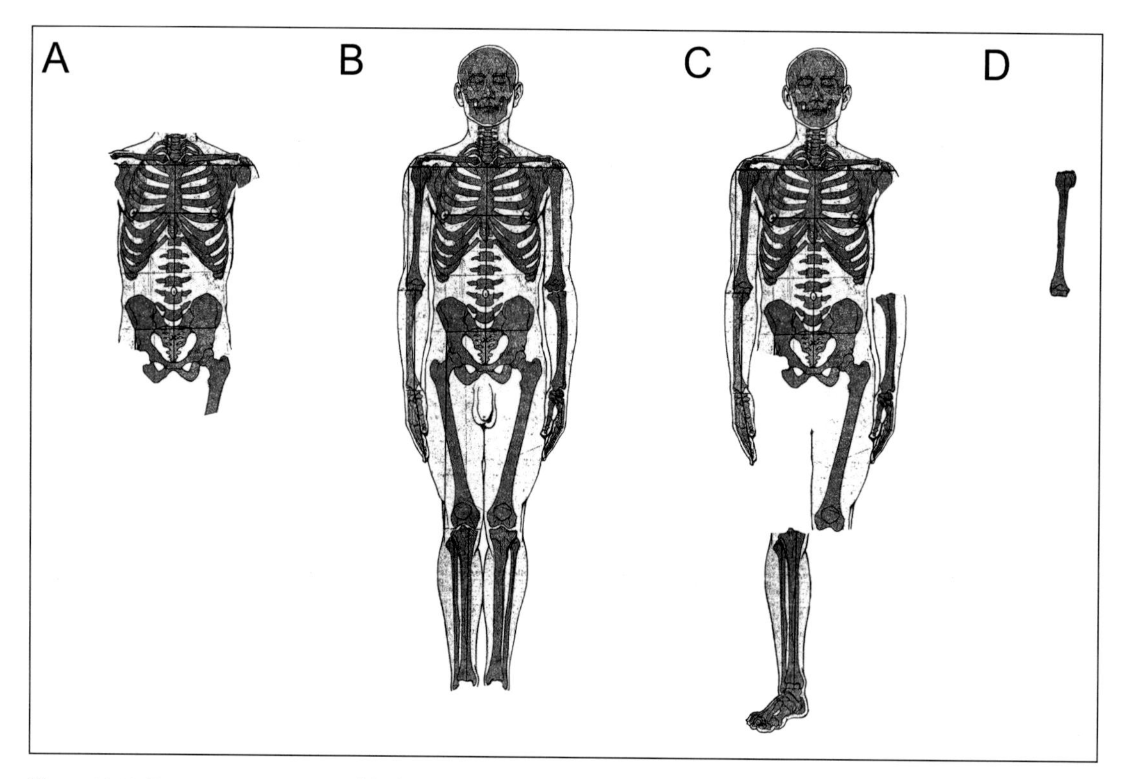

Figure 10.10. Four representations of differential preservation of skeletal elements.

Question 1: For Case A, can a reasonable estimate of forensic stature be made from these remains? If so, is it more appropriate to use an anatomical estimation method or to use a regression equation based on long bone lengths?

Question 2: For Case B, can a reasonable estimate of forensic stature be made from these remains? If so, is it more appropriate to use an anatomical estimation method or to use a regression equation based on long bone lengths?

Question 3: For Case C, can a reasonable estimate of forensic stature be made from these remains? If so, is it more appropriate to use an anatomical estimation method or to use a regression equation based on long bone lengths?

Question 4: For Case D, can a reasonable estimate of forensic stature be made from these remains? If so, is it more appropriate to use an anatomical estimation method or to use a regression equation based on long bone lengths?

Question 5: Based on this small sample of cases, would you think regression equations or anatomical methods for stature estimation are more commonly used?

Exercise 2

Materials Required: **Table 10.4** and a computer with internet access.

Directions: **Table 10.4** presents a selection of stature estimation equations for a male decedent with the following recorded long bone lengths:

Left humerus: 339 mm

Left femur: 487 mm

Left tibia: 398 mm

Question 1: Based on the data in **Table 10.4,** which skeletal element or combination of skeletal elements produces the best estimate of stature for these remains?

Question 2: Using the equation from your answer to Question 1 and the long bone lengths given above, calculate the 90 percent prediction interval for the stature of this decedent. Note

Table 10.4. Stature estimation equations generated for a male decedent based on available long bone lengths

Equation	90% Prediction Interval (in)	R^2
0.10862 × Humerus + 32.397	± 3.5	0.526
0.09370 × Femur + 23.647	± 3.1	0.622
0.09555 × Tibia + 30.219	± 3.0	0.662
0.05599 × (Humerus + Femur) + 23.025	± 3.0	0.646
0.06072 × (Humerus + Tibia) + 23.872	± 2.7	0.718
0.04990 × (Femur + Tibia) + 24.583	± 2.9	0.669
0.03816 × (Humerus + Femur + Tibia) + 22.188	± 2.7	0.696

that the equation provides a result that is already in imperial units (i.e., inches)—there is no need for unit conversion.

Question 3: Go to www.namus.gov. Click on the "Explore NamUs" button and then, on the next page, click on the link for "Missing Advanced Search" (located at the bottom of the box on the left labeled "Quick Search"). Under the heading "Description," check the box for "Male" and record the number of possible results as "*Value A.*" Then, enter the 90 percent prediction interval for stature that you calculated in Question 2 under "Height." You can enter the numbers in whole inches (round up or down as appropriate)—the page will automatically convert them to feet and inches for you. Once the prediction interval is entered, record the number of possible results as "*Value B.*"

Approximately what proportion of missing males are recorded as having statures that fall into your 90 percent prediction interval? *Hint:* Divide *Value B* by *Value A* and multiply by 100.

Exercise 3

Materials Required: None.

Scenario: Responding to a tip from an informant, law enforcement has searched a dry wash for the remains of a white male who was reported missing two years ago. According to his driver's license, this individual was 5 ft, 1 in tall, or had a stature of approximately 61 in. During their search, they discovered scattered skeletal elements, including a left humerus, a right tibia, and a right femur. These bones have been brought to you to assess whether they are consistent with belonging to the missing person.

Question 1: The left humerus measures 277 mm in length. Using Fordisc, you have generated the following stature estimation equation:

Forensic Stature = (0.122 × humerus length) + 28.144

The output of this equation is in inches and has an associated 90 percent prediction interval of ±3.8 in.

Based on this information, is the humerus consistent with belonging to the missing person?

Question 2: The right tibia measures 424 mm in length. Using Fordisc, you have generated the following stature estimation equation:

Forensic Stature = (0.09562 × tibia length) + 31.637

The output of this equation is in inches and has an associated 90 percent prediction interval of ± 3.4 in.

Based on this information, is the tibia consistent with belonging to the missing person?

Question 3: The right femur measures 375 mm in length. Using Fordisc, you have generated the following stature estimation equation:

Forensic Stature = (0.08929 × femur length) + 26.806

The output of this equation is in inches and has an associated 90 percent prediction interval of ±3.5 in.

Based on this information, is the femur consistent with belonging to the missing person?

Question 4: Is it likely that the three bones recovered by law enforcement belong to the same individual? If not, how many individuals are likely represented?

References

Auerbach BM. 2011. Methods for estimating missing human skeletal element osteometric dimensions employed in the revised Fully technique for estimating stature. *American Journal of Physical Anthropology* 145: 67–80.

Barclay J. 1824. *A Series of Engravings Representing the Bones of the Human Skeleton; with the Skeletons of Some of the Lower Animals, and Explanatory References,* Second Edition. Edinburgh, Scotland: Printed for MacLachlan and Stewart.

Bidmos MA, Manger PR. 2012. New soft tissue correction factors for stature estimation: results from magnetic resonance imaging. *Forensic Science International* 214: 212e1–212e7.

Braus H. 1921. *Anatomie des Menschen: ein Lehrbuch für Studierende und Ärzte.* Berlin: Julius Springer.

Didia BC, Nduka EC, Adele O. 2009. Stature estimation formulae for Nigerians. *Journal of Forensic Sciences* 54: 20–21.

Dixon AF. 1912. *Manual of Human Osteology.* New York, NY: William Wood and Co.

Fully G. 1956. Une nouvelle méthode de détermination de la taille. *Annales de Medicine Legale* 36: 266–273.

Fully G, Pineau H. 1960. Détermination de la stature au moyen du squelette. *Annales de Medicine Legale* 40: 145–154.

Galloway A. 1998. Estimating actual height in the older individual. *Journal of Forensic Sciences* 33: 126–136.

Giles E. 1991. Corrections for age in estimating older adults' stature from long bones. *Journal of Forensic Sciences* 36: 898–901.

Giles E, Hutchinson DL. 1991. Stature- and age-related bias in self-reported stature. *Journal of Forensic Sciences* 36: 765–780.

Giles E, Klepinger LL. 1988. Confidence intervals for estimates based on linear regression in forensic anthropology. *Journal of Forensic Sciences* 33: 1218–1222.

Jantz LM, Jantz RL. 1999. Secular change in long bone length and proportion in the United States, 1800–1970. *American Journal of Physical Anthropology* 110: 57–67.

Jantz RL, Ousley SD. 2005. *FORDISC 3.0: computerized forensic discriminant functions.* Knoxville: University of Tennessee.

Jantz RL, Hunt DR, Meadows L. 1994. Maximum length of the tibia. How did Trotter measure it? *American Journal of Physical Anthropology* 93: 525–528.

Jantz RL, Hunt DR, Meadows L. 1995. The measure and mismeasure of the tibia: implications for stature estimation. *Journal of Forensic Sciences* 40: 758–761.

Kobayashi M, Togo M. 1993. Twice-daily measurements of stature and body weight in two children and one adult. *American Journal of Human Biology* 5: 193–201.

Mahakkanukrauh P, Khanpetch P, Prasitwattanseree S, Vichairit K, Case DT. 2011. Stature estimation from long bone lengths in a Thai population. *Forensic Science International* 210: 279e1–279e7.

Meadows L, Jantz RL. 1995. Allometric secular change in the long bones from the 1800s to the present. *Journal of Forensic Sciences* 40: 762–767.

Morris H, McMurrich JP. (Eds.) 1907. *Morris's Human Anatomy: A Complete and Systematic Treatise by English and American Authors,* Fourth Edition. Philadelphia, PA: P. Blakiston's Son & Co.

Ousley S. 1995. Should we estimate biological or forensic stature? *Journal of Forensic Sciences* 40: 768–773.

Ousley SD, Jantz RL. 1998. The Forensic Data Bank: documenting skeletal trends in the United States. In: Reichs KJ (Ed.), *Forensic Osteology: Advances in the Identification of Human Remains,* Second Edition. Springfield, IL: Charles C. Thomas. p. 441–458.

Raxter MH, Auerbach BM, Ruff CB. 2006. Revision of the Fully technique for estimating statures. *American Journal of Physical Anthropology* 130: 374–384.

Steele DG. 1970. Estimation of stature from fragments of long limb bones. In: Stewart TD (Ed.), *Personal Identification in Mass Disasters.* Washington, DC: Smithsonian Institution. p. 85–97.

Trotter M. 1970. Estimation of stature from intact long limb bones. In: Stewart TD (Ed.), *Personal Identification in Mass Disasters.* Washington, DC: Smithsonian Institution. p. 71–83.

Trotter M, Gleser G. 1951. The effect of aging on stature. *American Journal of Physical Anthropology* 9: 311–324.

Trotter M, Gleser GC. 1952. Estimation of stature from long bones of American Whites and Negroes. *American Journal of Physical Anthropology* 10: 463–514.

Trotter M, Gleser GC. 1958. A re-evaluation of estimation of stature based on measurements of stature taken during life and of long bones after death. *American Journal of Physical Anthropology* 16: 79–123.

Willey P. 2016. Stature estimation. In: Blau S, Ubelaker DH (Eds.), *Handbook of Forensic Anthropology and Archaeology.* Second edition. London: Routledge. p. 308–321.

Willey P, Falsetti T. 1991. Inaccuracy in height information on driver's licenses. *Journal of Forensic Sciences* 36: 813–819.

Wilson RJ, Herrmann NP, Jantz LM. 2010. Evaluation of stature estimation from the Database for Forensic Anthropology. *Journal of Forensic Sciences* 55: 684–689.

11

Personal Identification

Learning Goals

By the end of this chapter, the student will be able to:

Describe different kinds of antemortem skeletal signatures and how they can provide identifying information.

Describe the basic differences between two national databases (NamUs and CODIS) that can be used to match unidentified remains with missing persons.

Identify anatomical features that can be used to establish a positive identification.

Apply population frequencies recorded for some traits to evaluate the strength of potential identifications.

Describe alternative ways of generating investigative leads from unidentified skeletal remains.

Introduction

The process of restoring a name to unidentified human remains is referred to as **personal identification**, and it is one of the most important tasks undertaken by a forensic anthropologist. Aside from its role in the investigation and prosecution of homicides, the identification of an unknown decedent is important for the creation of a death certificate and release of the body. Such documentation allows for the execution of wills, the inheritance of property, the transference of retirement benefits, and even remarriage of the surviving spouse. Equally important, the identification of an unknown decedent may help to bring closure to the family and friends that survive them (Komar and Buikstra, 2008; Christensen et al., 2019).

As with the determination of cause and manner of death, personal identification of the deceased is ultimately the responsibility of the medicolegal authority that has jurisdiction (Christensen et al., 2019). Identifications are often categorized based on the level of certainty associated with them. **Tentative** identifications are based on circumstantial evidence, such as when human remains are discovered with a photo ID or perhaps a name written on an associated article of clothing. **Presumptive** identifications are usually based on multiple, concordant lines of evidence and the absence of any contradictory information. Both tentative and presumptive identifications typically require further investigation before a positive identification can be established, although presumptive identifications can sometimes suffice for the generation of a death certificate. **Positive** identifications are made when a medicolegal authority is satisfied that human remains have been identified correctly (Komar and Buikstra, 2008). Although forensic anthropologists do not generally have the authority to establish a positive identification, their work can contribute to the process both by providing information that enables the inclusion and exclusion of missing persons profiles as potential matches for the decedent and by performing identification comparisons (Christensen et al., 2019).

It is often the case that human remains enter the medicolegal system without sufficient information to establish even a tentative identification. In such cases, a forensic anthropologist may be asked to examine the remains and generate a biological profile. The four main components of a biological profile (sex, age, ancestry, and stature) were discussed in detail in the preceding chapters. While it is often assumed that forensic anthropologists work exclusively with skeletal remains, they may also be asked to examine decomposed, burned, dismembered, and even relatively well-preserved decedents. For example, consider a scenario in which a body is recovered in a relatively "fresh" state but without any identifying information. Stature and sex may be readily discerned by measuring the body and examining the intact soft tissue. The estimation of age-at-death, however, is less straightforward and may require the involvement of a forensic anthropologist. In addition to the standard components of the biological profile, anthropological analyses may contribute information pertaining to the postmortem interval (see chapter 15) and the documentation of antemortem skeletal changes.

This chapter discusses the process of establishing a positive identification, beginning with an overview of the different kinds of antemortem skeletal changes that can contribute identifying information. National missing persons databases are discussed in the context of generating potential matches for an unknown decedent. This is followed by a description of common methods of scientific identification comparisons, such as comparative radiography, and the growing use of likelihood ratios to provide quantitative identifications. The chapter closes with a brief discussion of other methods that can contribute to the identification process, such as craniofacial superimposition and facial approximation.

LEARNING CHECK

Q1. Decomposed human remains are recovered from the driver's seat of a vehicle parked in a secluded location deep in the woods. Investigators find a wallet with a driver's license in the back pocket of the decedent's jeans. The name on the license matches the name on the vehicle's registration. Further, after anthropological analysis, the biological profile of the remains is found to be consistent with the information on the driver's license. Does the evidence support a tentative, presumptive, or positive identification?

A) Tentative identification
B) Presumptive identification
C) Positive identification

Q2. Skeletonized remains are recovered from a desert context. Aside from the clothes that they are found in, investigators also recover a cell phone. By contacting the phone's service provider, they are able to determine a name associated with that phone number. Is this an example of a tentative, presumptive, or positive identification?

A) Tentative identification
B) Presumptive identification
C) Positive identification

Antemortem Skeletal Changes

Antemortem skeletal changes include *osseous anomalies, pathological changes* to the skeleton resulting from disease or degeneration, evidence of *healed trauma,* and *medical interventions.* It is important for a forensic anthropologist to be familiar with the different kinds and mechanisms of antemortem changes that can occur in the skeleton as they can both affect the conclusions of anthropological analyses and contribute to the identification process. Some osseous anomalies, for example, have the potential to be mistaken as perimortem trauma. Similarly, healed trauma and medical interventions have the potential to be matched with antemortem medical records. While a thorough discussion of the range of antemortem skeletal changes is beyond the scope of this chapter, some examples are provided below.

Osseous anomalies are morphological variations of the skeleton. They are frequently categorized as either accessory bones, fusion anomalies, accessory foramina, or miscellaneous anomalies (e.g., Byers, 2017; Cunha, 2006). Perhaps the most common examples of accessory bones are Wormian bones (**Figure 11.1**). These are small ossicles that occur within cranial sutures. Their prevalence varies between populations and can also be affected by cultural practices such as head binding, where the skulls of infants are wrapped to alter the shape of their head by constraining the directions in which it can grow (Del Papa and Perez, 2007). Other examples of accessory bones include bipartite zygomatic bones (Hanihara et al., 1998) and the fabella—a small sesamoid bone that can sometimes be found posterior to the knee joint (Berthaume et al., 2019). Some individuals even have additional vertebrae and ribs.

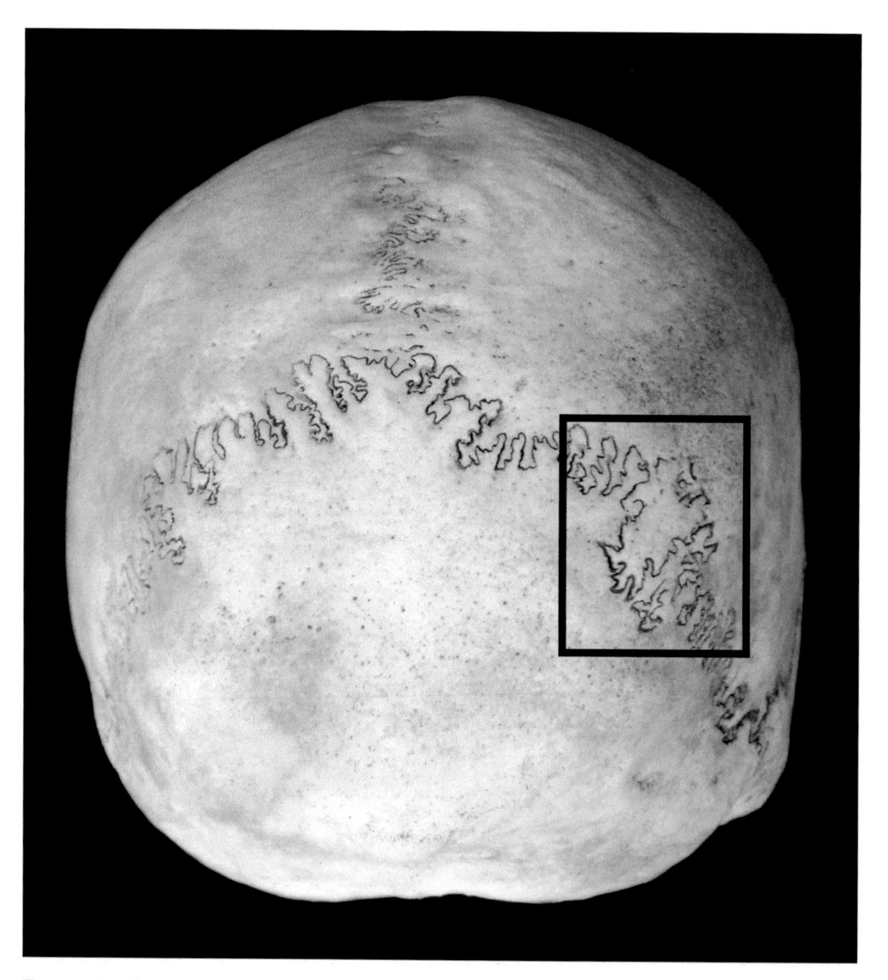

Figure 11.1. Cranium, posterior aspect. Black rectangle encloses a Wormian, or sutural bone.

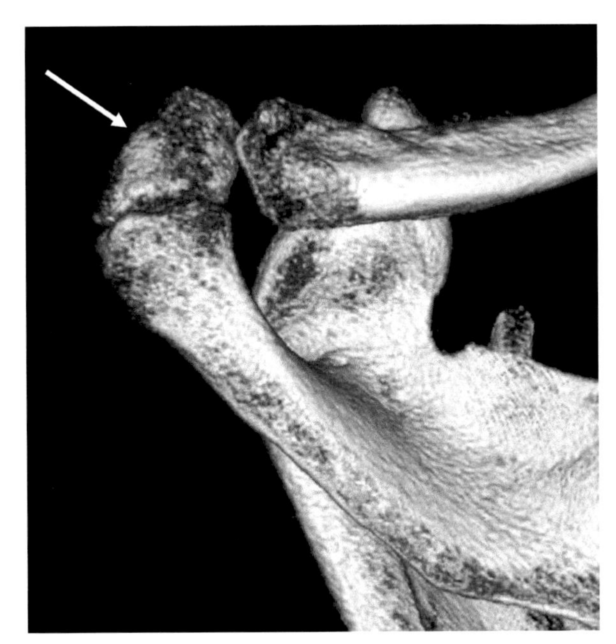

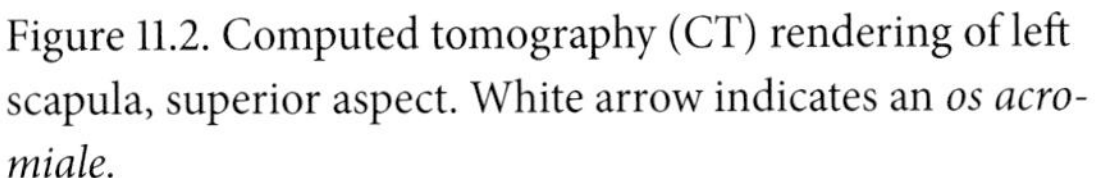

Figure 11.2. Computed tomography (CT) rendering of left scapula, superior aspect. White arrow indicates an *os acromiale.*

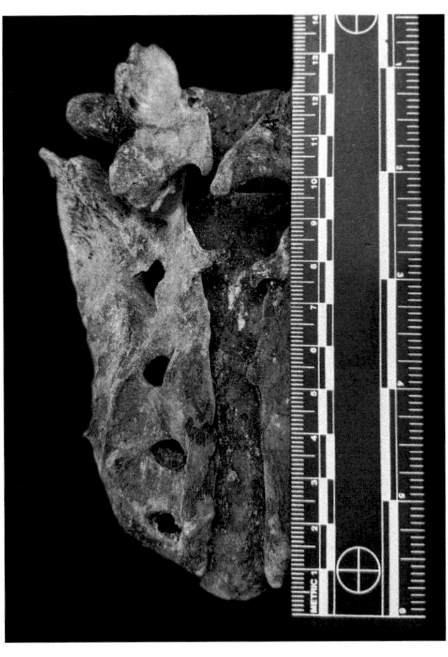

Figure 11.3. Posterior view of sacrum showing unfused neural arches (spina bifida).

Many seemingly "extra" bones are actually the result of fusion anomalies where different centers of ossification fail to unite during normal growth and development. An example of this is the os acromiale—a nonunion between the acromion process of the scapula and the rest of the bone (**Figure 11.2**). Radiographically, this anatomical variant has the potential to be misinterpreted as a fracture (Cunha, 2006). A more well-known example of a fusion anomaly is spina bifida, where the neural arches of the lumbar or sacral vertebrae fail to unite (**Figure 11.3**). These can be associated with neural tube defects and, consequently, have clinical manifestations that can potentially be linked to antemortem medical records (Barnes, 1994).

Other osseous anomalies have the potential to be confused with evidence of perimortem trauma. A classic example is the occurrence of a sternal foramen—a congenital perforation through the body of the sternum that is sometimes confused with a gunshot wound (**Figure 11.4**). **Figure 11.5** depicts an individual with dramatically enlarged parietal foramina. While foramina enlarged to this degree are a normal, if relatively rare, anatomical variant, it is easy to see how practitioners unfamiliar with such variants might interpret their appearance as evidence of trauma. While most osseous anomalies will be asymptomatic and are therefore unlikely to be included in medical records, their relative rarity makes them useful as individualizing characteristics should they happen to appear in antemortem medical imaging.

Unlike osseous anomalies, the majority of which reflect variations in the processes of growth and development and are relatively stable in their conformation, pathology can provoke dynamic responses within the skeleton. *Pathology,* in the context of forensic anthropology, refers to a change from the normal structure and function of an organ or tissue that adversely affects the functioning of an individual. Pathologies can be either congenital or acquired. Congenital pathologies are present at birth, such as genetic disorders. In contrast,

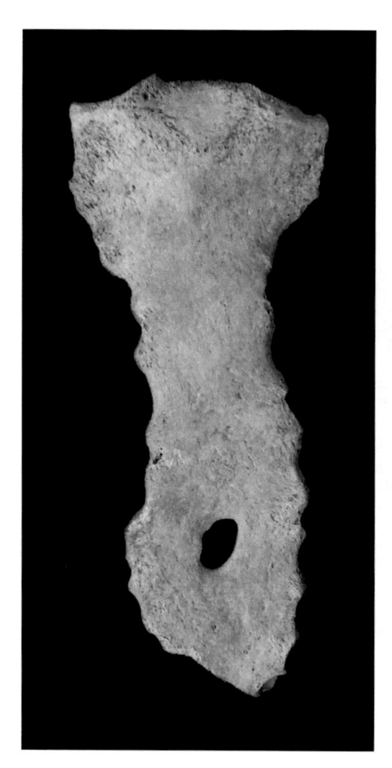

Figure 11.4. Sternal foramen.

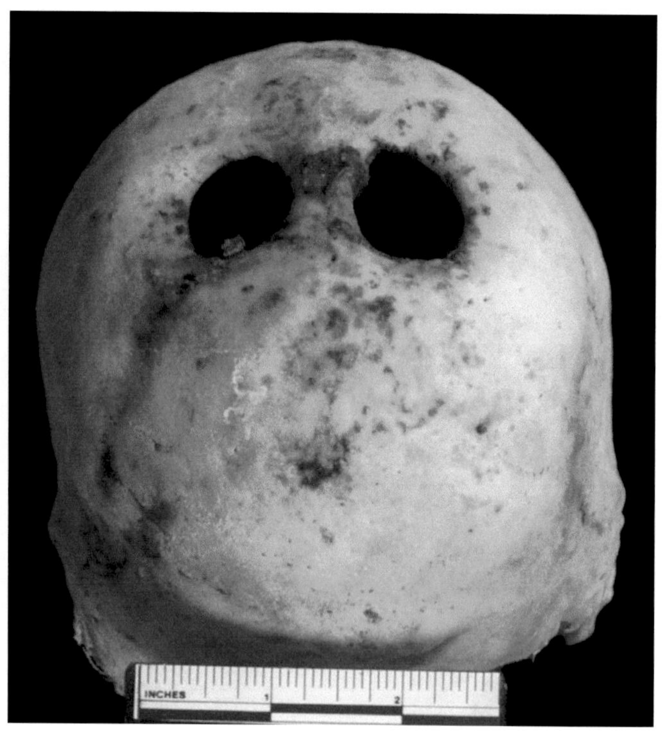

Figure 11.5. Cranium, posterior aspect. These enlarged parietal foramina are developmental anomalies and not the result of trauma.

acquired pathologies result from external factors. Infections, cancer, and trauma are all examples of acquired pathologies (Komar and Buikstra, 2008).

Recall from chapter 3 that maintenance of the skeletal system is mediated by different kinds of bone cells called osteoclasts (responsible for removing bone tissue) and osteoblasts (responsible for producing new bone tissue). These two kinds of cells also mediate skeletal responses to pathology. Reactions characterized by osteoclastic activity are termed *lytic* reactions, and generally result in the formation of pits (often referred to as cavitations), holes, and porosity in the bone (**Figure 11.6**). When mediated by osteoblasts, bone responses are called *proliferative* and involve the formation of new bone (**Figure 11.7**). Many skeletal reactions of pathological insults reflect varying combinations of both lytic and proliferative responses. For example, in the healing process initiated by a traumatic fracture, both osteoclasts and osteoblasts are implicated in the formation and remodeling of a bony callus (Baht et al., 2018). The limited number of ways that bone can react to stimuli often makes the diagnosis of specific pathologies from skeletal remains imprecise if not impossible. Few pathological processes leave observable traces on the skeleton and even fewer produce manifestations that are **pathognomic**, or indicative of a specific pathology. Students interested in the variety of conditions that can affect the skeleton are referred to the excellent and thorough reviews of the subject presented by Aufderheide and Rodríguez-Martín (1998) and Buikstra (2019).

One of the most common skeletal pathologies observed in forensic casework is degenerative joint disease, or *osteoarthritis* (**Figure 11.8**). Most instances of osteoarthritis are idiopathic, meaning that no immediate cause can be discerned, and are associated with the

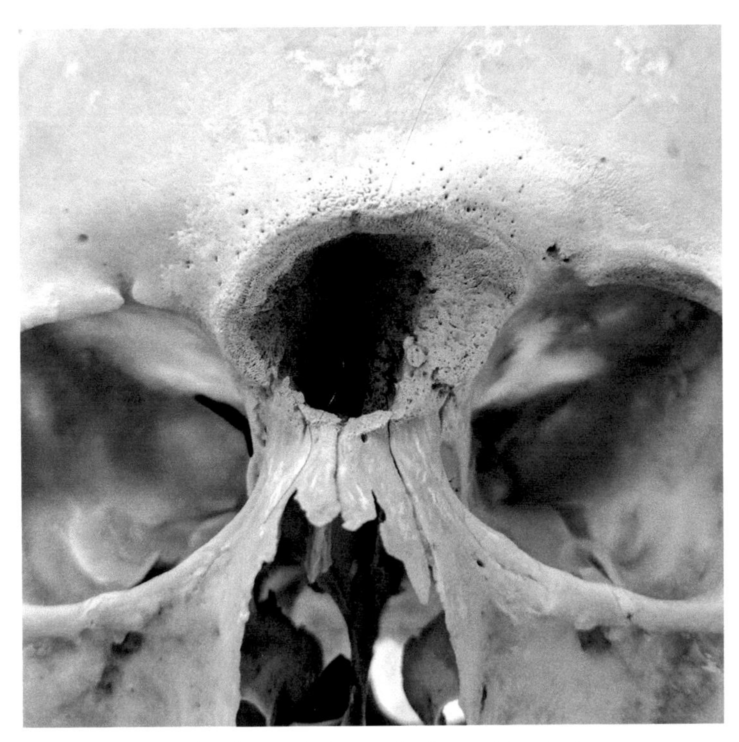

Figure 11.6. Example of a lytic lesion on the front of the skull.

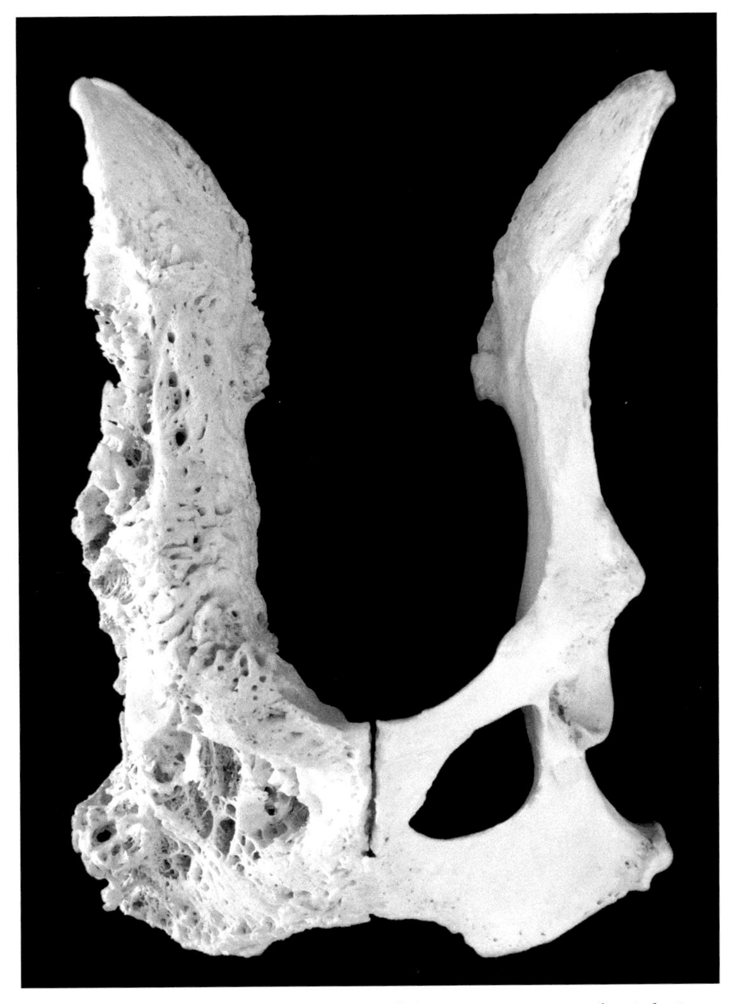

Figure 11.7. Pelvis of a dog. Compare proliferative reaction on the right innominate to the "normal" appearance of the left innominate.

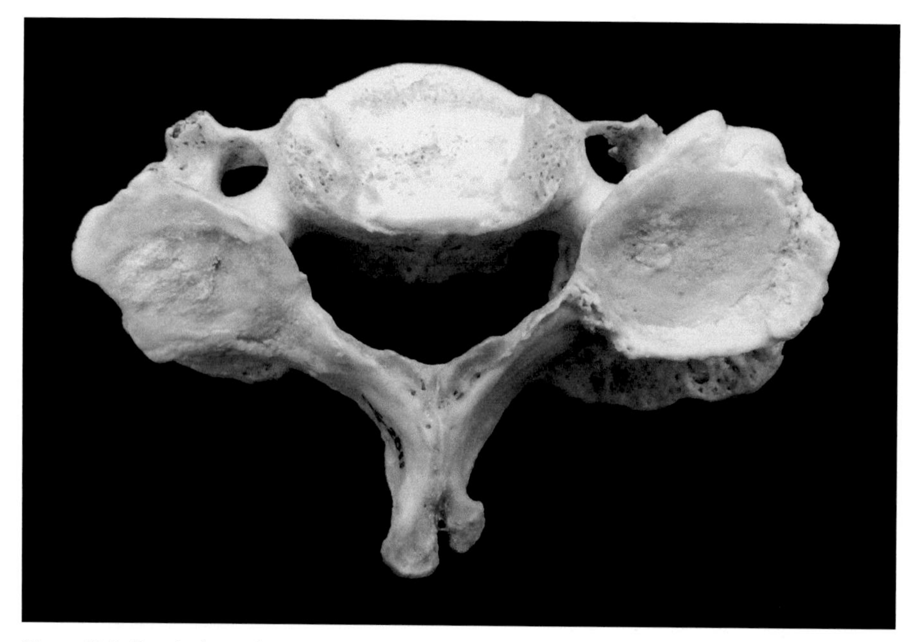

Figure 11.8. Cervical vertebra, superior aspect. The right articular facet is surrounded by bony outgrowth that is typical of osteoarthritis.

degenerative processes that characterize aging. Osteoarthritis can also result from other processes, such as trauma or infection (Aufderheide and Rodríguez-Martín, 1998). Generally, osteoarthritis is characterized by the formation of *osteophytes*, or smooth bony outgrowths around the margin of the affected joint (**Figure 11.9**). While the presence of osteoarthritic lesions is of little evidentiary value for the identification process, the pattern of osteophyte growth can sometimes be used as an individuating characteristic that may be observable in antemortem medical imaging. Further, if a decedent suffered from severe osteoarthritis, it is possible that their mobility would have been altered such that people who knew them in life

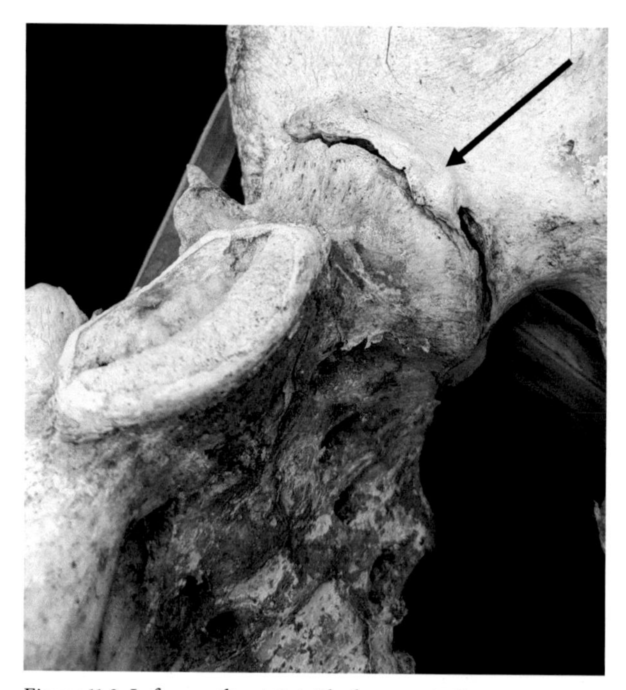

Figure 11.9. Left sacroiliac joint. Black arrow indicates osteophytes projecting across the joint surface.

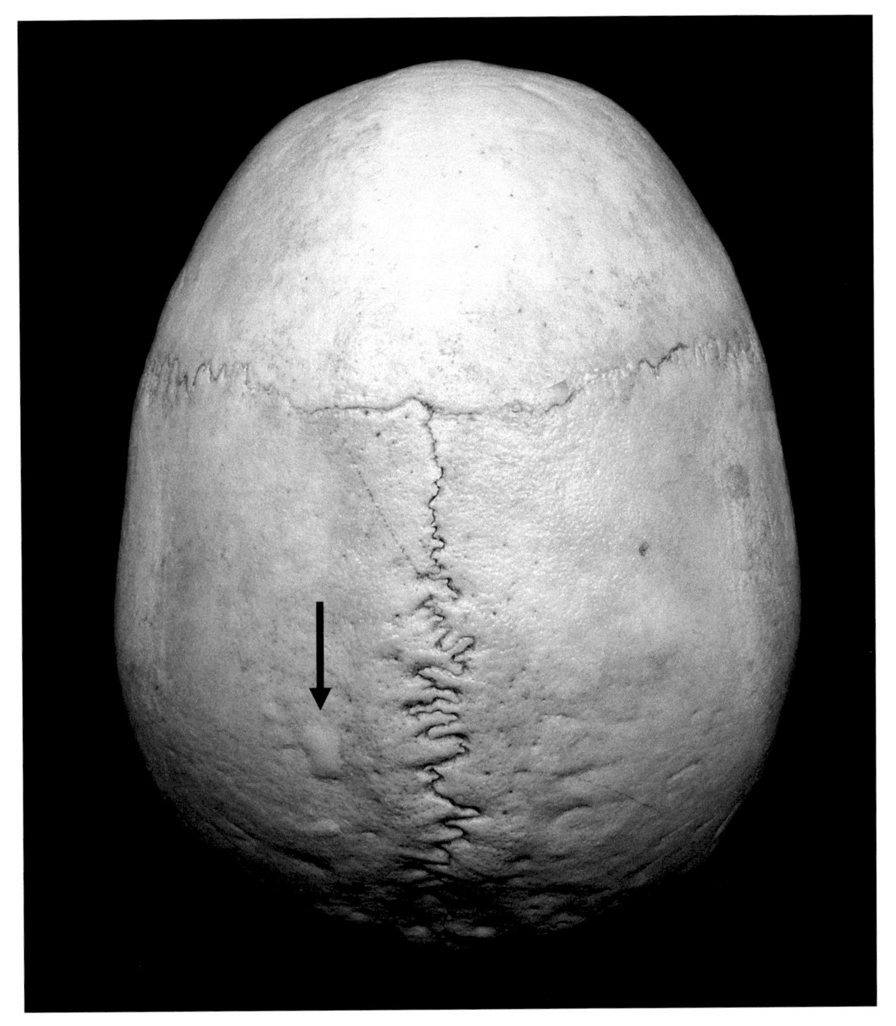

Figure 11.10. Cranium, superior aspect. Black arrow indicates an osteoma with two smaller satellite osteomas.

might describe them as having a "trick knee" or a "bad back." While vague, information such as this can help to flesh out a biological profile.

Other kinds of pathology may provide more specific and, consequently, potentially more useful information. *Button osteomas* (**Figure 11.10**) are slow-growing, benign neoplasms that are typically found on the external table of the cranial vault (Aufderheide and Rodríguez-Martín, 1998). While their presence is often obscured by hair in life, they are easily noted as a distinguishing feature when visible. Some pathologies can even point to a specific geographic area where a decedent spent a portion of their life. *Coccidioidomycosis,* better known as Valley Fever, is caused by a fungus that is endemic to the southwestern United States and especially common in parts of Arizona and California. It is estimated that approximately 100,000 people contract coccidioidomycosis each year. While the vast majority of people who contract Valley Fever never even know it, approximately 250 of those 100,000 cases will exhibit skeletal lesions resulting from a disseminated form of the infection (Aufderheide and Rodríguez-Martín, 1998; DiCaudo, 2006).

Perhaps the most useful antemortem skeletal changes for the purposes of identification are indications of *healed trauma* and *medical interventions* (**Figure 11.11**). Skeletal trauma frequently requires the attention of a medical professional. If a fracture is treated at a hospital, then it will be recorded in a patient's medical history and will likely be accompanied by

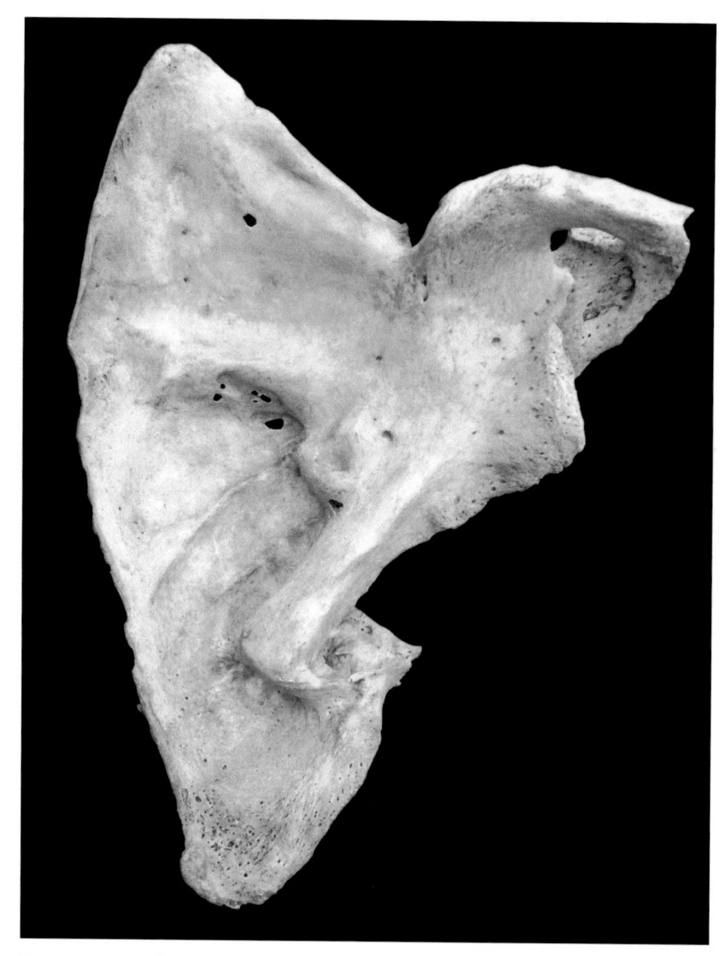

Figure 11.11. Left scapula with a healed and displaced fracture.

medical imaging, such as radiographs. If a fracture is severe enough, it may require surgical intervention and the use of a fixation device for stabilization. Medical procedures that result in surgically implanted devices may also provide information as to a decedent's identity. Examples include sternotomy wires (**Figure 11.12**), pacemakers, and even cosmetic implants.

While the medical records and antemortem imaging from medical procedures provide a rich dataset for use in the identification process, the devices themselves can also potentially be used to generate possible matches. Under the Safe Medical Devices Act of 1990 and the subsequent FDA Modernization Act of 1997, medical devices that are life sustaining, permanently implantable, or have the potential to generate health problems are required to be tracked by their manufacturer (Wilson et al., 2011). Many implanted devices will bear the name or logo of their manufacturer as well as serial and lot numbers and this information is typically recorded in the notes of the surgeon performing the procedure (**Figure 11.13**). In this way, certain medical devices (or certain pieces of composite devices) can potentially be linked directly to the person in which they were implanted and contribute to a positive identification. Both the serial number and the lot number are not necessarily unique, however, and it is possible that a batch of medical appliances will share the same information. In these cases, a medical device can still potentially generate a list of physicians or hospitals to which devices with that number were sold (Ubelaker and Jacobs, 1995; Wilson et al., 2011). Wilson and colleagues (2011) provide a relatively recent list of the contact information for several manufacturers of orthopedic devices as well as the typical locations of serial numbers for more commonly used implants.

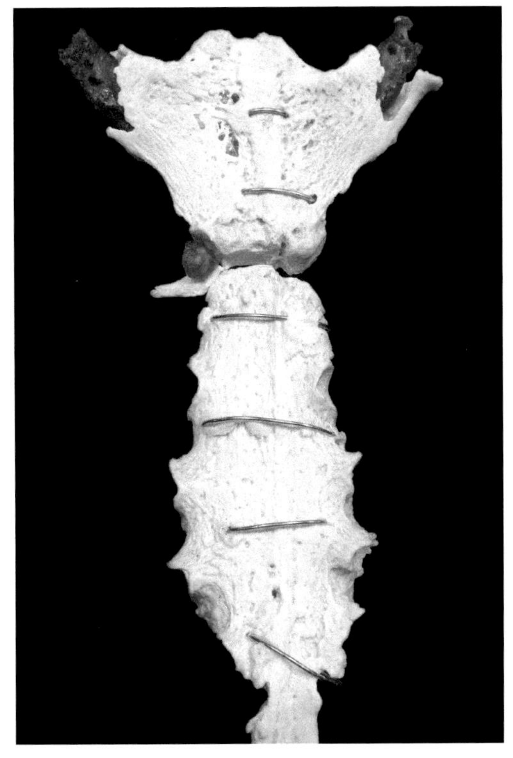

Figure 11.12. Healed midline sternotomy with wires in place.

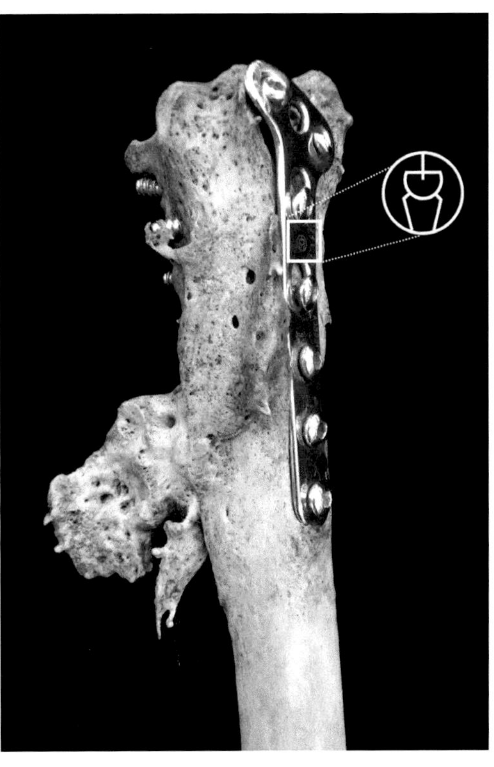

Figure 11.13. Surgical fixation with manufacturer's logo.

LEARNING CHECK

Q3. While examining an unidentified decedent, you notice that the intermediate and distal phalanges of their little toes are fused together (**Figure 11.14**). This is a relatively common, usually asymptomatic, osseous anomaly called pedal symphalangism. In which of the following ways is this observation most likely to contribute to establishing an identification?

A) It provides evidence of a disease process.

B) It suggests that the decedent engaged in a habitual activity.

C) It can provide individualizing information if it appears in antemortem medical imaging.

D) It indicates that the decedent had corrective surgery.

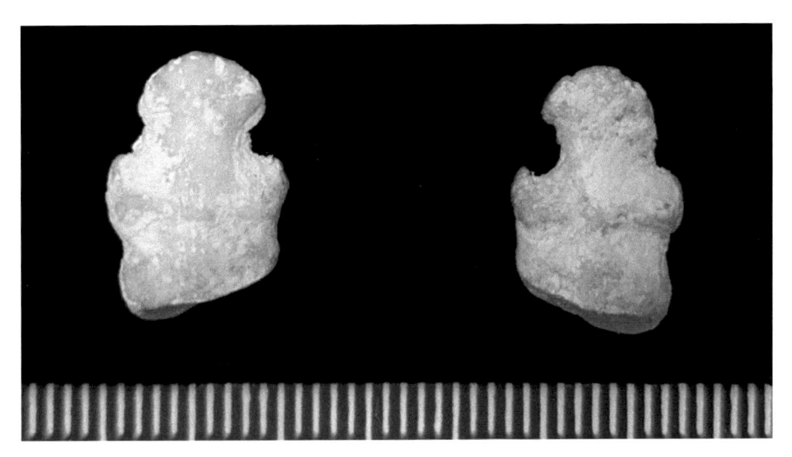

Figure 11.14. Bilateral pedal symphalangism.

Q4. In this picture of a proximal femur (**Figure 11.15**), is the bone reaction best considered as lytic, proliferative, or mixed?
A) Lytic
B) Proliferative
C) Mixed

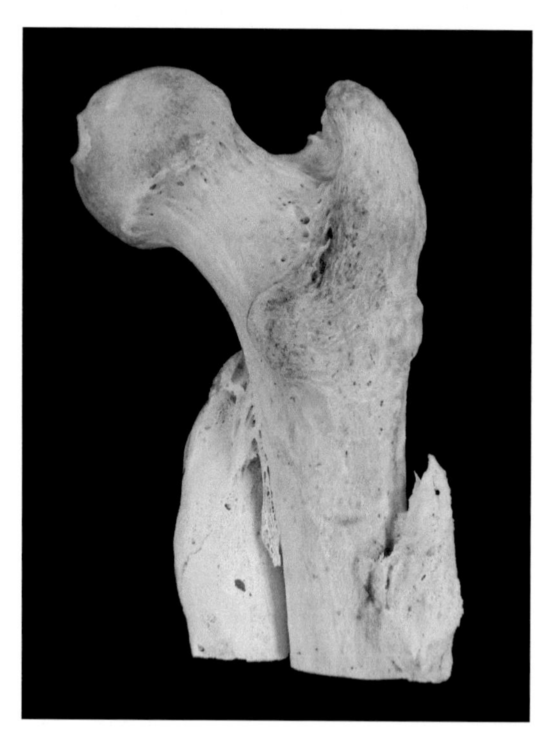

Figure 11.15. Proximal femur with extensive pathological changes.

Q5. Which of the characteristics shown in **Figure 11.16** has the *most* potential to help identify an unknown decedent?
A) A septal aperture, a relatively common, non-symptomatic osseous anomaly
B) Multiple healed rib fractures
C) Evidence of surgical intervention with generic hardware such as a piece of wire
D) Evidence of surgical intervention where the hardware bears a manufacturer's logo and a serial number

Missing Persons and Potential Matches

The biological profile, especially when augmented by the documentation of antemortem skeletal changes, serves as a filter that permits the inclusion or exclusion of missing persons as potential identifications for an unknown decedent. For example, if a decedent is estimated to be male, then missing persons files pertaining to females can be excluded from consideration. Similarly, if a missing persons file reports that the individual had a healed midshaft fracture on their left femur and no such fracture is discernible on the decedent, then that missing person can also be excluded as a potential match. In this way, the biological profile for an unknown decedent is compared to the profiles of missing persons drawn from reports and registries maintained by local law enforcement or included in national databases and a list of potential matches—missing person profiles consistent with the description of the decedent—is generated.

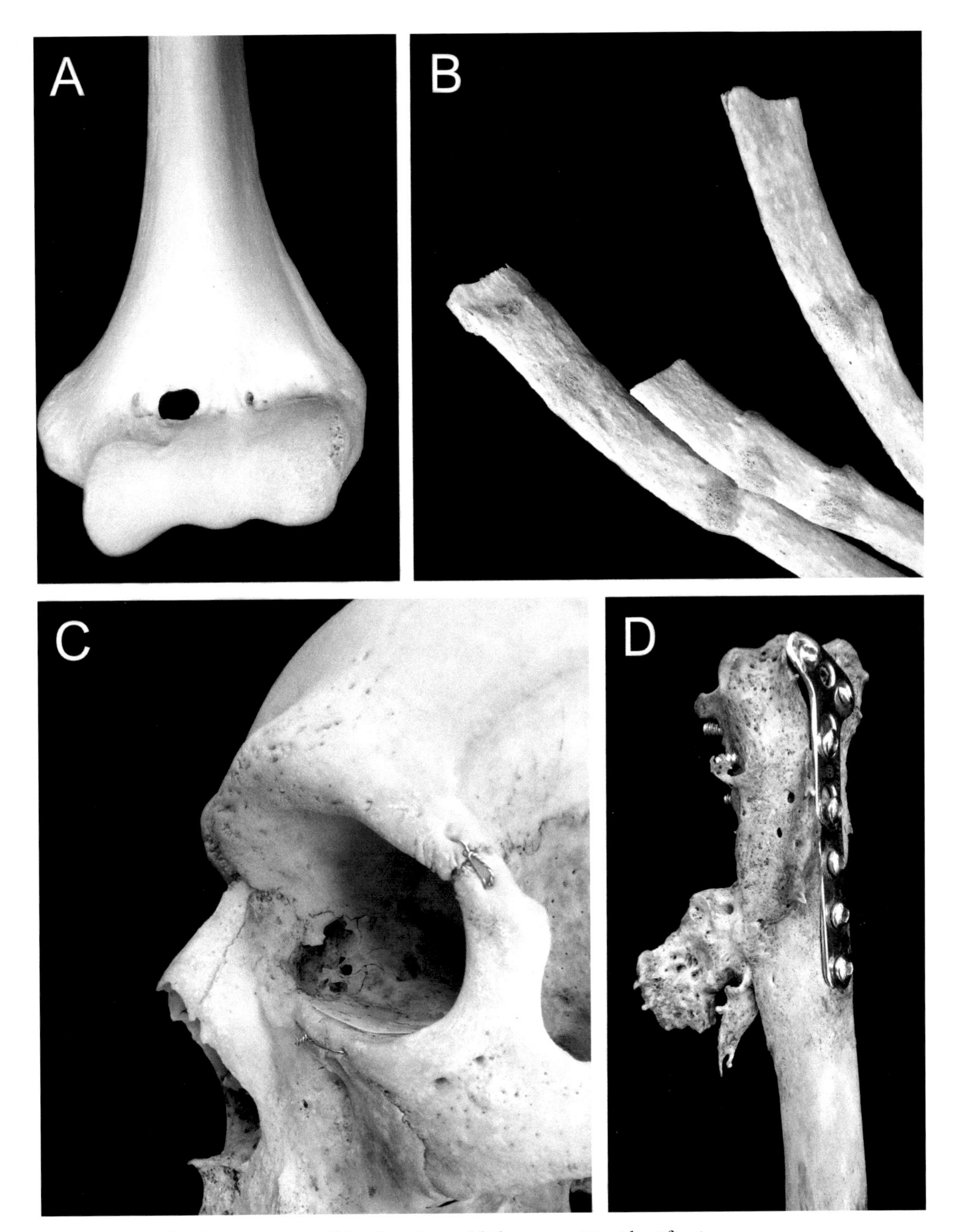

Figure 11.16. Skeletal signatures possibly of use in establishing a positive identification.

NamUs and CODIS

The National Missing and Unidentified Persons System, or **NamUs** (pronounced "name us"), consists of two linked databases. In one database, representatives of the medicolegal system enter data pertaining to unidentified decedents that are under their jurisdiction. Generally, this includes case information, demographic characteristics (such as those obtained from a biological profile), and the documentation of any associated personal items or potentially identifying characteristics such as tattoos, scars, or medical interventions. More detailed information may be available in the form of dental charts or radiographs, or indications as to whether fingerprint records are available or DNA samples have been taken. Paired with the unidentified persons database is a missing persons database. Here, information pertaining to missing persons can be entered by law enforcement personnel as well as concerned friends, family members, or other members of the public. When information is entered by members of the public, it is verified by NamUs personnel prior to being published in the missing persons database. Users can register with NamUs on their website (www.namus.gov) and conduct custom searches that compare profiles in both databases based on a set of user-specified parameters. The system will return a list of suggested matches between unidentified decedents and missing persons reports that can then be reviewed further and either excluded based on inconsistencies between the profiles or included as candidates for identification comparisons. It is important to note that the success of NamUs relies on accurate contributions to both databases—it cannot match a missing person to an unidentified decedent if information pertaining to one or the other is never entered into the system (Murray et al., 2018; Osborne-Gustavson et al., 2018).

Viable samples of DNA recovered from unidentified human remains can be submitted to the Combined DNA Index System (**CODIS**) which, in the United States, is administered by the Federal Bureau of Investigation. This database software is used by more than 190 crime laboratories across the nation and stores DNA profiles in one of several separate indexes. Separate indexes exist for DNA profiles obtained from unidentified remains, missing persons, family members of missing persons, offenders, arrestees, detainees, and forensic samples (those obtained from crime scene evidence). While CODIS has many applications, most relevant to forensic anthropology is the ability to search for matches and associations between a DNA profile obtained from unidentified remains and profiles stored in each of the separate indexes. Searches can be configured to look for direct matches or for associations that are consistent with a familial relationship. Potential matches identified in this way can then be assessed using more rigorous comparisons (Osborne-Gustavson et al., 2018).

LEARNING CHECK

Q6. A partial fibula is recovered from a wooded area. After an extensive search of the surroundings, no other skeletal remains or personal effects are recovered. Assuming that a successful DNA sample can be taken, does NamUs or CODIS have the most potential for generating potential matches for these unidentified remains?
A) NamUs
B) CODIS

Q7. Complete, mummified remains are recovered from a desert environment. Infrared photography of the skin documents the presence of several unique tattoos and there is still hair attached to the desiccated scalp. Combining these features with the biological profile of the remains, does NamUs or CODIS have the most potential for generating potential matches for these unidentified remains?
A) NamUs
B) CODIS

Establishing a Positive Identification

In recent deaths, if a decedent's appearance has yet to be affected by decomposition, their identity may be confirmed through *visual recognition* by someone who knew them during life. This, in fact, is how most positive identifications are established. While this is usually done through reference to the decedent's facial features, it may also rely on other distinguishing features, such as scars or tattoos (Christensen et al., 2019; Hurst et al., 2013). Evidence suggests, however, that visual recognition by individuals who did *not* know the decedent personally—such as when medicolegal personnel compare a decedent's face to the picture on a driver's license found with the body—is less accurate (Caplova et al., 2017). Visual recognition in such cases can be used to establish a tentative identification that will require further confirmation using a scientific comparison (Komar and Buikstra, 2008).

The use of scientific comparisons for establishing a positive identification relies on the availability of both antemortem and postmortem data. Antemortem information, such as photographs, medical and/or dental records, and any accompanying medical imaging, is obtained for missing persons of known identity whose profiles are consistent with that of an unknown decedent. A detailed comparison is then made between this and the postmortem data obtained through examination of the unidentified remains. Scientific comparisons typically result in one of three conclusions. *Identification* is achieved when antemortem and postmortem data are consistent in enough detail to conclude that they were derived from the same person. For a comparison to result in an identification, any discrepancies between the two sets of information must be explainable. For example, a healed humeral fracture observed postmortem but absent in an antemortem radiograph is an explainable difference in that the fracture could have occurred after the radiograph was taken. In contrast, an antemortem radiograph of a left maxillary first molar with an evident restoration is inconsistent with a postmortem assessment of remains in which that tooth is unrestored; such a discrepancy is unexplainable. Unexplainable discrepancies form the basis for *exclusion*, which indicates that antemortem and postmortem data are not derived from the same individual.

A scientific comparison may also conclude that there is insufficient information to warrant either the inclusion or exclusion of the individual associated with the antemortem information as the source of the postmortem data. This is typically the result of situations in which antemortem and postmortem datasets cannot be directly compared or when one or both are of poor quality (Christensen et al., 2019; Hurst et al., 2013). Methods of scientific comparison that are commonly accepted for establishing a positive identification include dactyloscopy (or fingerprint analysis), DNA, and radiographic comparisons (Christensen et al., 2019; Hurst et al., 2013; Komar and Buikstra, 2008). Dactyloscopy and DNA comparisons require training and certification that is typically beyond the scope of forensic anthropology. Many forensic anthropologists, however, routinely employ *comparative radiography* within the identification process.

Comparative radiography relies on the availability of antemortem imaging of individuals who are potential matches for an unknown decedent. Further, the anatomical regions captured in such imaging must be the same as those represented by the remains. An antemortem chest radiograph, for example, cannot be compared to a postmortem radiograph of the lower leg. An example of this potential disconnect is presented in **Table 11.1**, which compares the proportions by anatomical region of radiographs (X-ray images) taken of living patients in a medical teaching facility (Brogdon, 1998) to the proportions by anatomical region of radiographs used for identifications in forensic casework over a 14-year period in Michigan (Streetman and Fenton, 2018). Some of the discrepancies observed in **Table 11.1** are likely due to a dramatic increase in the use of computed tomography (CT) imaging for the head/neck and abdomen (Brogdon, 2011; Smith-Bindman et al., 2008; Viner, 2018).

Radiographic comparisons look for consistency between antemortem and postmortem images in characteristics such as bone contour, the configuration of anatomical landmarks, the arrangement of bone trabeculae, vascular grooves, and antemortem skeletal changes like

Table 11.1. Proportions of radiographs by anatomical region for antemortem imaging (AM) and postmortem comparisons (PM)

Anatomical Region	AM (%)[a]	PM (%)[b]
Chest/Thorax	51	32.2
Lower Extremity	11	27.7
Upper Extremity	10	11.6
Breast	8	-
Abdomen	7	18
Head/Neck	5	5.6
Pelvis/Hip	4	4.1
Other	4	0.7

[a] Adapted from Brogdon (1998).
[b] Adapted from Streetman and Fenton (2018).

those discussed earlier. While there is no threshold for the number of concordant features necessary to establish a positive identification, the use of anatomical structures or regions that either exhibit a complex shape or are relatively rare within the population at large is typically preferred (Brogdon, 2011; Cattaneo et al., 2006; Christensen et al., 2019; Hatch et al., 2014; Hurst et al., 2013). Despite skeletal remodeling over time, it has been demonstrated that distinguishing features of anatomical structures can persist relatively unchanged for several decades (Sauer et al., 1988).

Nevertheless, caution must be employed in radiographic comparisons as postmortem radiographs must be taken in such a way as to replicate as nearly as possible the projection, angulation, and position of any available antemortem images (Komar and Buikstra, 2008; Streetman and Fenton, 2018). Any discrepancies in these parameters can lead to distortion and impede or invalidate the comparison. Several studies have demonstrated the importance of experience and training for radiographic comparisons (Hogge et al., 1994; Koot et al., 2005; Stephan et al., 2011; Streetman and Fenton, 2018) and it is strongly suggested that anthropologists unfamiliar with these techniques consult with a forensic radiologist. While many anatomical structures have been used as the basis for establishing a positive identification (e.g., Adams and Maves, 2002; Angyal and Dérczy, 1998; Owsley and Mann, 1992; Rhine and Sperry, 1991; Riddick et al., 1983; Rougé et al., 1993; Stephan et al., 2011; Sudimack et al., 2002), two of the more widely used are the dentition and the frontal sinus.

LEARNING CHECK

Q8. Decomposing remains are recovered wrapped in plastic from a shallow grave (**Figure 11.17**). Image A depicts postmortem radiographs obtained of the right wrist. Law enforcement wants to know whether these remains are consistent with a local missing person, whose antemortem radiograph is presented in Image B. Based on this radiographic comparison, you can conclude that:

A) The antemortem and postmortem films are consistent, supporting a potential identification.

B) The antemortem and postmortem films are inconsistent, yielding an exclusion.

C) There is insufficient information to merit either an inclusion or an exclusion.

Q9. Skeletal remains are recovered from a dry wash. Image A in **Figure 11.18** depicts a photograph of the posterior aspect of the proximal right femur. After searching missing persons reports, law enforcement produces a potential match and provides you with Image B, an antemortem radiograph of the right proximal femur (the image is reversed due to different orientations). Based on this information, you can confidently say that:

A) The antemortem image is consistent with the remains, supporting a possible identification.

B) The antemortem image is inconsistent with the remains, excluding this missing person as a match.

C) There is insufficient information to merit either an inclusion or an exclusion.

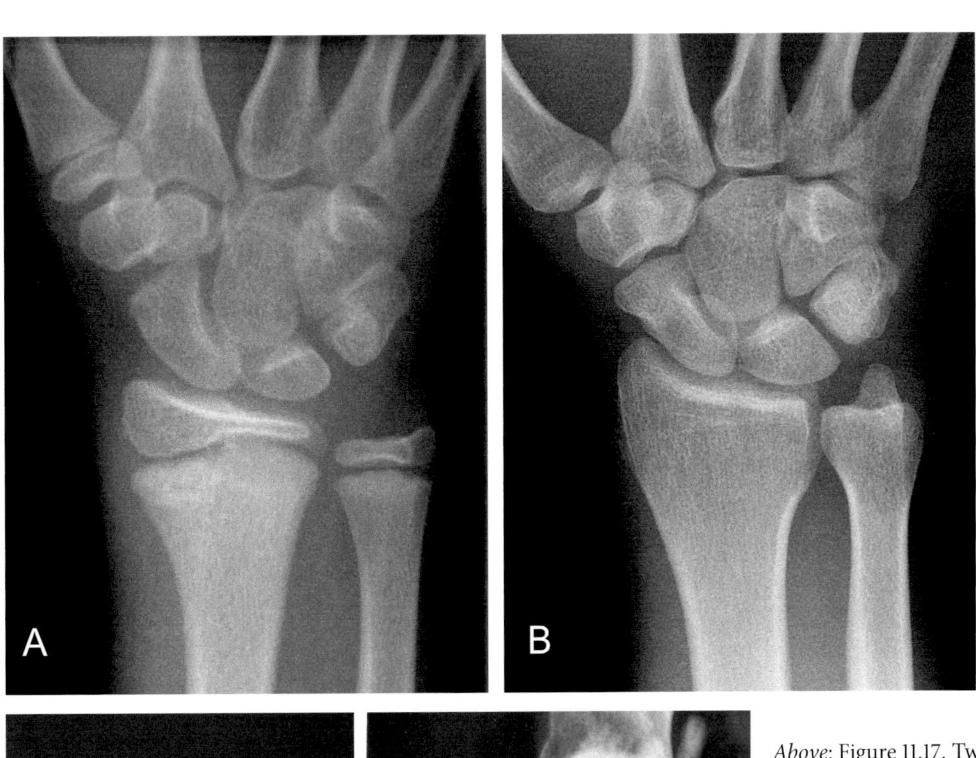

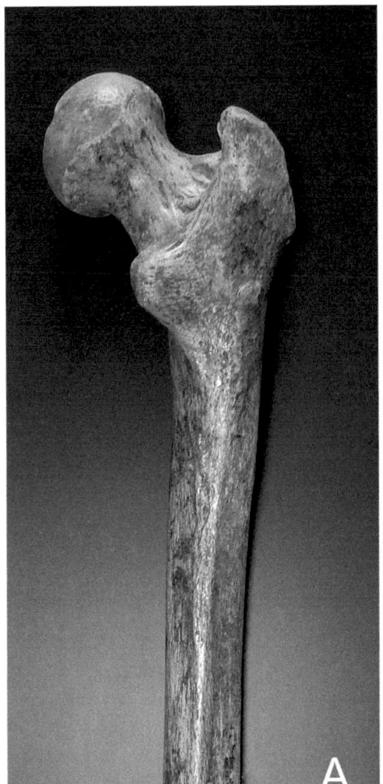

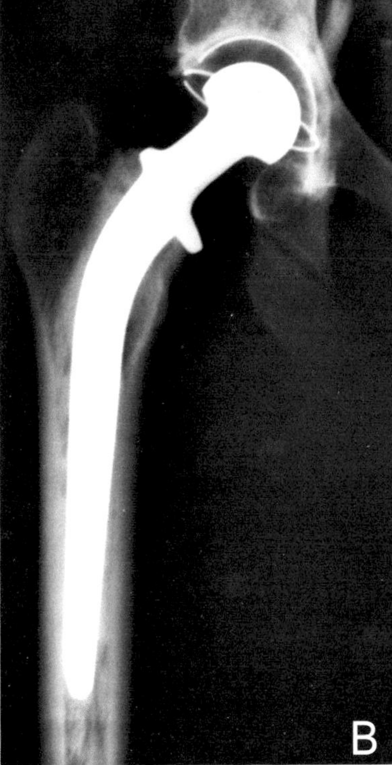

Above: Figure 11.17. Two radiographs of the wrist showing the bones of of the arm, carpals, and metacarpals.

Left: Figure 11.18. Real bone (*A*) and radiographic image (*B*) of human proximal femora.

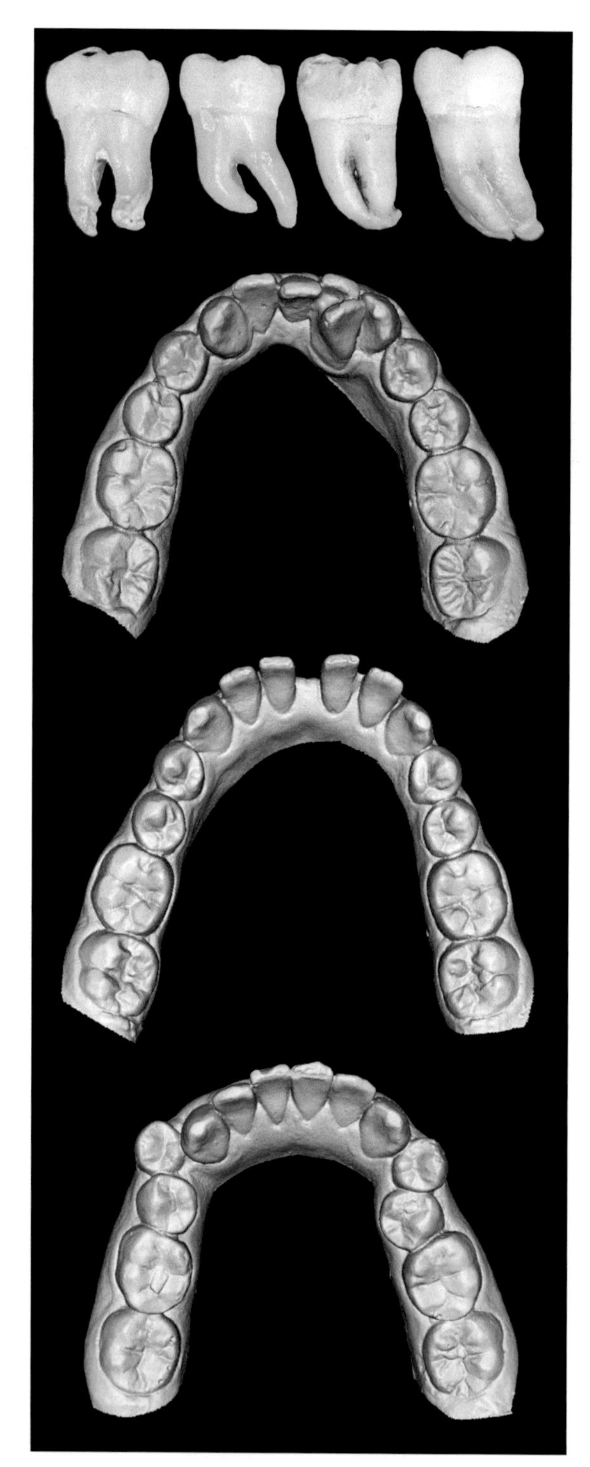

Figure 11.19. Variation in the mandibular dentition.

Dental Comparisons

Although not radiographic, the earliest recorded dental comparison for the purpose of identification occurred in AD 49. Agrippina the Younger, fourth wife of the Roman Emperor Claudius, arranged to have Lollia Pollina, a rival for her husband's attentions, first exiled and then executed. By the time Lollia's head arrived back in Italy, her face was no longer recognizable, and Agrippina is said to have inspected her rival's teeth and confirmed her identity by means of dental anomalies (Lipton et al., 2013). Today, most dental comparisons

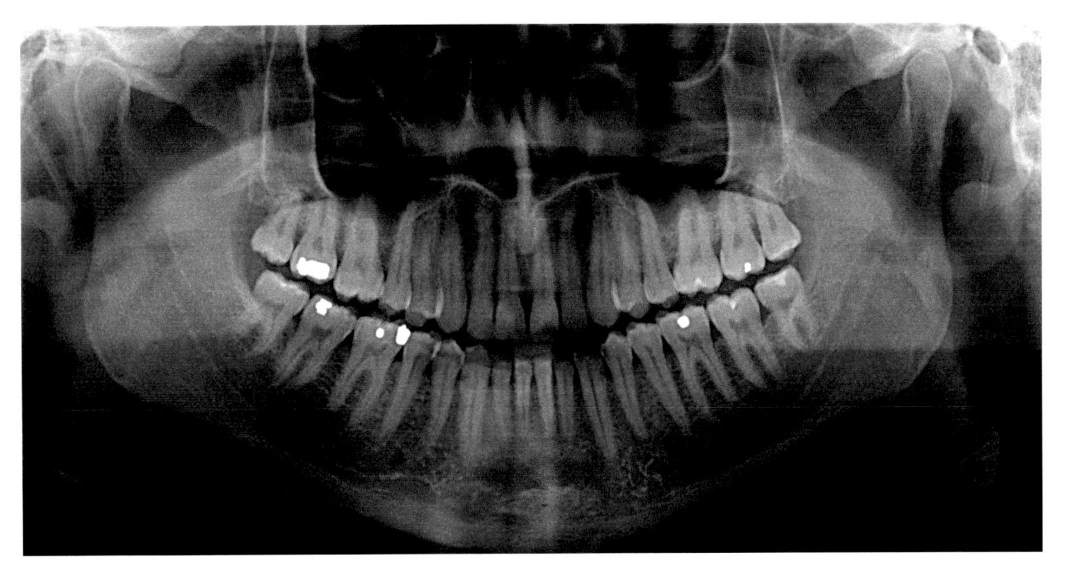

Figure 11.20. Panoramic dental radiograph.

are made using radiographs and, while more properly the domain of the forensic odontologist, they are common within forensic anthropology. This is largely due to two factors. First, owing primarily to the hardness of dental enamel, teeth often survive decomposition and taphonomic processes that damage other portions of the skeleton. They are even capable of surviving some fires relatively unscathed. Second, antemortem dental radiographs are more common than antemortem medical radiographs. Due to the increased likelihood of available antemortem and postmortem data, dental radiographic comparisons are often more cost-effective, more efficient, and frequently possible even when other scientific comparisons are not (Clement, 2013; Hurst et al., 2013).

The human dentition is astonishingly variable. Different individuals will exhibit slight differences in the rotation, position, and the morphology of both the crowns and roots of their teeth as well as their patterns of congenital absence and antemortem loss (**Figure 11.19**). In addition to this, the idiosyncrasies associated with dental work in terms of the number, type, placement, and shape of restorations and dental prosthetics dramatically increase the utility of the dentition for individualization. Other features that may be visible within panoramic dental radiographs include the contours of the nasal aperture and septum, the nasal conchae, and the morphological details of the mandible and eye orbits (**Figure 11.20**)(Clement, 2013; Forrest and Wu, 2010; Hurst et al., 2013). Even in the absence of available dental radiographs, the pattern of missing, filled, and unrestored teeth within an individual's dentition can be used to generate strong evidence in support of an identification (Adams, 2003a, b). While assessment of dental and osteological features is generally within the scope of a forensic anthropologist's training, the identification and documentation of specific kinds of dental restorations and implants is not. For this reason, forensic anthropologists are encouraged to work in conjunction with forensic odontologists for dental comparisons (Christensen et al., 2019).

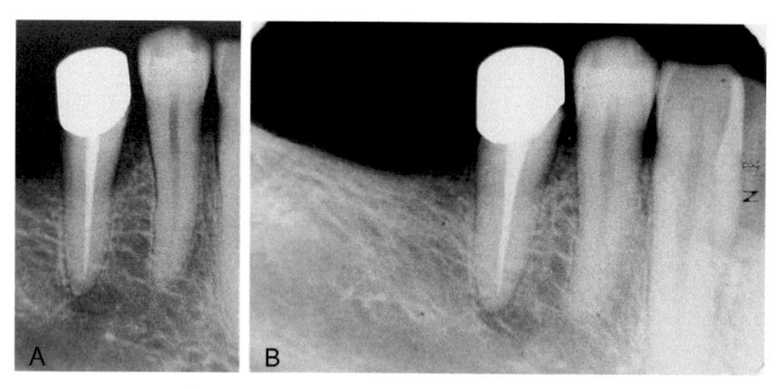

Figure 11.21. Radiographs of individual teeth showing dental interventions.

LEARNING CHECK

Q10. The radiograph on the left (**Figure 11.21A**) was obtained postmortem from an unidentified set of remains. The radiograph on the right (Image B) was included in a missing person's antemortem dental records. Based on this dental comparison, you can conclude that:

A) The antemortem and postmortem films are consistent and support an identification.

B) The antemortem and postmortem films exhibit unexplainable differences, and this missing person can be excluded.

C) There is insufficient information to merit either an inclusion or an exclusion.

Q11. The radiograph on the top (**Figure 11.22A**) was taken postmortem from an unidentified decedent. After finding a missing person whose biological profile was consistent

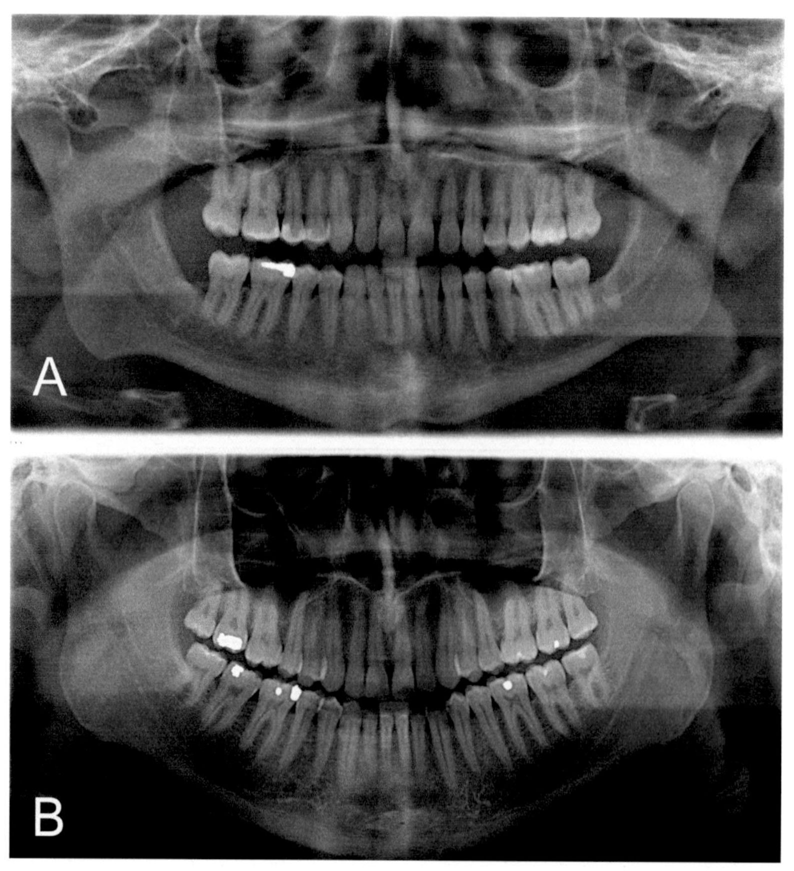

Figure 11.22. Two panoramic radiographs of the human dentition.

with that of the remains, antemortem dental radiographs were obtained for comparison (Image B). Based on this comparison, you can conclude that:

A) The two sets of radiographs represent the same individual.

B) The two sets of radiographs cannot represent the same individual.

C) The two sets of radiographs might represent the same individual, but you cannot be certain.

Frontal Sinus Morphology

In late June of 1925, two sets of human remains were found badly decomposed and dismembered in the Indus River. It was suspected that one of the bodies was that of an American tourist and real estate heir who had declared his intention to "disappear" while he attempted an illegal passage from India into Tibet. As it happens, this man had been operated on for left-sided mastoiditis and, during the procedure, radiographs of his frontal sinus and left mastoid process had been obtained. Nearly eight months after his disappearance, his alleged remains arrived in New York City where his identity was confirmed through radiographic comparison of the frontal sinus and the mastoid air cells (see **Figure 11.23**) (Culbert and Law, 1927). This case represents not only the first recorded positive identification by means of radiographic comparison, but also the first use of the **frontal sinus** for this purpose.

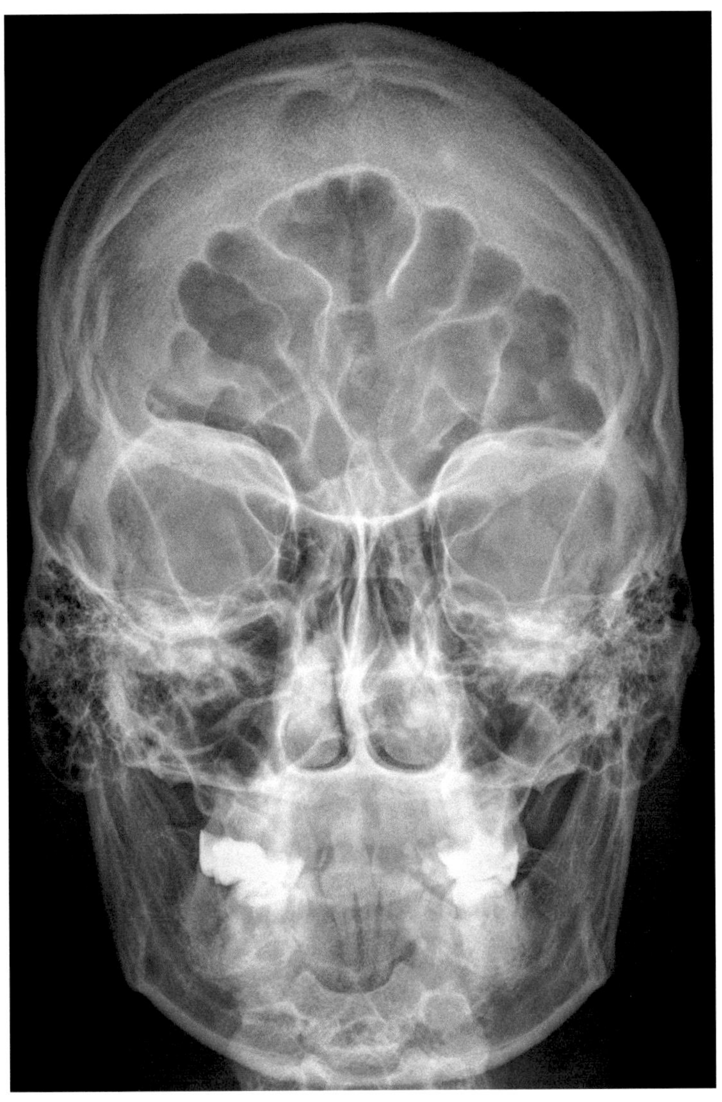

Figure 11.23. Radiograph with frontal sinus highlighted.

Although absent in some individuals, frontal sinuses begin to develop around the fourth or fifth fetal month. Typically, they can be seen in radiographs by around 5 years of age. The frontal sinus develops slowly until puberty and then rapidly until the approximate age of 20. At this point, barring bony remodeling resulting from trauma or disease, the shape of the frontal sinus stabilizes aside from potential thinning due to advancing age (Christensen and Hatch, 2018; Kirk et al., 2002). Although the growth and development of the frontal sinus limits its utility for radiographic comparisons to adults, its complex morphology is thought to be unique for every individual (Kirk et al., 2002; Ubelaker, 1984; Yoshino et al., 1989). Yoshino and colleagues (1989) developed a coding scheme for the description of frontal sinus morphology based on categorical assessments of size (area), asymmetry, contour complexity, and the presence or absence of partial septa and supraorbital cells. Using this scheme to assess a series of 100 skulls, it was found that 88 had unique code numbers and, further, the remaining 12 could be distinguished from each other using finer scale comparisons. This coding scheme was later refined by Cameriere and colleagues (2005) and used to demonstrate that the use of relatively simple codes for characterizing sinus morphology can result in low rates of false positive identification.

LEARNING CHECK

Q12. Compare the two images in **Figure 11.24**. Based on the shape of the frontal sinus, which of the following is a reasonable conclusion to draw from this comparison?

A) Image A and Image B represent the same individual.

B) Image A and Image B are not consistent with being derived from the same individual.

C) There is not enough information available from this comparison to determine whether Image A and Image B represent the same individual.

Likelihood Ratios

Radiographic comparisons are routinely accepted as a valid means of establishing a positive identification by medicolegal authorities, but this is not necessarily the case for courts. While forensic anthropologists are rarely required to provide expert testimony pertaining to the identification process (Christensen, 2005; Rogers and Allard, 2004; Steadman et al., 2006), a 2010 challenge to the admissibility of frontal sinus identifications within the Tennessee state court (Lesciotto, 2015) illustrates the need to ensure that radiographic comparisons meet

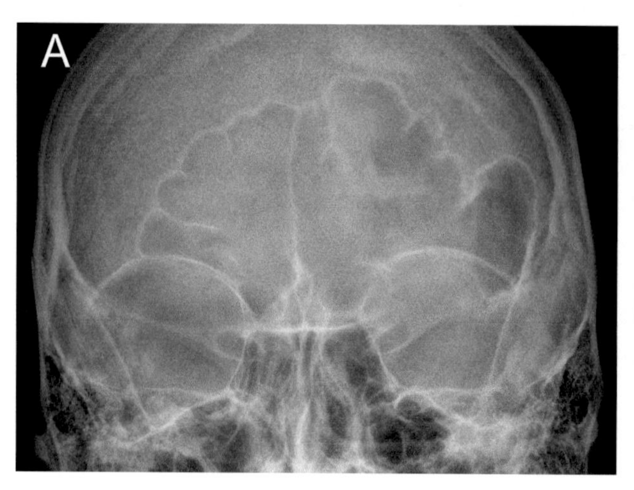

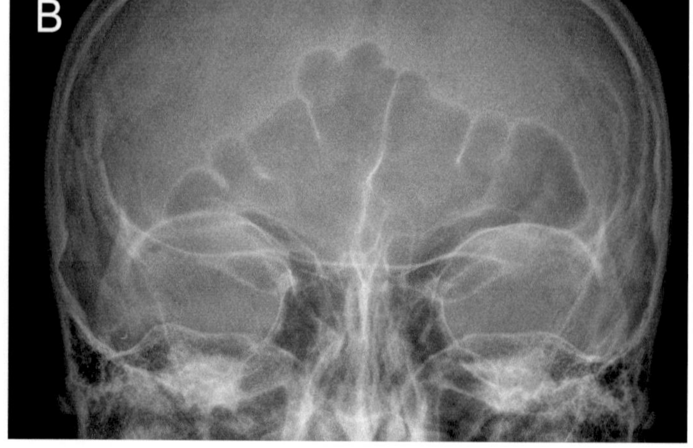

Figure 11.24. Two frontal radiographs showing the frontal sinus.

the criteria of objectivity and known error rates that were established by the *Daubert* ruling (Daubert, 1993; see chapter 2). Expressing the results of an identification comparison as a **likelihood ratio** is an increasingly common way of meeting this need.

A radiographic comparison is essentially a hypothesis test. The null hypothesis is that the antemortem and postmortem images are derived from the same person or, in other words, that the identification is correct. The alternative hypothesis is that the identification is incorrect and the antemortem and postmortem images represent two different individuals. Consider a scenario in which antemortem and postmortem radiographs both exhibit a sternal foramen, an osseous anomaly that occurs in 4.5 percent of the population (Christensen and Hatch, 2016). The likelihood (or probability) of this occurring under the null hypothesis is 100 percent, or 1. In contrast, if the antemortem and postmortem radiographs represent two different people, then the likelihood that they both have a sternal foramen is 4.5 percent, or 0.045. The ratio between these likelihoods is 1/(0.045), or 22.22. In other words, it is 22.22 times more likely that both radiographs exhibit a sternal foramen because they represent the same individual than it is if they are derived from different people.

The greater the magnitude of the likelihood ratio, the greater the strength of the evidence supporting the hypothetical identification. One benefit of likelihood ratios is that, if they are associated with independent traits, they can be multiplied to create a summary likelihood ratio that accounts for multiple points of comparison (Steadman et al., 2006; Christensen and Hatch, 2016). The use of likelihood ratios, however, is contingent on the availability of frequency data for the traits that are being considered. While efforts have been made to compile population frequency information for osseous anomalies in radiographic comparisons (e.g., Christensen and Hatch, 2016), many remain unknown. Likelihood ratios can also be used with skeletal features whose shape can be quantified. Christensen (2005), for example, used elliptic Fourier analysis (EFA) to create a numerical description of the shape of the frontal sinus and, using likelihood ratios, concluded that the odds of two different individuals sharing the same frontal sinus shape are around $10^{21.22}$ to 1. For a thorough presentation of the calculation and application of likelihood ratios within the context of forensic anthropology, readers are encouraged to refer to the work of Steadman and colleagues (2006).

LEARNING CHECK

Q13. According to Harris and colleagues (1987), the frequency of agenesis (i.e., a complete lack of development) of the frontal sinus varies from 5 percent to 15 percent across different populations. In examining a set of unidentified adult skeletal remains, you notice that the frontal sinus has never developed. After obtaining the antemortem medical images of an individual whose biological profile is consistent with those of the decedent, you realize that their frontal sinus is absent. If the frequency of agenesis of the frontal sinus is 10 percent in the population in which you work, what is the likelihood ratio associated with this observation?

A) (0.10)/1 = 0.1
B) (1–0.1)/1 = 0.9
C) 1/(0.1) = 10
D) 1/(1–0.1) = 1.11

Additional Contributions to Identification

While radiographic comparisons and surgical implants are the only methods that are readily available to forensic anthropologists for establishing a positive identification, it is often the case that no antemortem radiographs can be located for comparison. In such circumstances, other approaches can aid in the inclusion or exclusion of potential matches or the generation of new investigative leads.

Craniofacial superimposition involves the comparison of the structures of the skull to a portrait of the deceased (**Figure 11.25**). Within a forensic setting, the technique relies on the relationships between the soft tissue of the face and the underlying bony architecture to conduct a detailed comparison of the shape of the skull of an unknown decedent with an antemortem image of a potential match (Austin-Smith and Maples, 1994; Milligan et al., 2018). While considering variables including scale, lighting, positioning, and skull-to-camera distance, the skull is oriented to match the portrait as nearly as possible. The skull and the antemortem image are then superimposed using photography, videography, or, as is more common now, image analysis software (e.g., Austin-Smith and Maples, 1994; Dorion, 1983; Stephan, 2017; Ubelaker et al., 1992). Assessments of the accuracy of this technique indicate a high rate of identification errors resulting in the consensus that, at best, craniofacial superimposition can be used as a means of inclusion or exclusion, but not as the basis for establishing a positive identification (e.g., Austin-Smith and Maples, 1994; Christensen et al., 2019; Milligan et al., 2018; Dorion, 1983).

In the absence of any leads as to their identity, the remains themselves can be used to generate an idea of a decedent's appearance. In 1916, Giovanni Ramona was arrested for the murder of Dominick La Rosa after being identified by a woman startled by the ghost of La

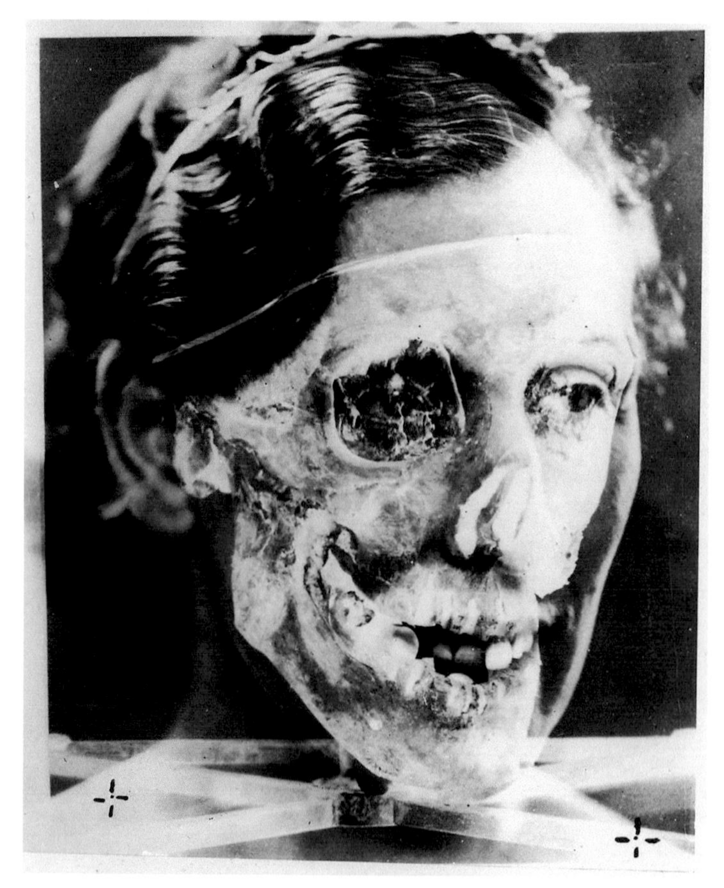

Figure 11.25. An early example of craniofacial superimposition.

Rosa. The "ghost" in this case was actually a wax face constructed upon a skull and adorned with a felt hat and some hair from the floor of a barber's shop ("Reconstruct Body," 1916). This represents the first recorded successful use of a technique now referred to as **facial approximation** (Stephan, 2013).

Techniques for facial approximation are typically talked about as belonging to one of two schools: the anatomical method or the tissue depth method. This, however, is a false dichotomy. While the anatomical method is ostensibly based on the reconstruction of individual muscles (Gerasimov, 1971), all techniques of facial approximation rely to some extent on mean tissue depths (Stephan, 2013). In contemporary practice, a replica of the skull is usually cast in plaster or acrylic or reproduced from a 3D scan. Tissue depth markers are placed at anatomical landmarks on the replica and used to guide the placement of clay or other modeling materials used to construct facial features upon the skeletal scaffolding (**Figure 11.26**) (Krogman and İşcan, 1986; Stephan, 2013). While the mean tissue depths used in facial approximation have historically been determined based on the estimated sex and population affinity of the decedent (e.g., Snow et al., 1970), a recent appraisal of the magnitude of measurement error involved in producing such figures indicates that mean tissue depths based on pooled data are more appropriate (Stephan and Simpson, 2008). One complication of contemporary facial approximation techniques is their reliance on *mean* tissue depths. Since tissue depths can never be negative, their distribution is positively skewed. Therefore, the use of mean tissue depths is a poor reflection of the actual tissue depths that characterize the population. While alternative measures have been proposed (e.g., Stephan et al., 2013), their utility is currently limited by small sample sizes (Stephan, 2014). While assessments of the resemblance of facial approximations to the individual that they represent yield recognition

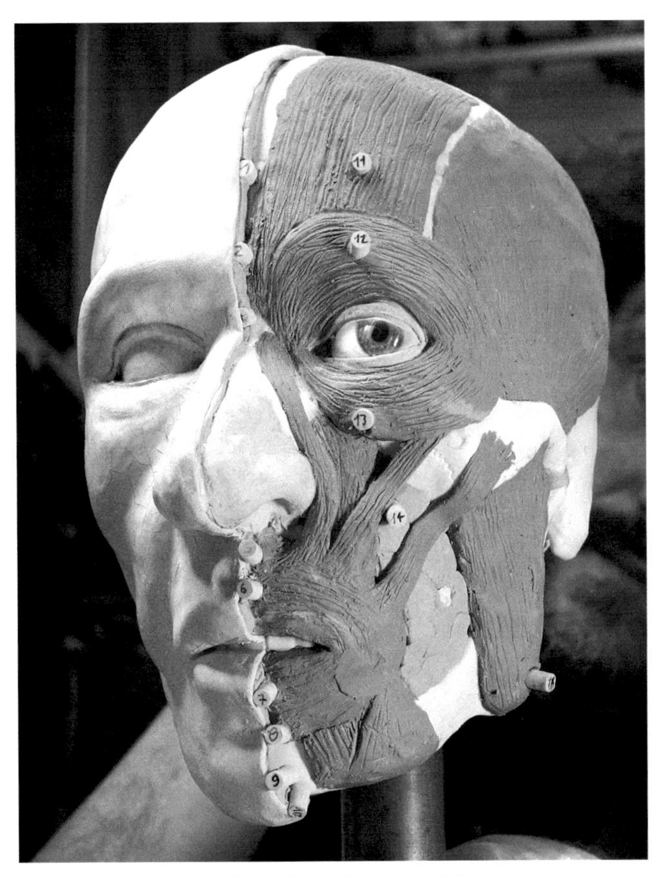

Figure 11.26. An example of three-dimensional facial approximation.

rates no better than would be expected by chance (e.g., Stephan and Cicolini, 2008; Stephan and Henneberg, 2001), this does not invalidate their contribution to the identification process in that the media release of a facial approximation helps to amplify public awareness of a case and generate new investigative leads (Christensen et al., 2019; Stephan, 2013).

End-of-Chapter Summary

This chapter provided an overview of the process of establishing a positive identification for unidentified decedents. Several kinds of antemortem skeletal changes, including osseous anomalies, pathologies, healed trauma, and surgical interventions, were described and discussed in terms of their potential for providing individualizing information. Combined with the main components of the biological profile (see chapters 6–10), this information can be used to generate potential matches between missing persons and the unidentified remains through consultation with local law enforcement and/or the use of national databases, such as NamUs and, where DNA samples have been obtained, CODIS. Potential matches are evaluated using methods of scientific comparison. For the forensic anthropologist, this usually involves radiographic comparisons. These are preferably made using morphologically complex anatomical features, such as the arrangement of the dentition or the frontal sinus. Of growing importance within forensic anthropology is the likelihood ratio. This provides a quantitative way of expressing the strength of the evidence in support of a possible identification. This chapter concluded with brief descriptions of approaches that can be used to aid the identification process when either there is insufficient evidence to permit a scientific comparison (e.g., craniofacial superimposition) or when potential matches have been exhausted and new investigative leads must be generated (e.g., facial approximation).

End-of-Chapter Exercises

Exercise 1

Materials Required: Computer with an internet connection.

Directions: Go to the NamUs website (www.namus.gov). Click on the link entitled "Explore NamUs"—this will take you to a page with a box on the left side labeled "Quick Search." At the bottom of this box, click on the link entitled "Missing Advanced Search." On the top right of the Missing Persons Search page that this link takes you to, the total number of missing persons records in the database is provided. Keep track of how this number changes as you perform the following steps.

1. In the box labeled "Demographics," use the drop-down menu under "Age" and select "Missing." This indicates the age at which the individual was reported missing rather than their age based on their year of birth. Enter a 10-year interval centered on your own age. For example, if you are 23 years old, enter the age as between 18 and 28.
2. In the box labeled "Description," select an option under "Sex."
3. In the box labeled "Description," select an option under "Race/Ethnicity."
4. In the box labeled "Description," under "Height" enter a four-inch interval centered on your own height. For example, if you are 5 ft 8 in tall, enter 5 ft 6 in and 5 ft 10 in.

Question 1: How does the number of possible results change as you enter each component of the biological profile?

Question 2: What is the final number of possible results after entering your information? How does this number compare to the numbers obtained by other students in your class? What do you think might account for the differences?

Question 3: Clear your selections and enter each component of the biological profile by itself. For example, select only an option under "Sex" or enter only an age range. Based on this, are there components of the biological profile that seem to be more useful for narrowing the number of potential results?

Exercise 2

Materials Required: **Figure 11.27** (or an example provided by the instructor) and a computer with internet access.

Directions: **Figure 11.27** depicts a mandible with the teeth numbered according to the Universal Charting System.

Go to www.odontosearch.com and click on the button labeled "Click Here to Begin Using OdontoSearch." On the following page, select "Universal" tooth numbering and "Generic" coding. Leave the default settings for Gender, Race, and Age. Using **Figure 11.27** as a reference, enter the information from this mandible into OdontoSearch using the following codes for tooth status:

"X" = a tooth that is missing antemortem.
"R" = a tooth that has been restored (e.g., with a filling or a crown).
"V" = a tooth that is unrestored.
"/" = a tooth that is unobservable.

Each tooth's status should be entered into the appropriate box. Once the data have been entered (make sure all of the maxillary teeth—the upper row of boxes on the page—are marked "/"), click "Search" at the bottom of the page.

Question 1: How many pattern matches were there between this mandible and OdontoSearch's database? What percentage of the population is this pattern expected to occur in?

Question 2: You notice that one of your potential matches for this decedent has an antemortem dental chart that matches the observed pattern in the mandible. What is the likelihood ratio for the hypothesis that the mandible shown in **Figure 11.27** belongs to this missing person?

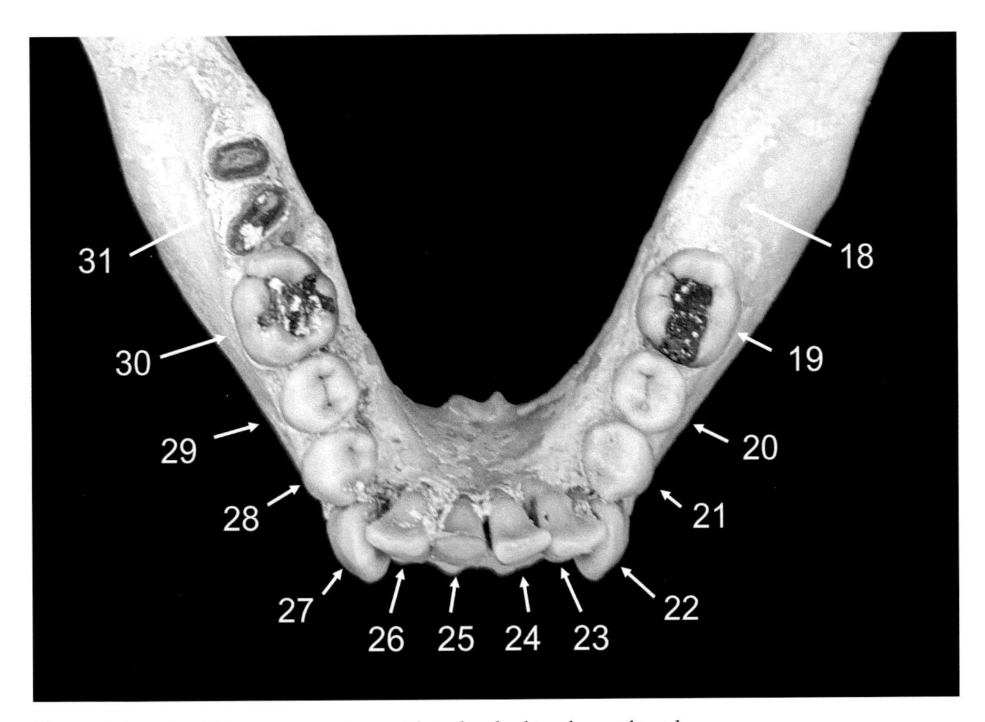

Figure 11.27. Mandible, superior view, with individual teeth numbered.

References

Adams BJ. 2003a. Establishing personal identification based on specific patterns of missing, filled, and unrestored teeth. *Journal of Forensic Sciences* 48: 487–496.

Adams BJ. 2003b. The diversity of adult dental patterns in the United States and the implications for personal identification. *Journal of Forensic Sciences* 48: 497–503.

Adams BJ, Maves RC. 2002. Radiographic identification using the clavicle of an individual missing from the Vietnam conflict. *Journal of Forensic Sciences* 47: 369–373.

Angyal M, Dérczy K. 1998. Personal identification on the basis of antemortem and postmortem radiographs. *Journal of Forensic Sciences* 43: 1089–1093.

Aufderheide AC, Rodríguez-Martín C. 1998. *The Cambridge Encyclopedia of Human Paleopathology.* Cambridge: Cambridge University Press.

Austin-Smith D, Maples WR. 1994. The reliability of skull/photograph superimposition in individual identification. *Journal of Forensic Sciences* 39: 446–455.

Baht GS, Vi L, Alman BA. 2018. The role of the immune cells in fracture healing. *Current Osteoporosis Reports* 16: 138–145.

Barnes E. 1994. *Developmental Defects of the Axial Skeleton in Paleopathology.* Niwot, CO: University Press of Colorado.

Berthaume MA, Di Federico E, Bull AMJ. 2019. Fabella prevalence rate increases over 150 years, and rates of other sesamoid bones remain constant: a systematic review. *Journal of Anatomy* 235: 67–79.

Brogdon BG. 1998. Radiological identification of individual remains. In: Brogdon BG (Ed.), *Forensic Radiology.* Boca Raton, FL: CRC Press. p. 149–187.

Brogdon BG. 2011. Radiological identification of individual remains. In: Thali MJ, Viner MD, Brogdon BG (Eds.), *Brogdon's Forensic Radiology,* Second Edition. Boca Raton, FL: CRC Press. p. 153–176.

Buikstra JE (Ed.). 2019. *Ortner's Identification of Pathological Conditions in Human Skeletal Remains,* Third Edition. London: Academic Press.

Byers SN. 2017. *Introduction to Forensic Anthropology,* Fifth Edition. London: Routledge.

Cameriere R, Ferrante L, Mirtella D, Rollo FU, Cingolani M. 2005. Frontal sinuses for identification: quality of classifications, possible error and potential corrections. *Journal of Forensic Sciences* 50: 770–773.

Caplova Z, Obertova Z, Gibelli DM, Mazzarelli D, Fracasso T, Vanezis P, Sforza C, Cattaneo C. 2017. The reliability of facial recognition of deceased persons on photographs. *Journal of Forensic Sciences* 62: 1286–1291.

Cattaneo C, De Angelis D, Porta D, Grandi M. 2006. Personal identification of cadavers and human remains. In: Schmitt A, Cunha E, Pinheiro J (Eds.), *Forensic Anthropology and Medicine: Complementary Sciences from Recovery to Cause of Death.* Totowa, NJ: Humana Press. p. 359–379.

Christensen AM. 2005. Testing the reliability of frontal sinuses in positive identification. *Journal of Forensic Sciences* 50: 18–22.

Christensen AM, Hatch GM. 2016. Quantification of radiologic identification (RADid) and the development of a population frequency data repository. *Journal of Forensic Radiology and Imaging* 7: 14–16.

Christensen AM, Hatch GM. 2018. Advances in the use of frontal sinuses for human identification. In: Latham K, Bartelink E, Finnegan M (Eds.), *New Perspectives in Forensic Human Skeletal Identification.* London: Academic Press. p. 227–240.

Christensen AM, Passalacqua NV, Bartelink EJ. 2019. *Forensic Anthropology: Current Methods and Practice,* Second Edition. London: Academic Press.

Clement JG. 2013. Odontology. In: Siegel JA, Saukko PJ (Eds.), *Encyclopedia of Forensic Sciences.* Waltham, MA: Academic Press. p. 106–113.

Culbert WL, Law FM. 1927. Identification by comparison of roentgenograms of nasal accessory sinuses and mastoid processes. *Journal of the American Medical Association* 88: 1634–1636.

Cunha E. 2006. Pathology as a factor of personal identity in forensic anthropology. In: Schmitt A, Cunha E, Pinheiro J (Eds.), *Forensic Anthropology and Medicine: Complementary Sciences from Recovery to Cause of Death.* Totowa, NJ: Humana Press. p. 333–358.

Daubert v. Merrell Dow Pharmaceuticals, Inc., 509 U.S. 579, 1993.

Del Papa MC, Perez SI. 2007. The influence of artificial cranial vault deformation on the expression of cranial nonmetric traits: its importance in the study of evolutionary relationships. *American Journal of Physical Anthropology* 134: 251–262.

DiCaudo DJ. 2006. Coccidioidomycosis: a review and update. *Journal of the American Academy of Dermatology* 55: 929–942.

Dorion RB. 1983. Photographic superimposition. *Journal of Forensic Sciences* 28: 724–734.

Forrest AS, Wu HY. 2010. Endodontic imaging as an aid to forensic personal identification. *Australian Endodontic Journal* 36: 87–94.

Gerasimov MM. 1971. *The Face Finder.* Philadelphia, PA: JB Lippincott.
Hanihara T, Ishida H, Dodo Y. 1998. Os zygomaticum bipartitum: frequency distribution in major human populations. *Journal of Anatomy* 192: 539–555.
Harris AM, Wood RE, Nortjé CJ, Thomas CJ. 1987. Gender and ethnic differences of the radiographic image of the frontal region. *The Journal of Forensic Odonto-Stomatology* 5: 51–57.
Hatch GM, Dedouit F, Christensen AM, Thali MJ, Ruder TD. 2014. RADid: a pictorial review of radiologic identification using postmortem CT. *Journal of Forensic Radiology and Imaging* 2: 52–59.
Hogge JP, Messmer JM, Doan QN. 1994. Radiographic identification of unknown human remains and interpreter experience level. *Journal of Forensic Sciences* 39: 373–377.
Hurst CV, Soler A, Fenton TW. 2013. Personal identification in forensic anthropology. In: Siegel JA, Saukko PJ (Eds.), *Encyclopedia of Forensic Sciences.* Waltham, MA: Academic Press. p. 68–75.
Kirk NJ, Wood RE, Goldstein M. 2002. Skeletal identification using the frontal sinus region: a retrospective study of 39 cases. *Journal of Forensic Sciences* 47: 318–323.
Komar DA, Buikstra JE. 2008. *Forensic Anthropology: Contemporary Theory and Practice.* Oxford: Oxford University Press.
Koot MG, Sauer NJ, Fenton TW. 2005. Radiographic human identification using bones of the hand: a validation study. *Journal of Forensic Sciences* 50: 263–268.
Krogman WM, İşcan MY. 1986. *The Human Skeleton in Forensic Medicine,* Second Edition. Springfield, IL: Charles C. Thomas.
Lesciotto KM. 2015. The impact of *Daubert* on the admissibility of forensic anthropology expert testimony. *Journal of Forensic Sciences* 60: 549–555.
Lipton BE, Murmann DC, Pavlik EJ. 2013. History of forensic odontology. In: Senn DR, Weems RA (Eds.), *Manual of Forensic Odontology,* Fifth Edition. Boca Raton, FL: CRC Press. p. 1–39.
Milligan CF, Finlayson JE, Cheverko CM, Zarenko KM. 2018. Advances in the use of craniofacial superimposition for human identification. In: Latham K, Bartelink E, Finnegan M (Eds.), *New Perspectives in Forensic Human Skeletal Identification.* London: Academic Press. p. 241–250.
Murray EA, Anderson BE, Clark SC, Hanzlick RL. 2018. The history and use of the National Missing and Unidentified Persons System (NamUs) in the identification of unknown persons. In: Latham K, Bartelink E, Finnegan M (Eds.), *New Perspectives in Forensic Human Skeletal Identification.* London: Academic Press. p. 115–126.
Osborne-Gustavson AE, McMahon T, Josserand M, Spamer BJ. 2018. The utilization of databases for the identification of human remains. In: Latham K, Bartelink E, Finnegan M (Eds.), *New Perspectives in Forensic Human Skeletal Identification.* London: Academic Press. p. 129–139.
Owsley DW, Mann RW. 1992. Positive personal identity of skeletonized remains using abdominal and pelvic radiographs. *Journal of Forensic Sciences* 37: 332–336.
Reconstruct body to solve murders. *New York Times,* 25 September 1916. p. 1.
Rhine S, Sperry K. 1991. Radiographic identification by mastoid sinus and arterial pattern. *Journal of Forensic Sciences* 36: 272–279.
Riddick L, Brogdon BG, Lasswell-Hoff J, Delmas B. 1983. Radiographic identification of charred human remains through use of the dorsal defect of the patella. *Journal of Forensic Sciences* 28: 263–267.
Rogers TL, Allard TT. 2004. Expert testimony and positive identification of human remains through cranial suture patterns. *Journal of Forensic Sciences* 49: 203–207.
Rougé D, Telmon N, Arrue P, Larrouy G, Arbus L. 1993. Radiographic identification of human remains through deformities and anomalies of post-cranial bones: a report of two cases. *Journal of Forensic Sciences* 38: 997–1007.
Sauer NJ, Brantley RE, Barondess DA. 1988. The effects of aging on the comparability of antemortem and postmortem radiographs. *Journal of Forensic Sciences* 33: 1223–1230.
Smith-Bindman R, Miglioretti DL, Larson EB. 2008. Rising use of diagnostic medical imaging in a large integrated health system. *Health Affairs* 27: 1491–1502.
Snow CC, Gatliff BP, McWilliams KR. 1970. Reconstruction of facial features from the skull: an evaluation of its usefulness in forensic anthropology. *American Journal of Physical Anthropology* 33: 221–228.
Steadman DW, Adams BJ, Konigsberg LW. 2006. Statistical basis for positive identification in forensic anthropology. *American Journal of Physical Anthropology* 131: 15–26.
Stephan CN. 2013. Facial approximation. In: Siegel JA, Saukko PJ (Eds.), *Encyclopedia of Forensic Sciences.* Waltham, MA: Academic Press. p. 60–67.
Stephan CN. 2014. The application of the central limit theorem and the law of large numbers to facial soft tissue depths: T-table robustness and trends since 2008. *Journal of Forensic Sciences* 59: 454–462.
Stephan CN. 2017. Estimating the skull-to-camera distance from facial photographs for craniofacial superimposition. *Journal of Forensic Sciences* 62: 850–860.

Stephan CN, Cicolini J. 2008. Measuring the accuracy of facial approximations: a comparative study of resemblance rating and face array methods. *Journal of Forensic Sciences* 53: 58–64.

Stephan CN, Henneberg M. 2001. Building faces from dry skulls: are they recognized above chance rates? *Journal of Forensic Sciences* 46: 432–440.

Stephan CN, Simpson EK. 2008. Facial soft tissue depths in craniofacial identification (Part I): an analytical review of the published adult data. *Journal of Forensic Sciences* 53: 1257–1272.

Stephan CN, Winburn AP, Christensen AF, Tyrell AJ. 2011. Skeletal identification by radiographic comparison: blind tests of a morphoscopic method using antemortem chest radiographs. *Journal of Forensic Sciences* 56: 320–332.

Stephan CN, Simpson EK, Byrd JE. 2013. Facial soft tissue depth statistics and enhanced point estimators for craniofacial identification: the debut of the shorth and the 75-shormax. *Journal of Forensic Sciences* 58: 1439–1457.

Streetman E, Fenton TW. 2018. Comparative medical radiography: practice and validation. In: Latham K, Bartelink E, Finnegan M (Eds.), *New Perspectives in Forensic Human Skeletal Identification.* London: Academic Press. p. 251–264.

Sudimack JR, Lewis BJ, Rich J, Dean DE, Fardal PM. 2002. Identification of decomposed human remains from radiographic comparisons of an unusual foot deformity. *Journal of Forensic Sciences* 47: 218–220.

Ubelaker, DH. 1984. Positive identification from the radiographic comparison of frontal sinus patterns. In: Rathbun TA, Buikstra JE (Eds.), *Human Identification: Case Studies in Forensic Anthropology.* Springfield, IL: Charles C. Thomas. p. 399–411.

Ubelaker DH, Jacobs CH. 1995. Identification of orthopedic device manufacturer. *Journal of Forensic Sciences* 40: 168–170.

Ubelaker DH, Bubniak E, O'Donnell G. 1992. Computer-assisted photographic superimposition. *Journal of Forensic Sciences* 37: 750–762.

Viner M. 2018. Overview of advances in forensic radiological methods of human identification. In: Latham K, Bartelink E, Finnegan M (Eds.), *New Perspectives in Forensic Human Skeletal Identification.* London: Academic Press. p. 217–226.

Wilson RJ, Bethard JD, DiGangi EA. 2011. The use of orthopedic surgical devices for forensic identification. *Journal of Forensic Sciences* 56: 460–469.

Yoshino M, Miyasaka S, Sato H, Tsuzuki Y, Seta S. 1989. Classification system of frontal sinus patterns. *Canadian Society of Forensic Science Journal* 22: 135–146.

12

Skeletal Trauma and Timing of Bone Injury

Learning Goals

By the end of this chapter, the student will be able to:

- Define trauma as applied within forensic contexts.
- Categorize the different types of external forces that can act on bone and differentiate the fracture patterns that result from each.
- Define the three time periods in which trauma or bone breakage is categorized.
- Differentiate ante-, peri-, and postmortem trauma using gross observations of the fracture site.
- Describe the process of bone healing and fracture repair.
- Differentiate the signatures of wet and dry bone fracture patterns.

Introduction

Trauma is defined as an injury or disruption caused to *living* tissue by any *external* force. It was only recently that forensic anthropologists began to consult on the interpretation of trauma visible on the skeletons of victims of violent crimes. However, this aspect of forensic analysis has become increasingly important over the last few decades. With improved understanding of fracture formation processes and a growing number of experimental studies documenting how fractures propagate throughout the skeleton, it is now expected that the forensic anthropologist provides a description of trauma during their investigation. The goal is to provide an accurate and complete description of the wounds, but not necessarily to offer a specific interpretation that may reach beyond what the data can support in a court of law. The *Daubert* criteria (chapter 2) must always be considered.

The ability to recognize and interpret trauma contributes to the death investigation in several ways. First, trauma analysis can provide information that is useful to the medical examiner for declaring the manner of death (e.g., homicide, suicide, accidental). Second, trauma analysis can help identify the nature of the force that caused the injury, which may lead to the murder weapon. Third, by identifying the number of wounds and their relationship to each other, the investigator can determine whether one or more implements were used. This can potentially speak to the number of individuals involved in committing a crime. Fourth, ascertaining the sequence of injuries speaks to the circumstances of the crime scene and addresses the fundamental question—what happened? Fifth, trauma analysis can be used to reconstruct the spatial relationship between the attacker(s) and the decedent, which can corroborate or falsify suspect or witness testimony. Sixth, the ability to recognize healed trauma, i.e., trauma not directly related to the moment of death, can provide information on the decedent's medical history that could be used to positively identify them (see chapter 11).

Finally, repeated, healed, or healing injuries may also speak to patterns of domestic violence and child abuse.

Forensic anthropologists classify trauma with respect to three broad classes of weapons: **projectile** gunshots (e.g., handguns), **blunt force** objects (e.g., bats), and **sharp force** objects (e.g., knives). However, these broad categories may fail to categorize all traumatic injuries. For example, a bullet that impacts the skull after losing much of its velocity will create a wound more similar to blunt force trauma than a typical penetrating gunshot wound. For this reason, forensic anthropologists also recognize the importance of *velocity* in determining the extent of a traumatic injury. Two broad classes of weapons are identified: **high-velocity projectiles** and **low-velocity objects**.

The area of direct impact is also a consideration. Sharp force trauma entails an object physically impacting the body with low velocity over a very small area, thus making a cut in the bone but not shattering it. Blunt force trauma entails an object impacting the body with low velocity over a relatively wide area, often shattering the bone but not penetrating it. High-velocity projectile trauma often combines penetrating wounds (bullet holes) and more significant destruction of the surrounding tissues. Whether a bone fractures when put under stress depends on the energy it must absorb from the foreign object. Objects moving with higher velocity impart more energy onto the skeletal tissues, resulting in more damage. Identifying the nature of the force (projectile, blunt force, sharp force) impacting the body is one of the primary goals of the forensic analysis.

In addition to determining the type of force impacting the body, the forensic anthropologist is interested in determining the timing of impact with respect to the individual's death. Three time periods are considered: **antemortem trauma** is that which occurred prior to death, **perimortem trauma** is that which occurred *around* the time of death, and a **postmortem alteration** is bone damage that occurred after death. These are not clearly distinguishable categories (as discussed below), and it is often very difficult to determine whether damage is perimortem or postmortem.

Bone Biology and the Biomechanics of Bone Fracture

Although all bone is composed of collagen and hydroxyapatite (see chapter 3), this fact alone does not explain or predict how any specific traumatic impact will result in a fracture. Bone is *heterogeneous,* meaning the bones in the body vary considerably in shape, thickness, and the distribution of cortical and trabecular bone (see chapters 3 and 4). As described by Symes et al. (2012), bone is also *anisotropic,* which means it responds differently to external forces depending on the direction the force is applied. In other words, for any given element of the human skeleton, the same magnitude of force applied from different directions will result in different fracture patterns due to the intrinsic qualities of the bone. For example, the ends of long bones have a thin outer layer and are mostly composed of spongy trabecular bone. The long bone ends, then, will respond differently to external stress than the middle of the shaft, which is mostly thick, cortical bone. The most common directional descriptions of force are presented in **Figure 12.1**.

Under tension, the ends of the bones are pulled apart, which stretches the collagen fibers along their length. Fractures caused by tension are often clean breaks that have few secondary fracture lines. Under compression, the ends of the bones are pushed together causing multiple breaks and secondary fracture lines that radiate from the point of impact. Under shearing, the force impacts the body at an oblique angle, which often causes the two fracture ends to be misaligned because they are pushed in opposite directions. Under torsion, the

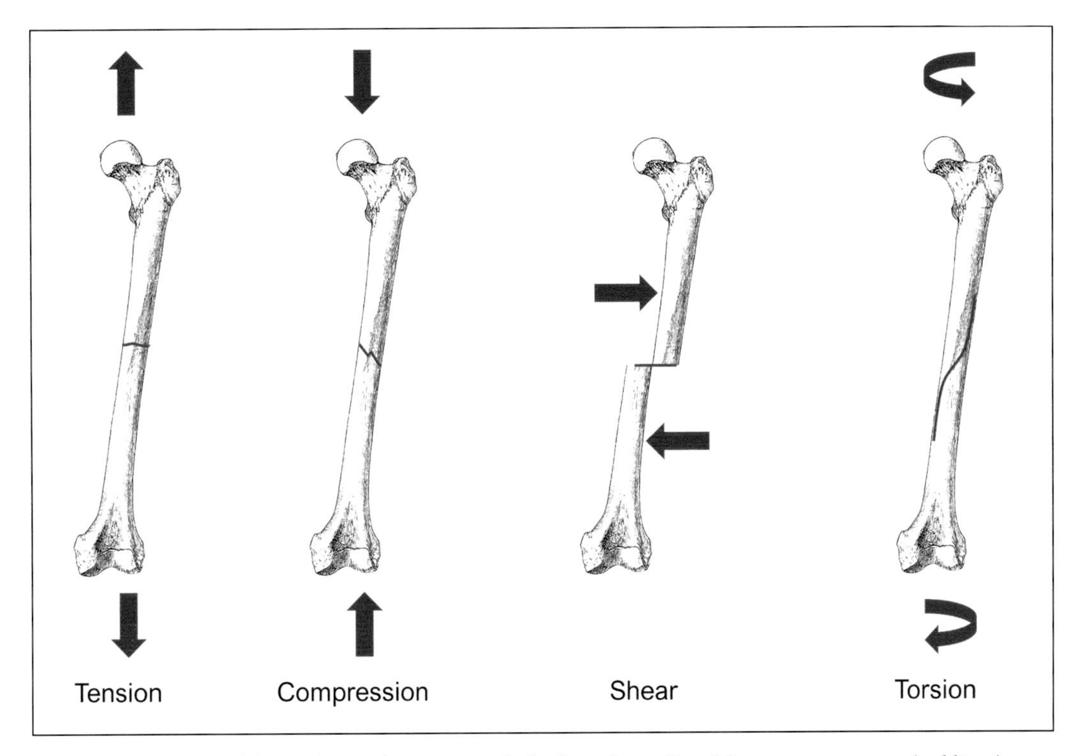

Figure 12.1. Directional forces (arrows) impacting the body with predicted fracture appearance (red lines).

force is also directed obliquely but with one end of the bone being stationary. This causes fracture lines to spiral down the axis of the bone shaft. In real-world situations, most fractures result from a combination of these directional forces. However, it is useful to consider tension and compression as complementary fracture mechanisms, which leads to the first axiom of interpreting skeletal trauma (Currey, 1970; Zephro and Galloway, 2014).

The **first axiom** for understanding bone fracturing patterns is that *bone is weaker under tension than under compression.* This means that bone will fracture first where tensile forces are experienced. The ability of bone to withstand compression is nearly twice that of the ability of bone to withstand tension.

Consider the diagram of a long bone shaft shown in **Figure 12.2.** A force is applied from the left side of the bone, striking it at a 90-degree angle. The side of the bone being struck is under compression because the bone is bending under the force of the blow. The small red arrows along the long bone shaft indicate the direction of bone movement under both compression and tension. The bone at the immediate point of impact is physically being squeezed together. The opposite side of the bone is under tension as the collagen fibers are being pulled apart. In other words, the bone is being stretched on the opposite side that the force is applied. *Because bone is stronger under compression than tension, it will first fracture on the opposite side of the bone from which the force was applied.* This fact alone allows the forensic anthropologist to determine the direction of force.

The bending of the bone shown in **Figure 12.2** demonstrates another factor that is critical for interpreting trauma (Berryman et al., 2012, 2013; Symes et al., 2012; Zephro and Galloway, 2014). Bone is a *viscoelastic* material, which means it responds to force differently depending on the speed at which a force impacts the body. *Static forces* impact the body at low velocity for a relatively long period of time (a blow to the head with a brick). *Dynamic forces* impact the body suddenly and quickly, but also dissipate rapidly (a bullet wound). Bone is better able to resist dynamic forces, but once the bone fails, it shatters without bending or deforming.

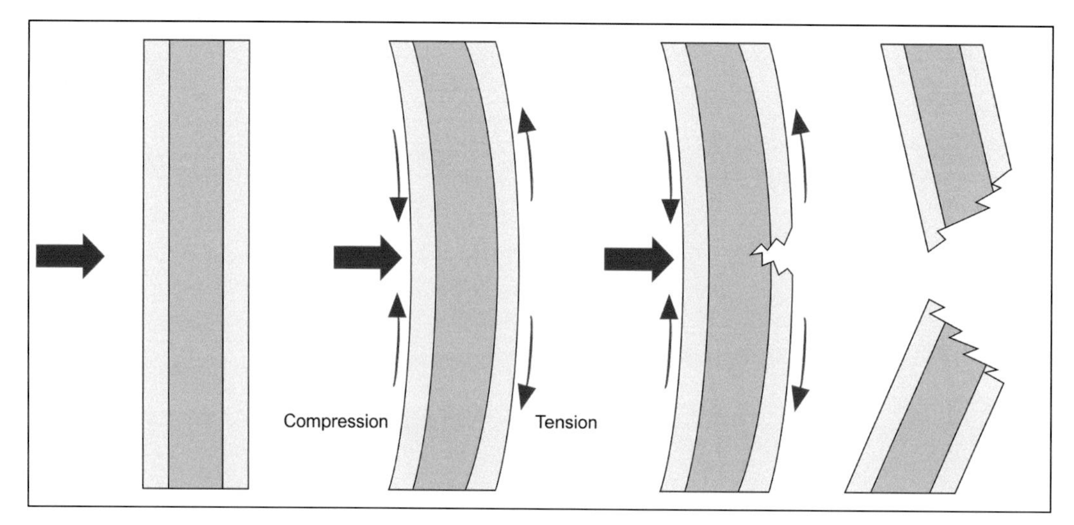

Figure 12.2. Response of a long bone shaft to a perpendicular force (thick red arrow) being applied. Smaller arrows indicate the direction in which the bone is moving under stress.

In other words, bones act like a brittle material (like a pane of glass) under a dynamic force. Comminuted fractures and secondary fractures (see below) are often the result of dynamic stress. To the contrary, bone reacts in a more pliable fashion under static stress. This means the bone will bend and deform prior to fracturing (like a piece of plastic). Static stress more often results in single fracture lines but not comminuted fractures.

When a static force is applied, bone undergoes three phases of change (Currey, 1970). During the first phase, the bone will bend in response to a force but will not be permanently deformed once that force is removed. This is called the *elastic phase*—the bone bends, absorbs the energy, and returns to its original state without any permanent change to its structure. Most minor blows to the body (a kick to the shin) fall into this category. As the magnitude or duration of force increases, the bone will continue to bend and deform to absorb the energy imparted. However, during this phase, when the force is removed the bone will *not* return to its original state. In other words, the bone is permanently deformed. This is called the *plastic phase.* The third and final phase is when the bone has been stressed beyond its capacity to absorb energy. The force exceeds the structural strength of the bone and gives way; the bone has *fractured.*

The **second axiom** for understanding bone fracturing patterns is that *bone responds as a brittle material under dynamic forces and as an elastic material under static forces.* Dynamic forces do not deform the bone but simply shatter it. Static forces deform the bone prior to causing a fracture. This helps to explain some of the differences between how bone reacts to high energy versus low energy trauma (see chapters 13 and 14).

Classifying Trauma

When enough force is applied to bone, a discontinuity (i.e., break) will form. If the discontinuity travels through the bone and physically separates it into two pieces, it is called a *fracture.* A *displaced fracture* occurs when the two continuous bone surfaces no longer meet. In **Figure 12.3,** the broken ends of the femur are displaced in the open wound example. If the bone is not "set" by a doctor, it will heal but will be permanently misshapen. *Comminuted* fractures are those in which multiple fractures occur at a site of injury. In comminuted fractures, the bone has fragmented into two or more pieces, indicating that a high level of energy was impacting the body. This is consistent with dynamic loading. A *butterfly fracture*

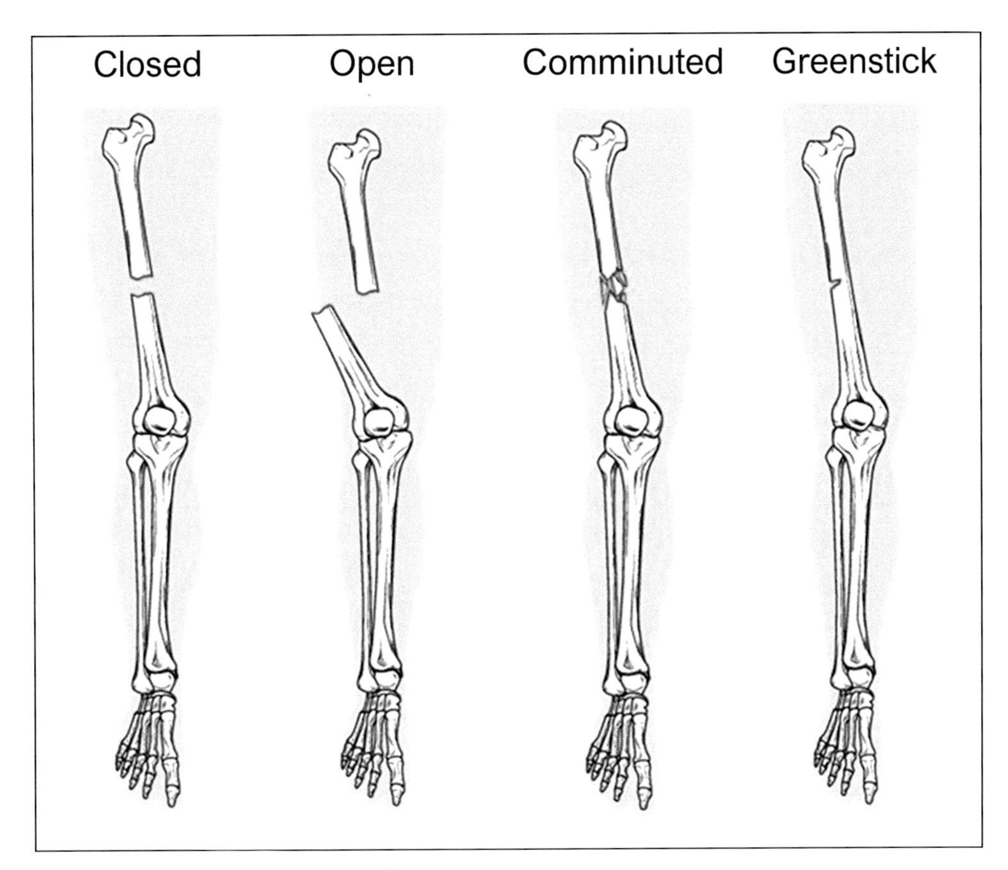

Figure 12.3. Different classifications of bone trauma.

is a special type of comminuted fracture of long bones, typically resulting from blunt force trauma. Fractures can also be characterized with respect to soft tissue injury. A fracture in which the bone does not break the outer surface of the skin is called *closed,* while a fracture that breaks the skin is called *open.* Open fractures have a much higher chance of secondary infection.

If the discontinuity does not completely separate the bone, it is called an *infraction* (or an *incomplete fracture*). All else being equal, fractures represent more severe injuries than infractions and suggest more energy has impacted the body. Infractions are more common in children because of the high collagen content of growing bones, whereas elderly individuals with low bone mineral density (osteoporosis) are more likely to experience a complete fracture. Infractions may also suggest that the energy impacting the body was either spread over a relatively wide area (such as being hit by a car), or that the amount of energy impacting the body was limited (being hit by a car at a low speed) (Galloway et al., 2014a). A *greenstick fracture* results when the bone is impacted at an angle or is bent resulting in tension on one side of the bone and compression on the other side. An incomplete fracture will form transversely and often travel longitudinally up and down the shaft of the long bone. These types of fractures are more common in children. A *depression fracture* is specific to the cranial vault. It involves an injury that usually crushes the trabecular bone and some of the outer cortical bone but does not completely separate the fragments. An example of a depression fracture is presented in **Figure 12.4**. There are dozens of additional clinically defined fracture categories; however, many of these are not relevant to forensic anthropology (see Galloway et al., 2014a).

Figure 12.4. Depression fracture of the cranial vault.

Secondary Fracture Lines

The preceding discussion focused on the formation of a *primary fracture* associated with the direct area of impact. However, discontinuities are often associated with *secondary fracture lines,* which usually originate from the point of impact or circumscribe the impact site. Secondary fracture lines result when the bone directly at the point of contact (bullet entry point) cannot absorb all of the energy imparted into the body through the formation of the primary fracture (bullet hole). There are two types of secondary fracture lines.

Radiating fractures disperse outward from the site of impact, making them useful for identifying the site of impact if the damage to the body has resulted in extreme fragmentation. Radiating fractures represent the energy traveling through the bone along the lines of least resistance—areas of low bone density, low bone thickness, high proportions of trabecular bone, cranial sutures, and the numerous foramina located throughout the body. **Concentric fractures** or **hoop fractures** form curved fracture lines that surround the site of impact. These lines are usually associated with high-velocity projectile and low-velocity blunt force trauma. Concentric fractures are associated with high levels of energy impacting the body.

Figure 12.5 shows both types of fractures resulting from a gunshot exit wound on the left side of the cranium. Radiating fractures are tinted in blue and concentric fractures are tinted in red. This figure nicely demonstrates what is meant by the term "concentric"—the fracture lines form like rings in a pond when a pebble is thrown into the water. These fracture lines represent the effects of the force of the bullet and the energy produced when it impacts the skull.

It is important to consider that radiating fractures have a single point of origin. This is a critical feature for distinguishing perimortem trauma with radiating fractures from damage that occurs to a cranium from natural processes once buried in the ground. Consider the cranium in **Figure 12.6**. At first glance, the damage and breakage to the side of this cranium may seem similar to blunt force trauma with extensive radiating fractures. However, note how haphazard the fracture lines are. There is no clear impact site that is the point of origin

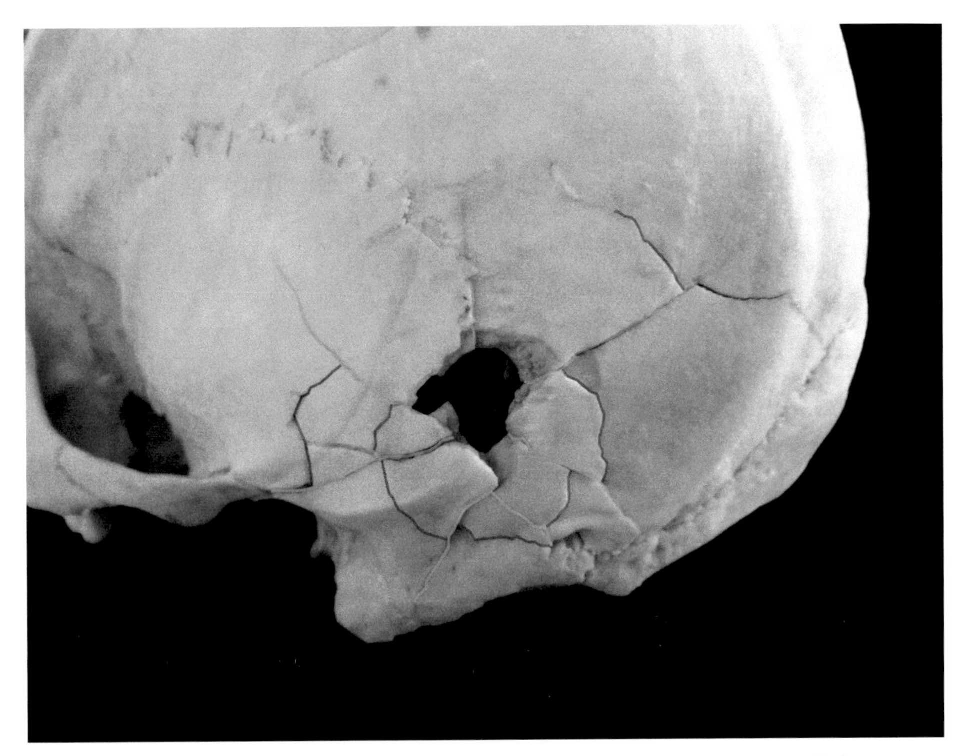

Figure 12.5. Gunshot exit wound resulting in both radiating (blue) and concentric (red) fractures.

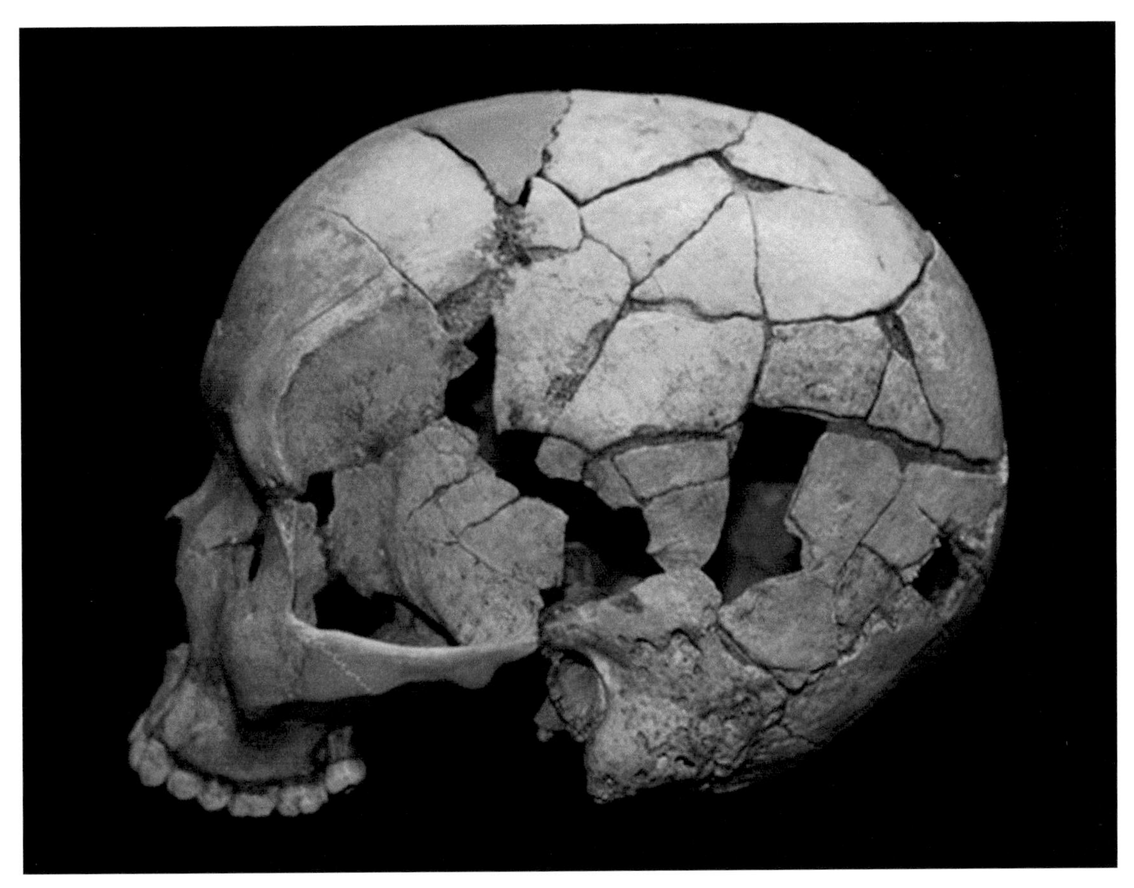

Figure 12.6. Lateral view of cranium depicting crushing due to soil compression.

for these different fracture lines. The fracture lines intersect at right angles to each other and the broken fragments of bone are square and rectangular rather than more triangular in shape. This damage is caused by soil compression and is not the result of trauma.

LEARNING CHECK

Q1. Describe the types of fractures seen in **Figure 12.7A** on the frontal bone.
A) Radiating
B) Concentric
C) Both

Q2. Describe the type of fracture(s) seen in **Figure 12.7B**. Note this is a blunt force trauma likely caused by something like a hammer.
A) Radiating
B) Concentric
C) Both

Q3. Describe the type of fracture(s) seen on the cranium in **Figure 12.7C**.
A) Radiating
B) Concentric
C) Both

Timing of Bone Injury

One of the most important considerations when assessing trauma is determining when a fracture occurred with respect to the time of the decedent's death (**Table 12.1**). Identification of **antemortem trauma** contributes to an investigation in two ways. First, if the trauma occurred well before the time of death, there may be medical records that can be matched to the body of an unknown decedent. That is, a fracture with a distinct pattern of healing could be compared to X-rays of an individual for whom a missing persons report has been filed. Second, trauma identified in various stages of healing may indicate a pattern of injury consistent with torture or domestic abuse. For children, observation of antemortem trauma may indicate a history of child abuse and neglect. The challenge with identifying antemortem trauma is that it is entirely based on observations of *bone healing* or signs of *secondary infection* in response to the traumatic insult. However, bone does not begin healing at the level of gross, macroscopic observation (i.e., with the naked eye) for days to weeks after an injury occurred. A decedent could have received a blow to the head causing blunt force trauma 1–2 weeks prior to being shot and killed. However, if the healing of the first incident had not begun at the time of the second incident then the timelines may not be distinguishable through forensic anthropological analysis.

Perimortem trauma is likely of greatest interest to the forensic anthropologist because it may directly relate to the cause of death. Observations of perimortem gunshot wounds, stab marks, or blunt force injuries would all be suggestive of how the decedent died. However, it is important to stress that in forensic pathology, the perimortem interval refers to the time within seconds or minutes of a decedent dying. In forensic anthropology, the time interval is much more coarsely defined and is usually measured in weeks, not minutes or seconds.

Postmortem breakage is likely less relevant to assessments of cause and manner of death.

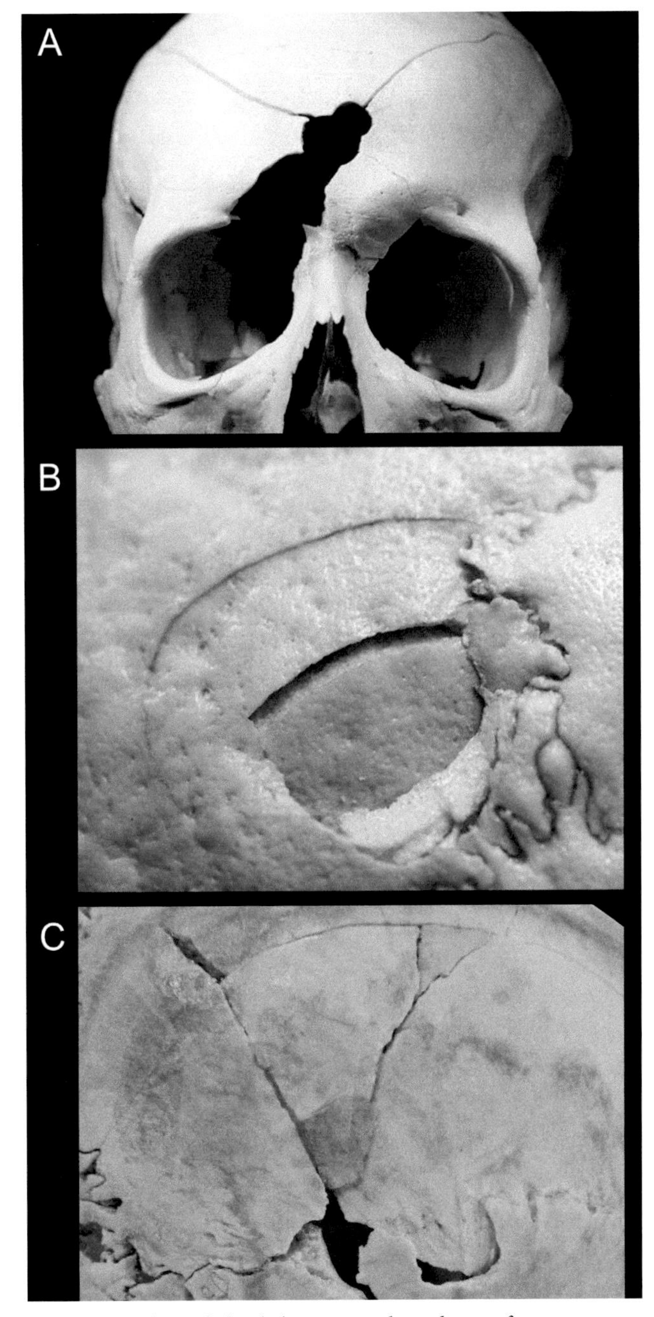

Figure 12.7. Three skeletal elements with evidence of trauma.

Table 12.1. Diagnostic criteria for differentiating antemortem, perimortem, and postmortem bone fracture/breakage

Time Period	Wet/Fresh Bone Response	Signs of Healing	Timeframe
Antemortem	Yes	Yes	Years to 2–3 weeks before death
Perimortem	Yes	No	2–3 weeks before death to weeks to months after death
Postmortem	No	No	Weeks to months after death

The challenge for the forensic anthropologist is differentiating perimortem trauma from postmortem damage, which is largely a function of the collagen content remaining in the bone at the time the break occurs. After death, bone loses moisture and the collagen component of bone begins to decay. This process of decay is a roughly continuous process, but it is also dependent upon the environment in which the body is decomposing. At a certain point after death, bone will begin to break like a dry, brittle material. Therefore, the difference between perimortem trauma and postmortem damage is whether the bone fractures with a *wet/fresh bone* response (indicating high levels of collagen or moisture) or with a *dry bone* response (indicating low levels of collagen or moisture). These criteria are summarized in **Table 12.1**.

Identifying Antemortem Trauma

The first step in determining the timing of bone injury is observing whether there is any evidence of osteogenic response at the site of fracture. The term "osteogenic" means "bone formation" and reflects the body's response to an injury. If the cellular processes were no longer functioning at the time the injury was sustained (i.e., the individual was not alive), then the healing process will not be evident at the site of fracture. Antemortem trauma can also be identified if the wound shows evidence of secondary infection. The healing process and the body's response to infection involve both osteoclasts (cells that remove bone) and osteoblasts (cells that make new bone). A general description of the healing process is as follows:

1) Soon after the trauma, blood from broken blood vessels will pool at the site of the injury resulting in visible bruising and swelling. The pooling of blood forms a hematoma, which is the body's initial attempt to stabilize the injury.
2) Within 5 to 10 days, the bone itself begins to heal as evidenced by *increased porosity or pitting* (small holes in the surface of the bone) surrounding the fracture margins. Osteoclasts will begin removing sharp fracture edges and resorbing dead pieces of bone within the fracture site. This results in the *rounding of the broken margins.*
3) As healing continues, osteoblasts will begin producing new bone to mend the fracture. Initially, a bony *callus* forms that is composed of woven bone, which is less structurally sound and more disorganized in appearance. This initial fracture callus will be replaced with mature bone that is more structurally sound as the wound is remodeled over the months following the injury.

In summary, antemortem trauma is evidenced by three macroscopic features of bone healing: increased porosity or pitting surrounding the fracture site, rounding of broken fracture margins, and the deposition of new bone (bony callus) to mend the fracture.

Examples of Antemortem Cranial Trauma

Figure 12.8A is a close-up showing the external surface of a cranial vault with active infection. The small holes represent porosity, which covers the entire area of the image. The larger holes are the parietal foramina; bony remodeling in response to the infection has obliterated the sagittal suture. Note the uneven appearance of the surface. **Figure 12.8B** shows the same individual from a different angle. Note that the left side of the image exhibits porosity and an uneven surface that is not smooth, while the right side of the image shows no porosity and is much smoother in texture (the texture that is visible is due to weathering). This image nicely demonstrates the difference between bone that is actively responding to a localized infection (left of the dashed line) and bone that is not (right side of the dashed line).

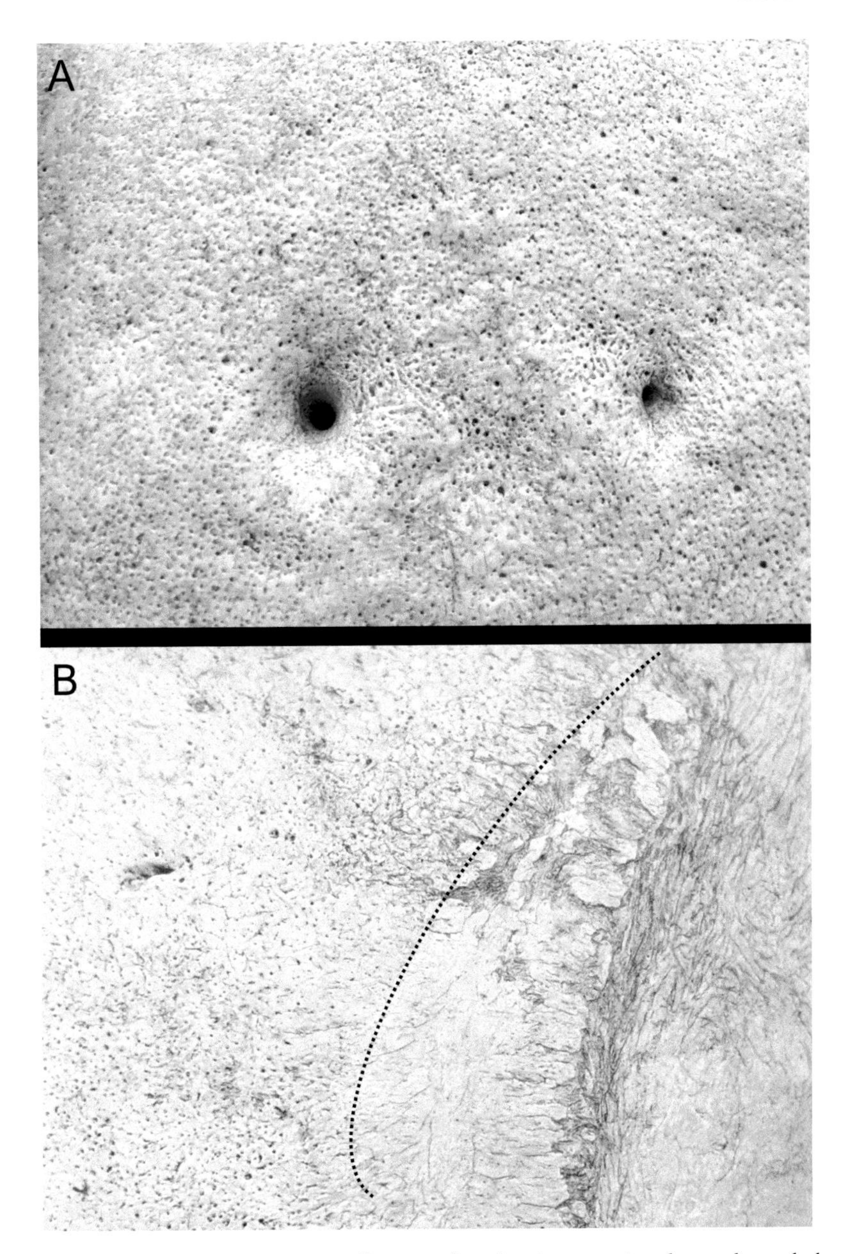

Figure 12.8. (*A*) Close-up image of bone surface showing porosity; the two larger holes are the parietal foramina. (*B*) A different view of the same individual illustrating active bone response (left of line) and smooth, but weathered, bone that is not reactive (right of line).

Figure 12.9A shows a single cranial gunshot wound that was fatal. There is no porosity surrounding the gunshot wound and the edges of the defect are sharp and crisp. The bone surrounding the bullet hole is smooth and even, indicating there is no remodeling of the margin. This would be considered perimortem trauma because no evidence of healing or active bony response is present. **Figure 12.9B** shows a bullet hole that has begun to heal. The black arrows indicate areas of active bone removal via osteoclastic activity. The red arrow indicates a smoothed fracture margin (also from osteoclastic activity). Note that the surrounding bone looks more porous and uneven in **Figure 12.9B** as the healing process has begun. **Figure 12.9C** shows evidence of more advanced healing. The white arrows indicate areas of new bone formation via osteoblastic activity. The body is attempting to close the hole by adding new bone to the fracture margins. The red arrow indicates an area that has been smoothed via osteoclastic activity, but no new bone has been laid down yet. The individual

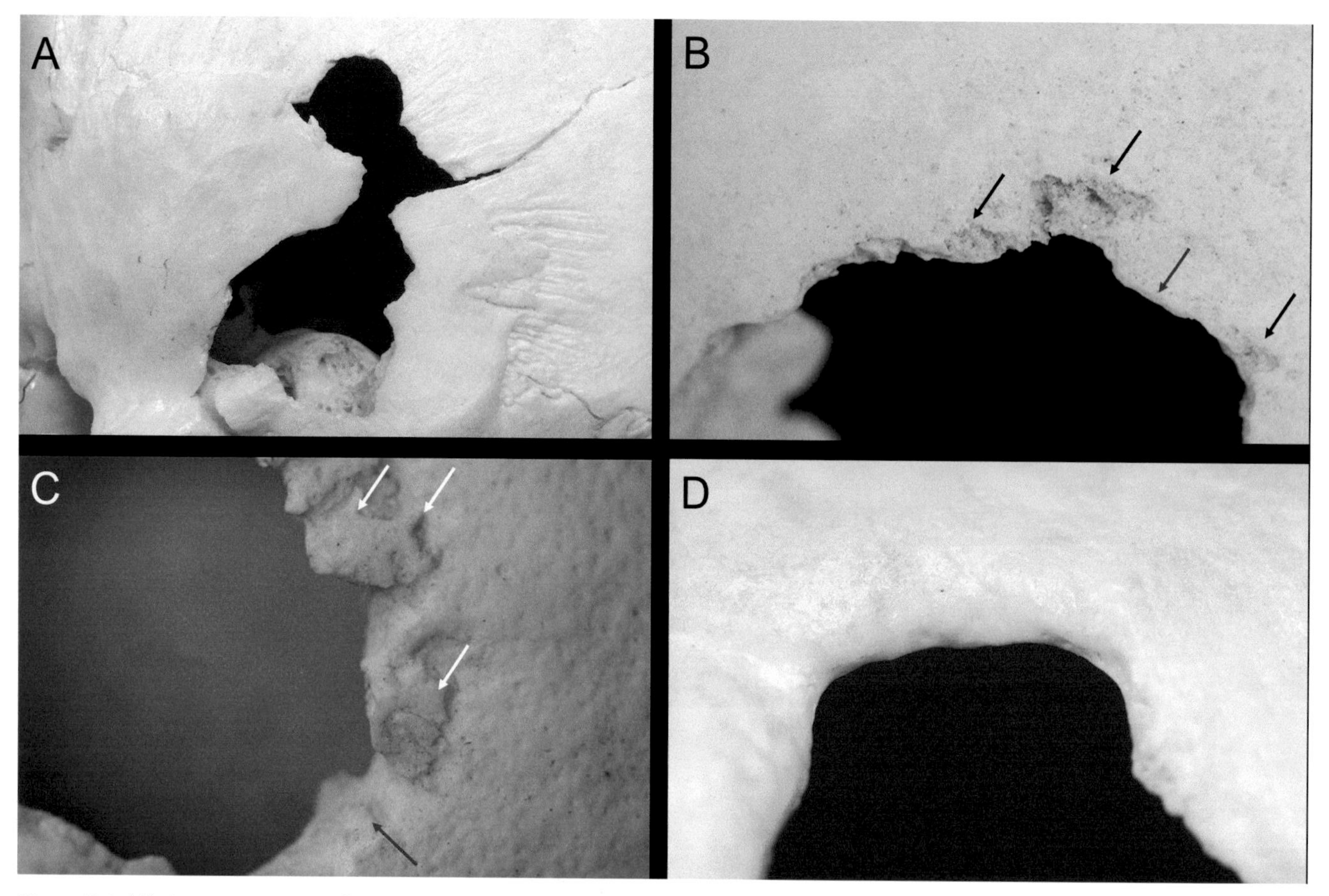

Figure 12.9. (*A*) A perimortem gunshot wound with no evidence of healing or remodeling of the fracture margins. (*B*) Gunshot wound showing active areas of bone removal (black arrows) and smoothing of the fracture margin (red arrow). (*C*) Gunshot wound showing a combination of osteoblastic (white arrows) and osteoclastic activity (red arrow). (*D*) Healed cranial fracture. Note the rounded and smooth fracture margins. The absence of porosity indicates the healing process has stopped.

in **Figure 12.9D** likely survived for several years after incurring this injury. Note how smooth and rounded the fracture margins are. The lack of porosity indicates the healing process has stopped. Although imperfect, the body has done what it can to repair the damage.

Antemortem Postcranial Callus Formation

For most postcranial fractures, the body will begin to quickly lay down new bone to mend the two halves of the fracture site. The result is a bony callus, initially composed of disorganized woven bone that will eventually be remodeled and replaced with mature bone. This process is shown in **Figure 12.10**. **Figure 12.10A** depicts an immature bony callus on a rib. Note the disorganized appearance of the woven bone. The fracture callus on the rib shown in **Figure 12.10B** is still dominated by woven bone but has begun to be remodeled—note the generally denser appearance of this bone relative to the woven bone. The clavicle in **Figure 12.10C** has progressed further in the healing process and much of the woven bone has been replaced by mature lamellar bone. Finally, in **Figure 12.10D**, the callus on this proximal radius has been completely remodeled. The body has likely stopped attempting to mend this fracture any further.

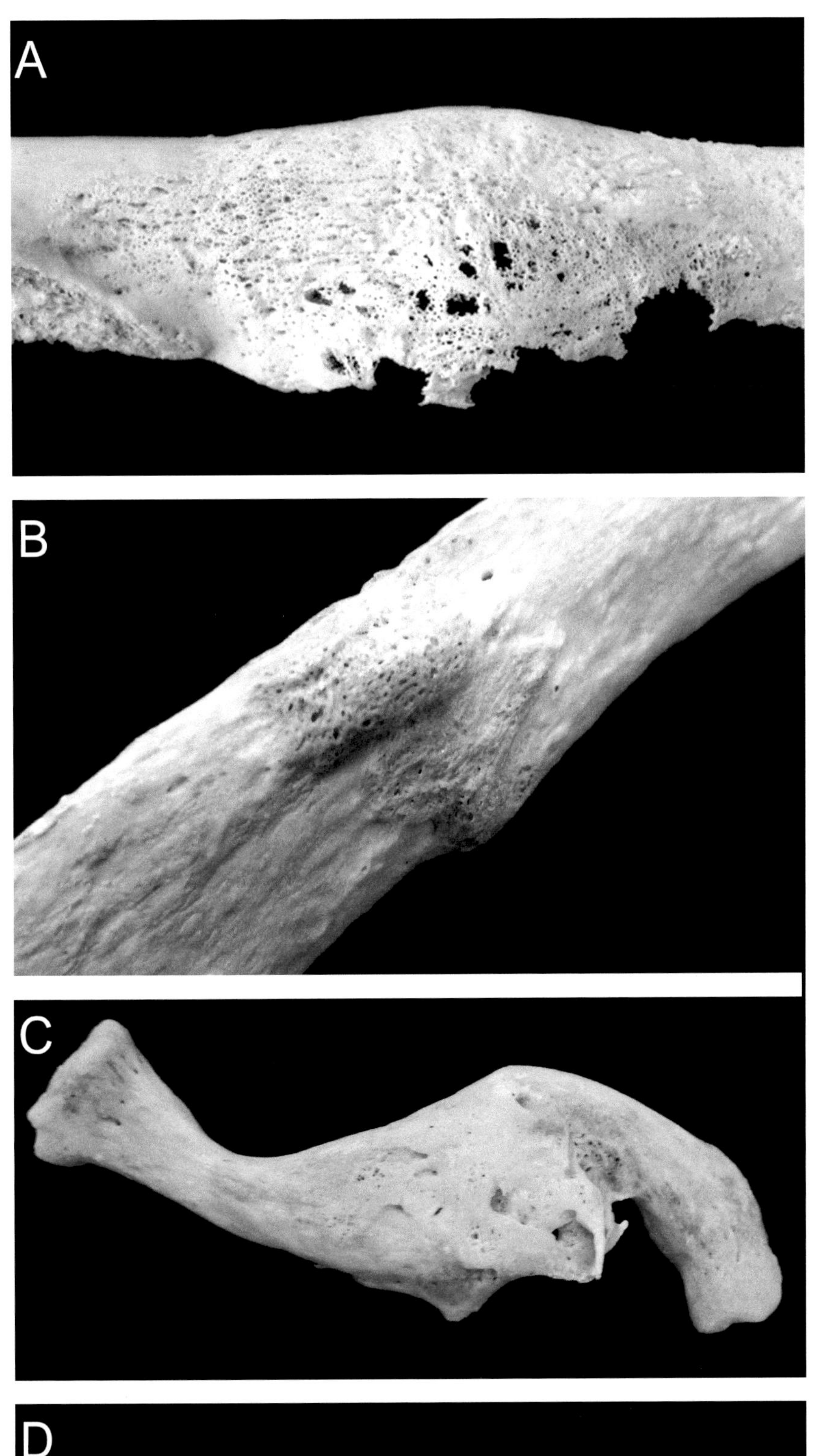

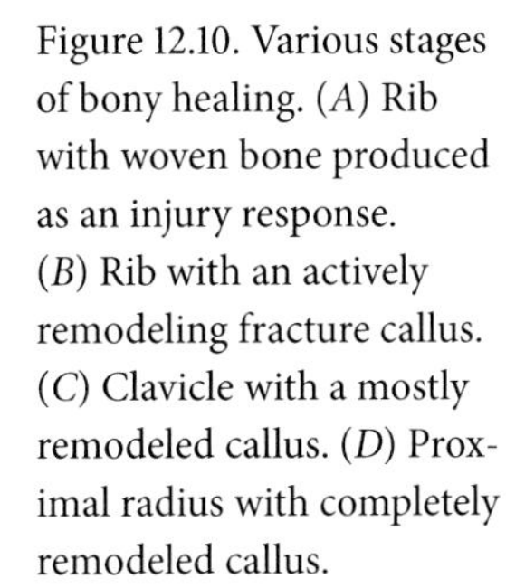

Figure 12.10. Various stages of bony healing. (*A*) Rib with woven bone produced as an injury response. (*B*) Rib with an actively remodeling fracture callus. (*C*) Clavicle with a mostly remodeled callus. (*D*) Proximal radius with completely remodeled callus.

Perimortem Trauma and Postmortem Alterations

If there is no evidence of healing, then the fracture could either be perimortem or postmortem. Recall that perimortem wounds are those in which the fracture occurred while the bone was in "a biomechanically fresh state (Christensen et al., 2014:349)." This is called a wet bone, green bone, or greasy bone response. The high collagen content of wet bone provides elasticity; the bone can better withstand stress and will deform prior to breaking. Postmortem fractures are those in which the break occurred when the bone was in a biomechanically brittle state. Dry bone is weaker than wet bone. It will not deform prior to fracturing, breaks under lower loads, and will tend to shatter into multiple pieces. Differentiating the two is one of the most difficult tasks in forensic anthropology (Galloway et al., 2014b; Symes et al., 2012, 2014; Ubelaker, 1991a; Ubelaker and Adams, 1995). Research on this topic is ongoing and the following discussion is meant to provide general guidelines and not hard and fast rules. The postmortem loss of elasticity in bone is a continuous process and not one that can easily be divided into discrete either/or categories. Still, experimental studies have identified several features that allow the forensic anthropologist to identify perimortem trauma with some consistency (**Table 12.2**).

Feature 1: Appearance of the fracture line on the external surface of the bone

One consideration is the way the fracture line propagates with respect to the overall shape of the bone (**Figure 12.11**). Perimortem fractures can exhibit a variety of appearances. Some fractures have acute changes in direction while others can be curvilinear in their path and encircle the shaft of a long bone in a spiral pattern. Perimortem fracture lines also tend to stop at the articular ends of long bones that are composed of trabecular bone, which dissipates the energy of the trauma more effectively without producing a fracture line. Postmortem fractures tend to be aligned either perpendicular to the bone shaft or longitudinally along the bone shaft. These fractures tend not to have a curved pathway around the bone and look more regular in their appearance. Postmortem fractures can also have a stepped pathway along the bone length. Postmortem fracture lines will continue through the articular ends of bones.

Table 12.2. Differences between perimortem trauma and postmortem breaks

Feature	Perimortem	Postmortem
Fracture Outline	Curved fracture lines, radial fracture lines, fracture is crisp in appearance	Straight fracture lines, breaks occur at right angles to shaft, fracture is ragged in appearance
Profile Angle	Angle of break is curved or oblique with respect to outer surface	Angle of break is perpendicular to outer surface or stepped
Profile Texture	Profile wall is smooth	Profile wall is rough and torn in appearance
Profile Color	Profile color will be the same as outer surface of the bone	Profile color could be different from the outer surface of the bone
Fracture Path	Fracture line will terminate at articular ends	Fracture line will continue through articular ends
Primary Impact Appearance	Hinge fractures and other adhering bone fragments may be present	Hinge fractures are not present, bone may shatter but individual fragments will not adhere to main bone segment

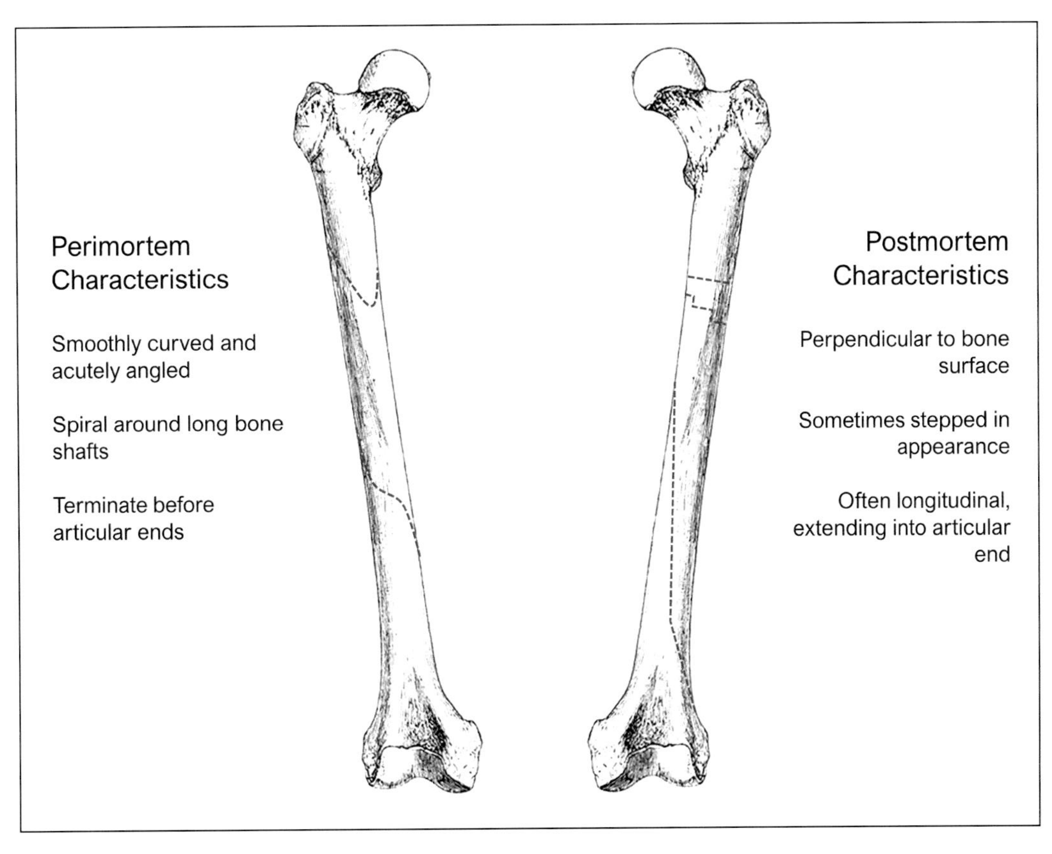

Figure 12.11. Common modes of perimortem and postmortem fracture propagation.

In perimortem fractures, the line itself tends to be sharp, clear, and well-defined. The bone tends not to shatter into multiple pieces. When fragmentation does occur, perimortem wounds are much more likely to have partially adhering bone and *hinge fractures* (**Figure 12.12**), in which the smaller fragments do not completely separate (i.e., they are infractions). In postmortem fractures, the line tends to be ragged in appearance. There is a greater tendency for the bone to shatter into multiple pieces, but these pieces will completely separate

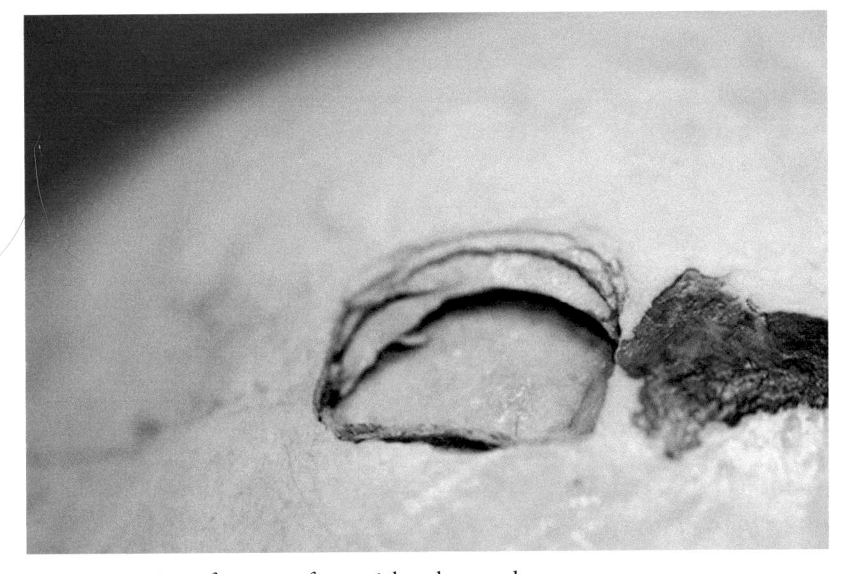

Figure 12.12. Hinge fractures of a cranial vault wound.

from the main body of the bone and not adhere. Hinge fractures are not seen with postmortem fractures.

Feature 2: Appearance of the exposed fracture edge

When a complete discontinuity results from trauma, the margin of the broken edge exhibits diagnostic criteria of peri- and postmortem trauma. Three observations are noteworthy: *fracture profile angle, fracture profile texture,* and the *fracture profile color.*

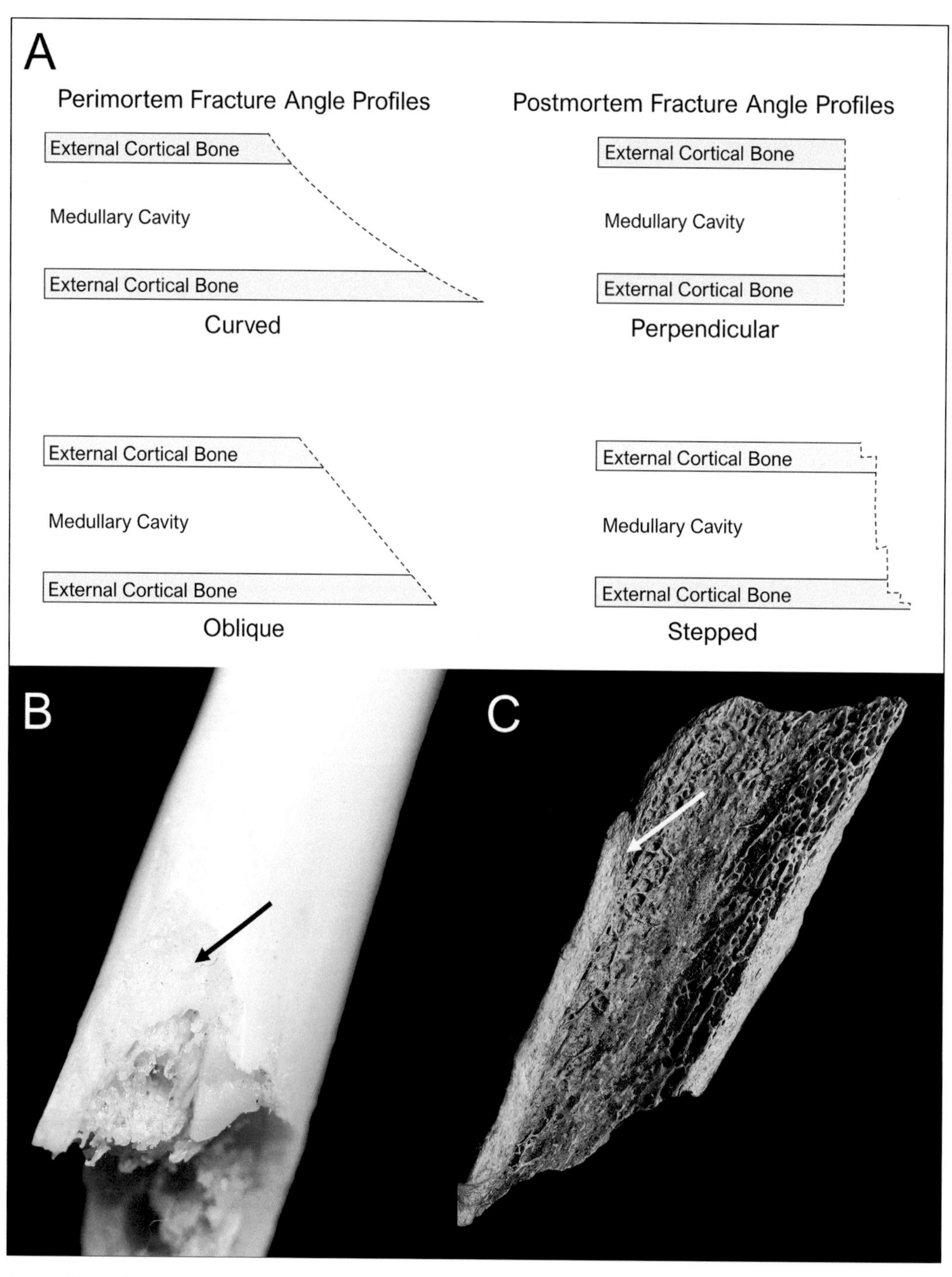

Figure 12.13. Differences between perimortem and postmortem fractures in fracture profile angle, texture, and coloration. (*A*) Diagram showing differences in fracture angles. (*B*) Perimortem fracture with oblique angle, smooth texture, and uniform coloration between the fracture margin and the exterior surface. (*C*) Postmortem fracture with perpendicular angle, rough texture, and different coloration between the fracture margin and the exterior surface.

The *fracture profile angle* refers to the shape of the broken edge. In perimortem trauma, the broken edge is obliquely angled or curved. In postmortem trauma, the bone breaks at a right angle to the shaft, and the fracture line is straight or perpendicular with respect to the outer surface of the bone. Postmortem breaks can also exhibit a stepped appearance (**Figure 12.13A**).

The *fracture profile texture* refers to the texture of the broken edge itself. In perimortem trauma the broken edge is smooth. In postmortem trauma the broken edge is rough and ragged. Postmortem fractures have the appearance of being "torn," reflecting the lack of elastic properties in the bone. Although difficult to photograph, the differences between a smooth and rough profile texture are best indicated by the black arrows in **Figures 12.13B** and **12.13C**.

The *fracture profile color* refers to whether the color of the fractured edge is the same color as the exterior cortical surface. In perimortem trauma, the edge and exterior surface of the bone are exposed to the same burial environment. As such, any changes in bone color that occur due to soil contact will be shared by both surfaces. In postmortem trauma, the exterior surface is exposed to burial color changes for a longer period of time than the fractured margin. Therefore, the margin should be lighter in color than the exterior bone surface. This is demonstrated in **Figure 12.13B** and **12.13C**. The perimortem trauma has a consistent color throughout the bone. The postmortem trauma is darker on the exterior and nearly white along the fracture edge margin. Note that it is possible for postmortem breaks to assume the same color as the exterior bone surface depending on when the break happened and how long the body has been in the soil.

LEARNING CHECK

Q4. Describe the timing of the gunshot wound to the head shown in **Figure 12.14A**.

A) Antemortem
B) Perimortem
C) Postmortem

Q5. Describe the timing of the fracture of this os coxa shown in **Figure 12.14B**.

A) Antemortem
B) Perimortem
C) Postmortem

Q6. Describe the timing of the fracture of this proximal tibia shown in **Figure 12.14C**.

A) Antemortem
B) Perimortem
C) Postmortem

Q7. Describe the timing of this trauma to the cranial vault shown in **Figure 12.14D**.

A) Antemortem
B) Perimortem
C) Postmortem

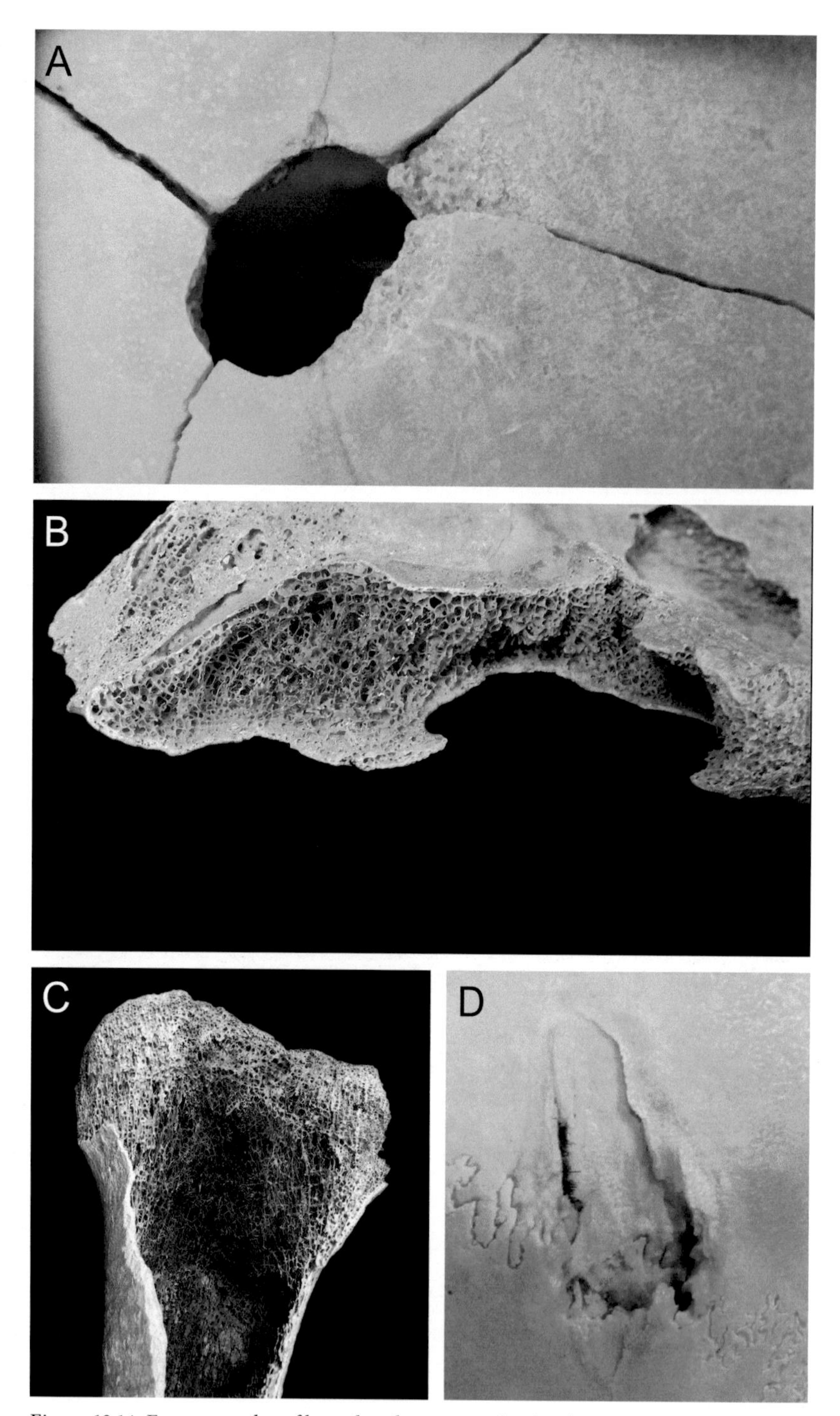

Figure 12.14. Four examples of bone breakage caused either by trauma or postmortem damage.

Q8. Describe the timing of the fractures on the femur shown in **Figure 12.15A**.
   A) Antemortem
   B) Perimortem
   C) Postmortem

Q9. Describe the timing of the tibia fracture shown in **Figure 12.15B**.
   A) Antemortem
   B) Perimortem
   C) Postmortem

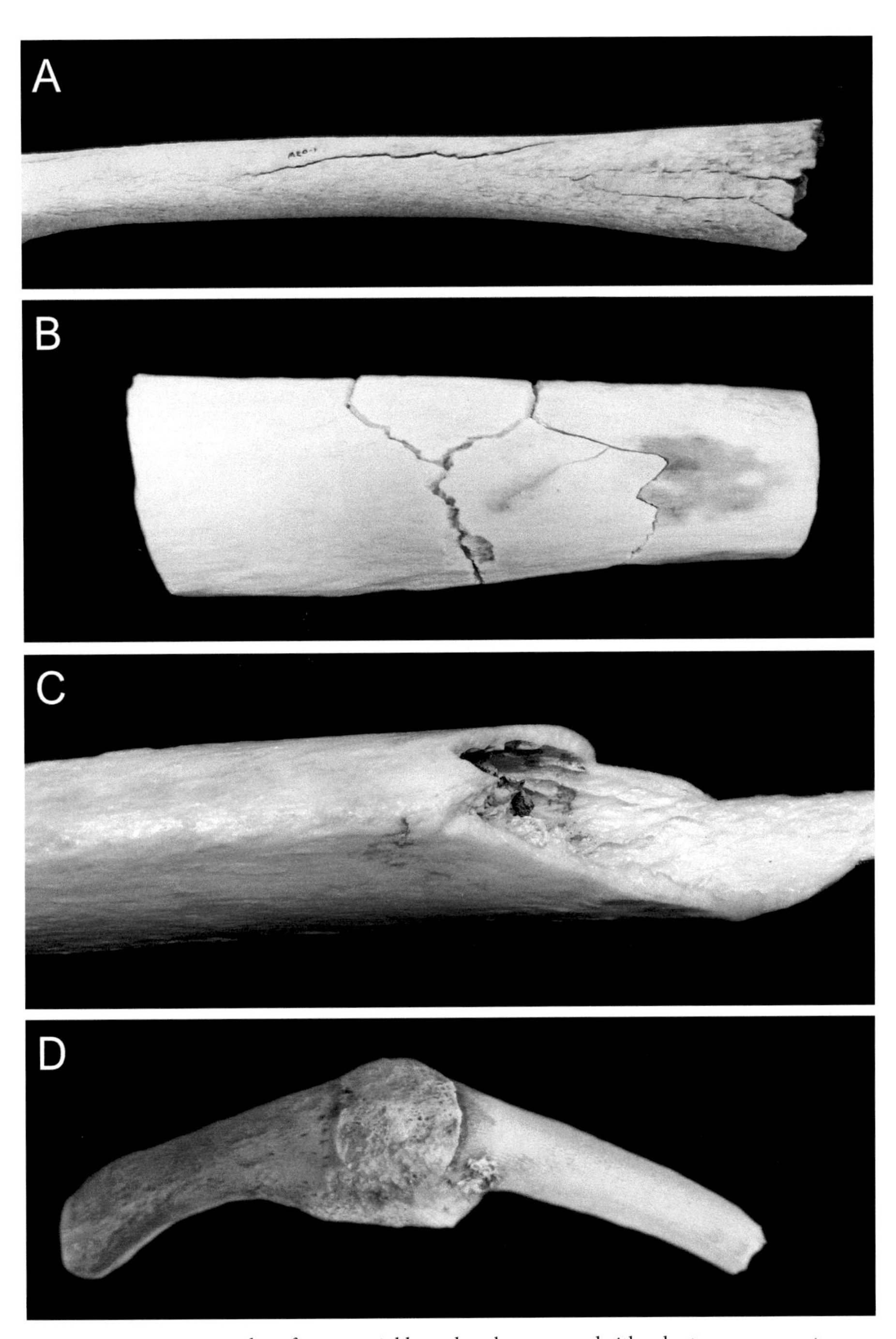

Figure 12.15. Four examples of postcranial bone breakage caused either by trauma or postmortem damage.

Q10. Describe the timing of the long bone fracture shown in **Figure 12.15C**.
A) Antemortem
B) Perimortem
C) Postmortem

Q11. Describe the timing of the rib fracture shown in **Figure 12.15D**.
A) Antemortem
B) Perimortem
C) Postmortem

Q12. Describe the timing of the fracture on the cranial vault shown in **Figure 12.16A**.
A) Antemortem
B) Perimortem
C) Postmortem

Q13. Describe the timing of the fractures of the cranial vault shown in **Figure 12.16B**.
A) Antemortem
B) Perimortem
C) Postmortem

Q14. Describe the timing of the injury to the cranial vault shown in **Figure 12.16C**.
A) Antemortem
B) Perimortem
C) Postmortem

Q15. What type of fracture is shown on the femur shown in **Figure 12.16D**?
A) Antemortem
B) Perimortem
C) Postmortem

End-of-Chapter Summary

This chapter introduced concepts relevant to the analysis of trauma in forensic anthropology. The formal definition of trauma was presented and the distinctions between the different types of fractures were summarized. A key point is that bone fails under tension before it fails under compression, which generates predictions about the direction of force impacting the body. In addition, the chapter discussed the differences between antemortem, perimortem, and postmortem fractures. Signs of healing were defined as a key identifier of antemortem trauma. Indicators of healing include porosity, rounding of the fracture margins due to osteoclasts removing the sharp edges of bone, and the addition of new bone in the form of a callus.

Differentiating perimortem and postmortem bone fractures requires an understanding of how bone breaks when wet (alive) or dry (dead). Perimortem fractures have sharp, clean edges with angled and smooth fracture profiles. Perimortem fractures orient obliquely with respect to the shaft or curve around the shaft. Postmortem breaks lack the sharp edges of perimortem fractures and often break at right angles to the long bone shaft. The color of the fracture margin may be different than the color of the bone surface. Postmortem fracture margins are rough and appear torn due to the brittle nature of dry bone. Differentiating perimortem and postmortem fractures takes considerable expertise and, in some cases, may not be possible.

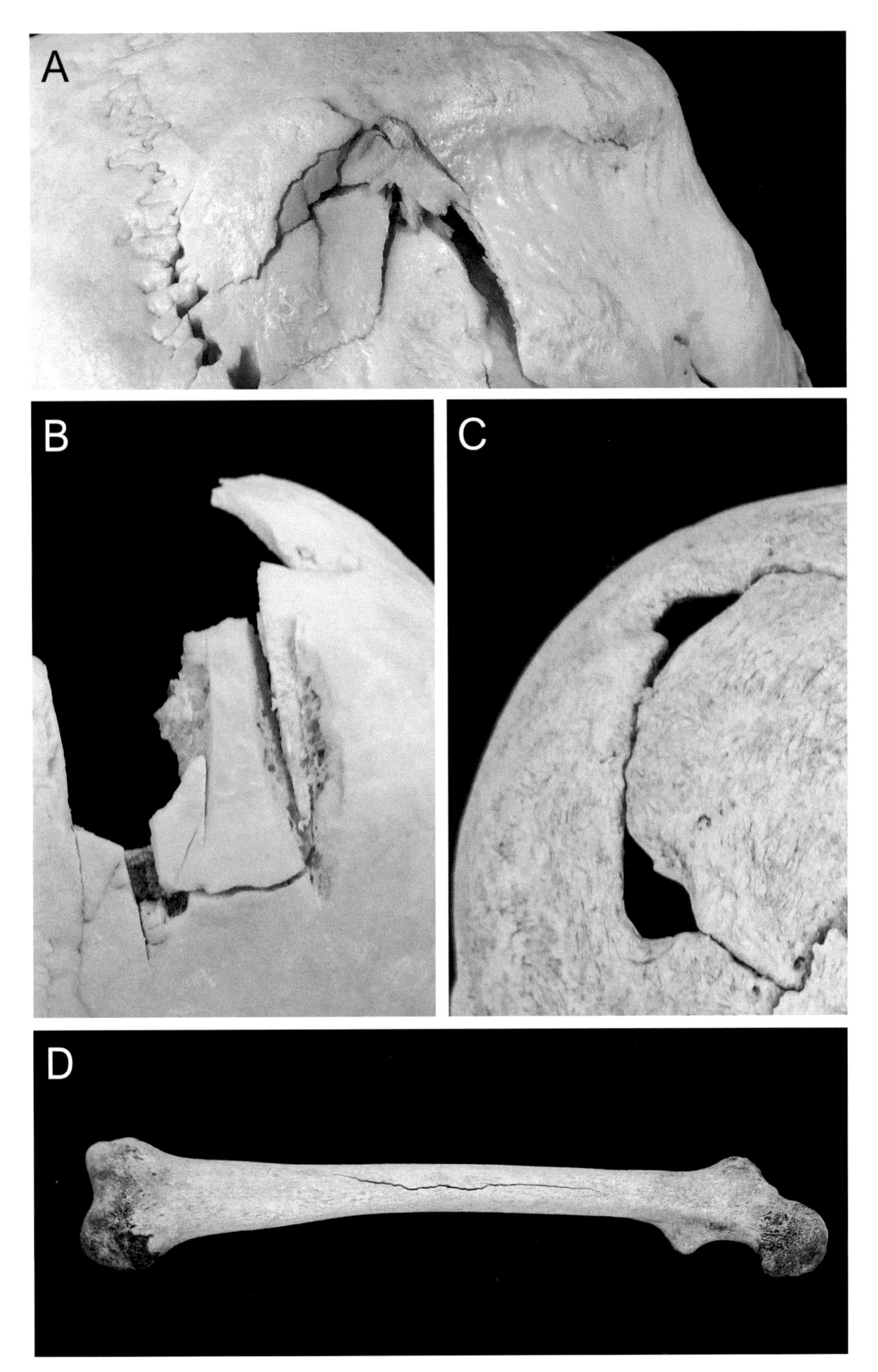

Figure 12.16. Four examples of bone damage caused either by trauma or postmortem damage.

End-of-Chapter Exercises

Exercise 1

Materials Required: **Figure 12.17**

Scenario: A colleague of yours has asked that you review their assessment of a set of human remains that were recovered from a shallow grave. They have included a photograph for your reference, shown in **Figure 12.17**. Their description and conclusions are as follows:

"There is a linear defect located on the occipital bone, just to the right of midline. The right side of this defect is associated with a semi-circular fracture line that surrounds an area where the bone is depressed inward. The edges of the linear defect are crisp and sharp-edged; there is no evidence of bony healing. The exposed bone along the defect margins (indicated by the white arrow) is the same color as the external surface of the bone."

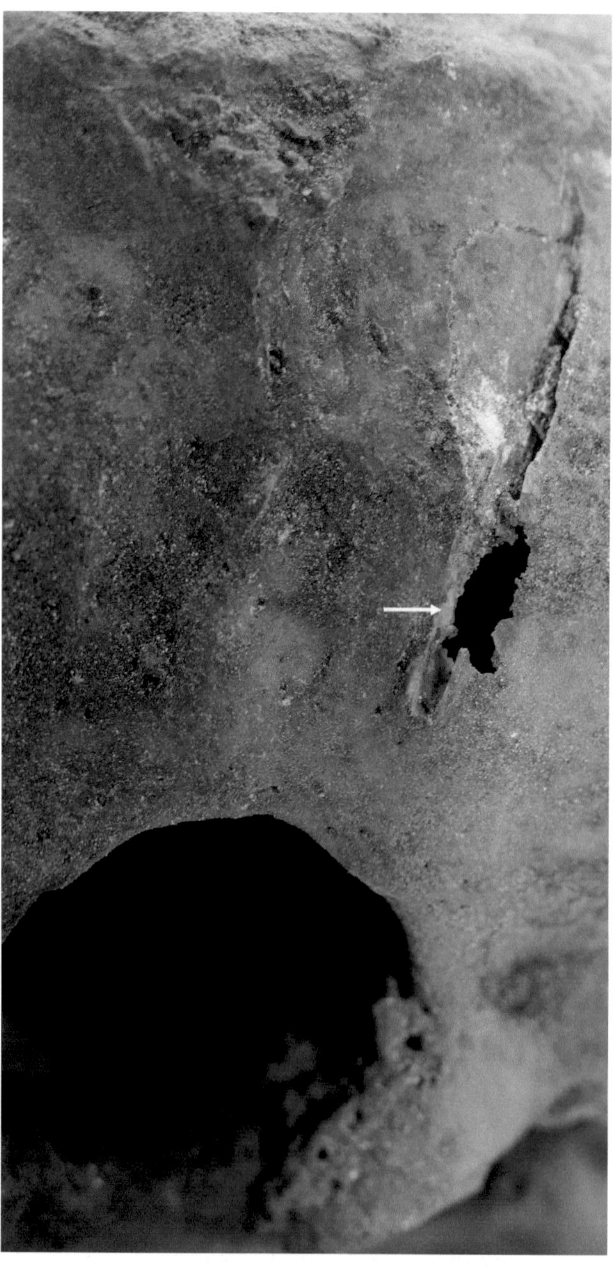

Figure 12.17. Close-up image of damage to skull (white arrow) possibly consistent with trauma.

Based on this description, your colleague has concluded that this defect is consistent with perimortem trauma.

Question 1: Do you agree with your colleague's description of the defect shown in **Figure 12.17**? Make any corrections you think are necessary.

Question 2: Do you agree with your colleague's conclusions regarding the timing of this defect? Why or why not?

Exercise 2

Materials Required: **Figure 12.18**

Directions: **Figure 12.18** depicts a femur that has been reconstructed after being fractured by a bullet. Use this image to answer the following questions.

Question 1: Which of the *numbers* on **Figure 12.18** refer to radiating fractures?

Question 2: Which of the *numbers* on **Figure 12.18** refer to concentric fractures?

Question 3: Based on your answers to Questions 1 and 2, which of the *letters* on **Figure 12.18** is most likely to represent the point where the bullet initially impacted the bone?

Exercise 3

Materials Required: **Figure 12.19**

Scenario: The hyoid is a U-shaped bone located in the neck. In younger individuals, it often appears as three separate pieces: the body and the two greater horns. While these often fuse together to form one bone later in life, this is not always the case. When one or both of the greater horns remains unfused to the body of the hyoid, the junction between them is smooth and appears much like a joint surface. This can be seen in **Figure 12.19A**. Unfused

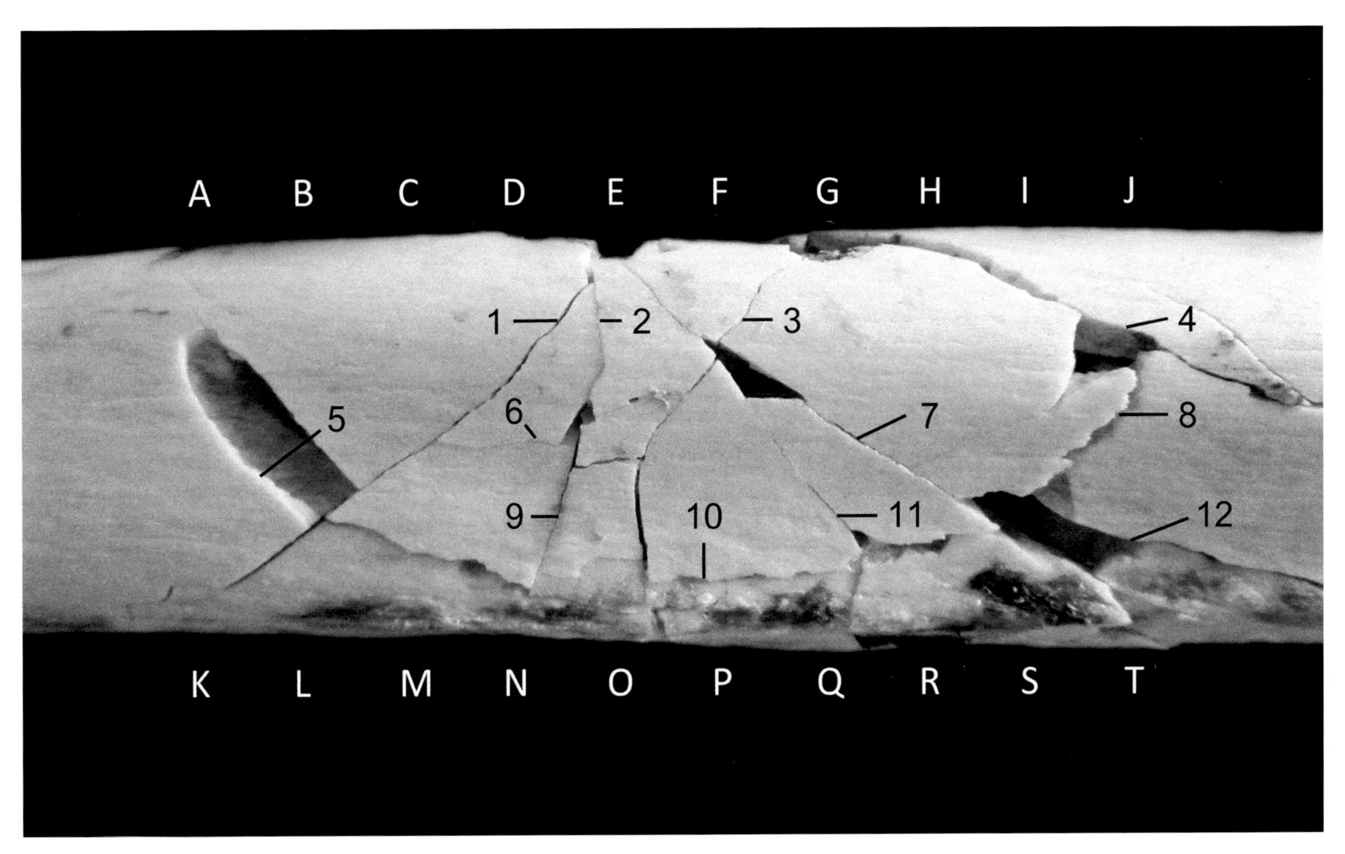

Figure 12.18. Long bone fracture lines and comminuted fragments.

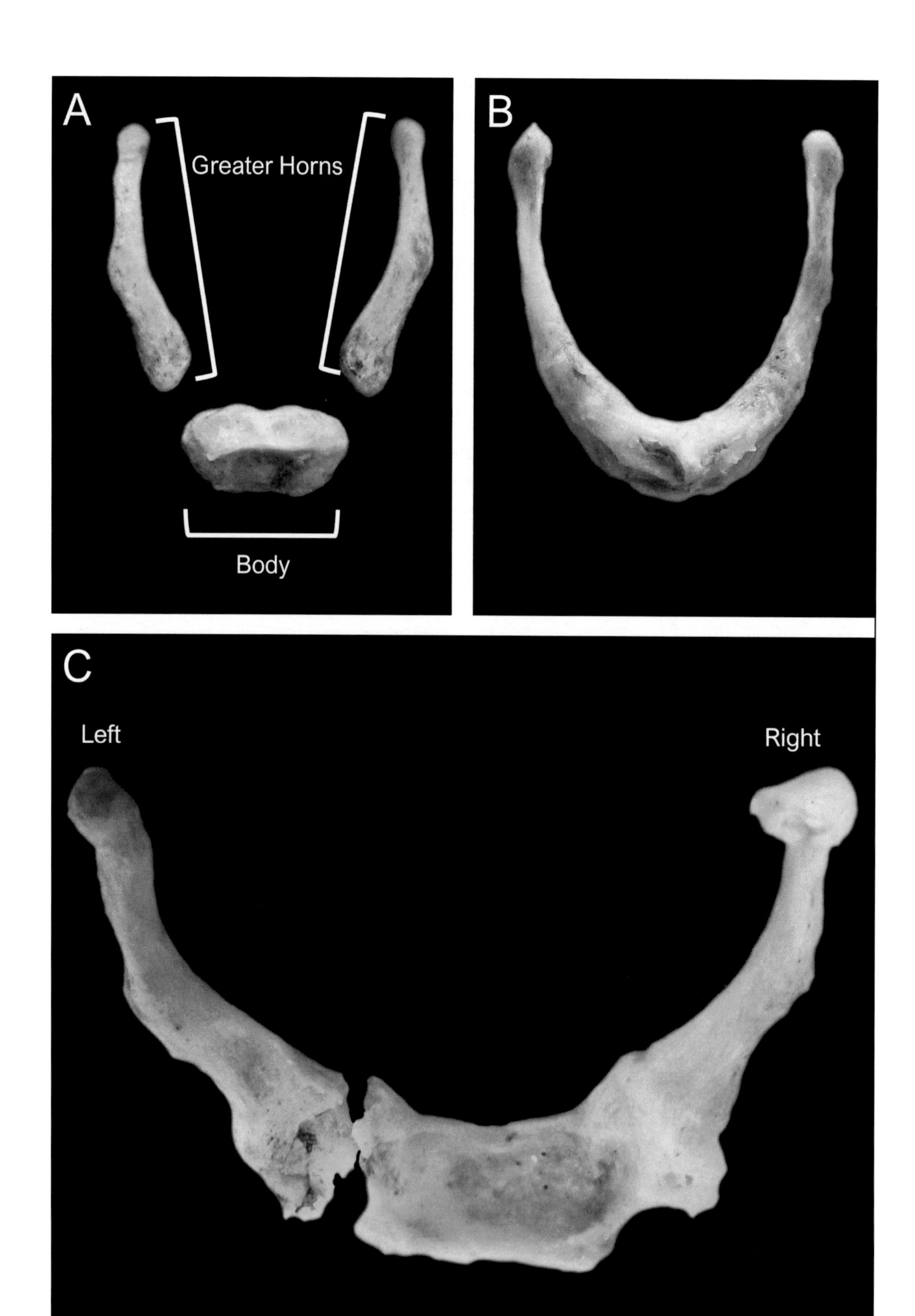

Figure 12.19. Three different hyoid bones.

greater horns are often mistaken as evidence of trauma. Hyoids that have fused together usually do so such that the contour of the bone across the junction between the body and the greater horns is smooth (**Figure 12.19B**).

You have been asked to assess the hyoid shown in **Figure 12.19C** for trauma. The view shown is looking at the inferior aspect of the hyoid, and the greater horns have been labeled according to their side.

Question 1: Is there any evidence of antemortem trauma on this hyoid? If so, please describe its features and location.

Question 2: Is there any evidence of perimortem trauma on this hyoid? If so, please describe its features and location.

Question 3: Is there any evidence of postmortem damage on this hyoid? If so, please describe its features and location.

Question 4: Based on your answers to the above questions, does the hyoid shown in **Figure 12.19C** provide evidence for multiple episodes of trauma? Please explain your answer.

References

Berryman HE, Lanfear AK, Shirley NR. 2012. The biomechanics of gunshot trauma to bone: research considerations within the present judicial climate. In: Dirkmaat DC (Ed.), *A Companion to Forensic Anthropology.* London: Blackwell. p. 390–399.

Berryman HE, Shirley NR, Lanfear AK. 2013. Low-velocity trauma. In: Tersigni-Tarrant MA, Shirley NR (Eds.), *Forensic Anthropology: An Introduction.* Boca Raton, FL: CRC Press. p. 271–290.

Christensen AM, Passalacqua NV, Bartelink EJ. 2014. *Forensic Anthropology. Current Methods and Practice.* Oxford: Elsevier.

Currey JD. 1970. The mechanical properties of bone. *Clinical Orthopedics and Related Research* 73: 210–231.

Dixon AF. 1912. *Manual of Human Osteology.* New York, NY: William Wood and Co.

Galloway A, Zephro L, Wedel VL. 2014a. Classification of fractures. In: Wedel VL and Galloway A (Eds.), *Broken Bones: Anthropological Analysis of Blunt Force Trauma.* Springfield, IL: Charles C. Thomas. p. 59–72.

Galloway A, Zephro L, Wedel VL. 2014b. Diagnostic criteria for the determination of timing and fracture mechanism. In: Wedel VL and Galloway A (Eds.), *Broken Bones: Anthropological Analysis of Blunt Force Trauma.* Springfield, IL: Charles C. Thomas. p. 47–58.

Symes SA, L'Abbé EN, Chapman EN, Wolff I, Dirkmaat DC. 2012. Interpreting traumatic injury to bone in medicolegal investigations. In: Dirkmaat DC (Ed.), *A Companion to Forensic Anthropology.* London: Blackwell. p. 340–389.

Symes SA, L'Abbé EN, Stull KE, Lacroix M, Pokines JT. 2014. Taphonomy and the timing of bone fractures in trauma analysis. In: Pokines JT, Symes SA (Eds.), *Manual of Forensic Taphonomy.* Boca Raton, FL: CRC Press. p. 341–365.

Ubelaker DH. 1991a. Perimortem and postmortem modification of human bone. Lessons from forensic anthropology. *Anthropologie* 29: 171–174.

Ubelaker DH, Adams BJ. 1995. Differentiation of perimortem and postmortem trauma using taphonomic indicators. *Journal of Forensic Sciences* 40: 509–512.

Zephro L, Galloway A. 2014. The biomechanics of fracture production. In: Wedel VL and Galloway A (Eds.), *Broken Bones: Anthropological Analysis of Blunt Force Trauma.* Springfield, IL: Charles C. Thomas. p. 33–45.

13

High-Velocity Projectile Trauma

Learning Goals

By the end of this chapter, the student will be able to:

Describe the basic characteristics of firearms and bullet technology.
Describe the relationship between firearm and bullet characteristics and the type of skeletal trauma that results.
Identify entrance and exit wounds based on wound beveling.
Describe the relationship between bullet wound dynamics and fracture propagation.

Introduction

In 2016, the FBI recorded over 15,000 homicides in the United States. Of these, over 11,000 (73 percent) were caused by firearms. Of the 11,000 firearms-related homicides, over 7,000 were caused by handguns. Keep in mind that these data are for homicides only and do not account for all violent crimes of which there were over 1.2 million reported in 2016 alone (the majority were aggravated assaults). Therefore, research on firearms-related trauma is of critical importance for bringing the perpetrators of homicide to justice. *Ballistics* is the study of the mechanics and physics of firing weapons and the movement of the discharged projectiles. Forensic pathologists have performed decades of research on the effects of projectiles on soft tissues with entire books dedicated to the subject (DiMaio, 2021). By comparison, the analysis of bullet wounds on the skeleton, i.e., high-velocity projectile trauma, is in its infancy.

High-velocity projectile trauma is a compressive injury caused by a bullet (metal) moving at high speeds (~325–425 meters per second) and impacting the skeleton over a very small area, thus focusing the energy on a spot roughly the size of the bullet diameter (Christensen et al., 2014). The velocity of impact is what distinguishes the effects of gunshot trauma from other types of trauma encountered in a forensics context (blunt and sharp force). Specifically, high-velocity projectile trauma impacts the skeleton with significant amounts of energy, specifically kinetic energy, which is defined as $(1/2)mv^2$ where m is the mass of the object in kilograms and v is the velocity in meters per second. This equation means that a light object (such as a bullet) moving with high velocity (measured in meters per second) transfers much more kinetic energy to the body than a heavy object (such as a brick) moving at a much slower velocity (measured in feet per second). The relationship between kinetic energy, mass, and velocity also indicates that an increase in velocity increases the kinetic energy transfer at a much higher rate than an increase in mass. It is this basic fact of physics that gives high-velocity projectile trauma the potential to cause catastrophic damage to the body (Berryman et al., 2012).

In addition, the physical properties of human bone (see chapter 3) cause it to respond like a brittle material when subjected to high amounts of energy. This means that gunshot wounds tend to shatter bone without significant *warping* (termed *plastic deformation*) prior to the bone "giving way" and fracturing. Under low energy stress, typical of blunt and sharp force trauma, the bone will bend and deform just prior to fracturing, which results in *fracture lines that may not align when reconstructed.* Projectile trauma generally shows limited warping of fractured pieces, except in cases where bullet velocity has slowed significantly before impacting the body (Berryman and Symes, 1998).

For these reasons, gunshot trauma usually results in a complete fracture of the bony element, often with *secondary radiating fractures* emanating from the site of impact, and *tertiary concentric fracture lines* surrounding the primary impact site, depending on the amount of energy imparted to the body. In modern contexts, projectile trauma is usually caused by bullets. Arrows, javelins, and spears are projected weapons, but their velocity is comparatively low, and hence their energy is low. Their wound characteristics will differ from firearms and be more similar to blunt or sharp force trauma.

The effects of gunshot wounds on bone vary considerably depending on a number of factors, including the distance between the shooter and the victim, the type of bullet fired, the caliber of the weapon, the velocity and angle of bullet impact, the presence of intervening materials, and the bullet behavior at the time of impact (was it rotating, wobbling, or tumbling). These are *extrinsic* factors (i.e., those outside of the body of the decedent). *Intrinsic* factors that affect all forms of skeletal trauma include the gross anatomy of the impacted skeletal element, the age of the victim, and the health of the victim, which determines how brittle their bones are.

Generally, gunshot wounds are identifiable by round or oval bullet entry and/or exit wounds that have a distinctive form of *beveling* that indicates the direction the bullet was traveling. High-velocity projectile trauma also has the potential to cause complete and

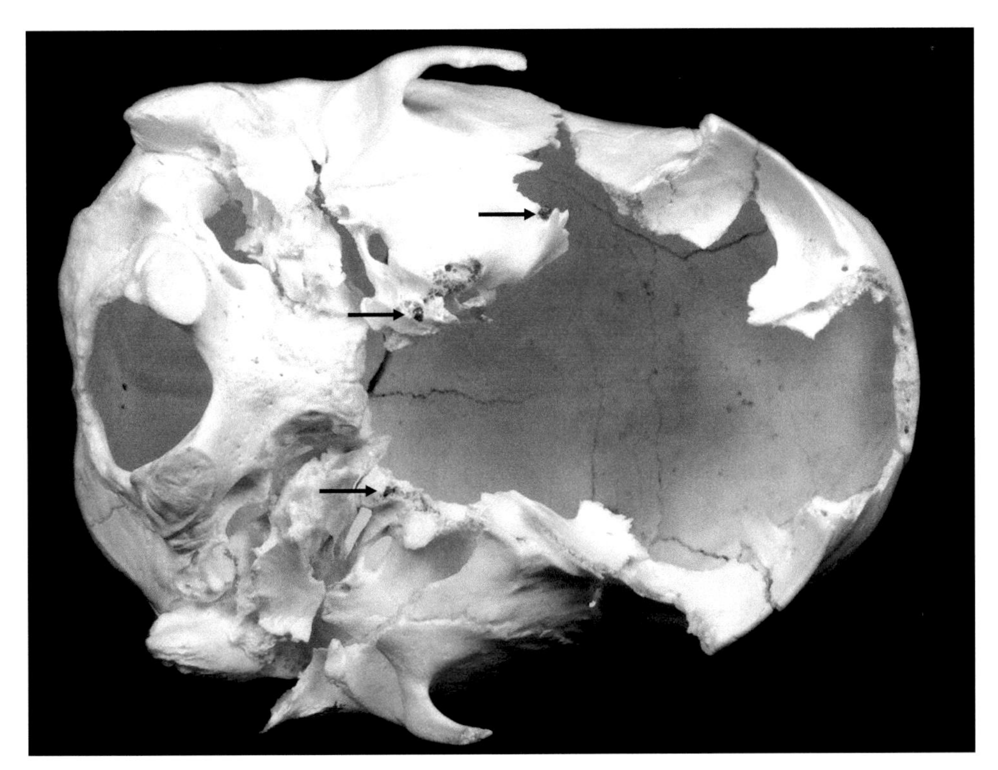

Figure 13.1. Severe destruction due to shotgun wound to the face. Note some pellets are still embedded in the base of the skull (black arrows). Inferior view, anterior is to the right.

catastrophic damage to the skeleton, as seen in **Figure 13.1**. This extent of damage is rarely seen in blunt or sharp force trauma.

Because the forensic anthropologist does not estimate cause or manner of death, their primary task is to describe the evidence of trauma observed on the skeleton. For projectile trauma, this entails identifying the entrance and exit wounds, linking these wounds to estimate the number of shots fired, and identifying the radiating and concentric fractures to reconstruct the sequence of shots. In some cases, statements can be made about the distance between the shooter and the victim as well as possible descriptive aspects of the firearm or bullet type; however, such statements often cause confusion in the courtroom and should be avoided.

Bullet Characteristics

Ballistics is a field of study that explores the physics of projectiles, from their launching and flight to their effects on the target. Although related to forensic anthropology, ballistics is a distinct field of study with its own expertise. This chapter will emphasize the minimum information needed to understand the effects of bullets and firearms on the human skeleton.

A basic firearms cartridge is pictured in **Figure 13.2,** which includes the following components:

1. Bullet: causes the damage and is propelled
2. Case: contains the gunpowder
3. Gunpowder: accelerates the bullet
4. Rim: forms a platform for holding the case in place
5. Primer: initial spark that ignites the gunpowder

Despite a basic design, there is tremendous variability in cartridge construction, which has effects on forensic investigations. First, cartridges vary in *size.* Ammunition size is measured by caliber, where larger numbers indicate larger bullet diameters (.22, .45, 9 mm). Shotgun shells are measured by gauge; a smaller number indicates a larger diameter of shell (12 gauge = 1/12 pound ball would fit in the diameter of the barrel). Second, cartridges vary by *profile,* which refers to the shape of the tip of the bullet. Most rifle bullets are pointed, while handgun bullets are blunt or flat. Some are hollow pointed, which are designed to fragment upon

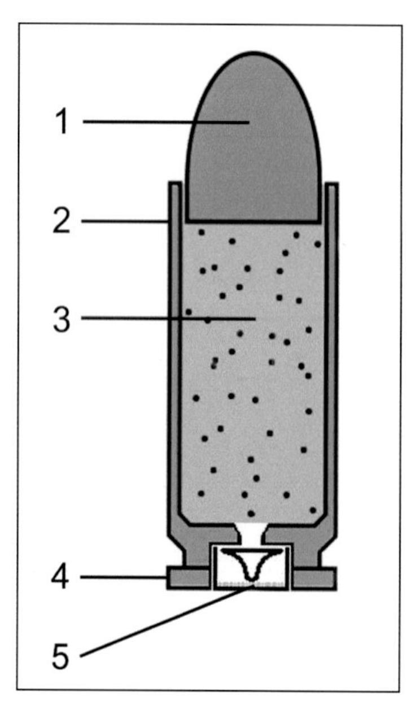

Figure 13.2. Components of a firearms cartridge.

impact to increase damage to the body. Third, cartridges vary by *internal composition*. For example, lead bullets deform easily, solid metal bullets deform less easily, and fragmenting bullets are meant to shatter upon impact. A hollow point bullet will deform significantly upon impact, even with relatively soft materials. Fourth, cartridges vary by *jacketing*, or the metal casing on the outside of the bullet. A full metal jacket will prevent a bullet from deforming upon impact.

Projectile Trauma and the Cranium

Ultimately, the bullet and firearm characteristics discussed above only provide guidelines for how one interprets damage to the skeleton. There are many variables to consider, and often only general statements can be offered. The amount of energy impacting the body matters most, and this depends on the size of the bullet, the bullet's construction, the velocity at which the bullet was fired, and the distance between the assailant and victim (which determines velocity at impact). Close-range gunshots cause more damage than distant gunshots. Larger caliber weapons cause more damage than smaller caliber weapons. Bullets that fragment or deform and stay in the body cause more damage than those that exit the body because all of their kinetic energy is released into the person's tissues.

In terms of the amount of damage caused, the following list is ordered from the least to the most amount of energy with which the bullet impacts the body.

A single entrance wound, no internal damage (no ricochet), no exit wound due to low velocity.
A single entrance wound with internal ricochet damaging the brain, no exit wound.
A paired entrance and exit wound of similar size, suggesting limited bullet deformation.
A paired entrance and exit wound, with the exit wound much larger than the entrance wound indicating bullet deformation or fragmentation.
Catastrophic damage due to high-velocity impact, the vault is in pieces.

Unlike most of the skeleton, the cranium is a round structure composed of thick, flat bones of the vault and thin bones of the face. Variation in thickness influences the behavior of bullets when they impact the cranium. In addition, because the cranial vault surrounds the brain, there is a larger distance the bullet travels between the entrance and potential exit during which deflection, deformation, and fragmentation can occur. No other bone in the body offers a similar geometry (essentially the cranial vault is a flat bone that forms a round structure approximately 7 inches in diameter), which makes cranial gunshot wounds distinct in their appearance. Decades of research has documented a relatively consistent series of bullet wound characteristics that consist of clearly defined entry and exit wounds, as well as radiating and concentric fractures.

Primary Fractures: Entry and Exit Wounds

Investigators are often interested in reconstructing the relationship between the shooter and the victim, as well as reconstructing the sequence of gunshots with a multi-shot fatality. Doing so requires understanding the mechanics of bullet hole formation. Cranial vault impacts are the focus here because they have received the most attention in the forensic anthropological literature.

As bullets pass through the body, they cause beveling within the wound, *where the beveling points in the direction the bullet is moving* (**Figure 13.3A**). As the bullet enters the skull, it will cause a wound that will be larger on the inside surface of the skull than on the outside

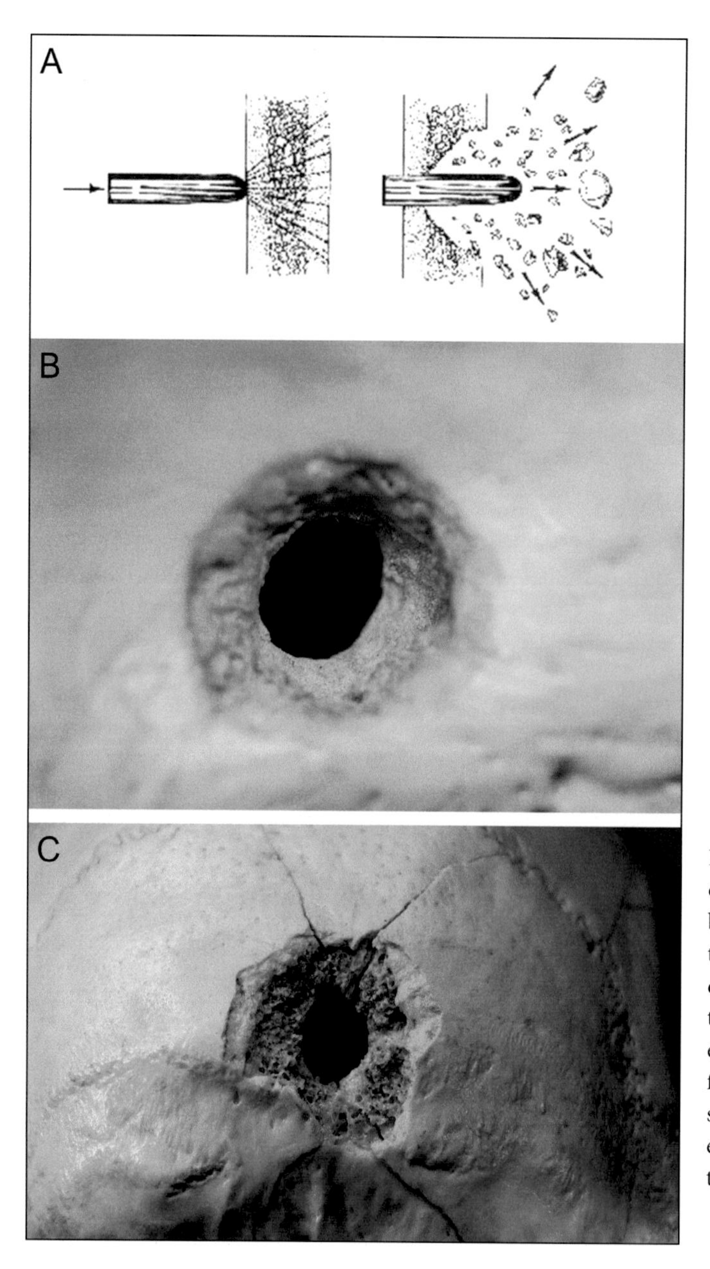

Figure 13.3. (*A*) Demonstration of bullet beveling characteristics as related to the direction of bullet travel. (*B*) Cranial entrance wound seen from the endocranial surface. (*C*) Cranial exit wound seen from the ectocranial surface.

surface of the skull (*inward* or *internal beveling*). **Figure 13.3B** shows the inside of the skull (endocranial surface). Note the bullet hole is cone-shaped and is larger on the endocranial surface. This is an entrance wound. As the bullet exits the skull, it will cause a larger wound on the outer surface of the skull than on the inner (*outward* or *external beveling*). **Figure 13.3C** shows an exit wound on the occipital bone with a much smaller inner margin hole than outer margin hole. Each bullet hole, therefore, is cone-shaped in cross-section with the *larger part of the cone facing the direction* in which the bullet was traveling. This rule applies to most of the thicker parts of the cranial vault (frontal, parietal, occipital). However, bullet hole formation is affected by the properties of the skeletal element being impacted. Thinner bone, such as that found in the area of the temporal bone, may simply punch through, as seen in **Figure 13.4**. Exceptions to the beveling rule have also been reported in unusual circumstances such as grazing wounds (Baik et al., 1991 Bhoopat, 1995; Coe, 1982; Peterson, 1991).

In addition to internal beveling, cranial entrance wounds typically display a round-to-oval wound shape, clear and sharp edges, and a smaller size that more closely approximates the caliber of the bullet. In addition to external beveling, cranial exit wounds typically display

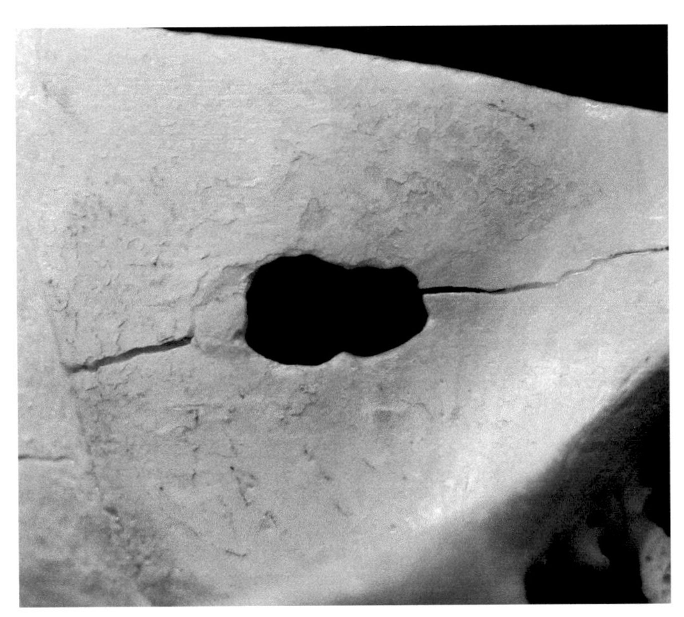

Figure 13.4. Gunshot wound to the left temporal bone.

a more irregular shape (oval, square, asymmetrical), and are generally larger than entrance wounds. In fact, with rare exceptions, linked entrance and exit wounds caused by the same bullet would almost always display a larger exit wound (**Figure 13.5**). There are two reasons for this. First, when the bullet impacts the outside of the cranium it immediately loses velocity. This causes a loss of axial rotation and a tendency for the bullet to tumble. A tumbling bullet that strikes the opposite side of the skull will do so with a higher surface area thus causing a larger exit wound. Second, the bullet may deform upon initial impact. Any deformation of the cylindrical shape of a bullet will result in a larger exit wound. One study found that the exit wound was between 1.14 and 5.76 times larger in diameter than the corresponding, matched entrance wound (Quatrehomme and Alunni, 2013).

It is important to note that not all entrance wounds are linked with an exit wound. If the velocity of the projectile is low enough, the bullet may only have enough energy to enter but not exit the skull. In such cases, the bullet could cause a "bone bruise" on the endocranial surface, which creates an impact injury similar to blunt force trauma (Smith et al., 1993). The bullet could also ricochet within the cranial vault (causing significant brain damage) or be

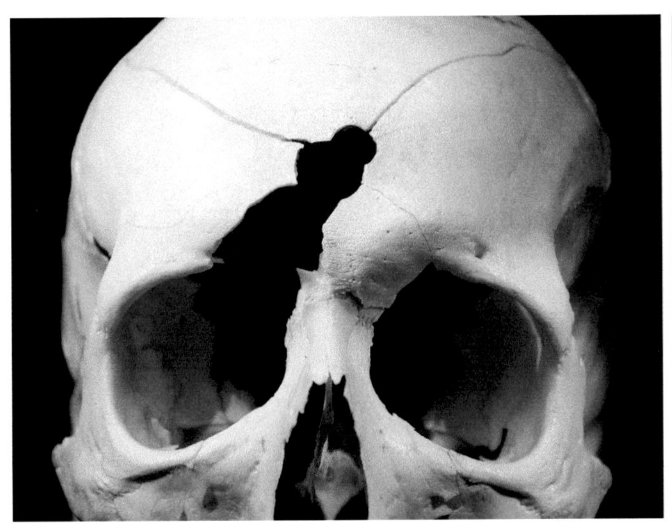

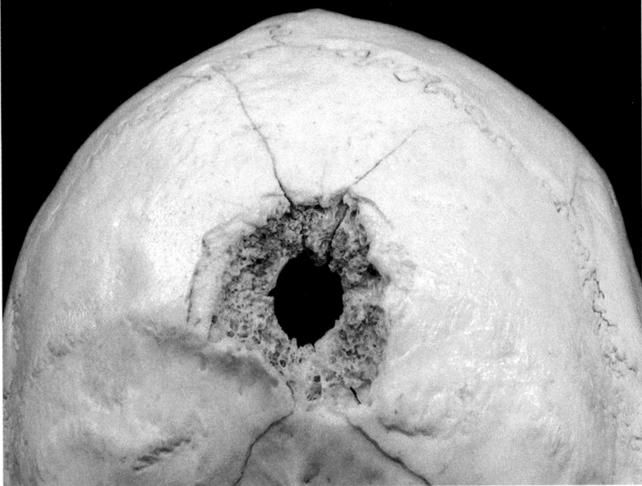

Figure 13.5. Entrance (*left*) and exit (*right*) wounds from the same bullet, note the difference in size.

designed to fragment upon initial impact. A bullet that does not exit the body releases all of its kinetic energy into the body, which causes more damage than a bullet that passes through the body.

Describing Primary Fractures

In analyzing bullet wounds, there are three factors to consider in preparing a descriptive report. These are *wound size, wound shape,* and a thorough description of *wound location* with respect to anatomical landmarks. For *wound size,* the maximum and minimum size of the bullet hole and the surrounding beveled region should be recorded. Information on wound size can help establish the direction of bullet travel through the body. Wound size is only weakly correlated with bullet caliber, however (Berryman et al., 1995), and research suggests only very general statements can be made about bullet size based on bullet hole measurements.

Wound shape is usually described as round, oval, square, or irregular. The shape of the bullet wound is determined by: 1) the angle of impact (perpendicular to the bone or at an angle), and 2) the bullet's rotational behavior as it impacts the body. A bullet with enough velocity to maintain its rotation around its long axis that impacts at a right angle will create a round bullet hole with symmetrical wound beveling. Bullet wounds that are oval indicate the bullet was wobbling on its axis as it impacted the body (**Figure 13.6**). This could indicate a deflection (when the bullet strikes another object prior to impacting the body) or that the distance between the shooter and victim was such that the bullet began to lose velocity and wobble. An *irregular* bullet wound indicates it was tumbling upon impact, suggesting a deflection. Because bullets lose velocity with distance, an irregular- or oval-shaped bullet wound may be useful in a forensic investigation for directing areas of search for additional evidence (gun casings) or comparing against a suspect's sworn statements.

Keyhole wounds are a special class of bullet wound shape. They result from a bullet impacting the skull at a shallow angle, which fractures the bullet into two or more fragments. These fragments have separate trajectories depending on the angle of impact and can create an entrance and exit wound at the same impact site (Dixon, 1982). Consider **Figure 13.7**.

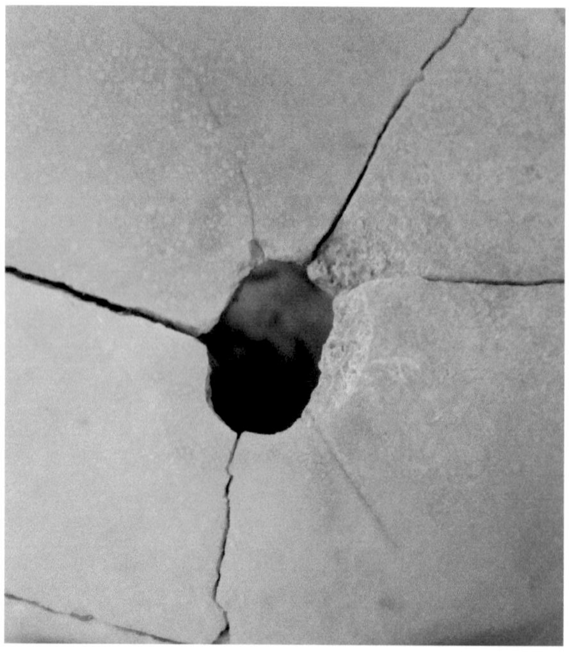

Figure 13.6. Cranial entrance wound. The oval outline of this defect suggests a loss of axial rotation prior to impact.

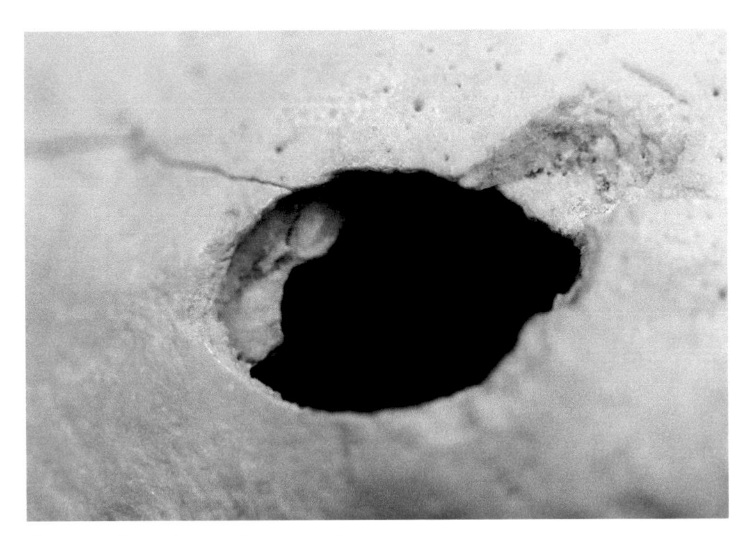

Figure 13.7. Keyhole bullet wound.

On the bottom and left side of the wound there is a round, clean entrance wound. However, toward the upper right of the impact site, a square fragment of bone was removed creating an area of damage with external beveling. Because keyhole wounds *combine an entrance and exit wound* in the same impact site, they indicate the direction the bullet was traveling and possibly the position of the assailant with respect to the victim. In **Figure 13.7,** the bullet was traveling from left to right, creating an entrance wound on the left side of the impact and an exit wound on the upper right side of the impact. In the surrounding soft tissues, there may be two apparent gunshot wounds that result from the bullet fragmentation (the initial impact entrance wound and the subsequent exit wound from bullet fragmentation). Observation of the underlying skeletal tissues that indicate a keyhole defect may help the forensic pathologist understand the soft tissue observations. Although usually associated with entrance wounds, keyhole lesions have also been described in association with exit wounds (Dixon, 1984a).

Secondary and Tertiary Fractures

In addition to primary fractures caused directly by the bullet's impact, there are two other types of fractures associated with gunshot wounds (**Figure 13.8**). *Radiating fractures* travel out from the primary impact site. In cross-section, radiating fractures show *no beveling* (unlike primary impact sites), and the fracture lines form *perpendicular* to the outer and inner surfaces of the vault. Radiating fractures will often stop at the cranial sutures, which dissipate the energy from the bullet. High levels of energy can cross the suture, however, or travel along it for some length before re-emerging as a fracture line once again. The physical separation of the suture line caused by the energy traveling through it is called a *diastatic fracture* (white arrows in **Figure 13.8**). Radiating fractures associated with exit wounds indicate considerable energy was still present in the bullet as it left the body. Because they radiate out from the site of impact, *radiating fractures also form before concentric fractures* (Smith et al., 1987). In fact, the formation of radiating fractures occurs faster than the speed at which bullets travel (Dixon, 1984b). This has been established through case reports demonstrating that radiating fractures associated with exit wounds will stop at radiating fracture lines caused by the corresponding entrance wound. It is estimated that fractures propagate at the velocity of 450 meters per second (Madea and Staak, 1988), while most firearms used in crimes in the United States project bullets at a velocity of 325–425 meters per second (prior to impacting any object, including the human body) (Christensen et al., 2014).

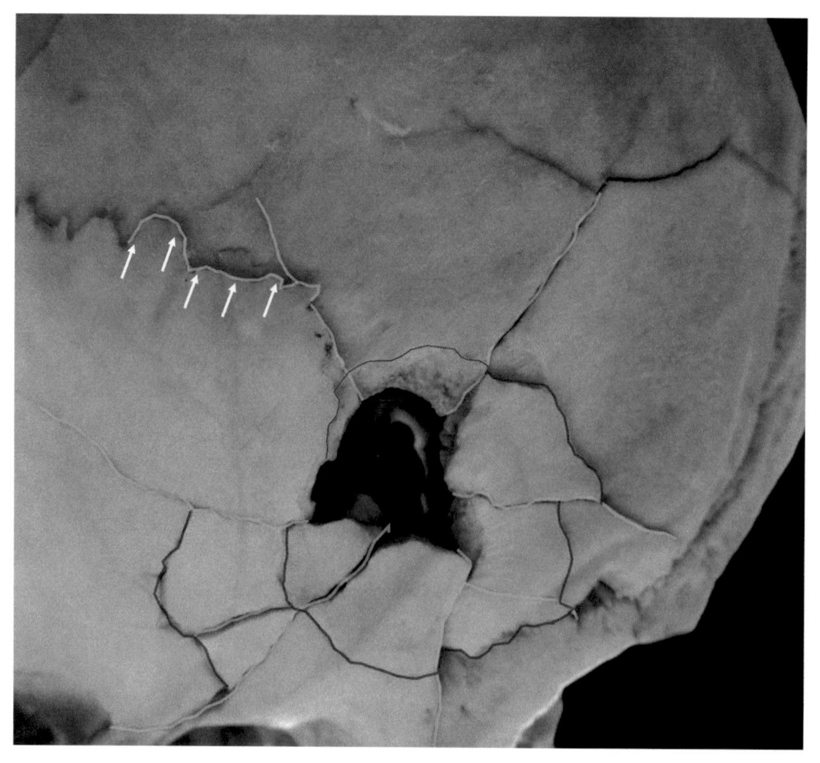

Figure 13.8. Reconstructed gunshot exit wound with radiating fractures (blue) and concentric fractures (red). Arrows indicate where a radiating fracture has become a diastatic fracture.

Concentric fractures develop as a result of an increase in intra-cranial pressure caused by the bullet impact, which results in the bone surrounding the primary impact site to *heave outward.* That is, the concentric fracture is formed from the cranial bones lifting upward. During the formation of a concentric fracture, the bone tears under tension beginning at the endocranial surface and the fracture line travels through the cross-section of the bone until it appears on the ectocranial surface (Hart, 2005; Smith et al., 1987). That is, concentric fractures form on the inside of the skull and move outward. This means that concentric fractures caused by gunshot wounds exhibit *external beveling.* This is not the case with radiating fractures caused by gunshots, or any type of fracture caused by blunt force trauma. Concentric fracture lines will terminate at radiating fracture lines because they form later in the fracture propagation sequence. As such, they are classified as *tertiary fractures.* Smith et al. (1987) report that concentric fractures are always found in association with radiating fractures because they represent the effects of maximum energy being dissipated by the skeletal tissues. That is, you will rarely have a gunshot wound with concentric fractures but no radiating fractures.

Linking Wounds and Wound Sequence

A thorough forensic analysis of high-velocity projectile trauma will seek to provide an accurate description of the primary, secondary, and tertiary fractures, their size and shape, their location on the skull, and their relationships with each other. The general process for providing this information in descriptive reports is as follows:

Reconstruct the skull if shattered. This is generally easy to do with gunshot trauma because the bones do not deform prior to shattering (which is not the case with low-velocity blunt force trauma).

Describe wound number and location.

Link entrance and exit wounds based on trajectories of bullet passage and wound size and shape. This provides information on the direction of fire.
Determine the minimum number of bullets fired.
Determine the sequence of wounds.

The identification of entrance and exit wounds is based on primary wound characteristics (size, shape, beveling). In the case of multiple gunshots (as in **Figure 13.9**), it may be difficult to link entrance and exit wounds based on wound characteristics alone. Furthermore, it may be difficult to reconstruct the wound sequence without reference to the pattern of radiating and concentric fractures (Dixon, 1984b). **Puppe's Rule** is useful in both cases. Puppe's rule states: "Fractures from subsequent impacts are arrested at pre-existing fracture lines . . . (Madea and Staak, 1988:321)." Consider **Figure 13.10**. Three gunshots are shown (A, B, C), each with three radiating fracture lines (labeled A1, A2, A3, etc.). Puppe's Rule states that gunshot

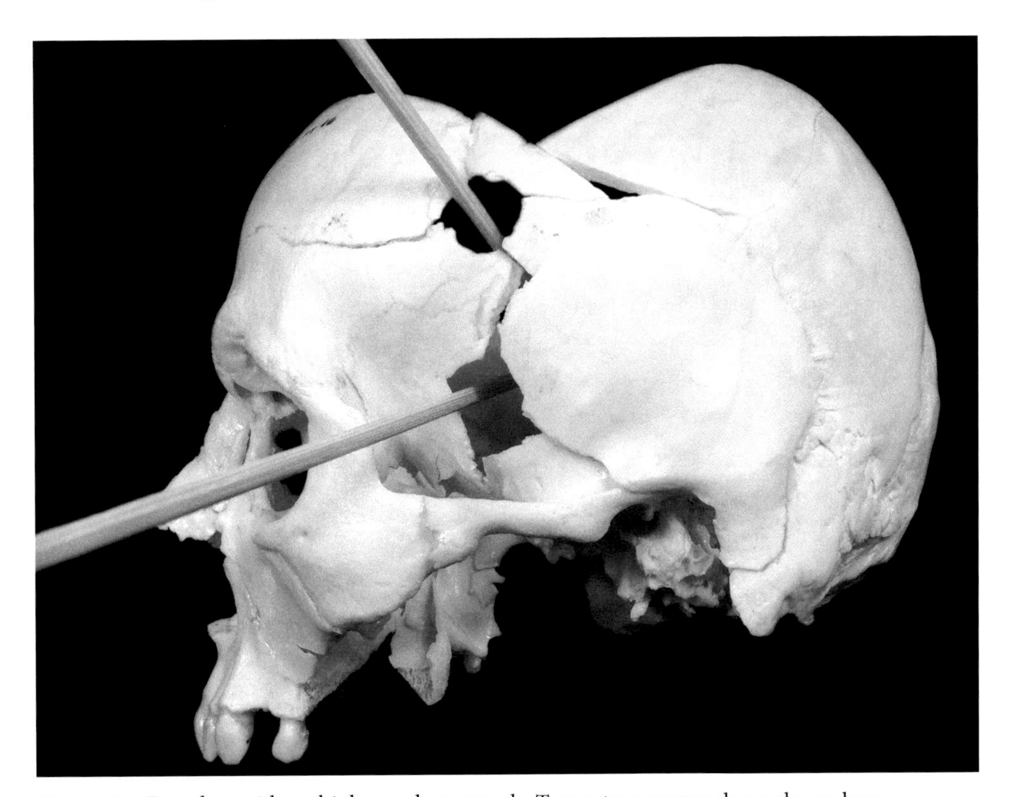

Figure 13.9. Decedent with multiple gunshot wounds. Two entrance wounds are shown here.

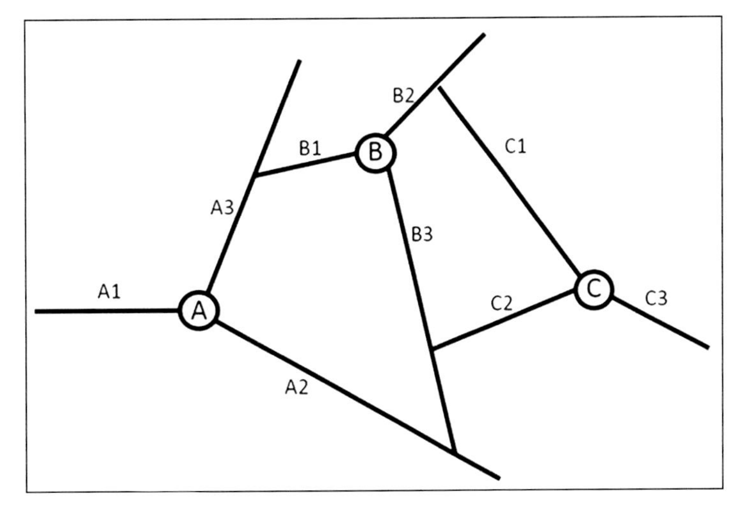

Figure 13.10. Puppe's Rule of fracture formation sequence.

A occurred first because radiating fracture B1 terminates in radiating fracture A3. Likewise, gunshot C occurred last because radiating fracture C1 terminates in radiating fracture B2. Using these simple principles, the forensic anthropologist can determine the sequence of multiple gunshot wounds. Another implication of Puppe's Rule is that the fracture lines associated with exit wounds will always terminate at the fracture lines associated with entrance wounds. Therefore, Puppe's Rule is useful for identifying entrance and exit wounds, in addition to determining their sequence. Note that in multiple gunshot cases, the secondary shots will also create less damage (fewer and shorter radiating fractures) due to the release of intracranial pressure caused by the first gunshot. This fact can also help establish wound sequence.

Wound Size and Bullet Characteristics

While it is tempting to assume that the size of a gunshot wound is related to the caliber of the bullet that produced it, this relationship is far from straightforward. The intrinsic characteristics of the bullet (size, material properties, jacketing), its movement (how stable it is moving and rotating around its axis), the angle at which it strikes the body (perpendicular or acute), and the intrinsic properties of the bone it impacts (skull vs. postcrania, cross-sectional properties of the impact site, ratio of cortical to trabecular bone, brittleness based on age or disease status) all affect wound size. Berryman and colleagues (1995), for example, found that a .22 caliber bullet produced defects ranging from .22 to .43 inches while a .38 caliber bullet produced defects ranging from .32 to .72 inches. A follow-up study by Ross (1996) produced similar data. Note that these results indicate that bullet wounds can be smaller than the bullet that created them—a phenomenon usually related to the angle of impact and whether a bullet enters the cranium at the site of a suture or a pre-existing fracture (Berryman et al., 1995; Fischer and Nickell, 1986; Paschall and Ross, 2017). The wide range of wound sizes that can result from the same caliber of bullet and the substantial amount of overlap in these ranges means that while it may sometimes be possible to distinguish small-caliber bullets from large-caliber bullets based on wound size, finer distinctions should not be attempted.

Postcranial Gunshot Wounds

There is much less research focusing on the mechanics of bullet hole formation in the postcranial skeleton. This may be because homicides typically involve gunshots to the head, but it also reflects the more varied shapes of the bones of the postcranial skeleton, which makes generalizations more difficult. In studying postcranial gunshot wounds, researchers have divided the skeleton based on bone shape. Flat bones are thought to respond to gunshot wounds in a manner similar to the cranial vault, where beveling allows an analyst to reconstruct the trajectory of the bullet (**Figure 13.11A**). Tubular bones, such as long bones, may respond differently to bullet impacts depending on whether the diaphysis or the metaphysis is affected. Gunshot wounds to the diaphysis may produce a clear entrance defect, often with embedded metal, or bullet wipe, around the impact site (**Figure 13.11B**, left). A clear exit wound, however, may be absent or obscured as projectiles tend to comminute the bone (**Figure 13.11B**, right). In contrast, the high proportion of trabecular bone in the proximal and distal ends of long bones will often result in a circular, punched-out defect associated with radiating and concentric fractures such as those seen in the cranial vault and flat bones (**Figure 13.11C**).

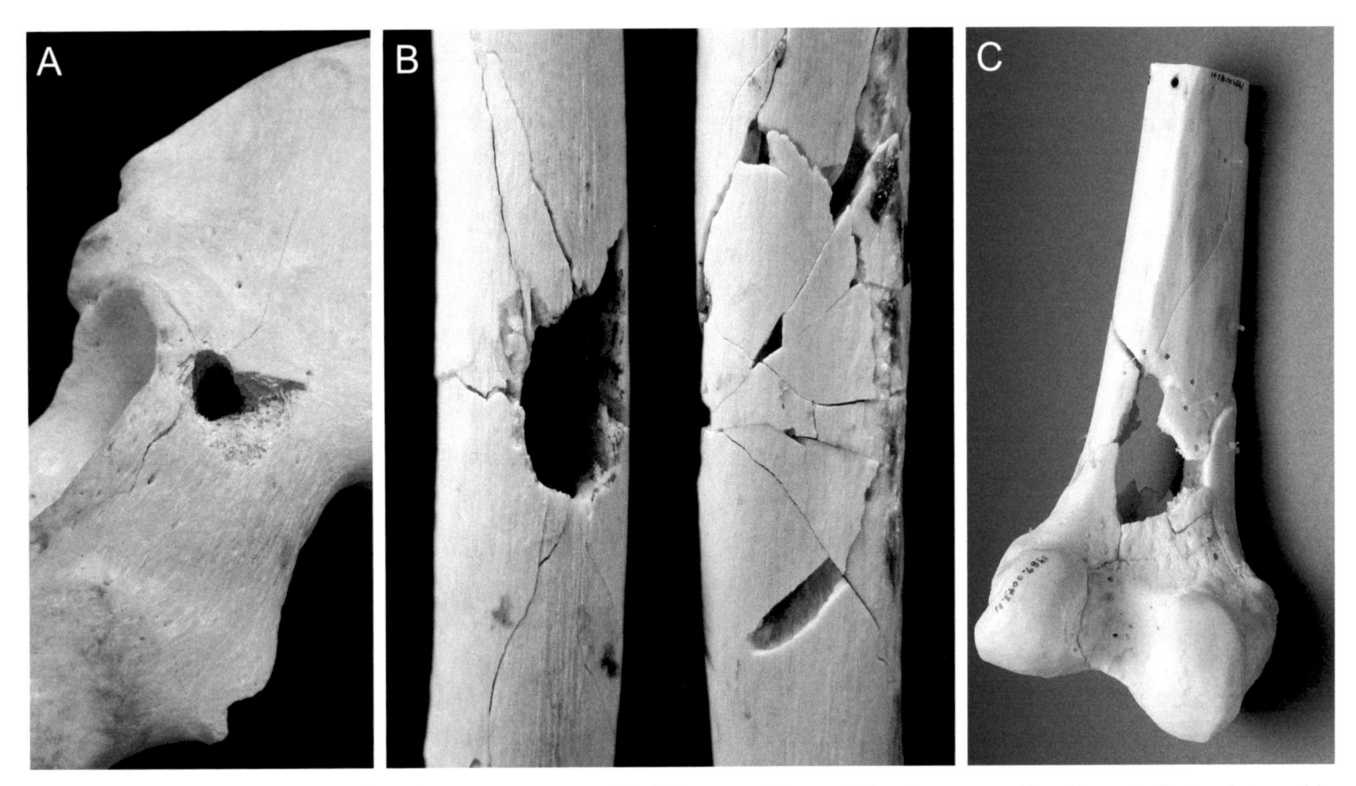

Figure 13.11. (*A*) Gunshot exit wound on the posterior aspect of the left os coxa. Note radiating fractures and beveling, similar to what would be seen in the cranial vault. (*B*) Entrance (*left*) and exit (*right*) wounds caused by the same bullet impacting the femoral diaphysis. Note the metal wipe in the left image as well as the comminution on the right. (*C*) Gunshot to the distal femur looking through entry and exit wounds, as well as radiating and concentric fractures.

LEARNING CHECK

Q1. Based on the appearance of the defect shown in **Figure 13.12A**, is this an entrance or exit wound?
- A) Entrance
- B) Exit

Q2. What type of fractures are associated with the defect shown in **Figure 13.12A**?
- A) Radiating
- B) Concentric
- C) Both

Q3. Based on the appearance of the defect shown in **Figure 13.12B**, is this an entrance or an exit wound?
- A) Entrance
- B) Exit

Q4. What type of fractures are associated with the defect shown in **Figure 13.12B**?
- A) Radiating
- B) Concentric
- C) Both

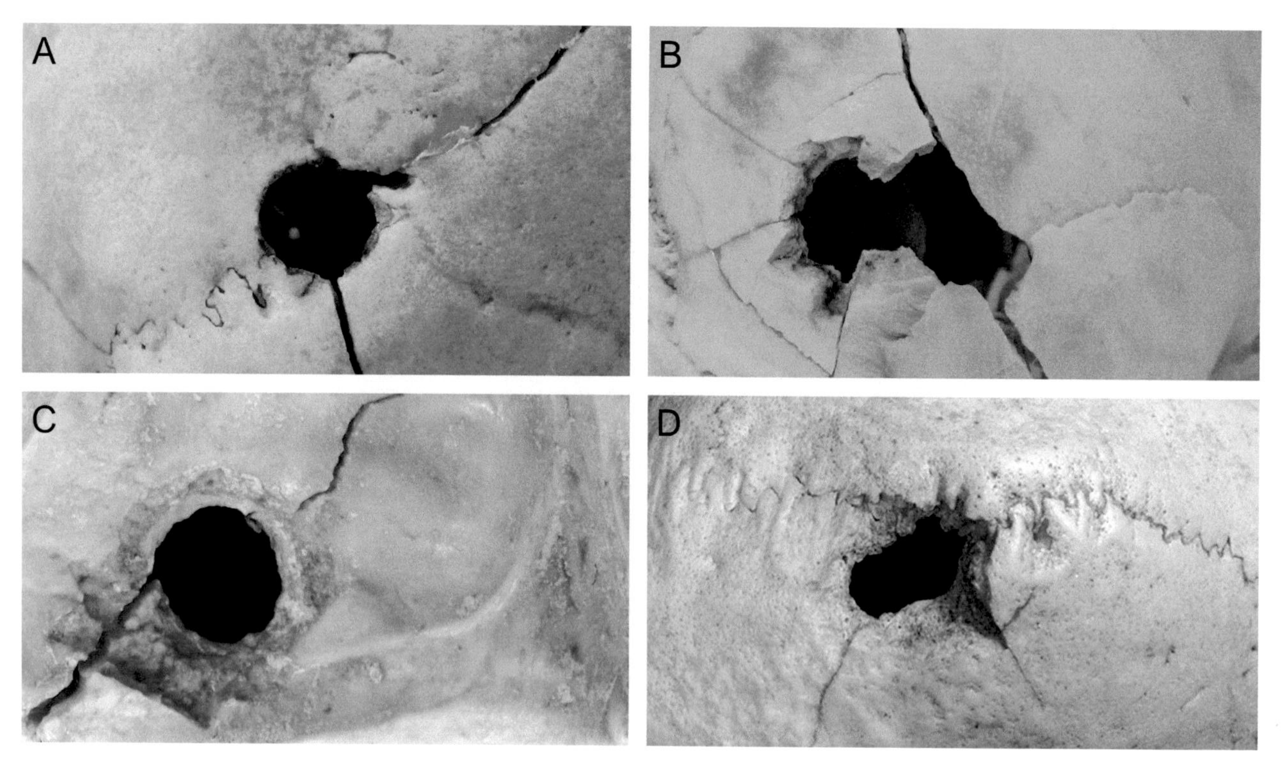

Figure 13.12. Four examples of cranial bullet wounds.

Q5. How would you characterize the defect shown in **Figure 13.12C**? This is the endocranial surface.

A) Entrance wound
B) Exit wound

Q6. What type of fractures are associated with the defect shown in **Figure 13.12C**?

A) Radiating
B) Concentric
C) Both

Q7. Based on the appearance of the defect shown in **Figure 13.12D**, is this an entrance or an exit wound?

A) Exit wound
B) Entrance wound

Q8. In which image can you see an example of internal beveling?

A) **Figure 13.12A**
B) **Figure 13.12B**
C) **Figure 13.12C**
D) **Figure 13.12D**

Q9. How would you characterize the defect shown in **Figure 13.13**? This is the endocranial surface.

A) Entrance wound
B) Exit wound

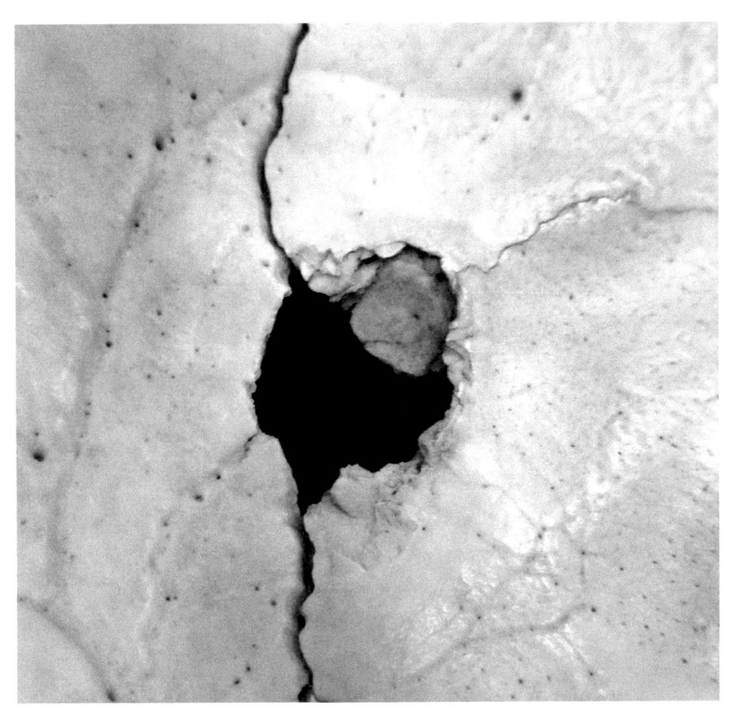

Figure 13.13. An example of a cranial bullet wound.

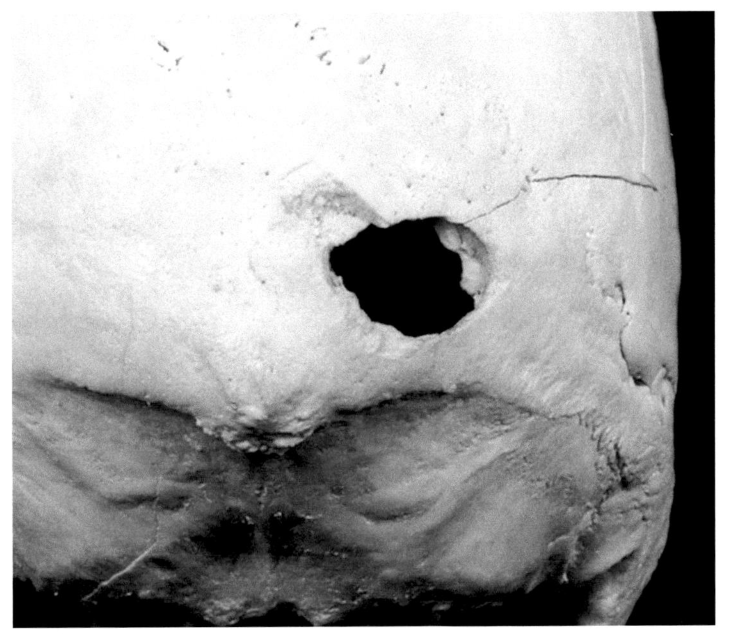

Figure 13.14. An example of a cranial bullet wound with radiating fracture.

Q10. Look closely at the bullet wound shown in **Figure 13.14**. Was the bullet moving from left to right, or from right to left? Consider what kind of bullet wound this is and examine all aspects of the impact site.

A) Left to right

B) Right to left

End-of-Chapter Summary

Projectile trauma is one of the most common forms encountered in a forensic investigation because firearms are often used in violent crimes. In this chapter, the basics of firearm and bullet technology were presented. In terms of skeletal trauma, it is important for the forensic anthropologist to differentiate entry and exit wounds via wound beveling patterns. Secondary and tertiary fracture patterns and wound beveling also allow the forensic anthropologist to reconstruct the sequence of events relating to the decedent's death. Although cause and manner of death are the purview of the forensic pathologist, a well-written descriptive report can provide important information for official declarations. Cranial trauma research is well established; however, the varied composition of postcranial skeletal elements makes generalizing about expected wound patterns difficult. The skeletal indicators of high-velocity projectile trauma result from the significant amount of energy imparted to the body, which causes the skeleton to respond like a brittle material that breaks before any bending or deformation occurs. This contrasts with low-velocity blunt force trauma, as discussed in the next chapter. This basic difference in velocity allows the forensic anthropologist to differentiate blunt and projectile trauma.

End-of-Chapter Exercises

Exercise 1

Materials Required: **Figures 13.15** and **13.16.**

Scenario: Skeletonized remains that were recovered from a ravine have been submitted to you for analysis. In examining the cranium, you have documented defects consistent with multiple gunshot wounds. Four views of the cranium are presented in **Figures 13.15** and **13.16,** with paired defects labeled such that "A1" indicates the entrance of Projectile A and "A2" indicates the associated exit. Use the information in **Figures 13.15** and **13.16** to answer the following questions.

Question 1: What are the major components of the trajectory (e.g., anterior-to-posterior, left-to-right, inferior-to-superior) of Projectile A?

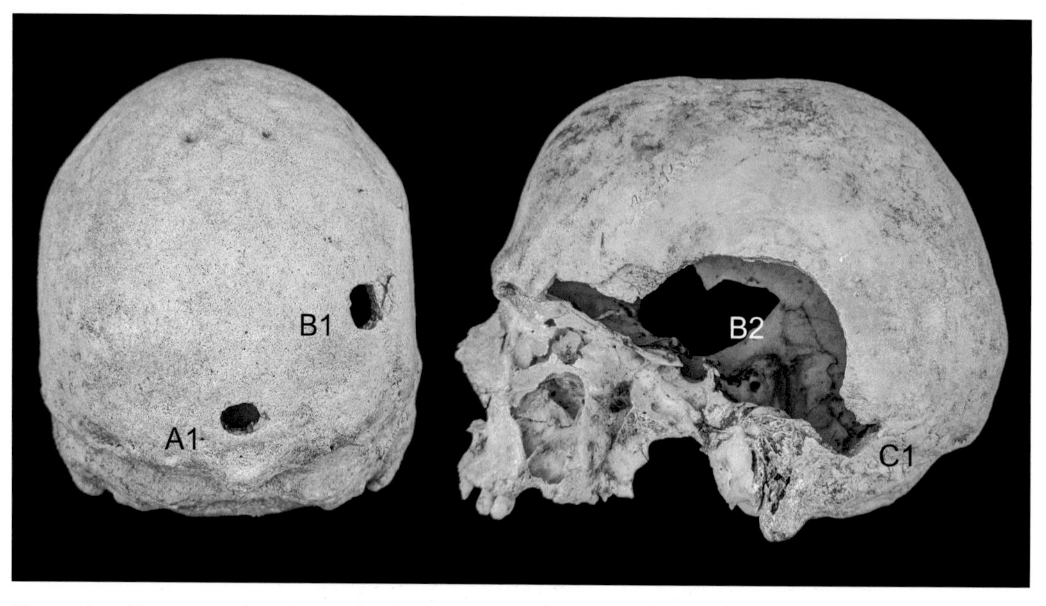

Figure 13.15. Two views of a cranium showing multiple gunshot wounds labeled A, B, and C.

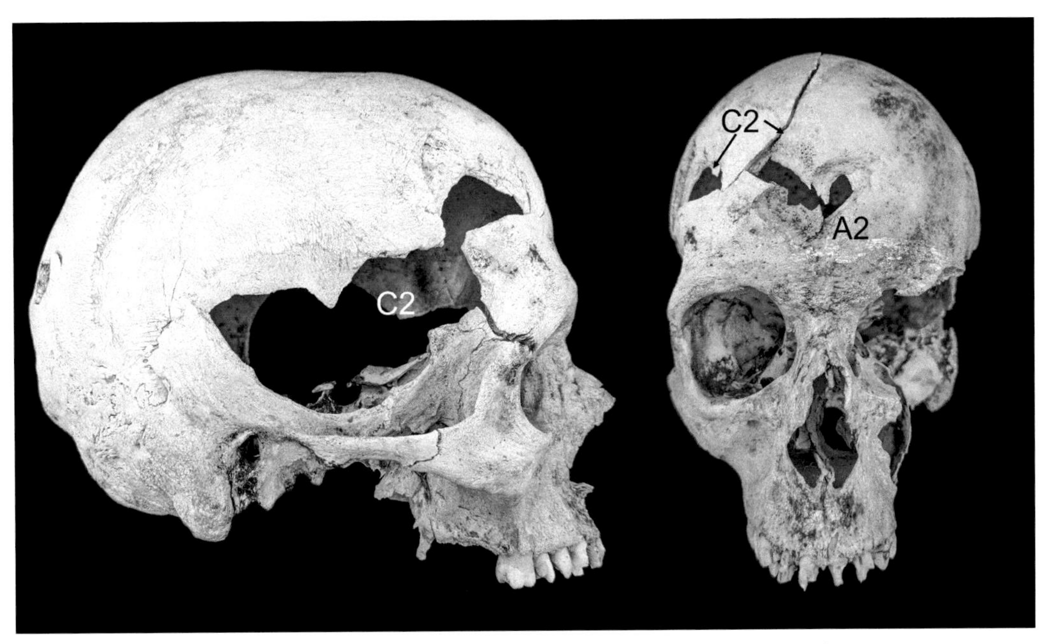

Figure 13.16. Two different views of the cranium in Figure 13.15 showing gunshot wounds.

Question 2: What are the major components of the trajectory of Projectile B?

Question 3: Using what you have learned in this chapter concerning secondary fracture patterns and Puppe's Rule, can you determine whether Projectile A preceded Projectile C? Please support your answer.

Exercise 2

Materials Required: **Figure 13.17**

Scenario: After pulling over to fix a flat tire, a motorist discovered skeletonized remains near a culvert passing under the road. After cleaning the remains, you noticed defects consistent with multiple gunshot wounds to the face and the lower leg. Your observations are shown in **Figure 13.17**. Use them to answer the following questions.

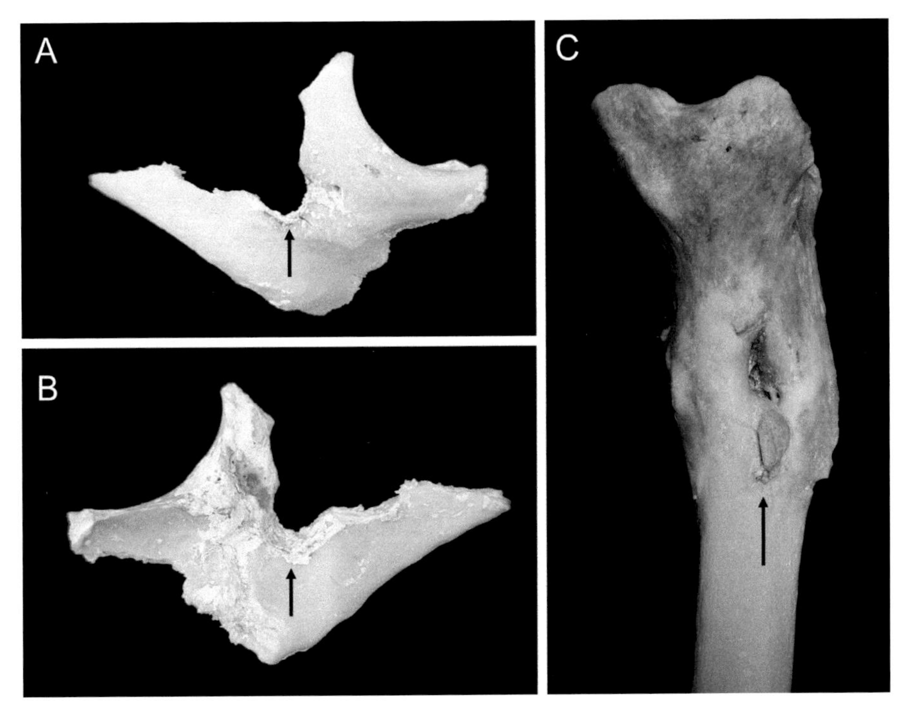

Figure 13.17. Multiple skeletal elements exhibiting gunshot wounds.

Question 1: **Figures 13.17A** and **13.17B** show a gunshot wound on the right zygomatic, indicated by the black arrows. **Figure 13.17A** shows the external surface and **Figure 13.17B** shows the internal surface. Based on these images, is this defect more consistent with an entrance or an exit?

Question 2: **Figure 13.17C** shows a gunshot wound to the distal tibia with the projectile indicated by the black arrow. Based on the appearance of this injury, did it occur at the same time as the injury to the zygomatic? Please support your answer.

Question 3: Which of the gunshot wounds shown in **Figure 13.17** has more potential to be used in the identification of the decedent? Why?

Exercise 3

Materials Required: **Figure 13.18**

Scenario: Construction workers repairing a bridge discovered skeletal remains eroding out of a riverbank. After observing defects in the cranial vault at the scene, investigators became concerned that the remains provided evidence of multiple gunshot wounds and therefore represented a homicide. They have asked you to consult. **Figure 13.18** shows the cranial vault in question; no other defects were observed.

Question 1: Look closely at **Figure 13.18,** noting characteristics of the bone surface surrounding the defects as well as the presence or absence of any fractures. Based on everything that you have learned in chapters 12 and 13, are the investigators right to be concerned? Please support your conclusion with specific observations.

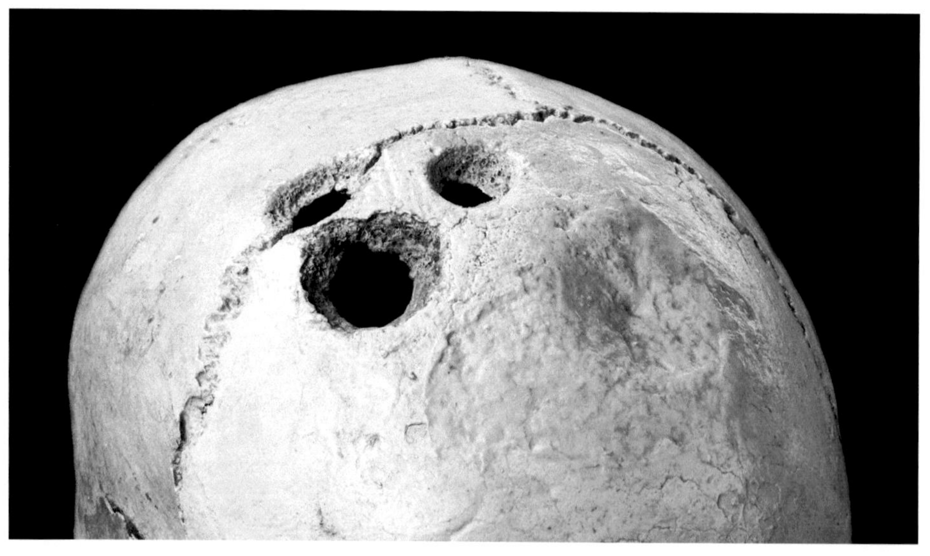

Figure 13.18. Superior aspect of a cranium with three holes evident.

References

Baik SO, Uku JM, Sikirica M. 1991. A case of external beveling with an entrance gunshot wound to the skull made by a small caliber rifle bullet. *American Journal of Forensic Medicine and Pathology* 12: 334–336.

Berryman HE, Lanfear AK, Shirley NR. 2012. The biomechanics of gunshot trauma to bone: research considerations within the present judicial climate. In: Dirkmaat DC (Ed.), *A Companion to Forensic Anthropology.* London: Blackwell. p. 390–399.

Berryman HE, Smith OC, Symes SA. 1995. Diameter of cranial gunshot wounds as a function of bullet caliber. *Journal of Forensic Sciences* 40: 751–754.

Berryman HE, Symes SA. 1998. Recognizing gunshot and blunt cranial trauma through fracture interpretations. In: Reichs KJ, (Ed.), *Forensic Osteology. Advances in the Identification of Human Remains,* Second Edition. Springfield, IL: Charles C. Thomas. p. 333–352.

Bhoopat T. 1995. A case of internal beveling with an exit gunshot wound to the skull. *Forensic Science International* 71: 97–101.

Christensen AM, Passalacqua NV, Bartelink EJ. 2014. *Forensic Anthropology. Current Methods and Practice.* Oxford: Elsevier.

Coe JI. 1982. External beveling of entrance wounds by handguns. *American Journal of Forensic Medicine and Pathology* 3: 215–219.

DiMaio VJM. 2021. *Gunshot Wounds: Practical Aspects of Firearms, Ballistics, and Forensic Techniques,* Third Edition. Boca Raton, FL: CRC Press.

Dixon DS. 1982. Keyhole lesions in gunshot wounds of the skull and direction of fire. *Journal of Forensic Sciences* 27: 555–566.

Dixon DS. 1984a. Exit keyhole lesion and direction of fire in a gunshot wound of the skull. *Journal of Forensic Sciences* 29: 336–339.

Dixon DS. 1984b. Pattern of intersecting fractures and direction of fire. *Journal of Forensic Sciences* 29: 651–654.

Fischer JF, Nickell J. 1986. "Keyhole" skull wounds: the problem of bullet-caliber determination. *Identification News* December: 8–10.

Hart GO. 2005. Fracture pattern interpretation in the skull: differentiating blunt force from ballistics trauma using concentric fractures. *Journal of Forensic Sciences* 50: 1276–1281.

Madea B, Staak M. 1988. Determination of the sequence of gunshot wounds of the skull. *Journal of the Forensic Science Society* 28: 321–328.

Paschall A, Ross AH. 2017. Bone mineral density and wounding capacity of handguns: implications for estimation of caliber. *International Journal of Legal Medicine* 131: 161–166.

Peterson BL. 1991. External beveling of cranial gunshot entrance wounds. *Journal of Forensic Sciences* 36: 1592–1595.

Quatrehomme G, Alunni V. 2013. Bone trauma. In: Siegel JA, Saukko PJ (Eds.), *Encyclopedia of Forensic Sciences.* Waltham, MA: Academic Press. p. 89–96.

Ross AH. 1996. Caliber estimation from cranial entrance defect measurements. *Journal of Forensic Sciences* 41: 629–633.

Smith OC, Berryman HE, Lahren CH. 1987. Cranial fracture patterns and estimate of direction from low-velocity gunshot wounds. *Journal of Forensic Sciences* 32: 1416–1421.

Smith OC, Berryman HE, Symes SA, Francisco JT, Hnilica V. 1993. Atypical gunshot exit defects to the cranial vault. *Journal of Forensic Sciences* 38: 339–343.

14

Low-Velocity Blunt and Sharp Force Trauma

Learning Goals

By the end of this chapter, the student will be able to:

- Identify the skeletal signatures of blunt force trauma.
- Identify the skeletal signatures of sharp force trauma.
- Differentiate blunt force, sharp force, and projectile trauma.
- Utilize a differential diagnosis decision tree to identify the cause of trauma.
- Explain the mechanics of how saw marks manifest on bone and identify signatures of different types of saws based on cutmark properties.
- Define the terms kerf, kerf wall, false start kerf, striae, and breakaway spur in relationship to cutmark analysis.
- Characterize the basic differences between hand and power saws and how they impact bone.

Introduction

In chapter 13 we discussed the effects of high-velocity projectile trauma on the skeleton. Here we consider low-velocity trauma, which includes a much broader range of weapons and objects, with two broad classes identified. **Blunt force trauma** includes blows with relatively heavy objects that impact over a wider area, thus crushing or fracturing the skeleton. **Sharp force trauma** includes blows with objects that have an edge, which focuses the area of impact resulting in cutting or stabbing injuries. Of the 15,000 homicides recorded by the FBI in the United States in 2016, 1,600 were caused by knives and other cutting objects. Approximately 500 were caused by blunt force objects, 650 were caused by fists and feet, and 100 were due to strangulation. In combination, then, approximately 1,200 homicides were caused by all forms of blunt force trauma. These data are in contrast with the 11,000 firearms-related homicides recorded during the same reporting period.

Blunt Force Trauma

Blunt force trauma results from an object impacting the body at a *low velocity* over a relatively *wide area* resulting in compression of the skeletal tissues and extensive fracturing. Examples of blunt force objects include fists, bats, bricks, and hammers. Falls are considered blunt force trauma (the object impacting the body is the earth itself) as are motor vehicle collisions and strangulations. The diagnostic criteria of blunt force trauma include: broad areas of *radiating and concentric fractures, plastic deformation*, and *cortical delamination* or *wastage* (in the case of multiple blows). The primary considerations in the analysis of blunt force trauma include:

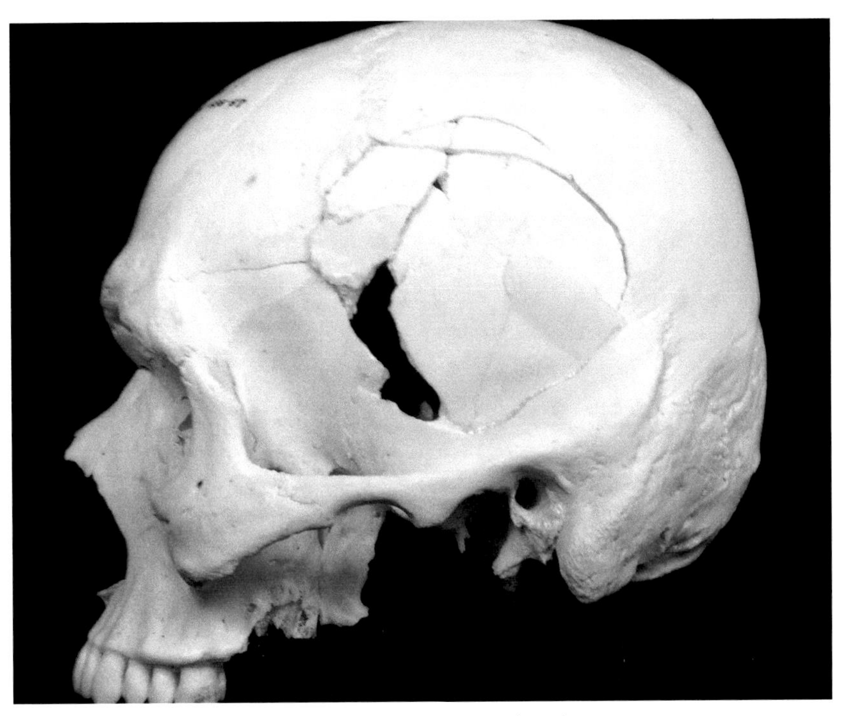

Figure 14.1. Blunt force trauma to the left lateral cranial vault.

identifying the exact point(s) of impact, recording the number and sequence of impacts, estimating the amount of energy imparted onto the body, and if possible, inferring characteristics of the weapon used to commit the act (Berryman et al., 2013). Because many blunt force injuries target the head, and because more research has been conducted on cranial fracture patterns, the discussion will initially focus on cranial vault injuries.

The diagnostic signature of blunt force trauma is crushing injuries over a relatively wide area (**Figure 14.1**). The larger and heavier the object, the wider the area of damage, which in some cases can be so catastrophic that it is difficult to determine whether a firearm or blunt force object was used. However, unlike high-velocity projectile trauma, all blunt force injuries result from *static loading* (i.e., slow loading) of the skeletal tissues. Recall that bone responds differently to static and dynamic forces (see chapter 12); static forces allow the bone to resist stress through an *elastic* (recoverable) and *plastic* (permanent, non-recoverable) phase of deformation *prior* to fracturing. Dynamic forces associated with high-velocity projectiles do not allow deformation prior to fracturing. Therefore, one of the key signatures of blunt force trauma is *plastic deformation,* particularly in the skull. This means that individual pieces of fractured bone may not fit back together after a blunt force impact because of the permanent deformation of the skull's contours.

Compare the reconstructed skulls in **Figures 14.2A** and **14.2B**. Note that the fracture edges do not touch in **Figure 14.2A**, which is a blunt force injury. This picture was taken after the skull was reconstructed (note the tape holding it together); the fracture margins will never touch again because the skull is permanently warped. However, the fractured pieces can easily be reconstructed in **Figure 14.2B** because there is no plastic deformation of the skull prior to it fracturing. This damage was caused by a firearm.

As a result of blunt force trauma, the skull is compressed from the outside as the energy from the blow is absorbed by the skeletal tissues. As the bones deform, they are pushed inward toward the brain, which creates tension on the interior surface of the cranial vault. Because bones are weaker under tension than compression, radiating fracture lines associated with blunt force trauma form on the inner surface of the cranial vault and travel outward

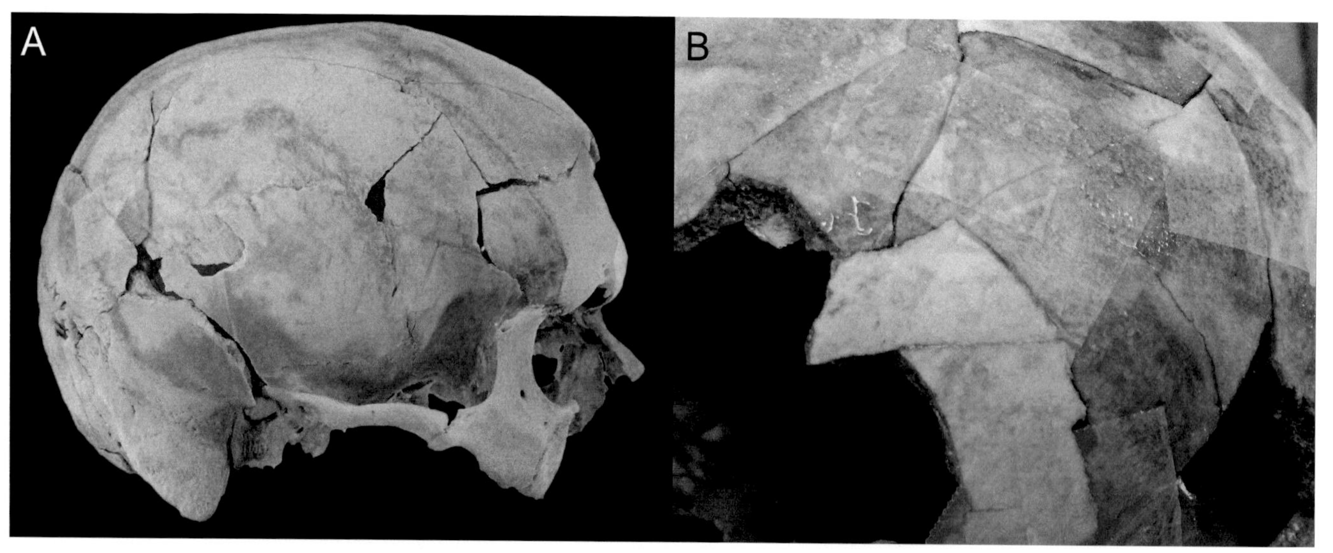

Figure 14.2. Two reconstructed crania. Compare the plastic deformation evident in (*A*) as a result of blunt force trauma to the lack of deformation in (*B*) resulting from a gunshot.

toward the outer surface of the skull. Radiating fractures are associated with the primary impact site and often (not always) create a series of triangular wedge-shaped pieces of bone that point toward the area of impact.

If these radiating fractures do not alleviate the stress of energy absorption, then concentric fractures may form surrounding the primary impact site. This is similar to a high-velocity projectile wound. However, there are two important differences. First, projectile trauma will create a penetrating primary fracture site associated with the bullet striking the skull. Second, concentric fractures resulting from blunt force trauma result from different biomechanical processes. In gunshot wounds, concentric fractures initiate on the inner surface of the skull and propagate outward; these fracture lines are externally beveled. However, with blunt force trauma, concentric fractures only form once the primary impact site is pushed further toward the brain, creating tension on the outer surface of the skull in the zone surrounding the primary impact site. *This means that concentric fractures in blunt force trauma are internally beveled* (Berryman et al., 2013; Hart, 2005). The beveling pattern of concentric fractures is another way to differentiate blunt force and projectile trauma.

Fracture propagation in blunt force trauma is not random. Because the fractures are absorbing and dissipating energy, the fracture lines will follow lines of least resistance, seeking weaker areas of bone and avoiding stronger ones. The stronger areas of bone are referred to as **buttresses**. The cranial vault has four areas of buttressing (mid-frontal buttress, mid-occipital buttress, anterior temporal buttress, posterior temporal buttress) that help predict fracture propagation, as shown in **Figure 14.3A**. **Figure 14.3B** demonstrates the behavior of energy within the cranial vault. Note how one radiating fracture (white arrows) moves anteriorly and then dives inferiorly to avoid the buttress in the temporal-frontal region. Note also how the smaller radiating fracture (black arrows) moves anteriorly and inferiorly to avoid the buttress of the mastoid process.

The preceding discussion assumes the blunt force object is sufficiently large and broad. However, if the object used in blunt force trauma is distinctive in shape, it may create tool marks—fracture discontinuities that match the shape of the object's striking surface. **Figure 14.4** depicts a case in which a hammer was used to strike the cranial vault, creating discontinuities that reflect the shape of the hammer head. Such tool marks are expected when the object has special shape characteristics that focus the energy into a narrower area. Instances

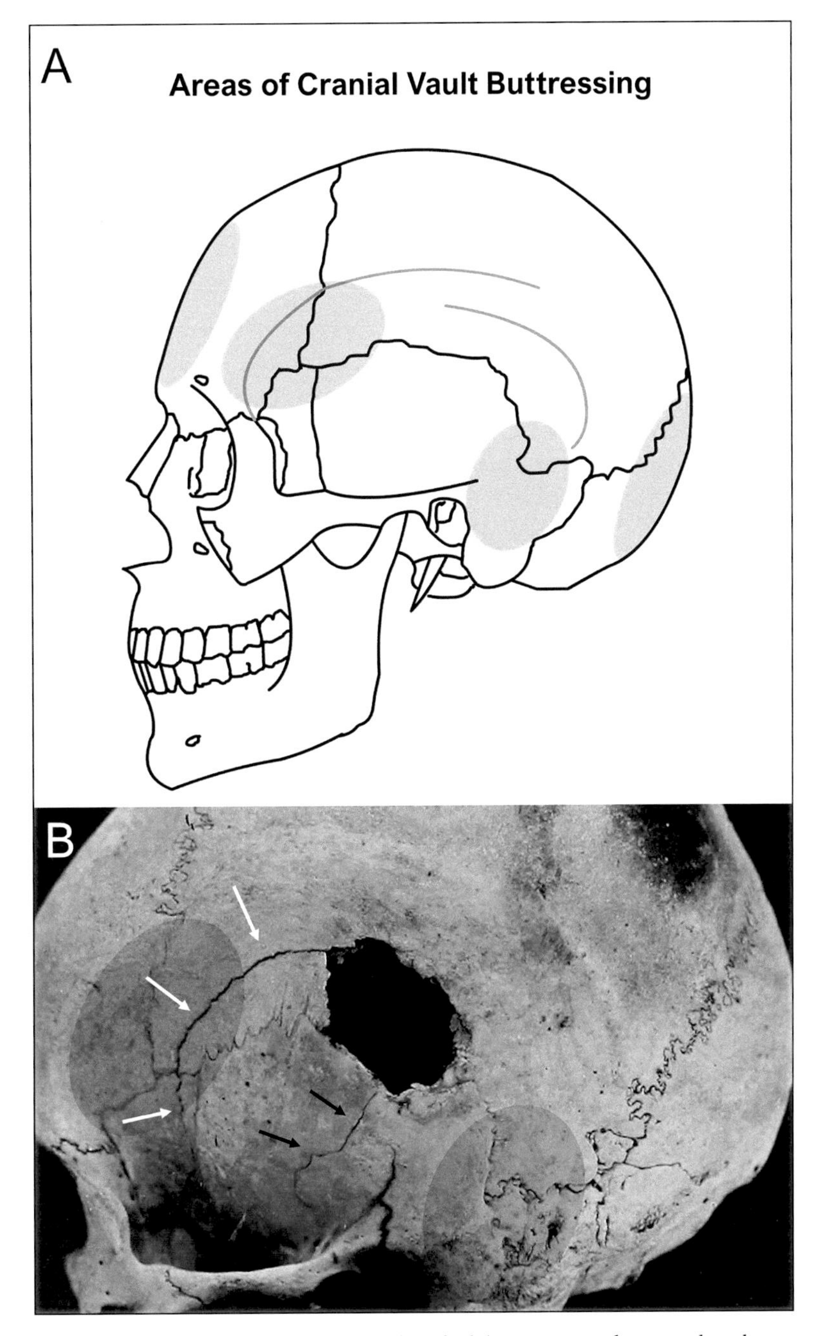

Figure 14.3. (*A*) Buttresses of the cranial vault. (*B*) Damage to the cranial vault with radiating fractures avoiding the buttressing of the vault.

of penetrating blunt force trauma are more common in elderly individuals where the bone is less elastic and more prone to react as a brittle material.

Blunt Force Trauma to the Face

Because the facial bones are so thin, direct strikes to the face cause characteristic breakage patterns that typically result in a portion of the face completely detaching from the vault. These types of blunt force fractures are called **Le Fort fractures**. Three classes of Le Fort fractures are recognized based on where the individual was struck (Le Fort, 1901). The energy of the blow is directed between the facial buttresses, areas of thicker bone that deflect energy to the thinner, weaker areas (**Figure 14.5A**). A Le Fort I fracture results from a blow received

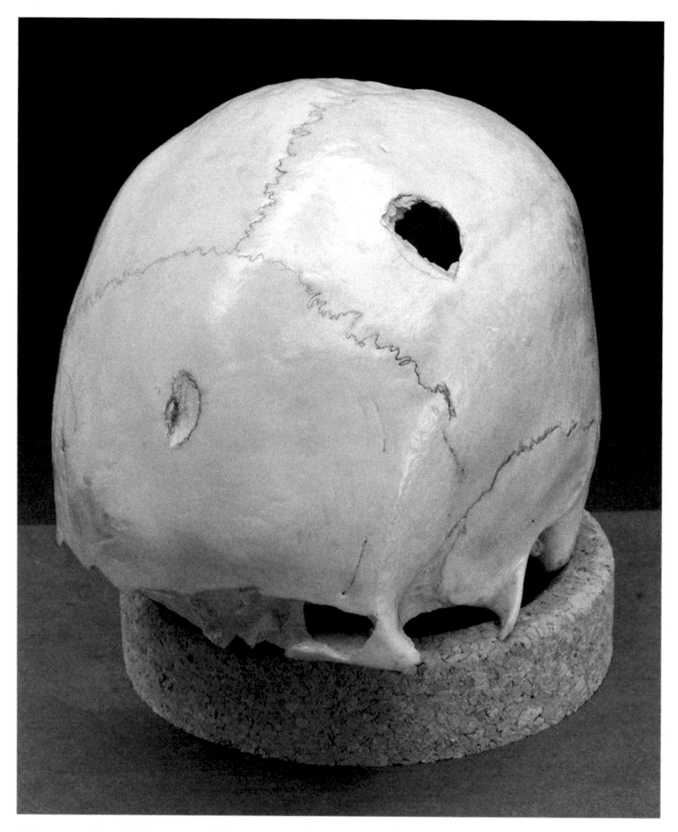

Figure 14.4. Hammer blows to the cranial vault.

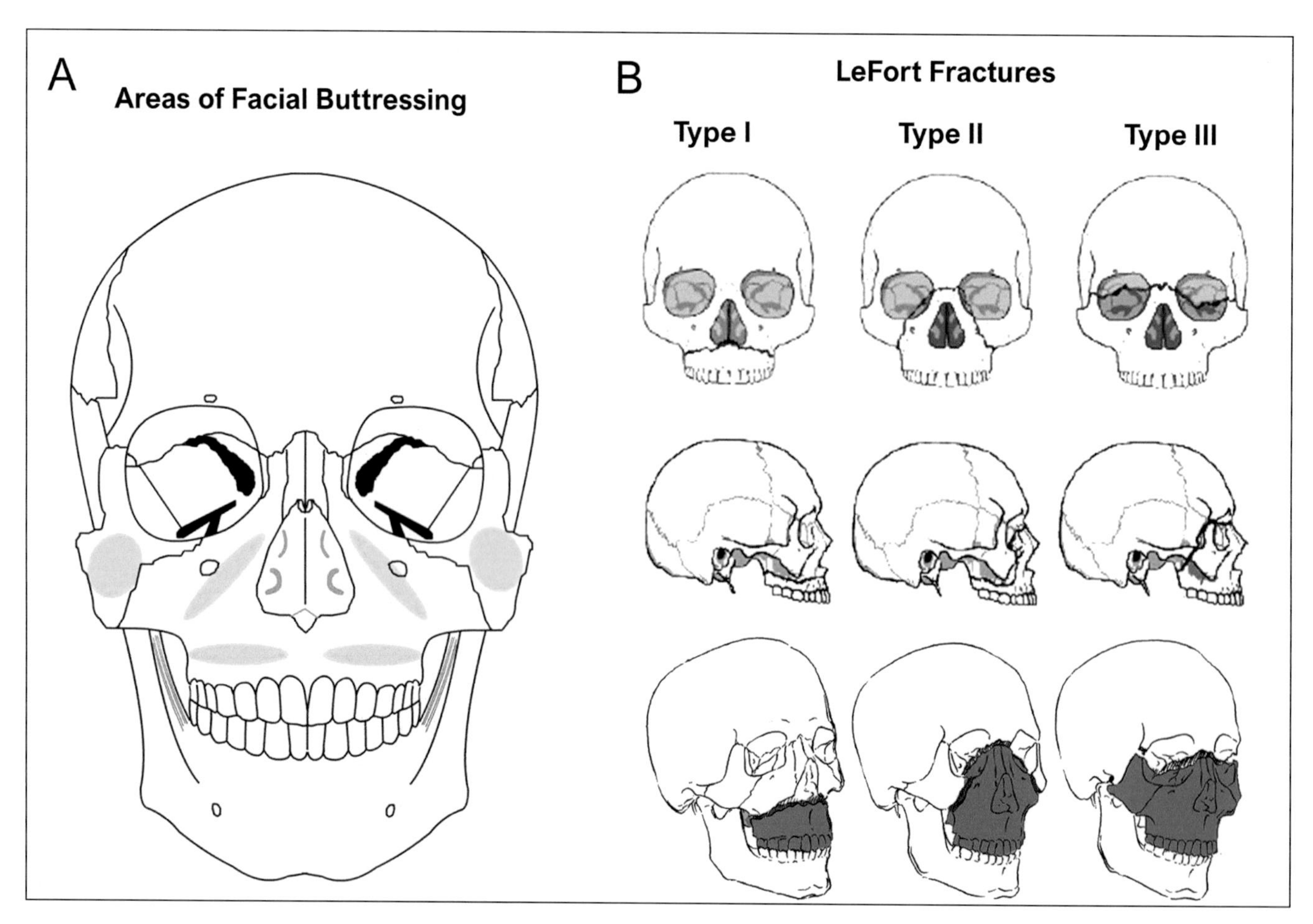

Figure 14.5. Comparison of the locations of the facial buttresses (*A*) with the portions of the facial skeleton involved in each of the three types of Le Fort fractures (*B*).

to the lower face in the area between the teeth and the lower part of the nose. This results in a separation of the maxilla at the alveolar process. A Le Fort II fracture is caused by a blow to the central part of the mid-face, resulting in separation of the mid-facial region. A Le Fort III results from a blow to the central part of the upper face, resulting in separation of the entire facial skeleton (**Figure 14.5B**).

Blunt Force Trauma to the Postcranial Skeleton

It is generally more difficult to predict blunt force trauma fracture patterns in the postcranial skeleton. Most bones of the axial skeleton are thin, with a high proportion of trabecular bone (ribs and vertebrae). Rib fractures can result from a punch to the chest and these usually appear as a single, incomplete fracture line that is vertically oriented (that is, perpendicular to the long axis of the rib) (**Figure 14.6A**). It is unlikely that blunt force trauma will result in rib comminution. Blunt force trauma to the vertebrae, shoulder, and pectoral girdles is rare, except in cases of falls or motor vehicle accidents (Berryman et al., 2013). Fractures of the

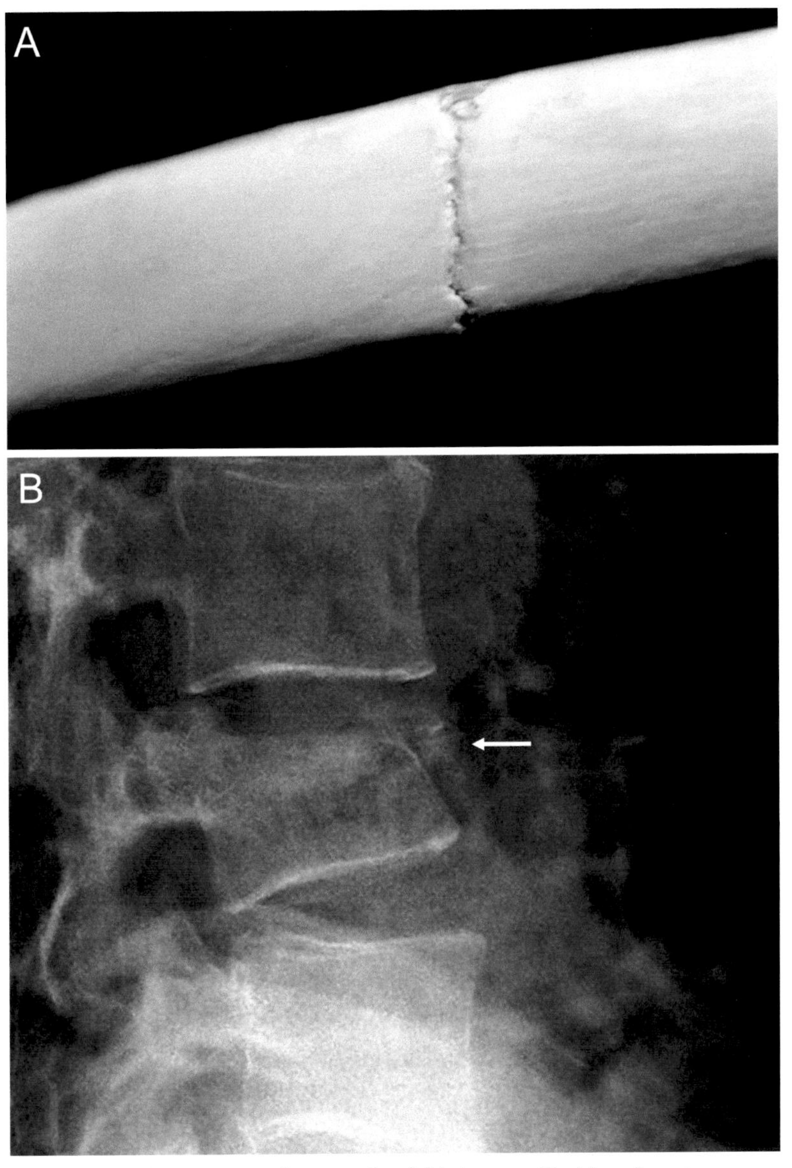

Figure 14.6. (*A*) Perimortem fracture of a rib likely caused by blunt force trauma. (*B*) Compression fracture of the spine showing collapse of the vertebral body.

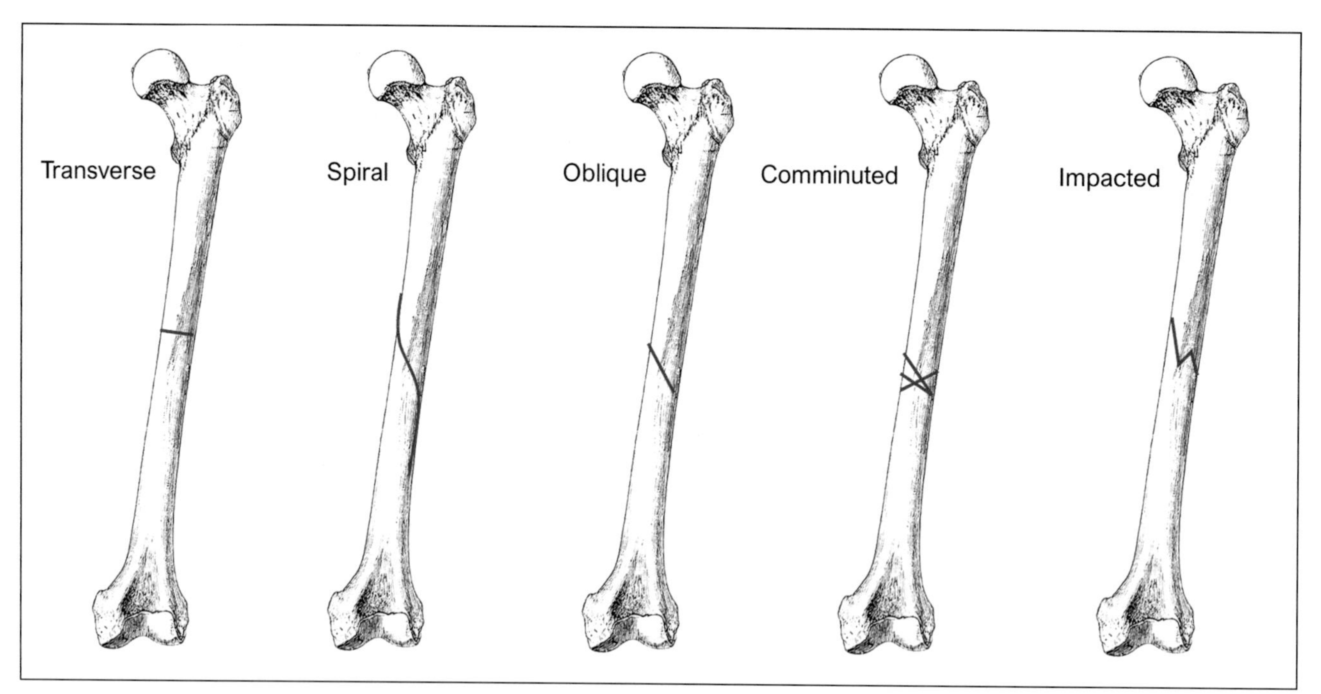

Figure 14.7. Common long bone fracture types (red lines).

spine are very common in falls and result from the body of the vertebrae collapsing under compression (**Figure 14.6B,** arrow).

Blunt force trauma to long bones will typically result in relatively clean, simple fractures without significant crushing or comminution. This is because the energy involved in an attack by a person is not sufficient to completely destroy the bone, but it is enough to fracture it. Long bones are tube-shaped elements that tend to show little plastic deformation prior to fracturing. Forensic anthropologists describe the shape of the fracture line, which can be used to infer the severity or direction of external force that impacted the body. Typical descriptive fracture types include transverse, spiral, oblique, comminuted, and impacted (**Figure 14.7**). Each can be associated with a specific direction of force. Spiral and oblique fractures, for example, indicate the impact was at an angle to the long bone. Comminuted fractures are associated with falls and motor vehicle accidents. Impacted fractures result from landing on one's feet after a fall.

Forces that impact a weight-bearing long bone at a perpendicular angle will create a specific type of fracture called a *butterfly fracture.* Butterfly fractures are valuable because they indicate the direction from which the force impacted the bone. Recall that bone is stronger under compression than tension. For this reason, when a long bone is struck at a perpendicular angle, the fracture will begin on the *opposite side of the bone from the point of impact* and travel back toward the impact site, often dividing and turning longitudinally along the shaft (**Figure 14.8**).

A **parry fracture** is a break in the ulna shaft, most often seen in the distal half of the bone (**Figure 14.9**). These fractures result from raising your arm above your head to defend yourself from a blow, resulting in a partial or complete fracture of the ulna shaft. Well-healed parry fractures may indicate a history of violent encounters, consistent with domestic abuse. Perimortem parry fractures indicate an attempt to defend oneself during the attack that led to one's death. Parry fractures are not to be confused with **Colles' fractures**, which affect the radius only and result from bracing yourself with an outstretched arm during a fall. Colles' fractures are the most common type of fracture in children.

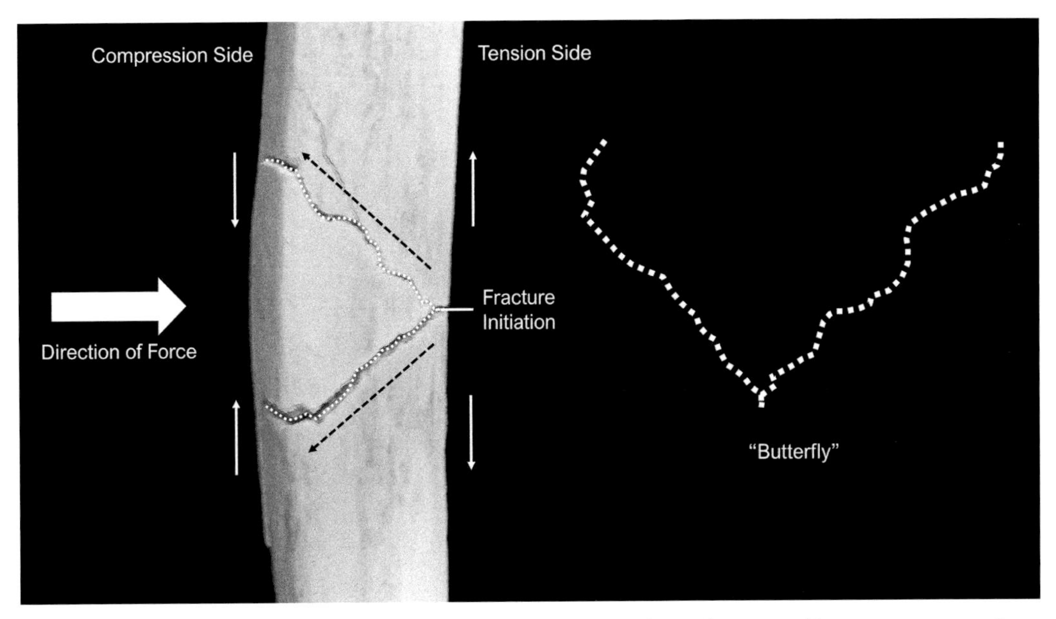

Figure 14.8. Butterfly fracture of a long bone. Small white arrows indicate direction of bone movement under application of force (large white arrow). Black arrows indicate direction of fracture propagation.

Strangulation

Strangulation represents a type of blunt force trauma that causes asphyxiation resulting from the slow, steady application of force in a compressive manner (Pollanen and Chiasson, 1996; Ubelaker, 1991b). Strangulation can be difficult to detect in skeletonized remains but may be indicated by a fractured *hyoid bone*. The hyoid forms from three different growth centers and, although they are generally understood to unite later in life, the hyoid frequently remains either entirely or partially unfused (**Figure 14.10**). Although a fractured hyoid *may* indicate strangulation, they have also been observed in cases involving "hangings, nonlethal throttling, sports injuries, motor vehicle accidents, and profuse vomiting" (Symes et al., 2012:355). Conversely, strangulation does not always result in damage to the hyoid—this is particularly the case when the body and the greater horns of the hyoid have not fused together (Fulginiti et al., 2021; Pollanen and Chiasson, 1996). There is thus no one-to-one correspondence between strangulation and hyoid fracture and other evidence should always be taken into account where available.

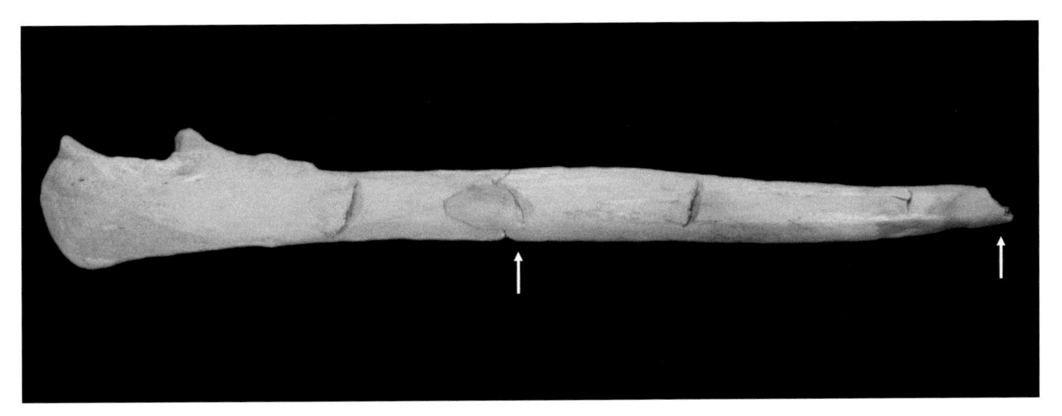

Figure 14.9. Posterior aspect of the ulna. White arrows indicate parry fractures.

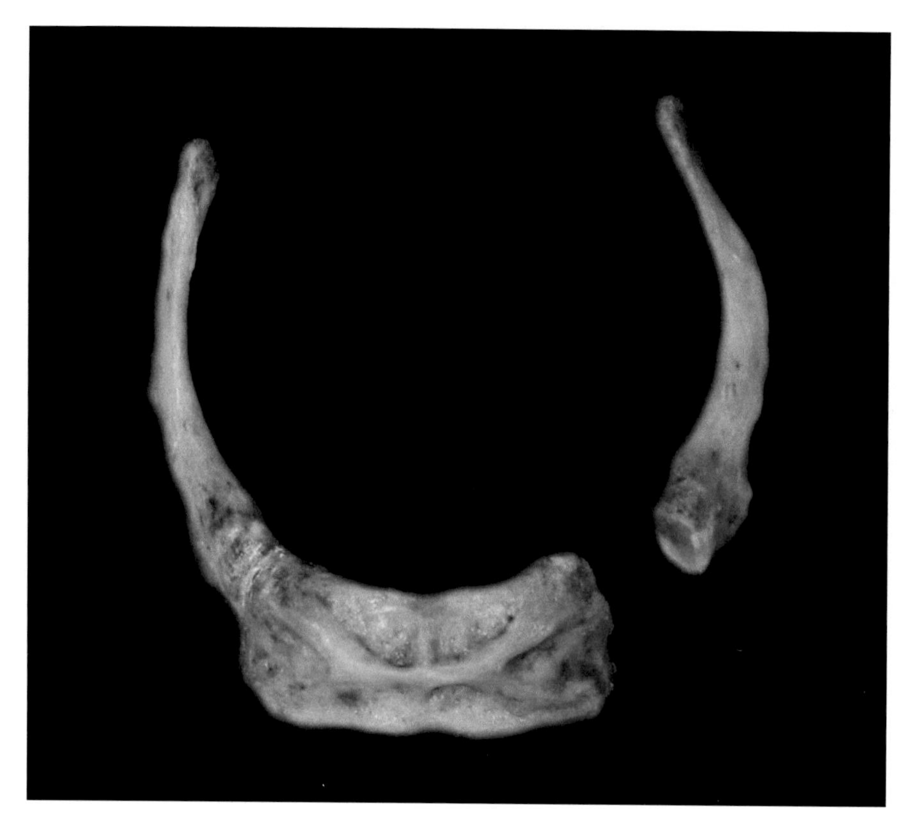

Figure 14.10. Hyoid with left greater horn unfused. Partial fusion of the hyoid can be mistaken for evidence of trauma.

Sharp Force Trauma

Sharp force trauma is an injury caused by compression or shearing forces applied to a *narrow area* of bone that results in a focal area of impact often resulting in a V-shaped cutmark. Examples of sharp force objects include knives, daggers, scissors, swords, machetes, and axes. Heavy sharp force objects (such as an axe) can produce wounds consistent with both blunt and sharp force injuries. They have an edge but are also heavy, which increases the amount of damage caused to the skeleton. A tell-tale sign of sharp force trauma is an injury that is linear with a well-defined straight edge. The sharp edge allows the object to sometimes puncture the bone or cleave off sections without significant additional damage. In general, sharp force trauma is much more likely to penetrate the bone than blunt force trauma. In this way, it is similar to high-velocity projectile trauma. A primary goal of the forensic analysis of sharp force trauma is to provide information about the murder weapon. Four criteria for doing so are discussed here (following Reichs, 1998): striations, secondary fracture lines, wastage, and hinging.

Striations are near microscopic lines that occur within the profile of the trauma impact site. They result from the direction of movement of the weapon with respect to the bone and can be oriented vertically (the object was moving perpendicular to the bone, such as in an up and down stabbing motion) or horizontally (the object was moving in a parallel fashion to the bone, such as in a slashing motion). In **Figure 14.11,** a section of parietal bone was sheared off with a sharp weapon. The wound cut has diagonally oriented striations (black arrows) indicating the direction of the blow. This suggests the decedent was attacked from his or her right side with the weapon moving from the right side of the image toward the left.

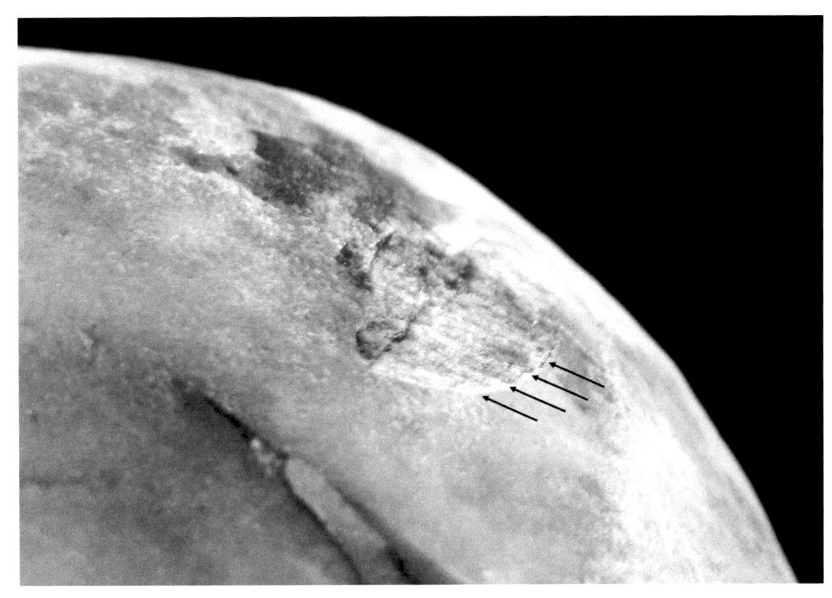

Figure 14.11. Closeup image of striations associated with a sharp force injury. The black arrows indicate the direction of movement.

Secondary fracture lines radiate out from the site of impact and indicate a higher level of energy was imparted onto the body. Their presence suggests the weapon used to commit the crime was heavier and/or wielded with enough force to create secondary fracture lines, like those seen with blunt force trauma and high-velocity projectile trauma (**Figure 14.12,** black arrow). Secondary fracture lines are useful for determining the sequence of blows because their propagation follows Puppe's Rule (see chapter 13).

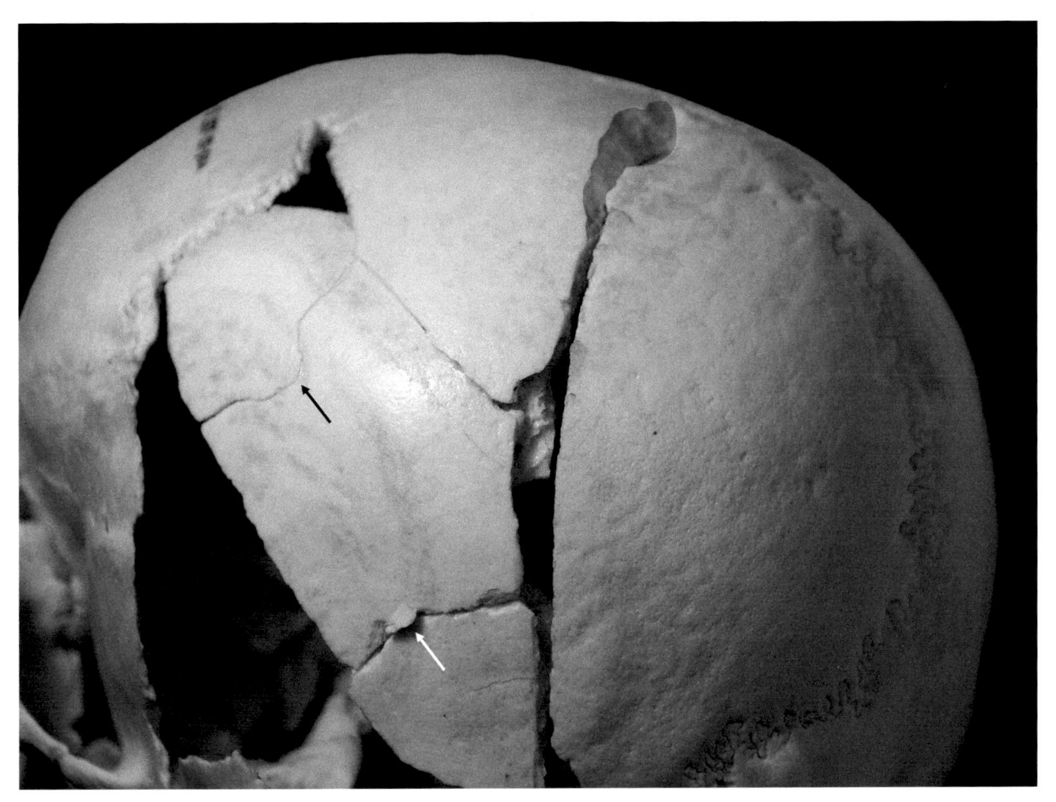

Figure 14.12. Sharp force trauma with a long implement, likely a machete. Note secondary fracture (black arrow), area of wastage (red tinting), and hinging (white arrow).

Wastage refers to pieces of bone that are removed from the primary impact site, similar to flaking/knapping with blunt force trauma (**Figure 14.12,** red tinting). In the case of sharp force trauma, wastage occurs when the object (usually heavy, such as an axe) is removed from the body after making the initial strike. The reverse motion detaches additional pieces of bone.

Hinging refers to the incomplete removal of bone fragments near the primary impact site (i.e., they are infractions). Hinging is diagnostic of a wet bone response and useful for differentiating perimortem and postmortem fractures (**Figure 14.12,** white arrow).

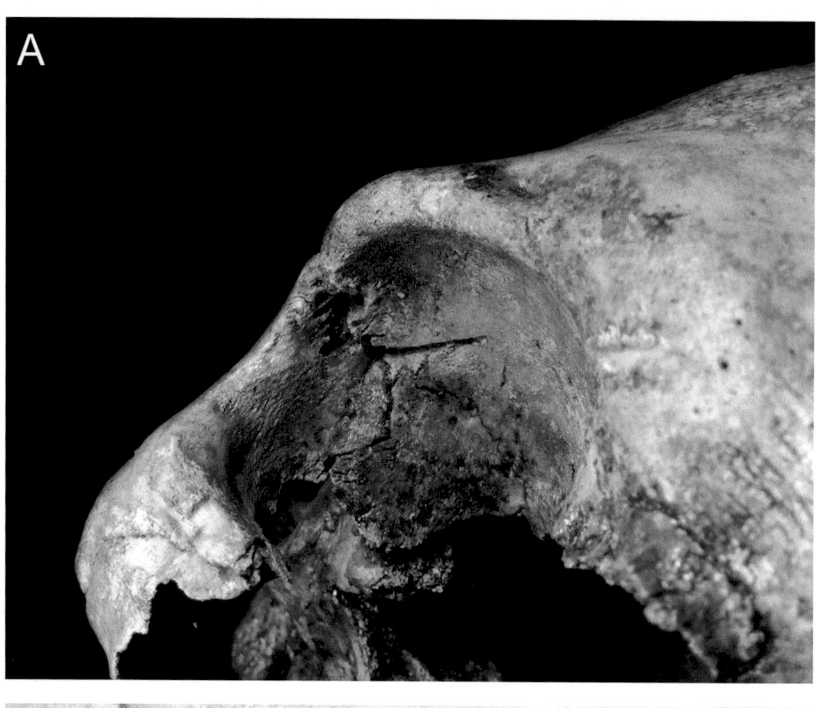

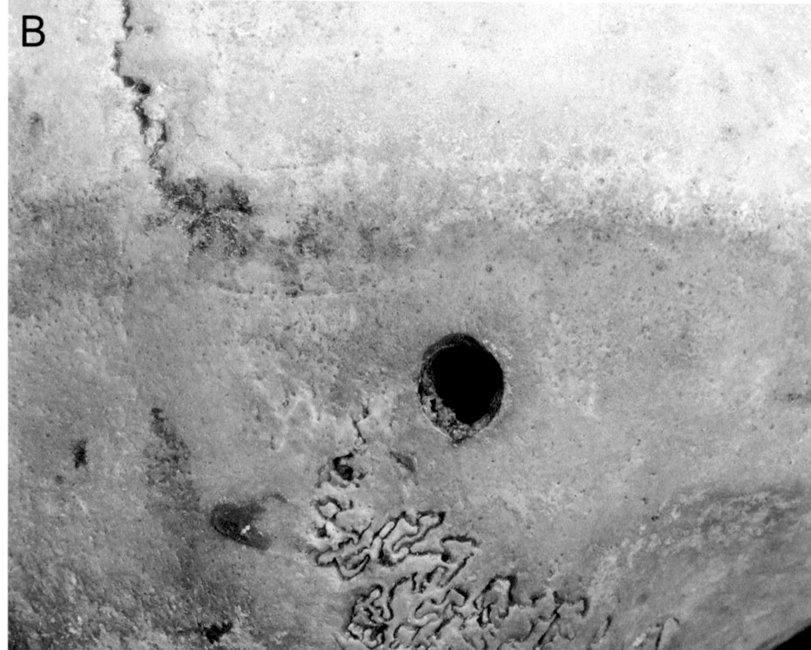

Figure 14.13. (*A*) Incision in the upper portion of the orbit. (*B*) Clean puncture wound to the skull. Note the lack of secondary fractures and the slight V-shape of the defect. Faint striations are visible on the lower right quadrant. These features are not observed in a gunshot wound.

Using these criteria, three different types of sharp force trauma can be described: incisions/cuts, punctures/stabs, and clefts/notches/chops.

Incisions/cuts are longer than they are wide and result from force being applied to the bone at an oblique angle. *Striations will be oriented horizontally* within the wound itself, and due to the low weight of most objects that cause incisions (knives, swords), there will be minimal secondary fracture lines, wastage, or hinging (**Figure 14.13A**).

Punctures/stabs result from objects being stabbed into the body at a perpendicular angle using objects such as closed scissors, knives, or ice picks. Punctures are approximately equal in their width and length, and the impact site can reflect the shape of the object used, especially in the cranial vault. Punctures have *vertically oriented striations* and limited wastage, although secondary fracture and hinging may be present if enough force was used to perform the initial strike (**Figure 14.13B**).

Clefts/notches/chops result from heavy instruments being used to attack an individual, which results in a *combination* of sharp and blunt force trauma characteristics. These include axes and other heavy objects. Because chopping instruments will usually impact the body at a perpendicular angle, they result in *vertically oriented striations*. The weight of the object means that secondary fracture lines, hinging, and wastage are much more likely to occur.

Note that hacking trauma tends to remove large pieces of bone, as opposed to blunt force trauma for which the damaged bone is often fragmented and pushed into the braincase. One of the primary differences between blunt and sharp force trauma is that sharp force objects (swords, machetes) produce long impact sites with clear, straight lines. This is rarely seen with a blunt force object.

Decision Tree for Differential Diagnosis

Differential diagnosis is a concept borrowed from the medical literature. The process of performing a differential diagnosis is to use the individual's personal information (the decedent in a forensics case) to establish a context (the crime scene) for comparing what is observed (the bone fracture patterns) to a known set of possible causes (experimental work on trauma).

One way to organize a diagnostic procedure is to create a **decision tree,** described as a series of Yes or No questions organized in a way to help decide upon a diagnosis. Information from the last three chapters was used to create a decision tree for trauma analysis. Please use **Figure 14.14** to answer the questions that follow.

This chart is relatively easy to use. Read each question or characteristic starting at the left, with the Yes and No responses guiding your path across the tree. The trauma diagnoses are in boxes at the right side of the tree.

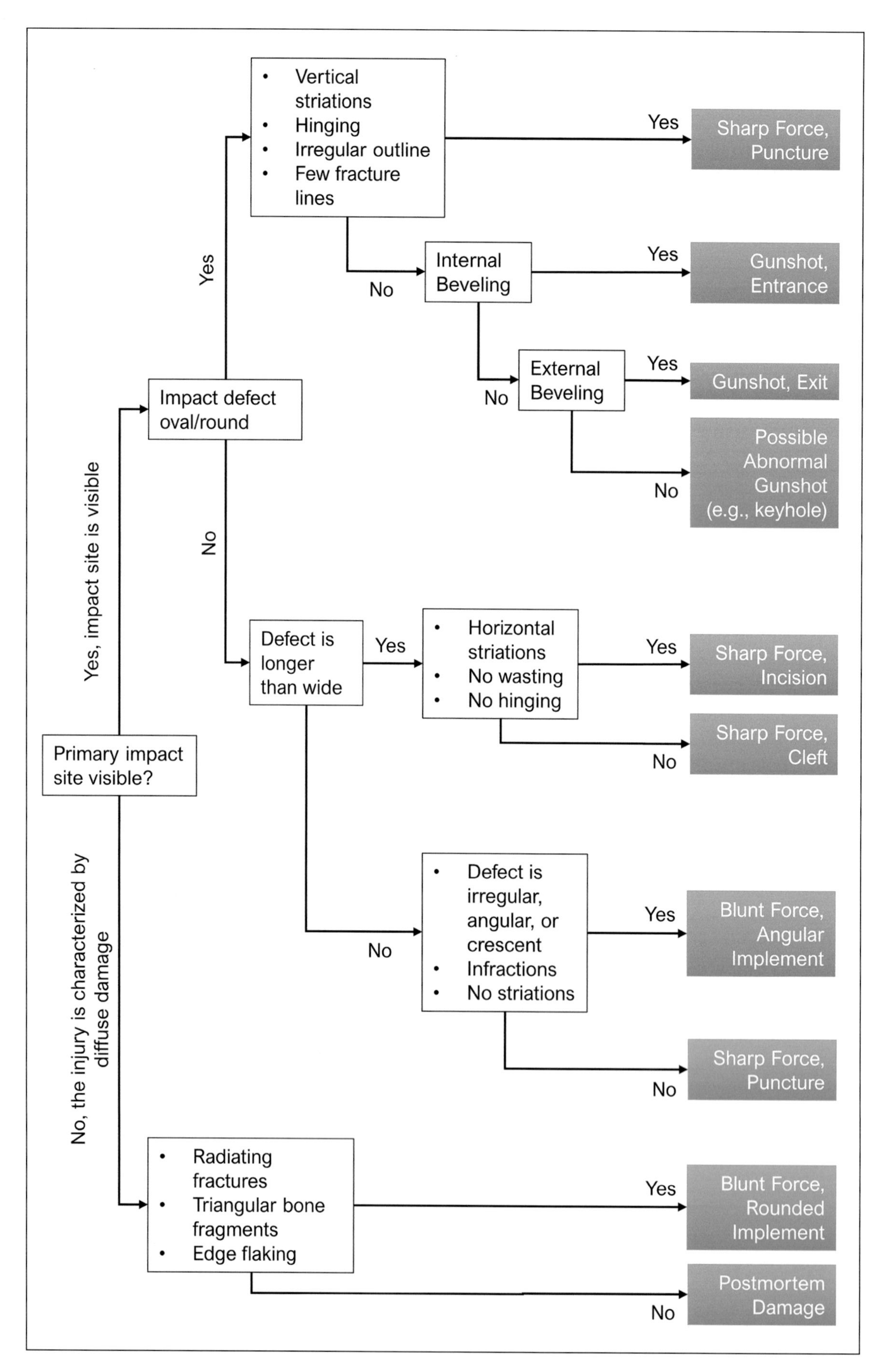

Figure 14.14. Decision tree for trauma analysis. Trauma diagnoses are in blue on the right side of the tree.

LEARNING CHECK

Q1. How would you characterize the damage seen on the nose of this individual? Note this was caused by a fist. Both views are of the same individual (**Figure 14.15**).

A) Sharp force trauma
B) Blunt force trauma
C) Projectile trauma

Q2. Now look very carefully at the bones of the nose in this individual shown in **Figure 14.15**. From what direction were they struck?

A) The blow came from the right side of this image.
B) The blow came from the left side of this image.

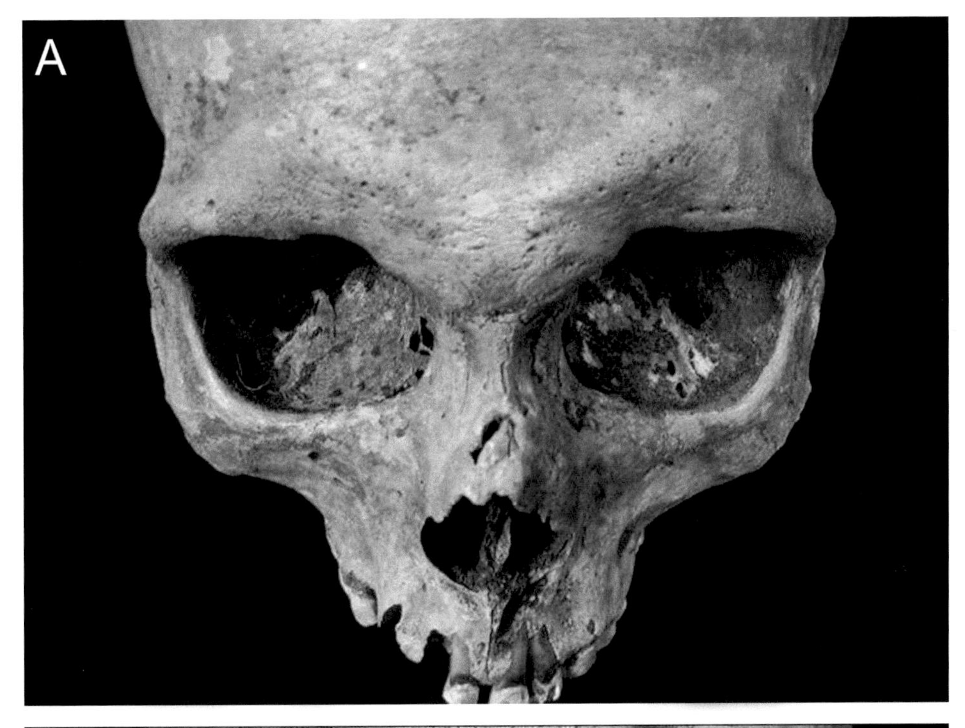

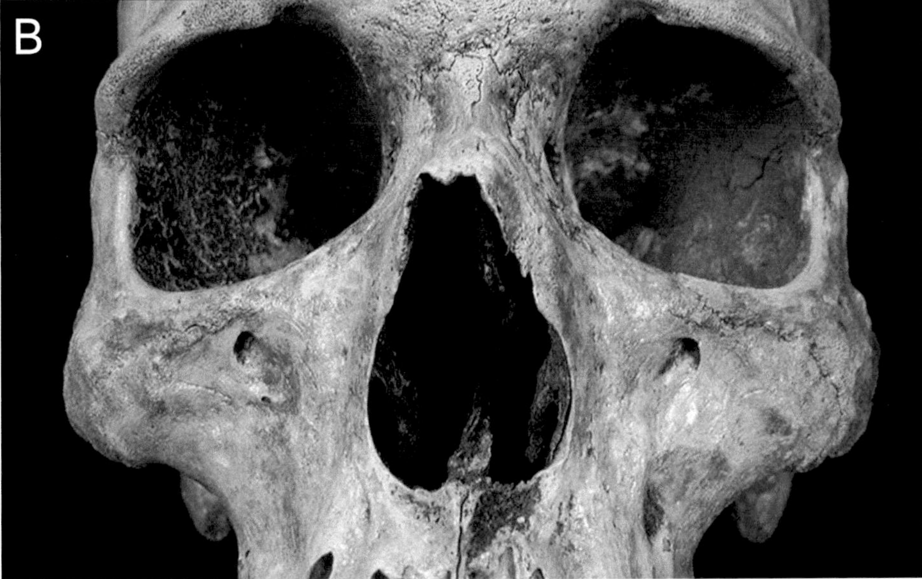

Figure 14.15. Two views of a cranium showing trauma to the nasal bones.

Q3. Two different weapon classes were used in the homicide shown in **Figure 14.16A**. Examine this image carefully and select two choices below. Look for key evidence of each trauma type as discussed in the last two chapters. Is there a bullet wound? Is there a straight fracture line associated with an edged weapon? Is there extensive comminution of the vault as caused by a blunt object?

A) Projectile weapon
B) Sharp force weapon
C) Blunt force weapon

Q4. Is there evidence of plastic deformation in the specimen shown in **Figure 14.16B**? Look closely at the primary impact site and any adhering bone fragments.

A) Yes
B) No

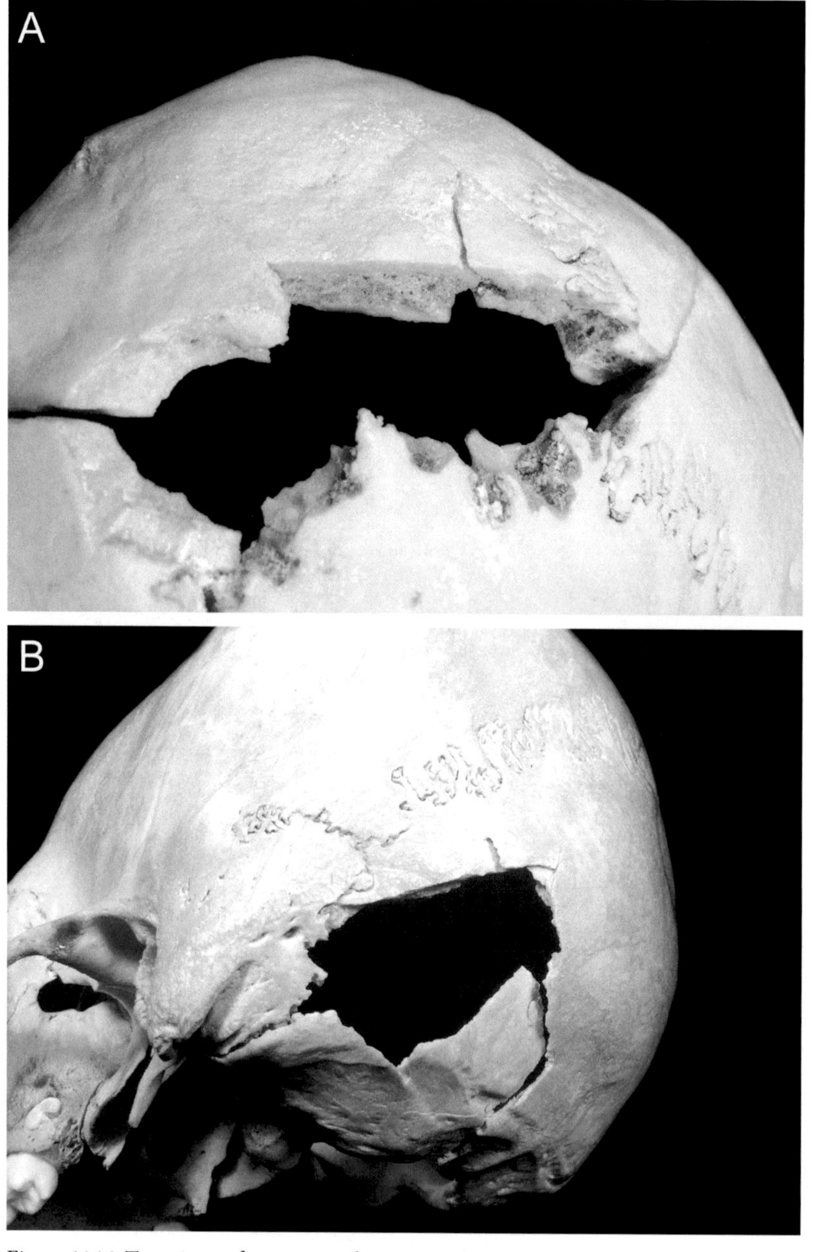

Figure 14.16. Two views of a cranium showing multiple wounds.

Q5. Based on the overall appearance of the wound on the skull shown in **Figure 14.17A**, how would you characterize it with respect to the type of trauma? Consider the area impacted and the evidence for healing.
A) Sharp force, perimortem
B) Sharp force, antemortem
C) Blunt force, perimortem
D) Blunt force, antemortem
E) Projectile, antemortem
F) Projectile, perimortem

Q6. Based on the overall appearance of the wound shown in **Figure 14.17B**, how would you characterize it with respect to the type of trauma?
A) Sharp force
B) Blunt force
C) Projectile

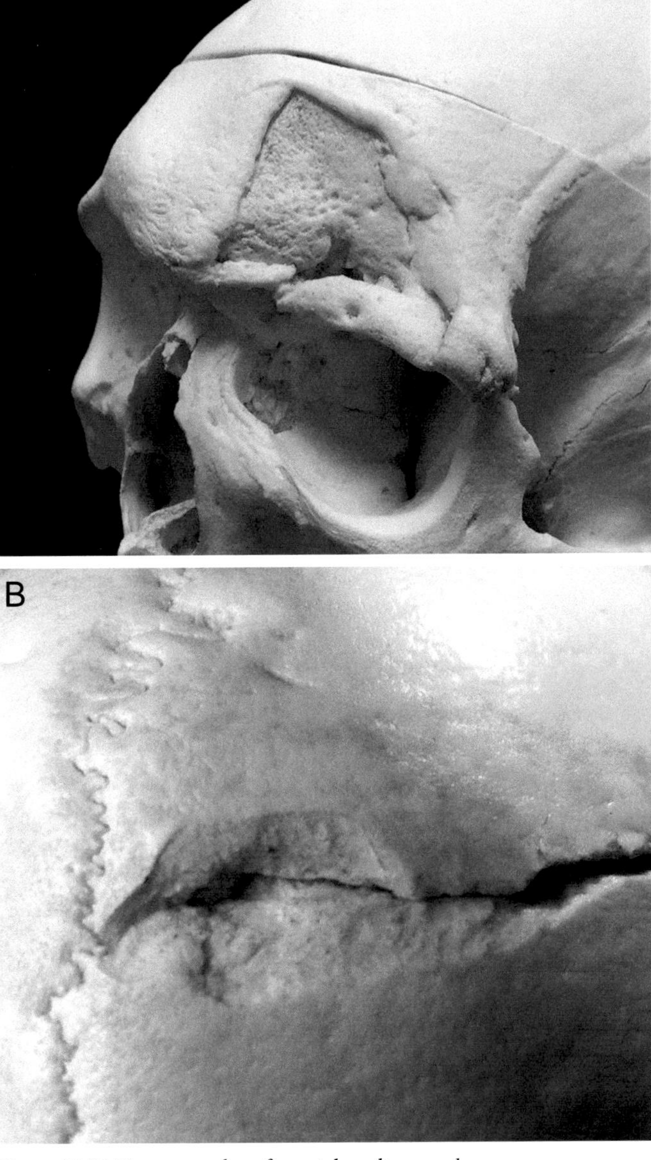

Figure 14.17. Two examples of cranial vault wounds.

Q7. **Figure 14.18A** depicts a perimortem injury. What feature of perimortem injuries is indicated by the black arrow?
 A) Wastage
 B) Hinge fracture
 C) Secondary fracture

Q8. How would you characterize the trauma to the mandible shown in **Figure 14.18B**?
 A) Perimortem sharp force
 B) Perimortem blunt force
 C) Antemortem blunt force
 D) Antemortem sharp force

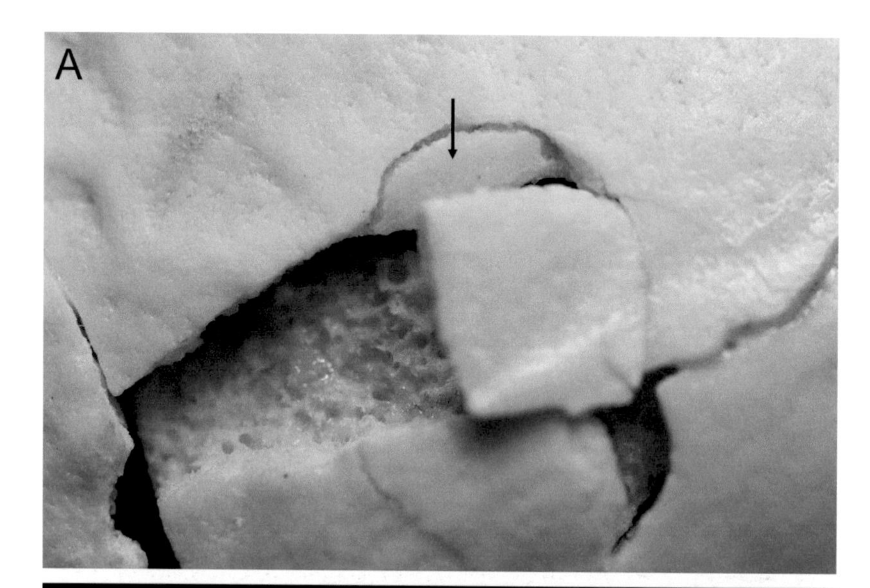

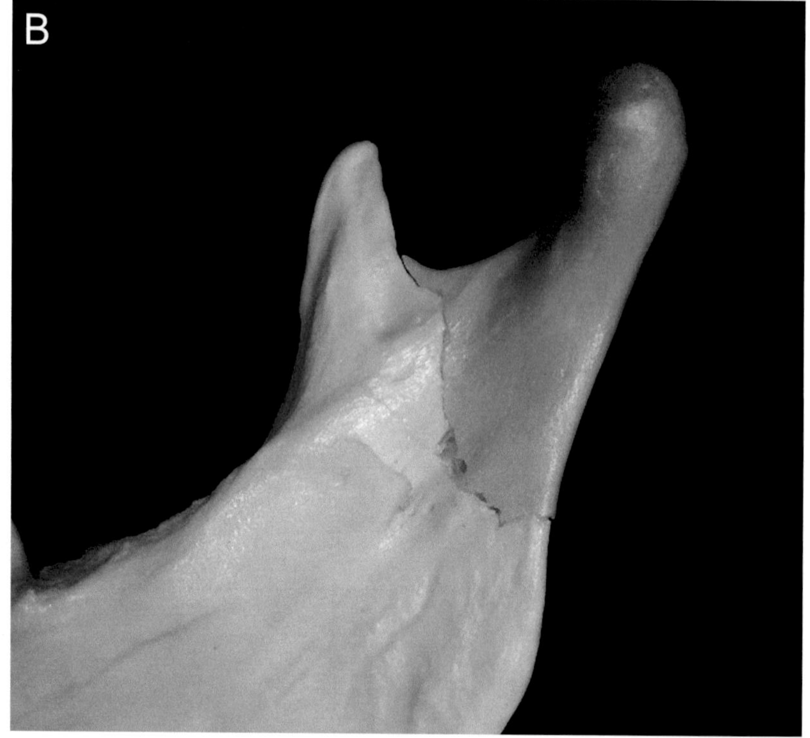

Figure 14.18. Two examples of perimortem trauma.

Q9. What type of Le Fort fracture is shown in **Figure 14.19A**? Compare to **Figure 14.5** and pay attention to the bones present on the side of the cranium.
A) Le Fort I
B) Le Fort II
C) Le Fort III

Q10. How would you characterize the trauma shown in **Figure 14.19B**? Focus on the primary impact site.
A) Sharp force
B) Blunt force

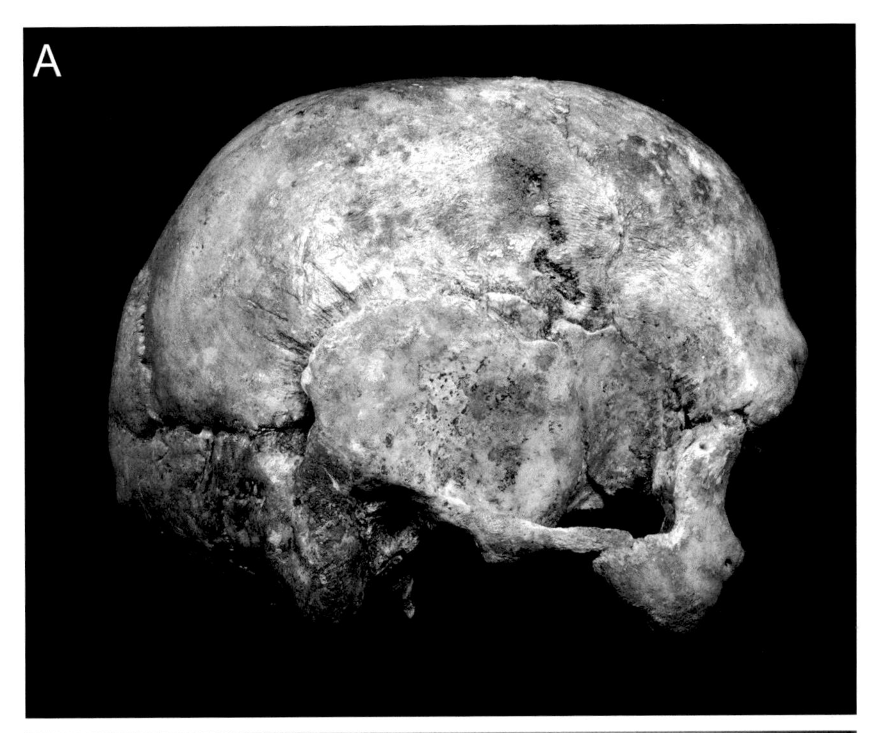

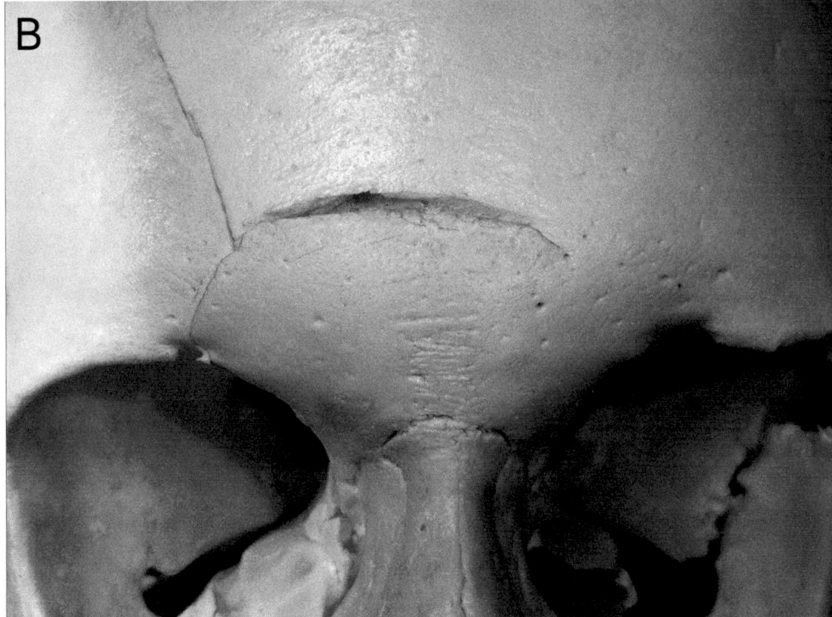

Figure 14.19. Two examples of cranial damage.

Wound Shape and Weapon Identification

In some cases, the shape of the weapon used to attack someone via blunt or sharp force trauma can be discerned from the wound characteristics. Although it is rare that length is evident from skeletal blunt force trauma, the weapon width can be seen in some cases. This tends to be more so with angular weapons that have a sharp edge. Angular weapons tend *not* to produce as many fracture lines because the energy is focused on a more focal area (as in sharp force), but they do tend to leave impact impressions with identifiable shape characteristics.

Consider **Figure 14.20**. The shape of the wound at the primary impact site in **Figure 14.20A** has a distinct angulation, which in this case reflects the type of weapon used (a tire iron or nail pull). A similar kind of wound is seen in **Figure 14.20B**. Note the crescent hinge fracture and the distinct 90-degree angulation. This type of injury could potentially be matched to a specific weapon. An ice pick or nail driven into the body will create circular wounds with a cone-shaped cross section, although this will not be evident if the bone is very thin (**Figure 14.20C**). Such defects typically lack the secondary fracture lines and beveling that characterize gunshot wounds. Aside from the shape of a defect, the patterning of wounds can also provide information on the type of implement used. The pairing of the defects shown in **Figure 14.20D**, for example, suggests a crowbar or the back of a claw hammer.

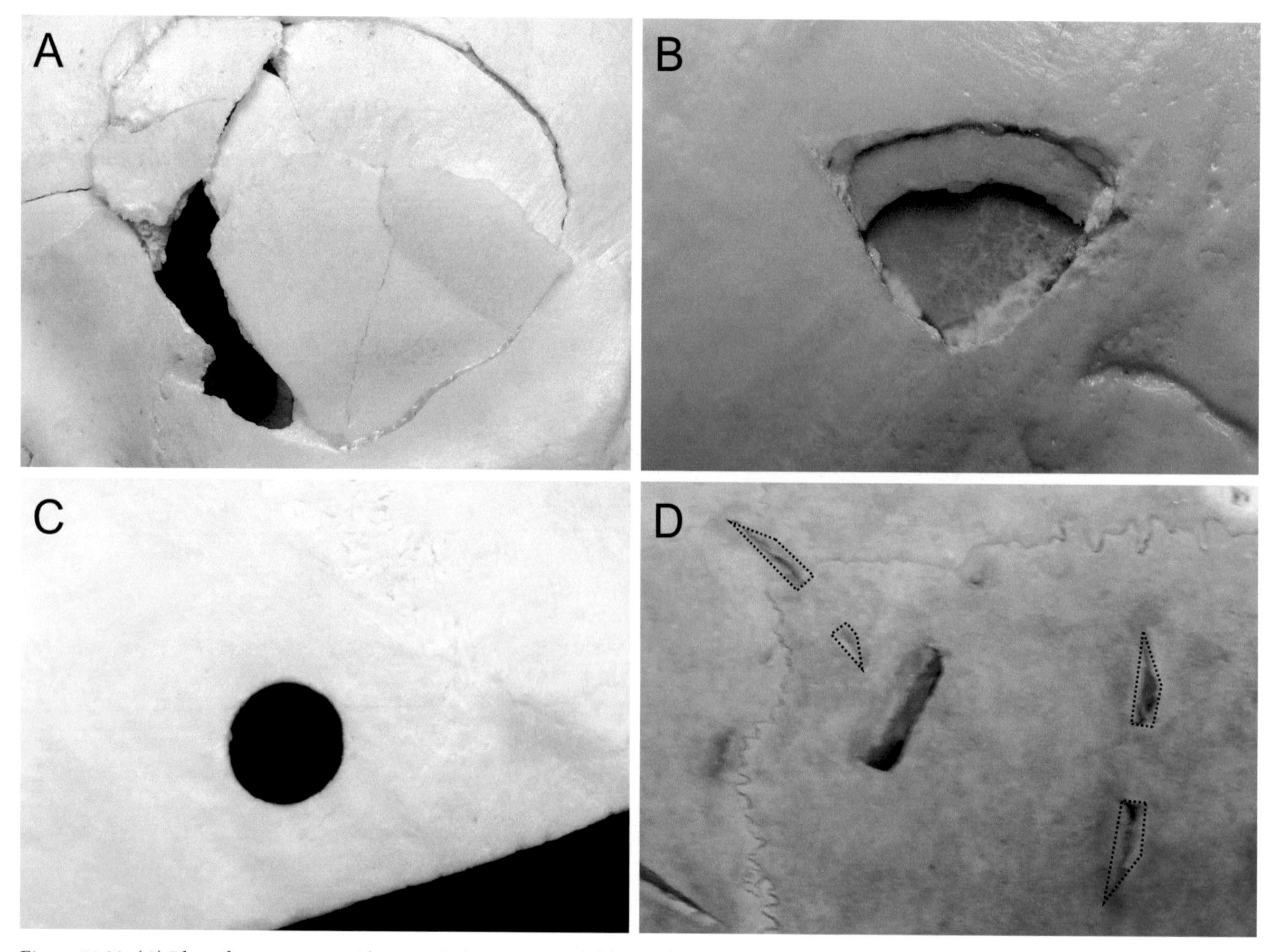

Figure 14.20. (*A*) Blunt force trauma with an angled instrument. (*B*) Blunt force trauma, focal injury with distinct shape. (*C*) Sharp force trauma where the wound shape is diagnostic of the weapon used. (*D*) Pairing of defects (dotted outlines) indicating the implement used.

LEARNING CHECK

Q11. Is there evidence of hinging present in **Figure 14.21A**?

A) Yes
B) No

Q12. In **Figure 14.21A**, what is the likely cause of this trauma? Consider the asymmetrical shape of the wound and what object is likely to cause this.

A) Keyhole gunshot wound
B) Blunt force trauma
C) Sharp force puncture

Q13. How would you characterize the weapon that caused the injury shown in **Figure 14.21B**?

A) Rounded object for blunt force trauma
B) Focused object like an ice pick
C) Hacking instrument like an axe

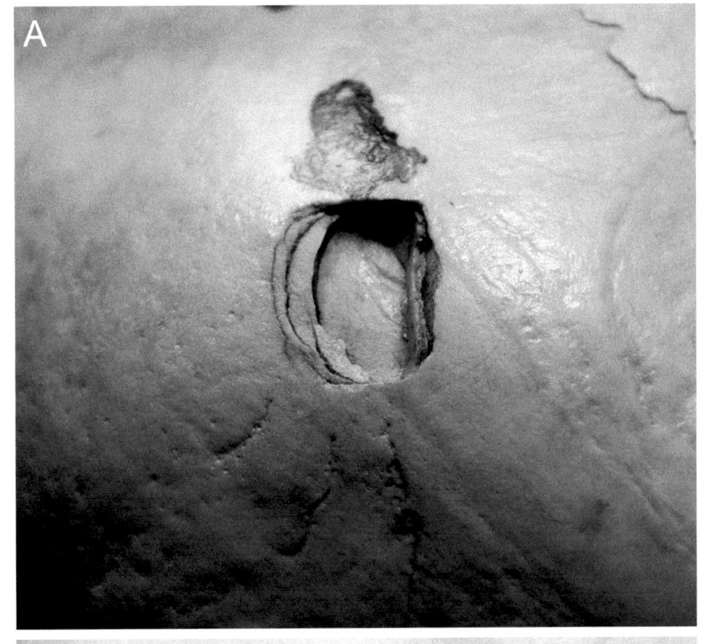

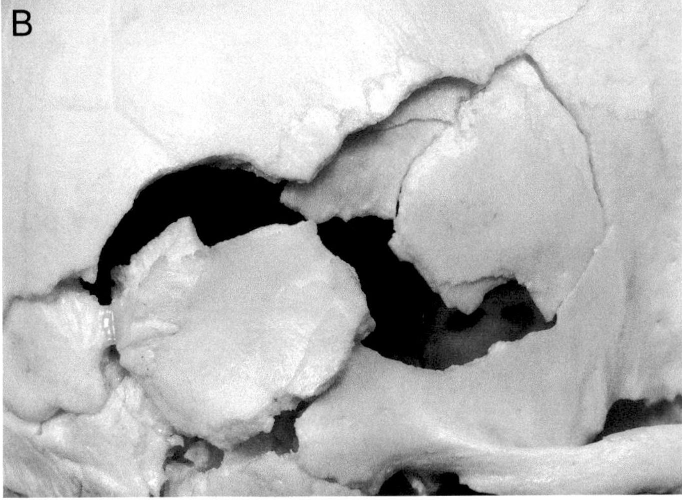

Figure 14.21. Two close-up views of cranial vault damage.

Dismemberment and Saw Marks

Dismemberment is the intentional separation of body segments, often performed to hide the identity of a victim. Because most dismemberments occur after the individual has died, they technically are part of the taphonomic history of a body, but some anthropologists categorize dismemberments as *sharp force trauma*. Dismemberment usually involves cuts to joints done by cutting (e.g., knives), chopping (e.g., axes), and chiseling implements (e.g., saws). The purpose is to reduce the size of the body to make it easier to hide, move or dispose of, or to disgrace the victim through desecration. However, when a body has been dismembered, it often creates more evidence for law enforcement to link back to the perpetrator because the dismemberment is usually done with a saw that adds to the evidence trail.

The principal goal of a forensic saw mark analysis is to identify the implement. Forensic anthropologists have attempted to generate predictive signatures for different types of saws, which include two broad classes—hand powered saws and power saws. There are many different types of hand saws with blades designed for specific tasks: cross-cuts, rips, metal, hack, concrete, etc. Power saws include bandsaws, chainsaws, and a variety of power tools used in the construction trades (circular saws, table saws). The shared feature among all saws is the presence of teeth that interact with the object being cut (**Figure 14.22**). Saw mark analysis evaluates features of saw marks on bones that can be used to infer three diagnostic characteristics:

1) The number of teeth per square inch and their size.
2) The angulation of the teeth and blade thickness.
3) The power source of the saw.

Dismemberment Cuts

As the saw teeth cut into bone, a **kerf** is formed. A kerf is the groove created in an object that is being cut **(Figure 14.23)**. Analyzing the characteristics of the kerf, the kerf wall, and the kerf floor is the primary goal of forensic cutmark analysis.

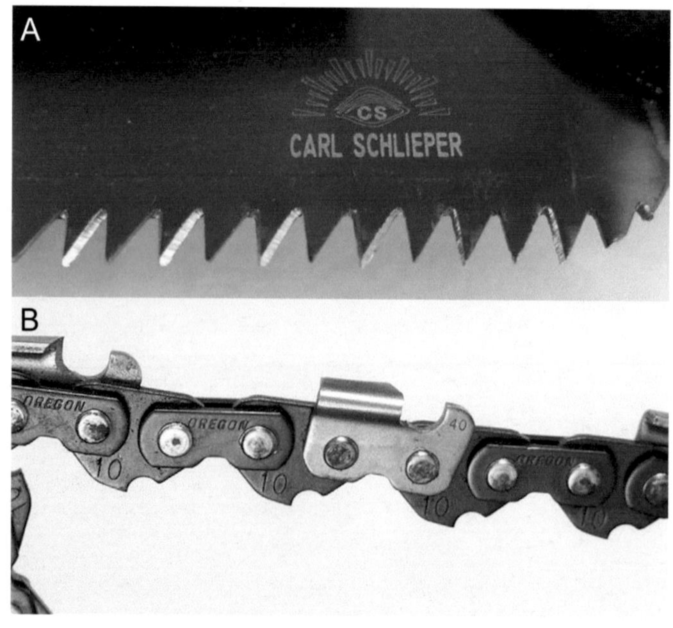

Figure 14.22. Two types of saw blades. (*A*) saw teeth on a cross-cut saw, (*B*) chainsaw blade.

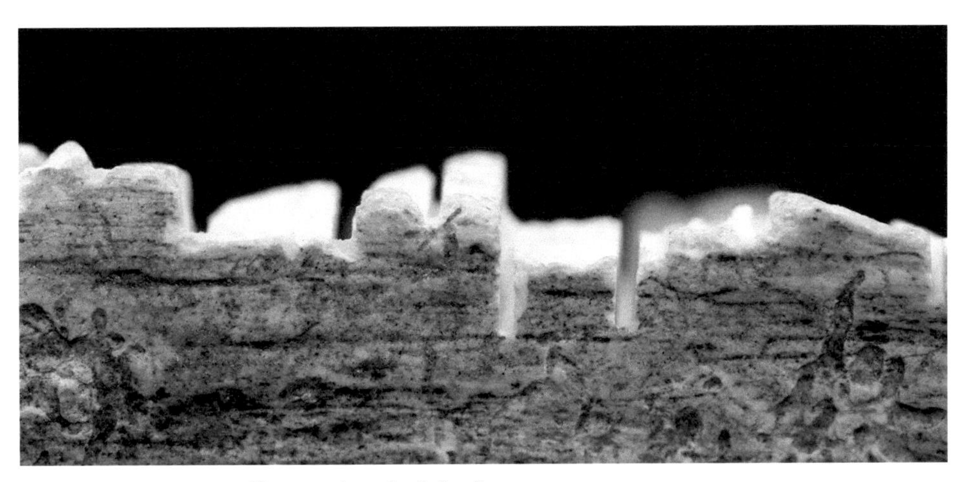

Figure 14.23. Fragment of bone with multiple kerfs present.

Saws cause three types of cuts on bone that are useful in a forensics investigation (**Figure 14.24**):

1. **Superficial false start scratches** are created when a saw blade bounces across the object and creates grooves or incisions on the bone surface.
2. **False start kerfs** are created during the push stroke. These represent aborted attempts to initiate a cut and are more common with hand saws. False start kerfs will often be parallel with the cut bone section.
3. **Sectioned bone cuts** are deep kerfs that have not completely divided the bone into two pieces. These are most useful for a forensic analysis because you can see the width of the blade, examine the walls of the kerf, and examine the floor of the kerf. Note in **Figure 14.24** that the cut does not completely divide the bone into two pieces.

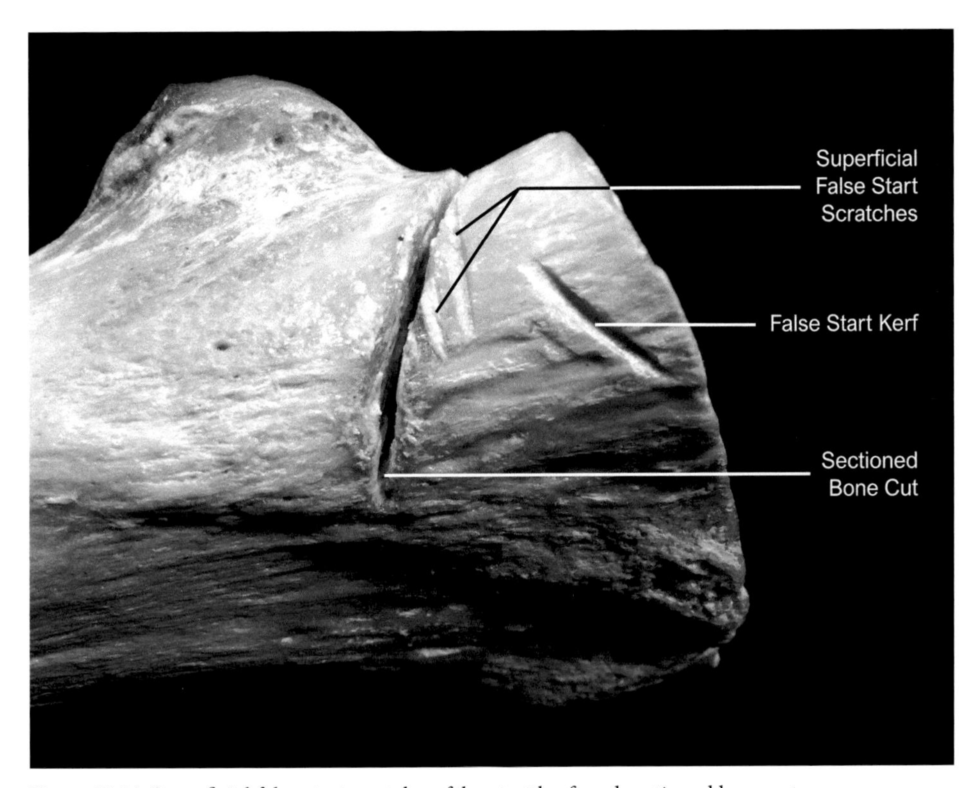

Figure 14.24. Superficial false start scratches, false start kerf, and sectioned bone cut.

Analysis of Saw Marks

The analysis of saw marks should focus on identifying five different features or aspects of the investigation.

1) Direction of Cut. Inferring the direction of cut may provide information on handedness and the relationship between the assailant and the victim. The direction of cut is based on the presence of false start scratches, which will be located on the surface of the bone where the cut initiated. In the absence of false start scratches some cuts will also have a **breakaway spur,** which is located on the opposite surface from where the cut was initiated (**Figure 14.25**). Breakaway spurs form because there is a tendency for the bottom of the cut to break away before the cut is complete due to the weight of the object. The size of the breakaway spur depends on the amount of force applied across the bone. Heavier objects usually have more leverage, and the more leverage applied, the larger the breakaway spur will be. For instance, the weight of a handheld circular power saw or chainsaw (offering leverage) often produces a larger breakaway spur than a saw which does not provide leverage.
2) Number of Teeth Present. Kerf walls contain information on the number of teeth per square inch and their size. The general rule is that kerf walls with rough, uneven appearances result from saws with fewer and larger teeth. Smooth kerf walls suggest the saw had both smaller teeth and more teeth per square inch. In **Figure 14.26A**, the large striations seen in the kerf wall suggest the saw had larger teeth. In addition, increasing the force applied to a saw will produce striae that are spaced further apart because greater force causes deeper swaths to be cut into the bone.

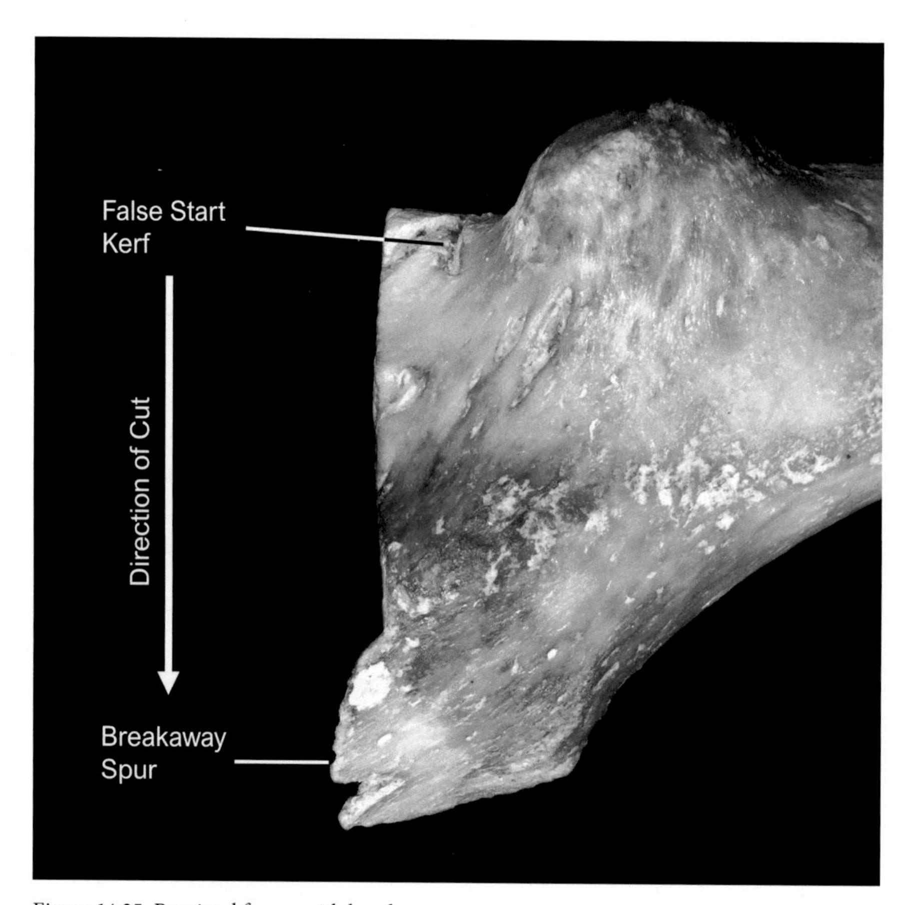

Figure 14.25. Proximal femur with breakaway spur.

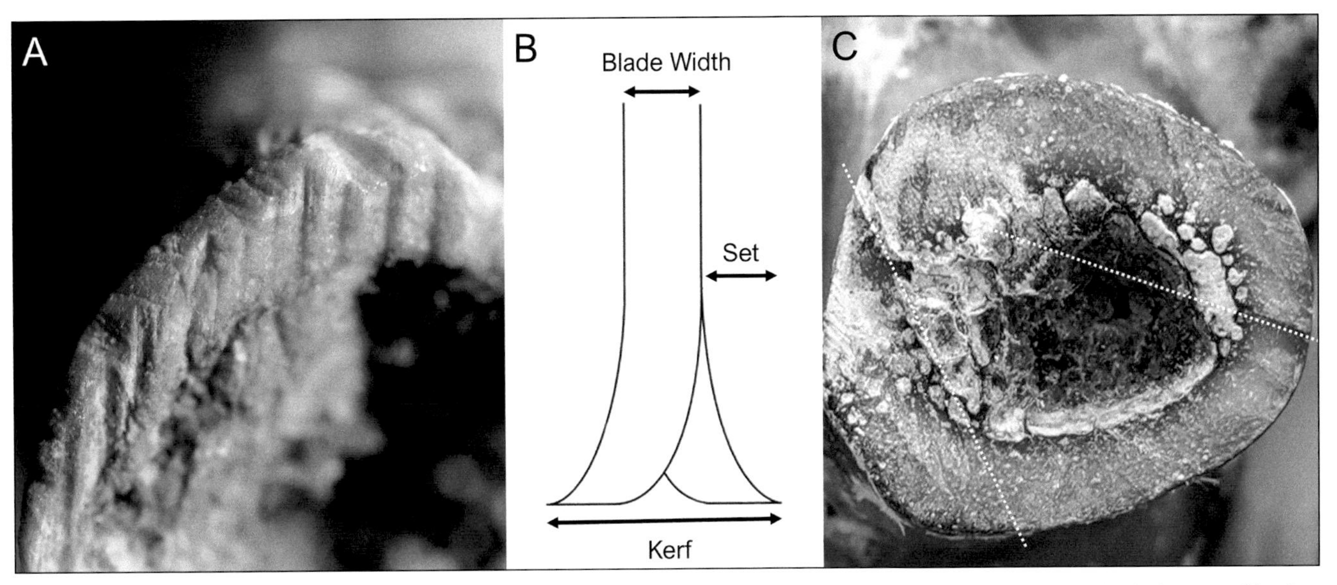

Figure 14.26. (*A*) Striations in a rough saw cut section. (*B*) Schematic demonstrating the relationship between saw blade thickness and kerf width. (*C*) Changes in striae direction on a cut section of long bone.

3) Width of the Blade. A saw's "**set**" is defined by the teeth that are bent laterally to one side of the blade. Set is represented by striations in the kerf wall. Tooth set creates a kerf wider than the saw blade, but no wider than 1.5 times the thickness of the blade. Estimation of blade width can usually only be done for intact kerfs. **Figure 14.26B** shows how the blade width will always be smaller than the resulting kerf because of the angulation of the teeth on the saw blade.
4) Shape of the Blade. This refers to whether the blade is on a fixed radius like a circular saw, or if the blade is a non-fixed radius such as hand saw or chainsaw. The forensic anthropologist can infer blade shape by the striae within the kerf wall. With a fixed radius blade, the striae will have the same orientation throughout the kerf wall. With a non-fixed radius blade the striae will vary in orientation. Consider **Figure 14.26C**. The dotted lines indicate striae orientation, which do not always orient in the same direction. This suggests a non-fixed radius saw was used. The change in orientation is needed to prevent the saw from binding and getting stuck within the cut wall. One must consider that these acts are being performed on a fully fleshed body and the bone itself is not visible when the decedent is being dismembered. So, in addition to the assailant changing their posture during the dismemberment, the body itself may be repositioned and moved during the act of dismemberment.
5) Energy Source. The goal is to distinguish between hand saws and power saws to help narrow the search for the object (which will have DNA preserved on it that can be linked to the crime scene and victim). Hand saws produce cruder, uneven kerf walls with striae that change direction as the saw position is adjusted during the cutting stroke. The striae are clear and deep. Power saws are easier to use and therefore tend to produce smooth kerf walls with smaller, finer striae that orient in the same direction. Chainsaws are the obvious exception. **Figure 14.27** demonstrates the differences between a cut created by a hand saw and by a power saw. **Figure 14.27A** is a hand saw. The cut wall is uneven, and the striae are deep and gouge the bone creating an uneven surface (although they do orient in the same direction). **Figure 14.27B** is more consistent with a power saw. Note how smooth-walled and even the cut surface is. Barely any striae are visible. This suggests a small-toothed power saw was used.

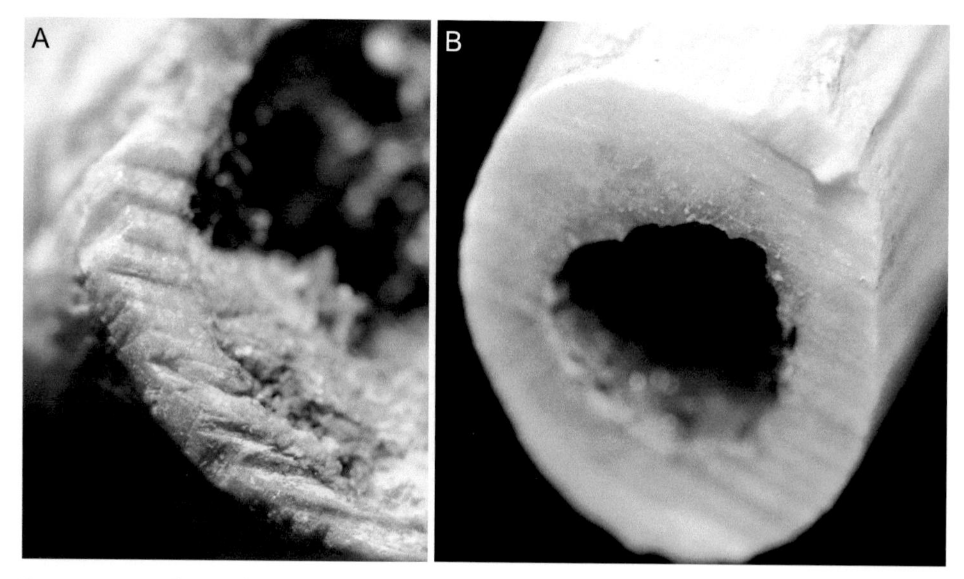

Figure 14.27. Difference between a hand saw (*A*) and power saw (*B*) in terms of the characteristics of the cut wall.

LEARNING CHECK

Q14. Based on the appearance of the kerf shown in **Figure 14.28,** would you say a hand saw or a power saw was used?
A) Power saw
B) Hand saw

Q15. What other feature of cutmark analysis would you be able to discern in **Figure 14.28**?
A) Width of the blade
B) Number of teeth present on the saw
C) Size of individual saw teeth

Q16. Assuming this cut is 1.5 mm wide, what is the *maximum* width of the blade used to make this cut?
A) 2 mm
B) 1 mm
C) 3 mm
D) .50 mm

End-of-Chapter Summary

This chapter discussed the different types of fractures that can impact the skeleton, with a focus on the direction of force and how this can be inferred based on wound characteristics. Blunt force trauma entails an impact to the body involving an object that affects a relatively wide area. Examples include bats, bricks, and hammers. Motor vehicle collisions and falls are also considered blunt force trauma. Sharp force trauma was defined with respect to the differences between incisions, punctures, and clefts. The characteristics of the wound help differentiate the types of sharp force objects used to attack a victim. Clefts combine some sharp force and blunt force characteristics because these types of injuries are caused by a heavy object with a sharp edge (such as an axe). In some cases, the wound shape can help determine the actual object used to commit the crime. A special case of sharp force trauma,

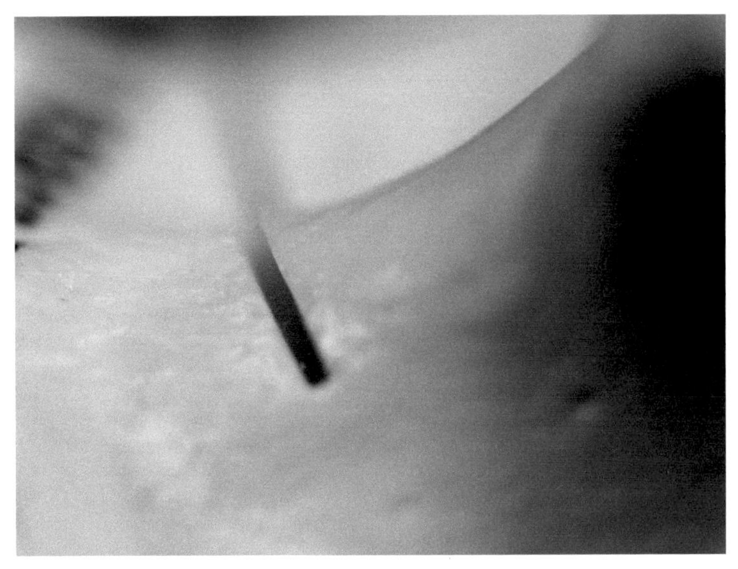

Figure 14.28. Close-up view of a cut in bone.

dismemberment, was also discussed. Analysis of saw cutmarks focusing on the location of cuts, kerfs, and kerf wall characteristics contributes to reconstructing the scene of dismemberment and identifying the saw type used in the act.

End-of-Chapter Exercises

Exercise 1

Materials Required: **Figures 14.14, 14.29,** and **14.30**

Scenario: Skeletal remains have been recovered from a shallow grave and submitted to you for analysis. You note damage to the cranial vault (**Figure 14.29**) as well as to the right radius and ulna (**Figure 14.30A,** anterior aspect; **Figure 14.30B,** posterior aspect).

Question 1: Using **Figure 14.14,** how would you classify the trauma that is shown in **Figures 14.29** and **14.30**?

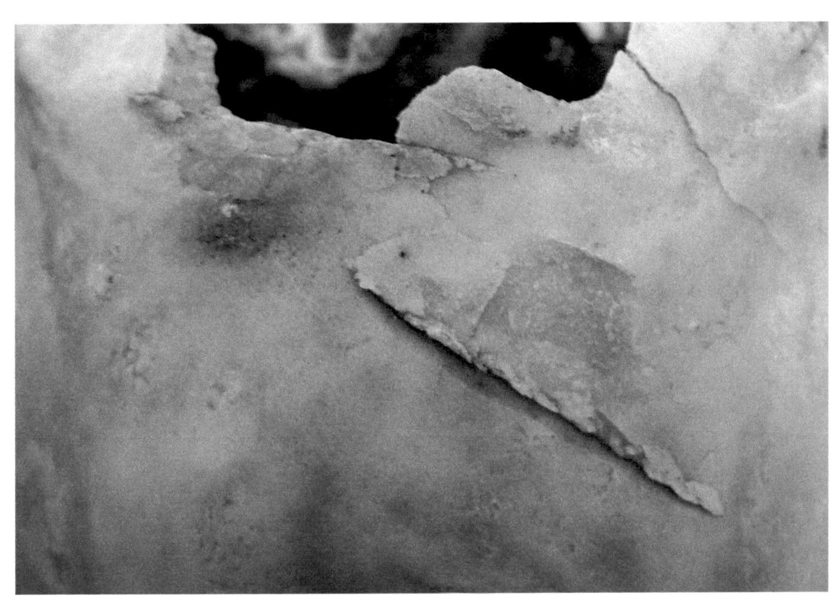

Figure 14.29. Close-up view of damage to the cranial vault.

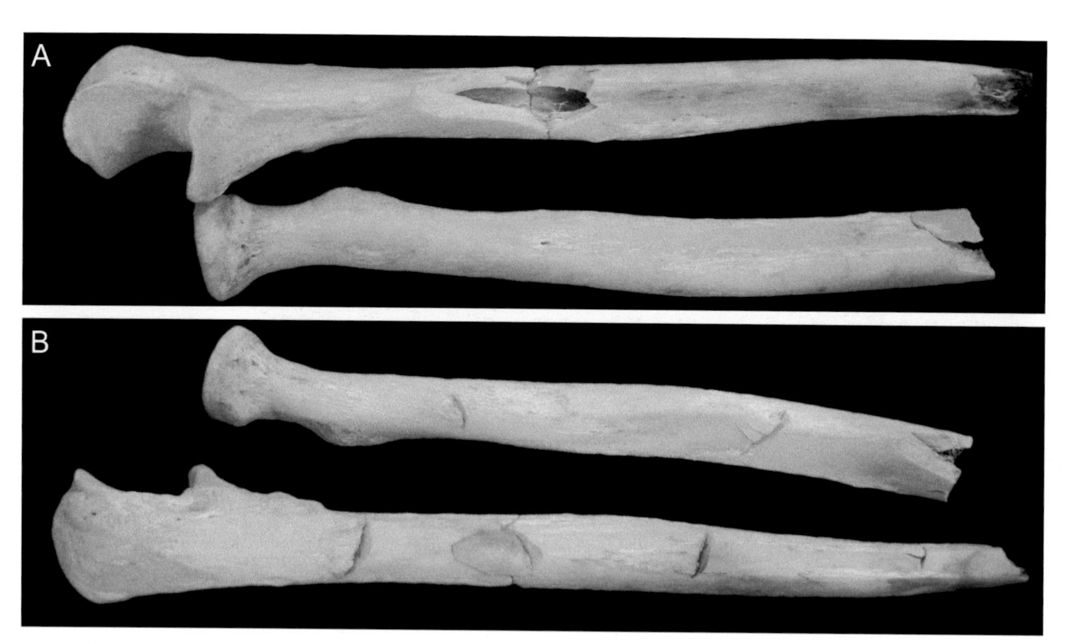

Figure 14.30. Two views of a radius and ulna. (*A*) Anterior aspect; (*B*) posterior aspect.

Question 2: How many separate impacts do you see on the cranial vault?

Question 3: How many separate impacts do you see on the bones of the forearm? Remember, these bones are articulated in life.

Exercise 2

Materials Required: **Figure 14.31**

Scenario: Multiple body segments were recovered from dumpsters spread across three neighboring cities. Eventually, enough segments were recovered that they could be reassembled and confirmed to belong to the same dismembered decedent. Use the body segments shown in **Figure 14.31** to answer the following questions.

Question 1: Figure **14.31A** shows the cut surface from a proximal femur. Look closely at the cut surface and its edges. What features can you observe that might tell you something about the number of teeth or the energy source of the saw used for this dismemberment? What features can you observe that can help you infer the direction of the cut?

Question 2: Look closely at **Figure 14.31B**. This is a picture of the anterior aspect of the left humerus. Based on what you can observe here, in which direction (e.g., posterior-to-anterior, medial-to-lateral) did the perpetrator *initially* attempt to cut through this bone? In which direction did the *final* cut through the bone proceed?

Question 3: Assuming that the reconstructed kerf in **Figure 14.31B** is 1.8 mm wide, what is the maximum width that the saw blade used for the dismemberment could be? Does anything make you doubt this number?

Exercise 3

Materials Required: **Figure 14.32**

Scenario: Over the weekend, investigators have recovered the skeletons of two individuals from separate locations. They are concerned that both individuals represent homicide cases and have asked your expert opinion. The cranial remains of these individuals are shown in **Figure 14.32.** Note that Individual A's cranium is shown from a superior aspect with the front of the cranium toward the bottom of the image.

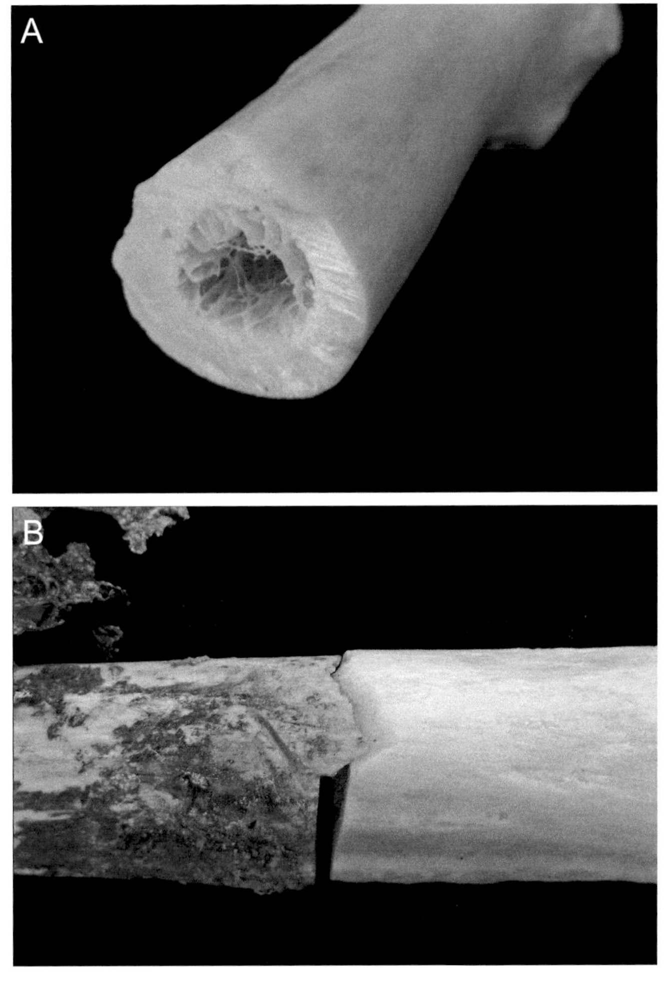

Figure 14.31. Two examples of long bones with cutmarks.

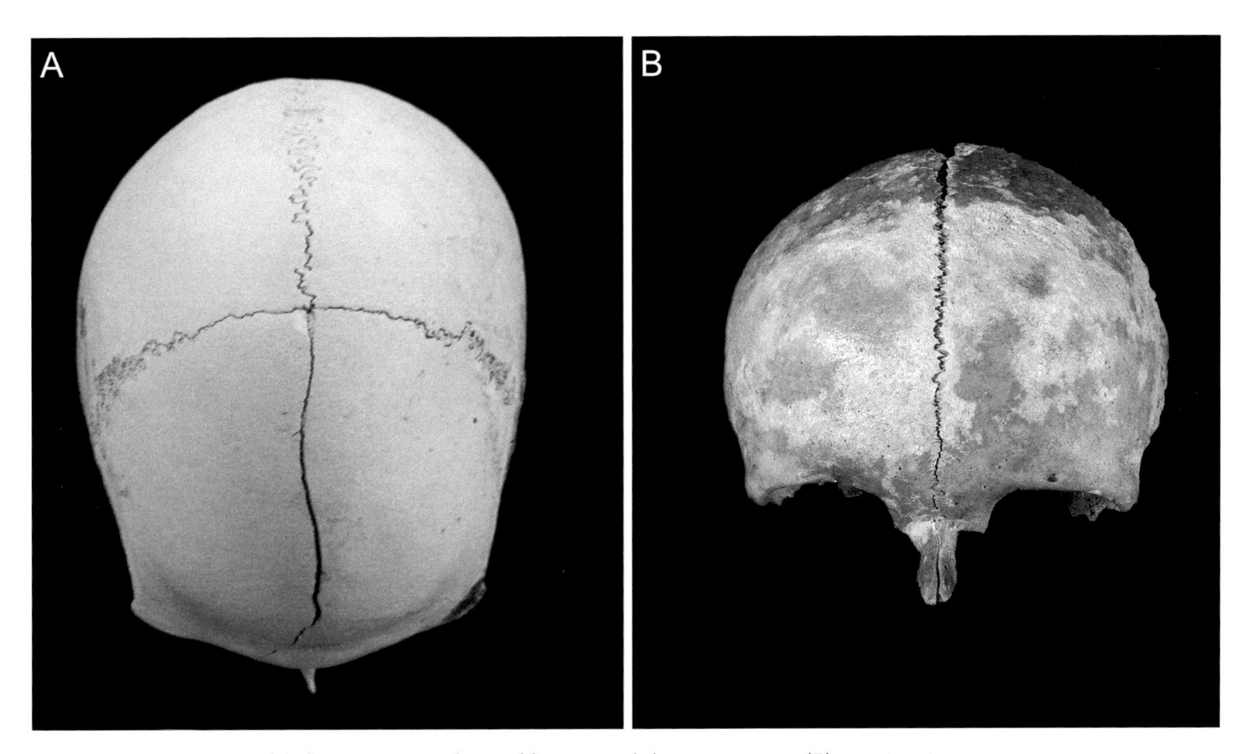

Figure 14.32. Two views of different crania with possible trauma. (*A*) superior view; (*B*) anterior view.

Question 1: Look closely at these two images, being sure to consider fracture characteristics and the behavior of radiating fractures with regard to cranial buttressing. In your opinion, are either of these cases likely to represent a homicide case? Please support your conclusion with observations.

References

Berryman HE, Shirley NR, Lanfear AK. 2013. Low-velocity trauma. In: Tersigni-Tarrant MA, Shirley NR (Eds.), *Forensic Anthropology: An Introduction.* Boca Raton, FL: CRC Press. p. 271–290.

Dixon AF. 1912. *Manual of Human Osteology.* New York, NY: William Wood and Co.

Fulginiti LC, Seidel AC, Bolhofner KL. 2021. Fusion and fracture: forensic implications of the hyoid bone. *Proceedings of the 73rd Annual Scientific Meeting of the American Academy of Forensic Sciences:* 33.

Hart GO. 2005. Fracture pattern interpretation in the skull: differentiating blunt force from ballistics trauma using concentric fractures. *Journal of Forensic Sciences* 50: 1276–1281.

Le Fort R. 1901. Étude expérimentale sur les fractures de la machoire supérieure. *Revue de Chirurgie Paris* 23: 208–227, 360–379, 479–507.

Pollanen MS, Chiasson DA. 1996. Fracture of the hyoid in strangulation. Comparison of fractured and unfractured hyoids from victims of strangulation. *Journal of Forensic Sciences* 41: 110–113.

Reichs KJ. 1998. Postmortem dismemberment: recovery, analysis and interpretation. In: Reichs KJ (Ed.), *Forensic Osteology. Advances in the Identification of Human Remains,* Second Edition. Springfield, IL: Charles C. Thomas. p. 353–388.

Symes SA, L'Abbé EN, Chapman EN, Wolff I, Dirkmaat DC. 2012. Interpreting traumatic injury to bone in medicolegal investigations. In: Dirkmaat DC (Ed.), *A Companion to Forensic Anthropology.* London: Blackwell. p. 340–389.

Ubelaker DH. 1991b. Hyoid fracture and strangulation. *Journal of Forensic Sciences* 37: 1216–1222.

15

Bone Taphonomy and Establishing the Postmortem Interval

Learning Goals

By the end of this chapter, the student will be able to:

- Describe the field of taphonomy and how it applies to forensic anthropology.
- Define the concept of postmortem interval.
- Define the stages of decomposition.
- Articulate the variety of factors that affect the rate of decomposition.
- Identify and differentiate between carnivore and rodent marks on bone.
- Identify the signatures of fire on bone and how the temperature of the fire can be assessed.
- Describe the process of bone weathering.

Introduction

Forensic taphonomy is the study of the postmortem changes to human remains that occurred after the individual died but before the body was recovered for analysis in the lab. Everything that affects the skeleton in the interim can be considered a taphonomic effect. Examples most relevant to a forensics context include: dismemberment (see chapter 14), intentional burning, animal scavenging and trampling, bone weathering and sun bleaching, bone sorting by moving water, soil conditions such as acidity and mineral content, and the effects of plant roots. The central question of forensic taphonomy is: *What happened to turn a living person into the skeleton currently being analyzed in the lab?* Understanding the range of factors that can affect what is observed on the human skeleton is important for several reasons, including differentiating natural (e.g., animal trampling, weathering) from trauma-induced signatures on bone, and documenting the environmental conditions in which a body decomposed. Taphonomy can also help establish if a body was dug up and moved or mishandled by law enforcement personnel.

Assessing the Postmortem Interval

Assessing the postmortem interval (PMI) is one of a suite of analyses that fall under the broad rubric of forensic taphonomy. The process of decomposition is part of the natural reduction of a body into simpler components once life's processes have ended. By studying the decomposition process, recording information on the stages the body goes through as it decomposes, the rate at which these stages are met, and the variety of factors that affect the rate of decay, forensic anthropologists can provide valuable information on PMI to the law enforcement community. However, it is important to stress that in fully fleshed and recently deceased bodies, the assessment of PMI falls under the purview of the forensic pathologist and will not enlist the expertise of a forensic anthropologist. It is only when the postmortem interval is long, resulting in partial or complete skeletonization, that a forensic anthropologist may contibute to estimating PMI.

In order to continue to refine our ability to estimate PMI, forensic anthropologists conduct research using body donation facilities where cadavers are used to record the process of decomposition under controlled conditions. The Anthropological Research Facility at the University of Tennessee-Knoxville is perhaps the most well known in the country (Vidoli et al., 2017) and has provided decades of data on human decomposition rates and processes (Bass, 1997; Rodriguez and Bass, 1983, 1985; Vass, 2011). This facility was founded by Dr. William Bass and continues to train forensic anthropologists in the methods of postmortem interval estimation. In recent years, other universities have established their own decomposition research facilities in recognition of the fact that temperature and climate significantly affect decomposition rates (Megyesi et al., 2005; Vass, 2011). Experimental research based on principles of scientific rigor is ongoing, with the goal of replacing the anecdotal evidence that was previously used to generalize about rates of human decomposition (Simmons, 2017) with scientifically tested methods that will stand up better in a court of law. Because experimental studies require access to human cadavers, the field has also used rabbits and pigs as model organisms because of their availability (rabbits) or similar body composition to humans (pigs).

The Relevance of PMI Estimation

Estimating the postmortem interval is important for several reasons. Foremost is that PMI provides an estimate of how long ago the individual died, which can help narrow the list of possible decedents for making a positive identification. This information is also useful for assessing suspect alibis and providing information on who was in the vicinity of the decedent at the time of their death. While this may generate a suspect list, it also provides a list of possible witnesses to the crime and helps corroborate witness testimony. Estimating PMI can also help in the case of a missing person's body that has not yet been located. Analysts can use information about the rate of decomposition to predict the state the body will be in when discovered. This could help guide the police in their search. PMI may also be important for establishing the order in which individuals passed away in multi-fatality cases, which could be critical for establishing rights of inheritance. Finally, accurately estimating the PMI is important for clerical reasons. It aids in filing the death certificate and providing data on when exactly an individual expired.

Dozens of methods for estimating PMI have been proposed over the years. However, many of these methods fail to satisfy the criteria needed for application by law enforcement. Methods need to be precise, repeatable, accurate, and expedient. Methods should generate quantitative data using statistical analyses that incorporate all relevant variables. Error rates should be known, and the method should always produce a range or interval of time rather than a single point estimate. For example, while complex biochemical methods can satisfy many of these criteria, they are not expedient nor are they cost efficient. The police need an assessment of PMI relatively quickly to begin narrowing the list of suspects. For this reason, it is better to produce a PMI estimate that is overly broad (bordering on useless) than to produce an estimate that is falsely specific and thereby steering law enforcement in an incorrect direction in their initial investigation. For example, if an overly confident analyst suggests a PMI of no more than 3 hours, law enforcement might use that confidence to rule out potential suspects that were in the area 5–6 hours ago, perhaps erroneously. In short, PMI as *exculpatory* evidence should be used with great caution, and it is for this reason that PMI alone is rarely the central argument in a legal proceeding.

Death and the Decomposition Process

Death, as they say, is a process not an event. As such, the "moment of death" is also considered somewhat difficult to define objectively. For this chapter's purposes, death will be defined as the irreversible process by which the cells of the body stop functioning and performing the tasks that make one "alive." The energy and nutrients that comprise the living organism are released back into the world through a complex process that involves the enzymes and bacteria found within the living body, as well as multiple external agents such as insects, animals, plants, and fungi that feed on the corpse.

Decomposition begins around four minutes after the final breath (Vass, 2001). The skin becomes pale (**pallor mortis**) within 15 minutes of death as blood stops circulating. The muscles initially relax, resulting in the loss of one's bowels and possibly regurgitation of the stomach contents (if the body has been moved). The jaw slackens and the eyelids open. If the eyes are not intentionally closed, a dark horizontal band will form on the sclera (whites of the eyes). This early indication of death is called *tache noire de la sclerotique,* and it is why coins are traditionally placed on the eyelids to prevent drying. As the cells die, they release their enzymes that begin the process of decomposition as the body begins to digest itself (**autolysis**). Within minutes, the odors released from the body attract blowflies to the corpse (if warm enough outside and not buried) that lay eggs within the warm, moist, and hidden parts of the body (mostly the orifices—mouth, nose, eyes, ears, anus, and genital openings).

Within a few hours, blood begins to settle (**livor mortis**), the muscle begins to stiffen (**rigor mortis**), and the body temperature drops toward the ambient air temperature (**algor mortis**). Within a few days, the bacteria within the body flourish due to their new food source (the corpse). Decomposition caused by bacteria is called **putrefaction**, which unleashes particularly foul odors that attract a greater variety of insect species to the body. Some feed on the corpse, others on the maggots that have now developed from the eggs initially laid by the first blowflies on the scene. As decomposition progresses, beetles and other scavengers (dogs, bears, coyotes) are attracted to the body and further remove soft tissue, separating the limbs from the trunk.

Once the skeleton is fully exposed, only a few carrion-eating beetles remain. The ligaments are the last remnants of soft tissue to decompose. Skeletonization exposes the bones to destructive elements that fragment, fracture, and weather them until they splinter. The teeth are the last of the physical body to be returned to the earth (except if fossilized). This entire process generally takes a few weeks (under ideal situations) to a few years, and this variation causes tremendous difficulty for estimating the PMI accurately (**Figure 15.1**).

Intrinsic Mechanisms of Decomposition

The cessation of breathing starves the cells of much-needed oxygen. As a result, cellular and metabolic processes fail, initiating a series of complex biochemical reactions that ultimately release the energy stored in the protein, fats, and carbohydrates that comprise the body. **Autolysis** and **putrefaction** are two complementary processes. Both are *intrinsic* mechanisms of decomposition; that is, both reflect processes that unfold from within a body. In addition,

Figure 15.1. Kusōzu, the death of a noble lady and the decay of her body. Wellcome Collection.

although both autolysis and putrefaction begin immediately after death, their effects are only normally visible after the first 48–72 hours. Both processes are also accelerated by warmer temperatures, including fever and active infection at the time of death.

Autolysis is defined as, "the post-mortem self-digestion and degradation of the cells and organs by intracellular enzymes" (Forbes et al., 2017:27). In other words, autolysis is the process of the body eating itself from the inside out. Without oxygen, cells accumulate acid and waste, swell, and eventually rupture, releasing metabolic enzymes that break down surrounding tissues. By definition, autolysis does *not* include the action of bacteria; it is only caused by enzymes found naturally within the body. Autolysis is first noticeable in the pancreas, kidneys, liver, stomach, and brain and last to develop in the muscles. The first visible signs are a clouding of the eye, skin slippage (when large sections of skin fall off the body), the loss of hair and nails, and *marbling* of the veins as the blood begins to break down.

Putrefaction is defined as, "a microbial-driven process . . . that degrades and liquefies tissues" (Forbes et al., 2017:27). During life, our bodies keep the bacteria that live in our digestive system in check. After death, however, these mechanisms are no longer functioning, and autolysis provides an abundance of resources (your body's tissues) for these bacteria to thrive. Bacteria multiply rapidly causing further destruction of the corpse as the soft tissues liquefy. Putrefaction begins in the gut and spreads through the blood vessels as bacteria migrate throughout the body. Marbling of the blood vessels is one of the earliest signs of putrefaction. The color of the skin changes from green to purple/brown to black as putrefaction progresses. Like autolysis, putrefaction also results in skin slippage of the hands, neck, and abdomen, as well as the loss of hair and nails. Unlike autolysis, bacterial fermentation of the liquefied tissues produces excessive amounts of foul-smelling gas. This results in bloating of the face, abdomen, and scrotum. As internal pressure increases, decomposition liquid will exit through the mouth and nostrils. Eventually the internal pressure is so high, and the skin so weak, that the body ruptures. Remaining soft tissue is predominantly black in color. Higher ambient temperature, fever, penetrating body trauma, and obesity all accelerate putrefaction. Individuals who die while taking antibiotics or die from considerable blood loss will experience slower rates of putrefaction.

Extrinsic Mechanisms of Decomposition

Many types of animals will feed on a corpse, including family pets (Colard et al., 2015). For bodies exposed outside, a number of wild carnivores (coyotes, bears, raccoons, opossums, pigs) will help reduce the corpse to a skeleton by removing soft tissue. However, insects by far have the greatest effect on the rate at which the body is reduced to a skeleton. On a warm sunny day, the blowfly (identifiable by its green iridescent body) will find the corpse within minutes. Flies will swarm the corpse once decomposition is fully underway (~48 hours) and lay thousands of eggs (*oviposition*) that hatch into maggots that can quickly consume the corpse while generating enough heat to further catalyze putrefaction and autolysis. As decomposition commences, other species of fly will appear, with carrion-eating beetles being the final group of insects to colonize the body.

Many factors affect the rate of insect activity on a corpse. Temperature, season of the year, time of day, rain conditions, and sun/shade conditions are important variables. Simply having access to the corpse is also critical. Bodies that are buried, decomposing inside a structure, or wrapped tightly in plastic will prevent insect colonization. A body found inside a house during the winter months may exhibit no insect activity. It is important to note that insect colonization preferences vary by species and many species are particular about the conditions under which they will deposit eggs. If the window of opportunity for oviposition is missed, certain species may never colonize a corpse. Because many fly species are only active during the day, a death at night can delay colonization and affect PMI estimates.

Factors That Affect the Rate of Decomposition

Several factors that affect the rate of decomposition have already been mentioned, however, there are many others that significantly complicate estimation of PMI. It is important to stress that these factors do not generally affect the process of decomposition. Given enough time, all bodies will eventually be reduced to a skeleton. *Therefore, these factors only affect the rate at which the body decomposes.* The potential list of factors is lengthy, and researchers are not in 100 percent agreement that all are relevant. Experimental research is ongoing.

All researchers, however, agree that temperature is the single most important variable that affects the rate at which the body decomposes. Higher temperatures result in quicker decomposition because autolysis, putrefaction, and insect activity are all accelerated at higher temperatures. For this reason, researchers no longer think that time is the dominant variable affecting decomposition rates; temperature is. Researchers now use the concept of **Accumulated Degree Days (ADD)** when discussing PMI estimation, which is simply the sum of the average daily temperatures (in Celsius) that a corpse experiences (Megyesi et al., 2005; Vass, 2011; Vass et al., 1992). As a very general rule, the time it takes to reduce a body to a skeleton = 1285/average daily temperature. This means that if the average daily temperature is 40°C (104°F), the body will skeletonize in approximately 32 days (1285/40=32). A more complex (and realistic) formula is presented in Vass (2011) that accounts for a greater variety of mitigating factors (see below).

If temperature is the primary variable that affects the decomposition rate, then one must consider all factors that can affect the temperature of a corpse. *Intrinsic* factors are those that relate to the body itself, and as mentioned earlier, the relevant variables include: age, body mass, body weight, percent body fat, muscularity, internal body temperature at time of death (mitigated by exercise, fever/infection, being involved in a struggle prior to death, exhaustion, hypothermia), whether there is penetrating trauma to the body, drug or alcohol use, and the position of the body (stretched or flexed). *Extrinsic* factors are those that relate to the microenvironment in which the corpse is found. Relevant variables include whether the body was inside or outside, if outside whether it was buried, how deeply it was buried, if left on the surface whether it was in the sun or the shade or whether it was windy during the first few days after death (wind increases evaporation and lowers body temperature), and whether the body was clothed, naked, or wrapped in materials that prevent insect colonization. Insect colonization increases the internal body temperature due to maggots feeding on the corpse, which in turn increases the rate of autolysis and putrefaction, which provides more food for those same insects.

As a general rule, corpses decompose faster if left outside than inside. For corpses left outside, surface remains decompose faster than buried remains. For buried remains, shallower burials decompose faster than deeper burials. Being buried lowers the temperature of the body, limits the access of insects and carnivores, and prevents the diffusion of gases during putrefaction. For buried remains, the pH of the soil also affects the decomposition rate. Bodies submerged in water decompose at an even slower rate than land burials.

Once a corpse has become skeletonized (**Figure 15.2**), a different set of factors come into play. Unlike the fats, proteins, and carbohydrates that comprise soft tissues, the skeleton is composed of collagen and hydroxyapatite (reviewed in chapter 3). Autolysis, putrefaction, insect activity, and carnivore feeding have all stopped, yet the body continues to deteriorate until the bones are turned to dust. At this point only the forensic anthropologist is involved in estimating PMI, which is primarily concerned with bone weathering. Intrinsic factors that

Figure 15.2. Complete skeletonization with breakdown commencing.

determine the rate at which the skeleton deteriorates include age (the bones of children may deteriorate faster), diseases that weaken bones, and the distribution of cortical and trabecular bone (with the latter breaking down more easily). Extrinsic factors include temperature (temperature variation hastens breakdown), groundwater (wet/dry cycling hastens breakdown), soil type (sand vs. clay), soil pH (acidic soils destroy bone), and sun exposure (which increases temperature variation). Animal trampling and plant roots can cause further damage and breakdown.

PMI Estimation beyond 48 Hours

For the first 48 hours after death, a forensic pathologist will rely on the *mortis triad* to estimate PMI: *rigor mortis* (the stiffening of muscles), *livor mortis* (the pooling of body fluids), and *algor mortis* (the decline in body temperature). After the first 48 hours, however, these standard approaches are less reliable and estimates of PMI must be based on knowledge of the rate of autolysis, putrefaction, skeletonization, and skeletal breakdown to produce rough estimates of PMI based on documented stages of decomposition. The earliest methods relied on crude stages of decomposition and less rigorous estimates of the timing of advancement through each stage that were highly dependent on climate and season. These methods have been replaced with empirical assessments of **Accumulated Degree Days**.

Stages of Decomposition

Over the years, researchers have developed several systems for organizing the decomposition process into discrete stages that can be compared against case reports to generate time intervals for each stage. The most commonly used staging scale was proposed by Bass (1997) based on cases from Tennessee and modified by Galloway (1997) for cases from southern Arizona. Descriptions of decompositional changes associated with each stage are presented in **Table 15.1** with exemplar images provided in **Figure 15.3**.

Table 15.1. Stages of decomposition

Stage	Description
1) Fresh	No discoloration except for lividity and incipient marbling of veins; no insect activity except for possible blowfly eggs near orifices (fine sawdust)
2) Early Decomposition	
a) Pre-Bloating	Skin slippage; hair and nail loss; pink discoloration of skin; darker discoloration of face and fingers; flies and maggots present near eyes, nose and mouth; marbling of veins is darker with decomposition fluid oozing from orifices
b) Bloating	Abdomen bloated and discolored green; beetles present
c) Post-Bloating	Body ruptures; discolored brown to black
3) Advanced Decomposition	Abdomen collapsed with skin sagging due to extensive maggot activity; remains are still moist; beetles are still active on corpse
4) Skeletonization	Skeleton largely exposed with some remaining flesh and ligaments; beetles dwindling; bones are greasy and yellow due to remaining fat; eventually drying to white
5) Skeletal Decomposition	Bones are sun bleached; cracks forming along long bones; cortical bone peeling away. In later stages, the long bone ends are destroyed and the shafts exhibit large splinters.

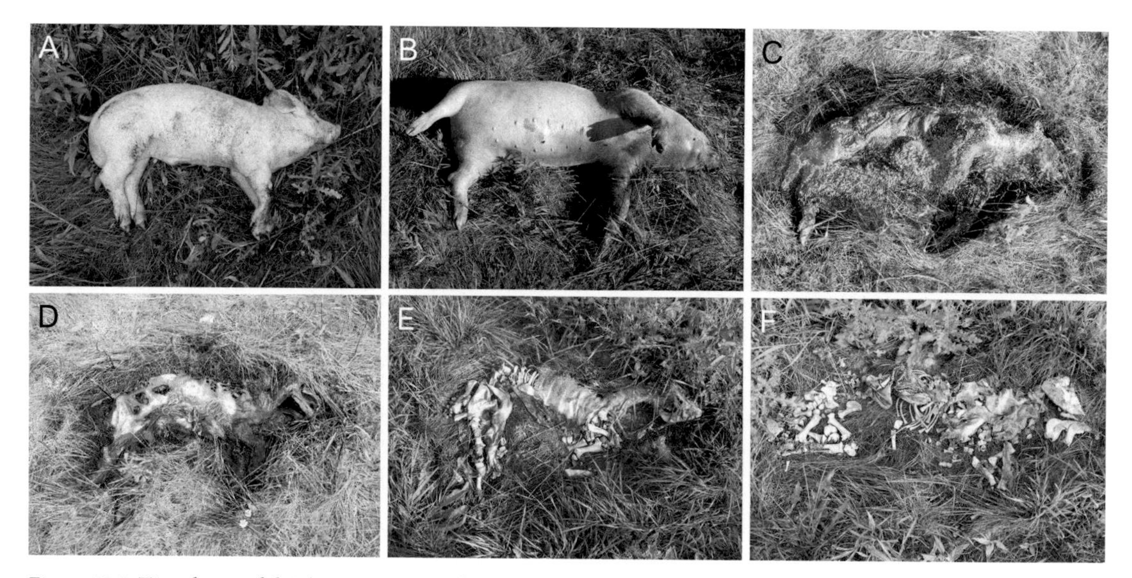

Figure 15.3. Time lapse of the decomposition of a pig. (*A*) Fresh stage of decomposition. Autolysis occurring but not evident. Insect eggs likely laid in eyes, nose, and mouth though not visible. (*B*) Bloating stage of decomposition. Note the distended abdomen and discoloration patterning indicating putrefaction is under way. Body green to red in color; (*C*) Advanced decomposition stage. Maggot mass is extensive with body largely reduced. Putrefaction extensive with brown-black tissue present; (*D*) Advanced decomposition stage. Maggot activity is decreasing with only a few visible here. Internal organs are largely consumed; (*E*) Partial skeletonization and the beginning of the dry stage of decomposition. There is limited insect activity, likely only including beetles. Note the decomposition process has killed the plants underneath the body; (*F*) Skeletonization. Only remnant, partially dried skin is present. Disarticulation of the skeleton is occurring.

The timing of each stage for a human corpse is presented in **Table 15.2**. The Bass (1997) data from Tennessee should be considered a very general rule of thumb; they are based on a small sample of human cadavers observed decomposing under controlled conditions. Galloway's (1997) data are more realistic and are based on nearly 200 retrospective case reports from the Desert Southwest. The wide ranges are discouraging, but this table includes PMI data from a wide range of decomposition scenarios known to alter the decomposition rate: buried remains, surface remains, indoor remains, winter deaths, etc. The "Normal Range" indicates the time in which the majority of cases reached that particular stage. Note, the case of extremely rapid skeletonization (7 days) occurred inside a non-air-conditioned house during the late summer. **Table 15.2** indicates just how important additional contextual data are for assessing PMI using decomposition stages. **Table 15.3** combines information from the mortis triad with the early and advanced stages of decomposition and skeletonization.

Accumulated Degree Days

Decomposition follows a fairly standard trajectory (**Table 15.3**) but the rate of decomposition is heavily affected by environmental conditions, temperature being among the most important. Because temperature varies so much throughout the year and from region to region, even within the U.S., researchers have attempted to define formulae that are more quantitative in their consideration of how environmental variables affect the rate of decomposition, which then affects estimates of PMI. Researchers interested in developing quantitative methods of PMI estimation work backward from known cases where they can record information on decomposition stage at the time of discovery. The research method takes known case data, records information on decomposition stage, generates data on the environmental conditions the body experienced, and then generates formulae to predict PMI based on past cases. The benefit of this approach is that, if your training sample is large enough, you can

Table 15.2. Climate-specific rates of decomposition

Decomposition Stage	Bass (1997) Warm, Humid	Galloway (1997) Warm, Dry
Fresh	1st Day	*Mode* = 1 day *Normal Range*: (1–2 days) *Total Range*: (1–7 days)
Early Decomposition (Fresh to Bloated)	1st Week	*Mode* = 3 days *Normal Range*: (2–10 days) *Total Range*: (1 day to 4 months)
Advanced Decomposition (Bloated to Decay/Putrid)	1st Month	*Mode* = 4 months *Normal Range*: (10 days to 4 months) *Total Range*: (3 days to 3 years)
Skeletonization	1st Year	*Mode* = 9 months *Normal Range*: 3 months to 3+ years) *Total Range*: (7 days to 3 years)
Skeletal Decomposition	1st Decade	*Normal Range*: (2 months to 3+ years) *Total Range*: (2 months to 3+ years)

Table 15.3. Short- and long-term changes in the human body after death

Time after Death	Changes Observed
0 minutes	Circulation and breathing stop Blood drains from surface of skin (pallor) Muscles relax
2 hours	*Tache noire de la sclerotique* Rigor mortis begins Algor mortis begins Livor mortis evident
4–6 hours	Livor mortis well developed
12 hours	Livor mortis fixed, blood will not respond to movement or skin pressure
24–48 hours	Livor mortis released due to blood re-liquefying
48–72 hours	Rigor mortis releases Marbling of veins easily seen Hair and nails detach
96 hours	Skin begins to slip off body Putrefaction well under way
Days to months	Discoloration spreads and intensifies Bloating of abdomen followed by release of gases Soft tissues liquefy
Months to years	Gradual loss of remaining soft tissue Skeletonization leading to complete skeletal deterioration

also generate ranges of error based on past casework, which satisfies the *Daubert* criteria (see chapter 2).

Central to these methods is the concept of **Accumulated Degree Days** (ADD). Megyesi et al. (2005:621) define ADD as representing "heat energy units available to propel a biological process such as bacterial growth or fly larvae growth." ADD is calculated by adding the daily temperatures a corpse was exposed to over the course of the decomposition process, which is the estimate of PMI investigators are interested in knowing. For example, if the average daily temperature was 30 degrees Celsius for a week, then the ADD for that corpse would be 30*7 or 210. This means that higher temperatures accumulate ADD more rapidly. This also means that temperatures below freezing (0 degrees Celsius) do not add to the total because the decomposition process usually stops entirely in freezing temperatures. On the other hand, research indicates that soft tissue decomposition was usually completed at 1285 ADD (Vass, 2011). In other words, after 1285 ADD the corpse is usually skeletonized. This means that in colder climates (temperature around 10 degrees Celsius), a corpse would take about 128 days to skeletonize (1285/10 = 128.5). In warmer climates (temperature around 30 degrees Celsius), a corpse would skeletonize in around 42 days (1285/20 = 42.8).

Megyesi et al. (2005) developed a method for estimating PMI using the concept of ADD and what they term the Total Body Score (TBS). The TBS is calculated using tables they provide for three parts of the body—the head and neck, the trunk, and the limbs. They differentiate the three sections of the body because each undergoes decomposition at different rates, with decomposition starting in the trunk. Using tables providing a point scoring system for each stage of decomposition, the researcher can define a TBS for a corpse, which can vary from 3 to 35. The TBS is then entered into a formula to estimate ADD, which is then divided by temperature data from where the body was found to arrive at a PMI stated in "number of days."

Vass (2011) provided a similar solution to estimating PMI that differentiates surface and buried corpses and also adds humidity as a primary factor determining the rate of decomposition. The formula provided by Vass (2011) is easier to use in that it directly estimates PMI in days rather than ADD, which must then be converted to PMI using additional temperature data. Vass (2011) developed two formulae for predicting PMI—one for surface remains and one for buried remains. We focus on the surface formula here because it is simpler to apply. The formula is:

$$(1)\ \text{PMI Aerobic} = \frac{1285*(\frac{decomposition}{100})}{.0103*\text{temperature}*\text{humidity}}$$

Use of the formula requires:

1) An estimate of soft tissue decomposition, ranging from 1 to 100, which reflects the total amount of soft tissue decomposition that has occurred expressed as a percentage. This is entered as a whole number and assumes the corpse has not been eaten by animals or been subjected to other conditions that affect soft tissue decomposition. A corpse that is mostly decomposed might be scored as a 90, or 90 percent of soft tissue has decomposed.
2) An estimate of the average temperature in Celsius that the corpse has been exposed to. This could be the temperature on the day of discovery or an average of multiple days if temperature fluctuated significantly during the time in which the corpse likely decomposed.
3) An estimate of the relative humidity, ranging from 1 to 100 percent. This could be the humidity on the day of discovery or an average of multiple days if humidity fluctuated significantly during the time in which the corpse likely decomposed.

Example: A corpse was discovered on the surface with 80 percent of the soft tissue decomposed on a 20-degree day at 60 percent relative humidity. Use of the Vass (2011) formula suggests a PMI of:

$$= \frac{1285*(\frac{80}{100})}{.0103*20*60}$$

$$= \frac{1028}{12.36}$$

$$= 83 \text{ days}$$

LEARNING CHECK

Q1. Based on the information shown in **Figure 15.4** and the decomposition stages presented in **Table 15.1,** what stage of decomposition is this pig in?
A) Fresh (no evidence of autolysis or putrefaction)
B) Pre-bloating (minor changes due to putrefaction but no abdominal bloating)
C) Bloating (abdomen bloated, discoloration spreading but still light in color)
D) Advanced decomposition (putrefaction evident throughout the body)

Q2. A corpse was discovered on the surface with 20 percent of the soft tissue decomposed on a 10-degree day at 10 percent relative humidity. Use of the Vass (2011) formula suggests a PMI of:
A) 300 days
B) 250 days
C) 100 days
D) 400 days

Q3. A corpse was discovered on the surface with 40 percent of the soft tissue decomposed on a 45-degree day at 90 percent relative humidity. Use of the Vass (2011) formula suggests a PMI of around:
A) 6 days
B) 12 days
C) 18 days
D) 24 days

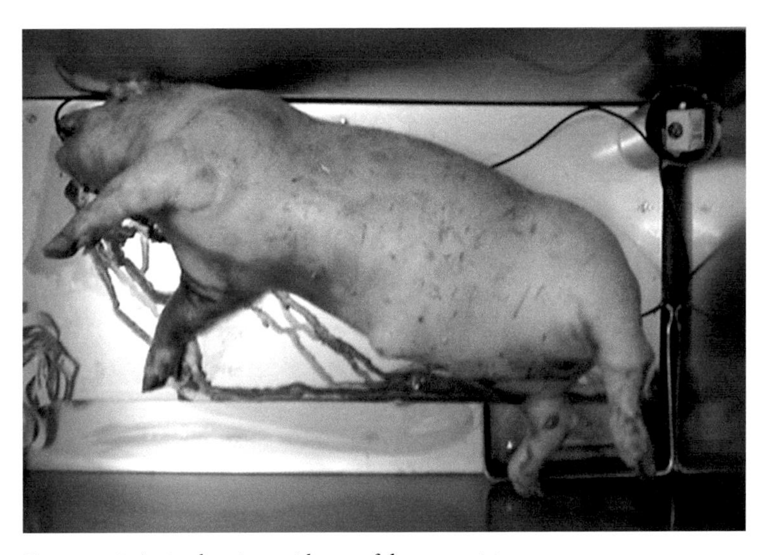

Figure 15.4. A pig showing evidence of decomposition.

Q4. Based on the Vass (2011) formula, which environmental variable has a greater effect on the rate of decomposition?
A) Temperature
B) Humidity
C) They both have the same effect.

PMI Estimation after Skeletonization

Methods for estimating PMI after complete skeletonization are limited, imprecise, and subject to very broad ranges. Nevertheless, this is an area where forensic anthropologists alone can provide some insight into time-since-death. With the soft tissue all but gone, the forensic anthropologist is left with just the skeleton to use as a source of information. Even after the soft tissue decomposes, the skeleton, composed of a much more durable matrix of collagen and hydroxyapatite (see chapter 3), continues to break down. At first, the collagen is lost, which is the remaining organic component of bone. The loss of collagen and the elasticity it provides is important for differentiating the timing of bone trauma, as discussed in chapter 12. Once the collagen is gone what remains is a brittle matrix of bone.

There are many factors that contribute to skeletal breakdown, including scavenging by animals, trampling by animals, plant and root destruction, and interactions between the skeletal tissues and the surrounding burial soils. All things being equal, the more taphonomic changes observed on a skeleton the longer the PMI is likely to be; however, these provide only very imprecise assessments.

Methods have been developed for estimating PMI based on bone weathering. Weathering is defined as "the process by which original microscopic organic and inorganic components of a bone are separated from each other and destroyed by physical and chemical agents . . ." (Behrensmeyer, 1978:153). Weathering is caused by fluctuations in temperature and humidity levels. Bones that experience temperature (excessive heating or cooling) or moisture (flooding followed by rapid drying) extremes weather more quickly.

Studies of weathering in different environments emphasize the distinction between the macroenvironment and the microenvironment. The macroenvironment refers to the broad climatic regime the remains are found in (arid desert, temperate, tropical; high or low latitude, high or low altitude), whereas the microenvironment refers to the specific location of the remains in relationship to moisture sources and sun exposure. For example, buried remains experience much lower rates of weathering because the burial environment maintains a more consistent temperature and moisture level, and of course, limited sun exposure. Surface remains directly exposed to the sun will weather much more quickly due to the temperature variation throughout the day. Even for surface remains, bone will weather mostly on the side exposed to the sun. This means differential weathering will likely be apparent on the upper (facing the sun) and lower (facing the ground) surfaces of the bone. The less consistent the microenvironment, the more severe weathering will be.

Most forensic anthropologists use the scoring stages of Behrensmeyer (1978) to assess the degree of weathering observed in a skeletonized body. These are summarized in **Table 15.4**. In applying these scoring criteria, a bone should be assigned the highest score based on the highest degree of weathering that covers at least 1 cm^2 of the bone's surface. Examples of different weathering stages are presented in **Figure 15.5**. It is rare to encounter stage 4 and 5 weathering in a forensics context because of the years of exposure required to develop this degree of breakdown.

Table 15.4. Weathering stages as defined by Behrensmeyer (1978)

Stage	Description
0	No signs of weathering; no cracking or flaking; bone fresh and may feel wet or greasy
1	Initial cracks appear oriented with the fiber structure of the bone, i.e., along the lengths of long bones; mosaic cracking on articular surfaces
2	Outer layer of bone begins to lift and form flakes, often at the margins of existing cracks
3	There are patches of heavily weathered bone with an overall fibrous appearance; the weathering is shallow but the concentric layers of bone are missing
4	Bone is coarse with a fibrous appearance with varying size splinters falling away
5	The bone has splintered in place with original shape of bone uncertain

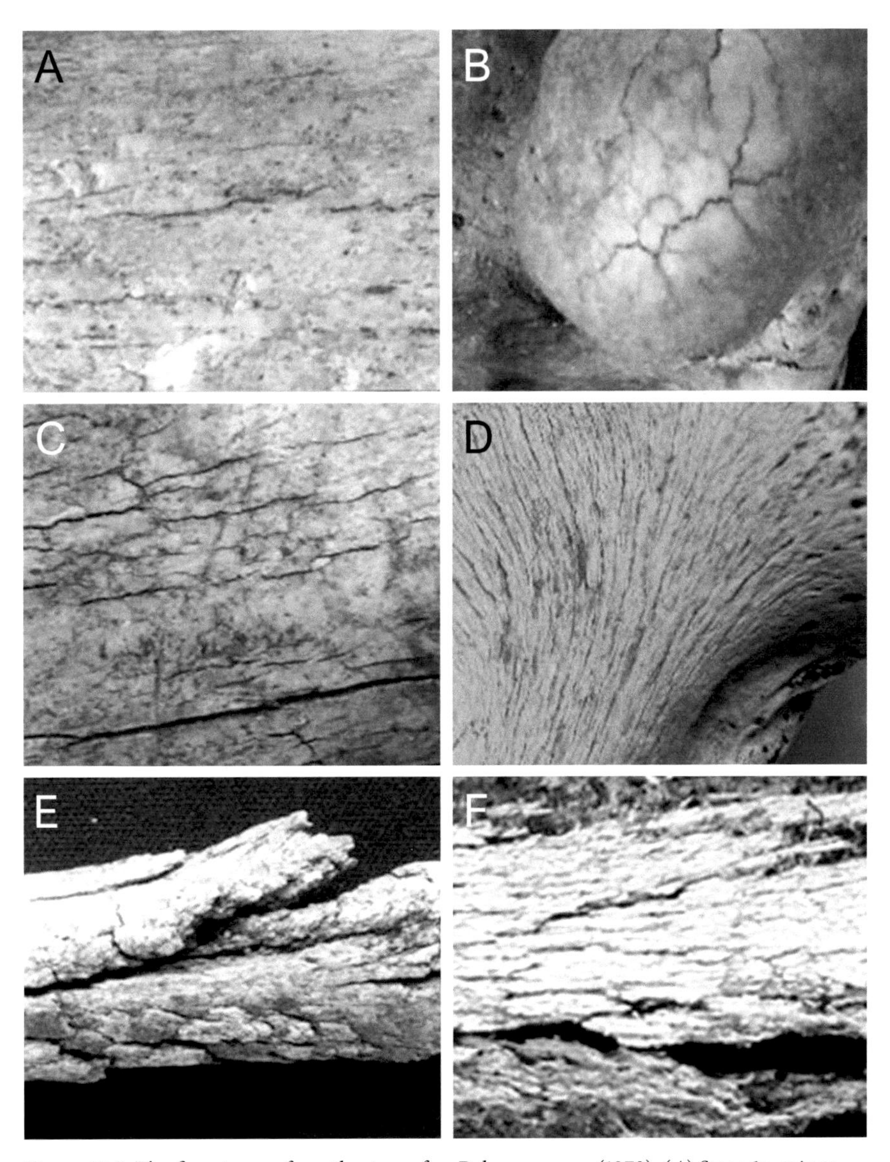

Figure 15.5. The five stages of weathering, after Behrensmeyer (1978). (*A*) Stage 1a, minor cracking on a long bone; (*B*) Stage 1b, mosaic cracking on an articular end; (*C*) Stage 2, more significant cracking with minor lifting of edges of cracks; (*D*) Stage 3, widespread cracking with fibrous bone texture; (*E*) Stage 4, uniform coarse fibrous appearance with large splinters; (*F*) Stage 5, bone is splintering in place and falling apart, loss of bone shape evident.

Converting observations of weathering into PMIs is challenging. Ross and Cunningham (2011) summarize the results of several studies across different climate zones. For arid sites, bones were generally unweathered (stage 0–1) until after 4 years, exhibited initial weathering (stage 1) between 4 and 8 years, and more advanced weathering (stages 2–3) after 10–15 years. To the contrary, open tropical sites advanced through the weathering stages more rapidly, reaching stages 1–2 in 2–4 years, stages 3–4 in 4–10 years, and stage 5 in 10–12 years. Closed tropical sites (rainforests with high tree coverage) had much slower rates of weathering and remains could be unweathered 15 years after death. This indicates the importance of the microenvironment for determining the rate of weathering.

LEARNING CHECK

Q5. Using **Figure 15.5** and **Table 15.4,** identify the stage of weathering of the specimen shown in **Figure 15.6A.**
- A) Stage 1
- B) Stage 2
- C) Stage 3
- D) Stage 4

Q6. Using **Figure 15.5** and **Table 15.4,** identify the stage of weathering of the specimen shown in **Figure 15.6B.**
- A) Stage 1
- B) Stage 2
- C) Stage 3
- D) Stage 4

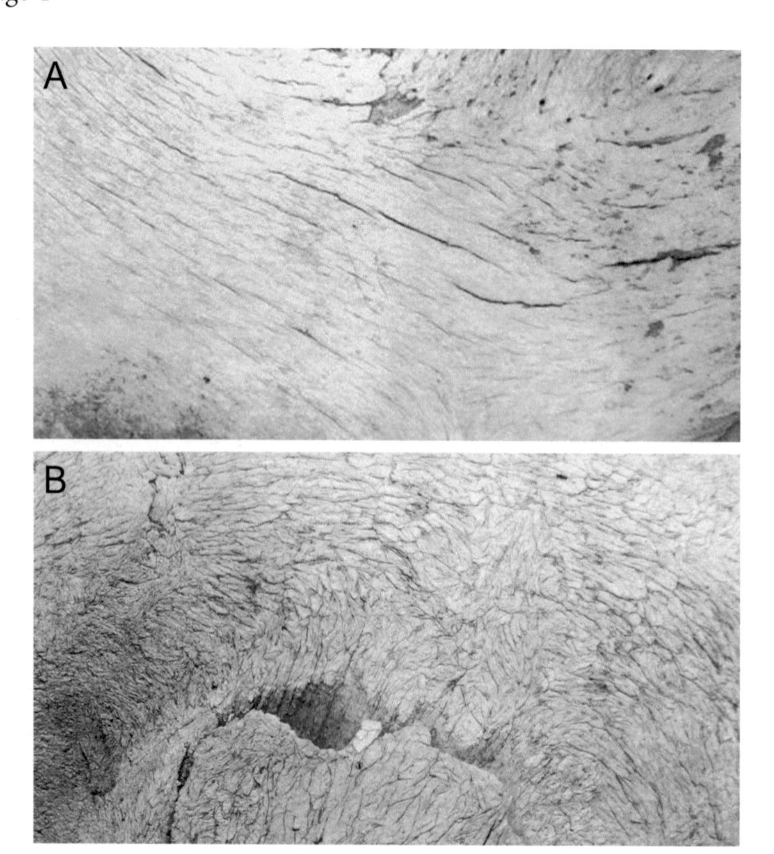

Figure 15.6. Close-up views of bone surfaces showing varying degrees of weathering.

Other Taphonomic Observations

Animal Scavenging

Animal scavenging has three main effects on a crime scene and forensic analysis: bone scattering across the crime scene, bone breakage through trampling on the remains, and bone destruction by chewing the remains. When a human body is scavenged by animals, there is a specific pattern that is typically followed. The soft tissue of the head and neck are consumed, then the thorax and internal organs, then the upper limbs are separated from the thorax, and finally the lower limbs are separated from the thorax. The result of continued scavenging by carnivores is that the skeletal elements are scattered over a relatively wide area.

It is important to be able to differentiate carnivore activity from other factors that may affect a body. Four kinds of postmortem bone changes are attributed to carnivores. *Punctures* are areas of bone that have collapsed under the force of carnivore tooth pressure (**Figure 15.7**). If the teeth only leave a depression and do not penetrate the bone, this is referred to as a *pit*. *Scoring* refers to the scratches caused by teeth moving across the surface of the bone. When these scratches are deep enough, they are called *furrows*. Rodents can also do extensive damage to bone because they chew on skeletal remains to supplement their calcium and to sharpen and wear their incisors. Rodent gnawing produces marks that are diagnostically shallow, parallel and paired, reflecting the shape and configuration of their incisors (**Figure 15.8**).

Animal scavenging can produce changes to bones that are remarkably similar to trauma. Punctures and pits, for example, may be misinterpreted as sharp force puncture wounds and scoring and furrows can be mistaken for cuts or incisions (see chapter 14). A key consideration when assessing evidence for dismemberment, carnivore scavenging, or sharp force

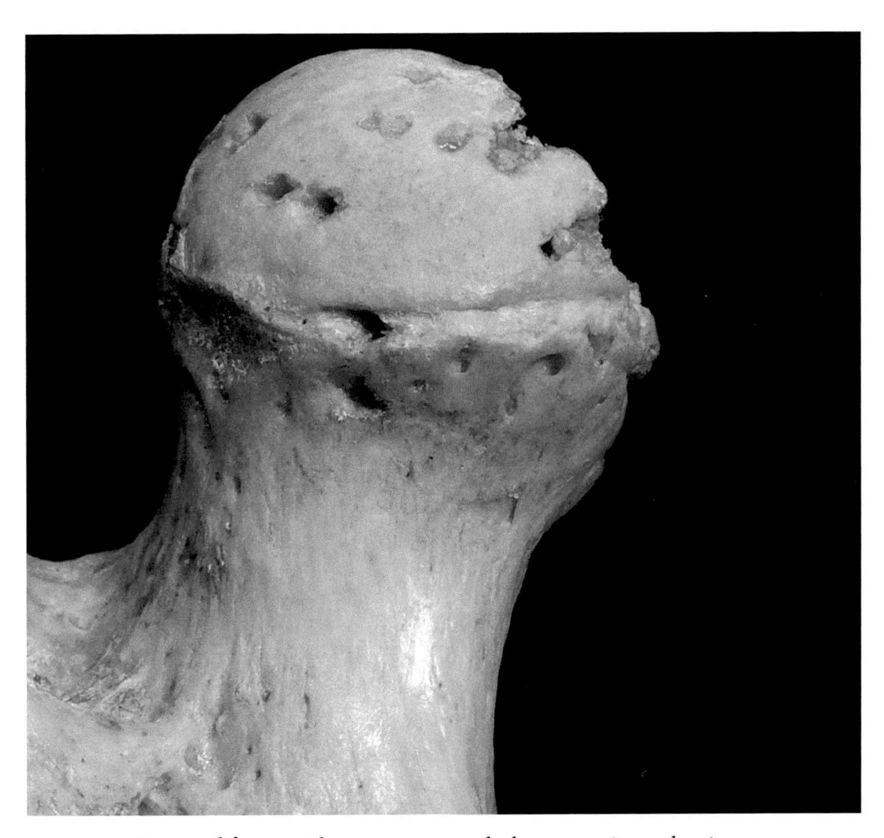

Figure 15.7. Proximal femur with puncture wounds due to carnivore chewing.

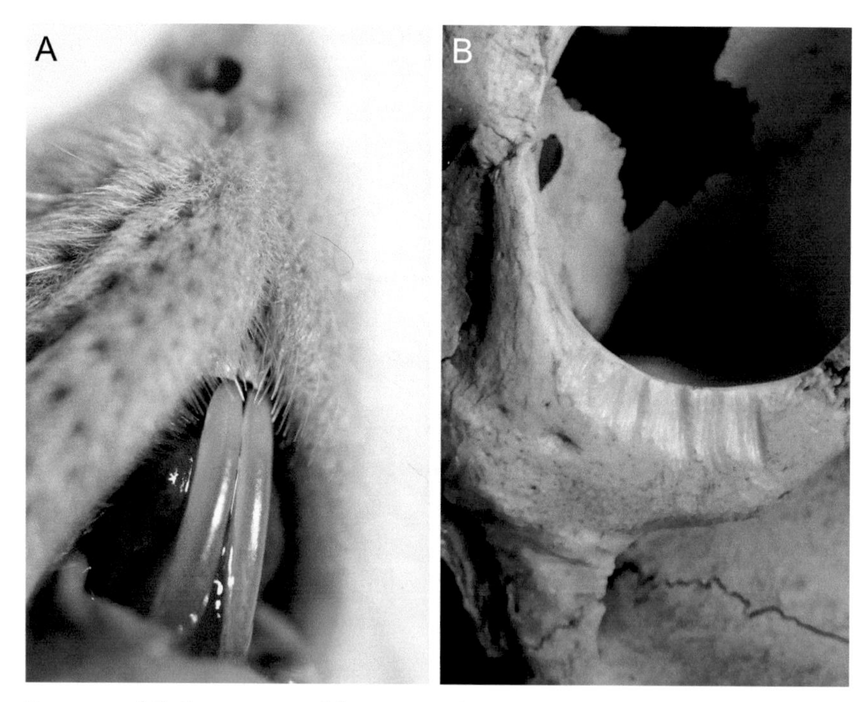

Figure 15.8. (*A*) Close-up view of the incisors of a rodent. (*B*) Rodent gnawing on the lateral aspect of the orbit.

trauma is the shape of the profile of the mark created in the bone. Metal and stone tools tend to create cuts with a V-shaped cross-section. Tooth marks on bone will have a flat or U-shaped cross-section that is more rounded. In **Figure 15.9A**, a tool made the cutmarks, which is evident in the V-shaped profile of the cut. In **Figure 15.9B**, the mark made in the bone has a slightly rounded bottom, consistent with a carnivore tooth.

Fire Damage

When bone is exposed to fire, it undergoes a series of color changes, shrinks, and then ultimately fragments into very small pieces that may defy forensic identification. This is often the specific purpose of burning a body—to hide evidence of a crime or to minimize the evidence that is recovered, including making it more difficult to identify the decedent. In its natural state, bone is yellowish-brown in color. As the temperature that a bone is exposed to increases, its color changes to a darker yellow-brown, and ultimately becomes black once all the fat in the bone is burned away (**Figure 15.10A**). With further heating, the color changes to dark gray, then light gray, and finally white (**Figure 15.10B**). Once a white state is reached, the bone consists only of calcium salts and is very brittle, often showing cracking similar to bone weathering.

The effect of fire damage on the forensic investigation varies with the temperature the remains were subjected to. For example, bone will shrink due to loss of moisture content as the temperature of the fire increases. If the fire is less than about 700 degrees Celsius there is limited shrinkage. Fires between 700 and 800 degrees Celsius will result in a 1–2 percent decrease in size. Fires that are over 800 degrees Celsius will result in a 10–15 percent reduction in size, a large enough difference to potentially impact assessments of sex or stature. When heated further and for longer periods of time, the bone breaks down to smaller pieces, most of which are white in color.

In addition to its potential effects on other analyses, evaluation of fire damage can provide further information that may be relevant in a death investigation. Consider **Figure 15.11.**

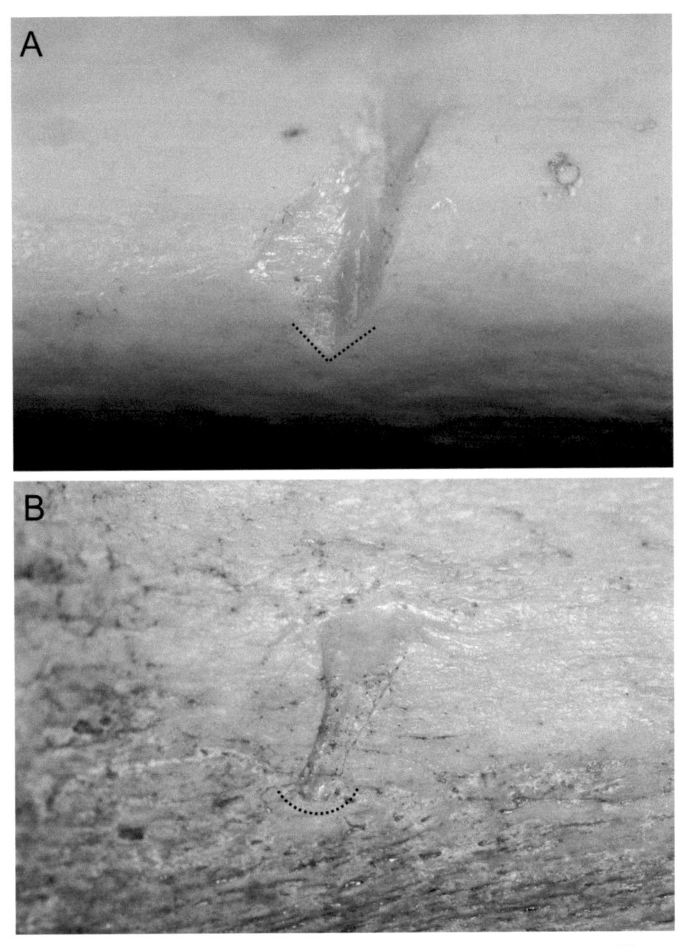

Figure 15.9. Cutmarks caused by a tool (*A*) vs. tooth marks caused by a carnivore (*B*).

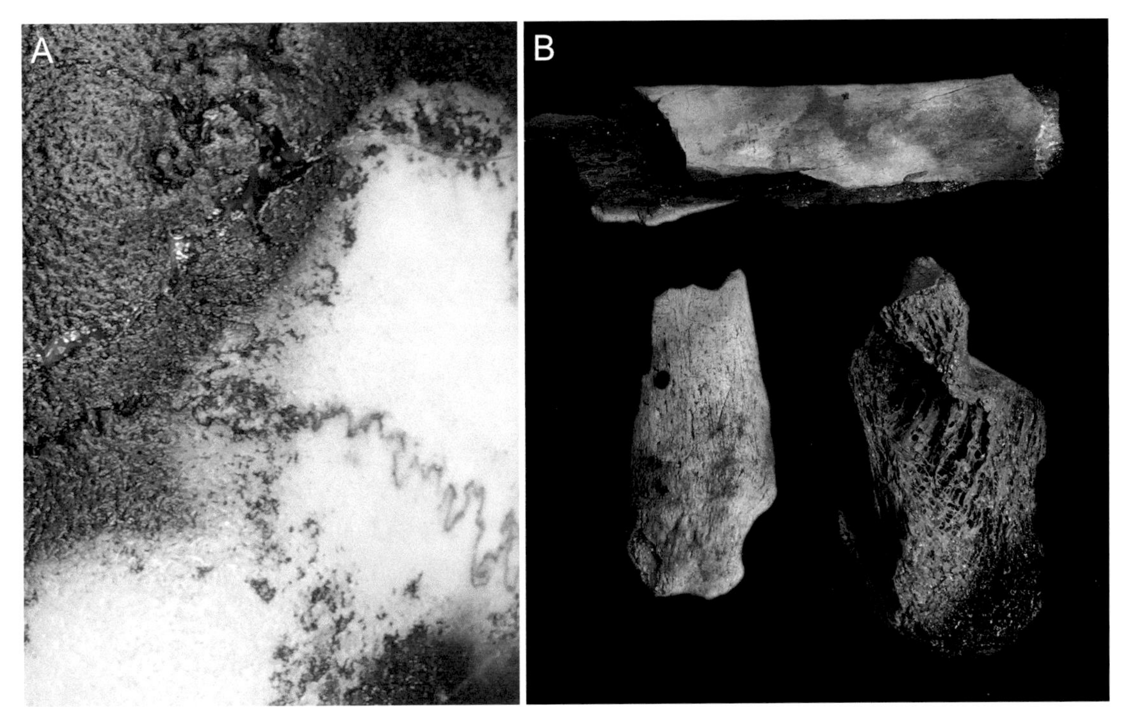

Figure 15.10. (*A*) Close-up view of a skull that has been severely burned. Note the transition in the color of the unburned portion from lighter yellow to darker yellow to tan to black. (*B*) Small fragments of bone showing color variation consistent with higher temperature fires.

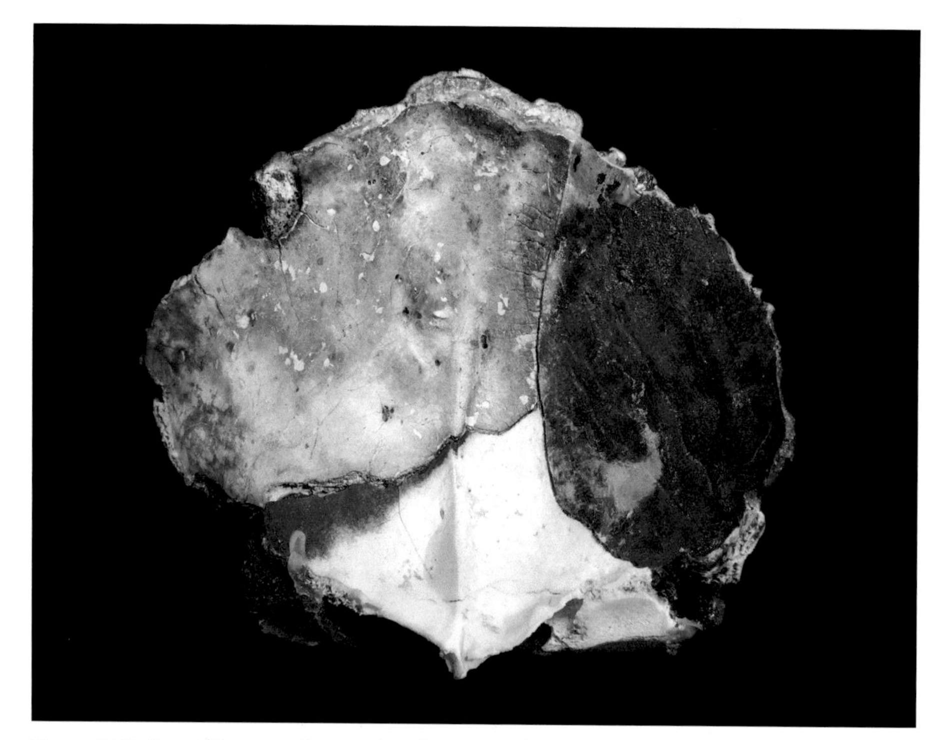

Figure 15.11. Frontal bone, endocranial surface. Note the variation in color even within the same element (which has been reconstructed). This suggests the bone was fragmented prior to burning.

This is the endocranial surface of the frontal bone that shows high variation in the degree of burning, which corresponds to where the bone was broken. In this reconstructed view it is clear that the bone was broken prior to it being burned. In other words, the fracturing was not the result of the fire itself, which means the fracture patterns may have an explanation that is relevant to the investigation.

Burial and Soil Changes

Simply being buried and in contact with soil will cause significant changes in bone color and appearance. Generally speaking, bone will stain a darker color depending on the soil composition and mineral content in the local area. The entire bone can change color or changes can be highly localized in the form of discrete surface stains. The most common staining is caused by manganese, which presents as patches of black. The frontal bone shown in **Figure 15.12A** is heavily stained from being buried for an extended period of time. Note the variation in color, from tan to darker brown, to near black. The white areas are recently exposed areas of bone, indicating that the staining seen here is superficial and does not penetrate very deeply into the surface. In **Figure 15.12B**, the darker brown bone was still buried at the time of discovery and has been stained by touching the underlying soil. There is a sharp transition from the brown surface to the lighter, yellower surface that represents the soil line. Once exposed, sun bleaching commences. Patterns of soil staining like this may reveal changes in the position of the body prior to its recovery.

Plant roots also cause damage to the skeleton. Roots will invade any hole or crack in the skeleton and push the bones apart. Roots will also mark the surface of the bone, as seen in **Figure 15.12C,** which is referred to as root etching.

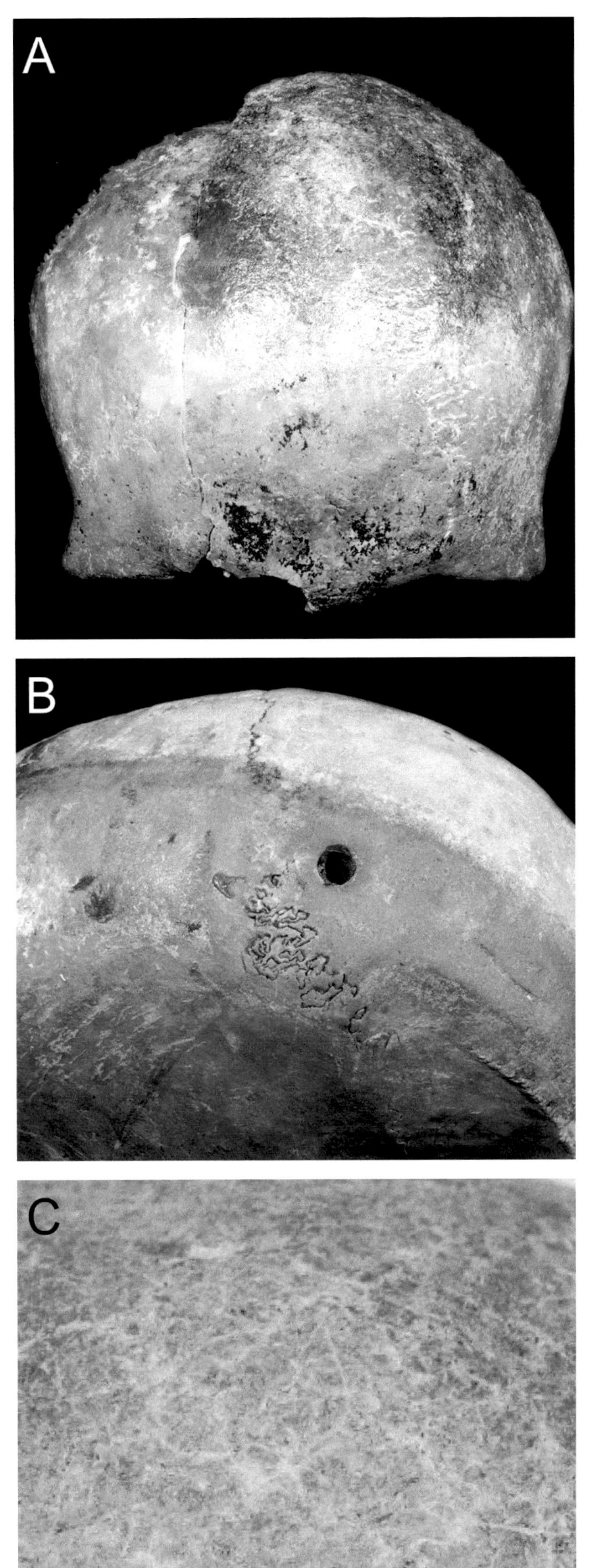

Figure 15.12. (*A*) Heavily stained frontal bone. (*B*) Skull with differential coloring due to partial exposure at time of discovery. (*C*) Root etching on a skull.

Figure 15.13. Two excavation score marks caused by tools during excavation.

Excavation and Recovery Damage

Damage also occurs to bones during the process of recovery and after the bones have been taken to the lab. For example, metal tools score bone more easily than softer wood or bamboo tools. For this reason, it is not recommended to use metal tools when excavating a body. These scoring marks are easily identified because they will be a different color (lighter) than the surrounding bone (**Figure 15.13**). Once removed from the burial environment, bones will also start to deteriorate if the environment they are kept in changes (in humidity). Often, bones will break upon excavation because they are **friable** and lack collagen. These factors need to be considered when analyzing the body for signs of trauma.

LEARNING CHECK

Q7. Based on what we have discussed here, what is the likely cause of the damage seen in the bone shown in **Figure 15.14A**?

A) Carnivore
B) Rodent
C) Excavator
D) Fire

Q8. Based on the overall appearance of the damage seen in **Figure 15.14A**, would you characterize this damage as occurring after the bones were removed from the ground?

A) Yes
B) No

Q9. Based on what is seen in **Figure 15.14B**, how was this individual oriented with respect to the ground?

A) Face down
B) Left side down
C) Right side down

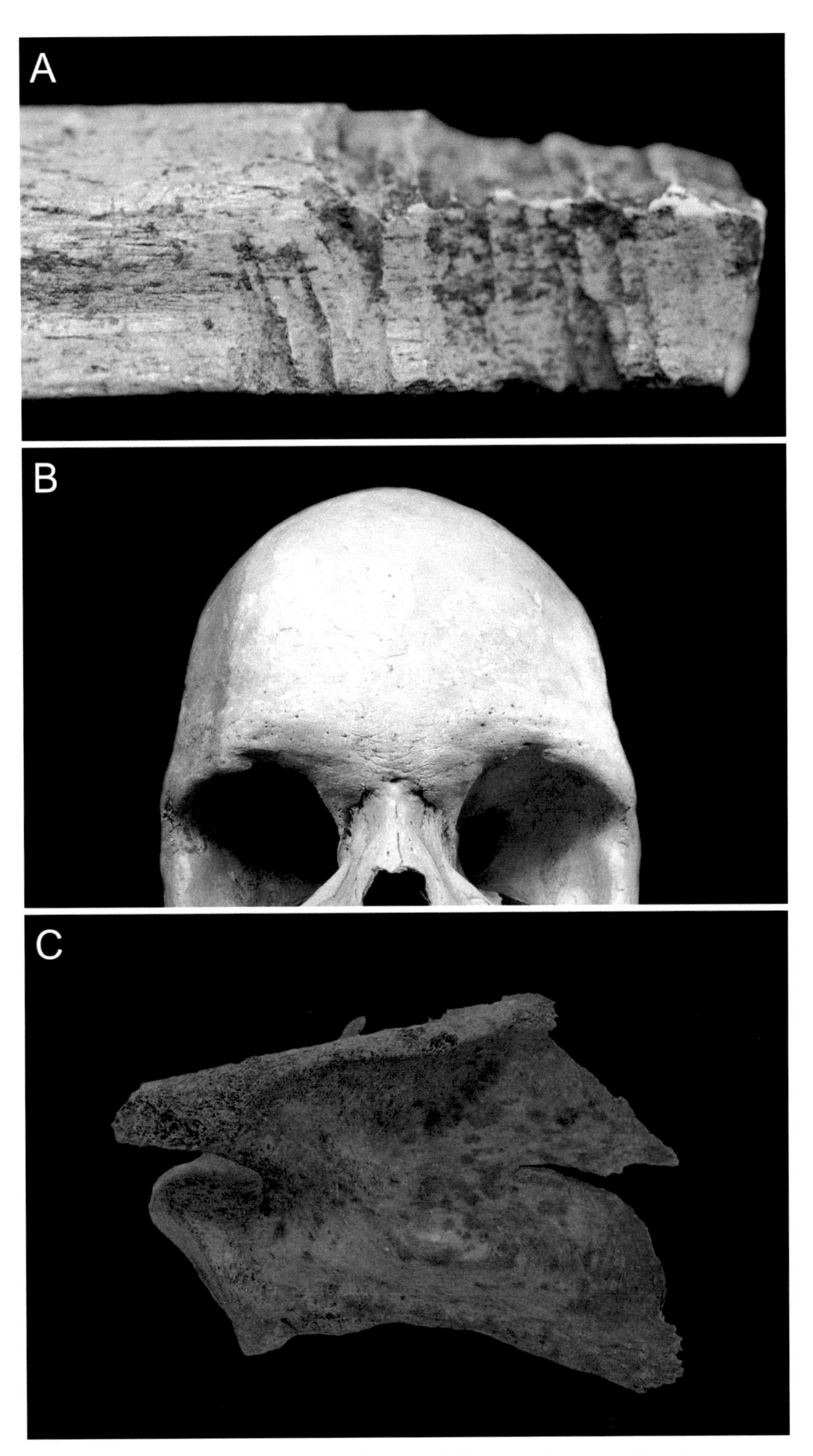

Figure 15.14. Three skeletal elements showing different taphonomic alterations.

Q10. Based on what is seen in **Figure 15.14C**, what is the likely cause of the discoloration of this scapula?

A) Low temperature fire

B) High temperature fire

C) Root etching

D) Soil staining

End-of-Chapter Summary

This chapter discussed taphonomic effects on the human body with an emphasis on the multiple ways to estimate PMI. The key takeaway from this chapter is the recognition that both time and temperature contribute significantly to the expected state of decomposition of a corpse. For short PMIs, less than 48 hours, the mortis triad is used by forensic pathologists to estimate PMI. After 48 hours, forensic anthropologists can estimate PMI using rough estimates of time frames in which a corpse reaches a certain stage of decomposition. The concept of Accumulated Degree Days provides a quantitative measure of PMI based on past case studies. For advanced stages of decomposition after the corpse has completely skeletonized, the forensic anthropologist can use information on bone weathering to describe the continual process of tissue breakdown. For longer PMIs, accuracy of estimates decreases and the estimate ranges become very broad. This chapter also discussed many factors that affect the appearance and condition of the human body after death, including animal scavenging, weathering, fire damage, and changes caused by the burial environment.

End-of-Chapter Exercises

Exercise 1

Materials Required: **Table 15.2**

Scenario: A colleague of yours is working in southern Arizona and has asked for your help in estimating the postmortem interval for a series of decedents. **Table 15.2** provides data linking stage of decomposition to PMI from a retrospective series of cases from the same area (Galloway, 1997). Using this information, answer the following questions.

Question 1: The first case that your colleague asks you about was recovered while actively undergoing putrefaction. During which postmortem interval is this stage of decomposition most often observed?

Question 2: The second case has been completely skeletonized. If you ignore the one occurrence of skeletonization at 7 days, this state of decomposition suggests a PMI of at least how long?

Question 3: Investigators would like to know if the third case, which was just beginning to bloat when recovered, could be the remains of an individual reported missing six months ago. Assuming the individual died at approximately the same time, how likely is it that the missing person and the third decedent represent the same individual?

Exercise 2

Materials Required: None

Scenario: A colleague has asked you to review their estimates of postmortem interval for two cases that they have been working on. They have provided you with the following brief descriptions of the context of recovery and their calculations of PMI:

Case 1: The remains of this decedent were discovered lying on the bottom of a wash in the desert in April. There was evidence of carnivore activity on the body and the left upper and lower extremities were not recovered. Decomposition was estimated to be 30 percent, the average temperature over the past month at the site of recovery was calculated to be 30°C, and the average humidity over the same period was 27 percent. PMI was estimated using the formula of Vass (2011) for aerobic environments (e.g., surface recoveries) as follows:

$$\text{PMI} = \frac{1285 * \left(\frac{20}{100}\right)}{0.0103 * 30 * 27} = 30.8 \text{ days.}$$

Case 2: The remains of this decedent were discovered lying on the ground in a pine forest in September. The body was relatively intact and decomposition was estimated to be 60 percent. The average temperature over the past month was calculated to be 74°F, and the average humidity over the same period was 44 percent. PMI was estimated using the formula of Vass (2011) for aerobic environments as follows:

$$\text{PMI} = \frac{1285*\left(\frac{60}{100}\right)}{0.0103*74*44} = 23 \text{ days.}$$

Use these descriptions and the discussion of Vass (2011) presented in this chapter to answer the following questions.

Question 1: Is the estimate of PMI presented for Case 1 reasonable? Why or why not? If the estimate is unreasonable, can it be easily corrected? If so, provide a corrected estimate of PMI.

Question 2: Is the estimate of PMI presented for Case 2 reasonable? Why or why not? If the estimate is unreasonable, can it be easily corrected? If so, provide a corrected estimate of PMI.

Question 3: Do you have any concerns regarding the accuracy and/or precision of the estimates of PMI produced using Vass's (2011) formula? How could they be improved?

Exercise 3

Materials Required: **Figure 15.15**

Scenario: In cleaning out a storage area, law enforcement personnel have discovered a set of skeletal remains contained in a cardboard box. They cannot locate any associated documentation and have transferred them to you for analysis. Images of these remains are presented in **Figure 15.15.** Based on what you can observe, answer the following questions.

Question 1: **Figure 15.15A** shows the left os coxa with the sides of the bone labeled. Based on the appearance of this bone, what position was the body likely in when it was originally recovered?

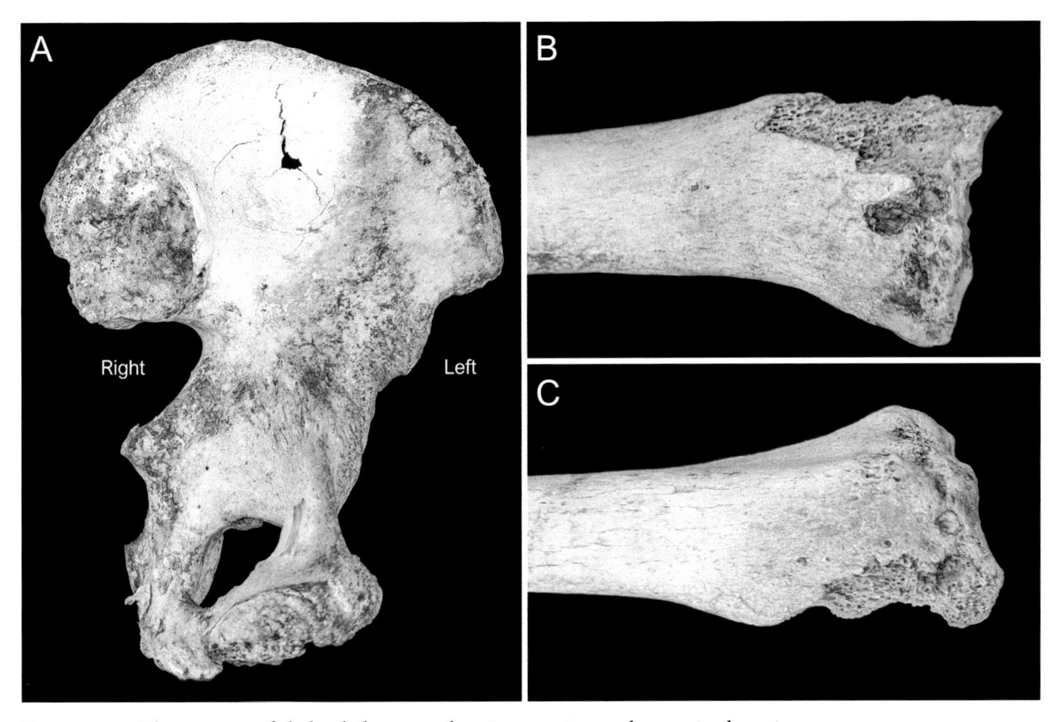

Figure 15.15. Three views of skeletal elements showing varying taphonomic alterations.

Question 2: **Figures 15.15B** and **15.15C** show the anterior and posterior aspects of the left proximal radius. Based on these images, is there any evidence that these remains were subject to animal scavenging? If so, please describe it.

Question 3: Considering the surface coloration of the elements shown in **Figure 15.15** and the fact that they are all from the left side, do you think it likely that the remains were still articulated when they were originally recovered? Why or why not?

References

Bass WM. 1997. Outdoor decomposition rates in Tennessee. In: Haglund WD, Sorg MH (Eds.), *Forensic Taphonomy. The Postmortem Fate of Human Remains.* Boca Raton, FL: CRC Press. p. 181–186.

Behrensmeyer AK. 1978. Taphonomic and ecological information from bone weathering. *Paleobiology* 4: 150–162.

Colard T, Delannoy Y, Naji S, Gosset D, Hartnett K, Bécart A. 2015. Specific patterns of canine scavenging in indoor settings. *Journal of Forensic Sciences* 60: 495–500.

Forbes SL, Perrault KA, Comstock JL. 2017. Microscopic post-mortem changes: the chemistry of decomposition. In: Schotsmans EMJ, Márquez-Grant N, Forbes SL (Eds.), *Taphonomy of Human Remains: Forensic Analysis of the Dead and the Depositional Environment.* New York, NY: John Wiley & Sons. p. 26–38.

Galloway A. 1997. The process of decomposition: a model from the Arizona-Sonoran Desert. In: Haglund WD, Sorg MH (Eds.), *Forensic Taphonomy: The Postmortem Fate of Human Remains.* Boca Raton, FL: CRC Press. p. 139–150.

Megyesi MS, Nawrocki SP, Haskell NH. 2005. Using accumulated degree-days to estimate the postmortem interval from decomposed human remains. *Journal of Forensic Sciences* 50: 618–626.

Rodriguez WC, Bass WM. 1983. Insect activity and its relationship to decay rates of human cadavers in East Tennessee. *Journal of Forensic Sciences* 28: 423–432.

Rodriguez WC, Bass WM. 1985. Decomposition of buried bodies and methods that may aid in their location. *Journal of Forensic Sciences* 30: 836–852.

Ross AH, Cunningham SL. 2011. Time-since-death and bone weathering in a tropical environment. *Forensic Science International* 204: 126–133.

Simmons, T. 2017. Post-mortem interval estimation: an overview of techniques. In: Schotsmans EMJ, Márquez-Grant N, Forbes SL (Eds.), *Taphonomy of Human Remains: Forensic Analysis of the Dead and the Depositional Environment.* New York, NY: John Wiley & Sons. p. 134–142.

Vass AA. 2001. Beyond the grave—understanding human decomposition. *Microbiology Today* 28: 190–192.

Vass AA. 2011. The elusive universal post-mortem interval formula. *Forensic Science International* 204: 34–40.

Vass AA, Bass WM, Wolt JD, Foss JE, Ammons JT. 1992. Time-since-death determinations of human cadavers using soil solution. *Journal of Forensic Sciences* 37: 1236–1253.

Vidoli GM, Steadman DW, Devlin JB, Meadows Jantz L. 2017. History and development of the first Anthropology Research Facility, Knoxville, Tennessee. In: Schotsmans EMJ, Márquez-Grant N, Forbes SL (Eds.), *Taphonomy of Human Remains: Forensic Analysis of the Dead and the Depositional Environment.* New York, NY: John Wiley & Sons. p. 463–475.

Illustration Credits

Figure 1.1. Image by Chris Stojanowski. [burned_human_bone_fragment] Catalogue No. 90.8.2. Courtesy of Maxwell Museum of Anthropology, University of New Mexico.

Figure 1.2. Image by Bitjungle [CC BY-SA 4.0, all images under CC BY-SA 4.0 are licensed under the following terms: https://creativecommons.org/licenses/by-sa/4.0/legalcode], via Wikimedia Commons.

Figure 1.3. Image by Chris Stojanowski.

Figure 1.4. Image by Chris Stojanowski.

Figure 1.5. Modified after image by University of Liverpool Faculty of Health [CC BY 2.0, all images under CC BY 2.0 are licensed under the following terms: https://creativecommons.org/licenses/by/2.0/legalcode], via Wikimedia Commons.

Figure 1.6. Image by Chris Stojanowski.

Figure 1.7. Images by Chris Stojanowski. Images A and B, [mastoids] Catalogue Nos. 77.11.14 and 77.11.1. Courtesy of Maxwell Museum of Anthropology, University of New Mexico; Images C and D, Texas State Donated Skeletal Collection 2013.053 and 2011.016. Courtesy of Forensic Anthropology Center at Texas State University.

Figure 1.8. Image by Andrew Seidel.

Figure 1.9. Image by Andrew Seidel.

Figure 1.10. Image by Chris Stojanowski.

Figure 1.11. Image by Chris Stojanowski.

Figure 1.12. Image adapted with the written permission of Joseph Hefner from Hefner (2007).

Figure 1.13. Image by Chris Stojanowski.

Figure 2.1. Image modified after Imklee238@asu.eduage by Arbeck (Own work) [CC BY 4.0], via Wikimedia Commons.

Figure 2.2. Image modified after Arbeck (Own work) [CC BY 4.0], via Wikimedia Commons.

Figure 2.3. Image modified after Arbeck (Own work) [CC BY 4.0], via Wikimedia Commons.

Figure 2.4. Image by Joaquim Alves Gaspar, vectorised by Antilived [CC BY 2.5, all images under CC BY 2.5 are licensed under the following terms: https://creativecommons.org/licenses/by/2.5/legalcode], via Wikimedia Commons.

Figure 2.5. Image by Chris Stojanowski.

Figure 2.6. Image by Realmastery [CC BY-SA 4.0, all images under CC BY-SA 4.0 are licensed under the following terms: https://creativecommons.org/licenses/by-sa/4.0/legalcode], via Wikimedia Commons.

Figure 2.7. Images modified after Sobotta (1909), Figures 14 and 15 [public domain], via Wikimedia Commons.

Figure 3.1. Images modified after image by Patrick J. Lynch [CC BY 2.5], via Wikimedia Commons.

Figure 3.2. Images by Edoarado [CC BY-SA 3.0, all images under CC BY-SA 3.0 are licensed under the following terms: https://creativecommons.org/licenses/by/3.0/legalcode], via Wikimedia Commons.

Figure 3.3. Image by Chris Stojanowski.

Figure 3.4. Image by Andrew Seidel.

Figure 3.5. Images modified after Sobotta (1909), Figures 37 and 40.

Figure 3.6. Image by Sarah Nabih [CC BY-SA 4.0], via Wikimedia Commons.

Figure 3.7. Image by BruceBlaus [CC BY 3.0, all images under CC BY 3.0 are licensed under the following terms: https://creativecommons.org/licenses/by/3.0/legalcode], via Wikimedia Commons.

Figure 3.8. Image A modified after image by Pbroks13 [CC BY 3.0], via Wikimedia Commons; Image B modified after image by Internet Archive Book Images [No restrictions], via Wikimedia Commons; Image C by patrick siemer from san francisco, usa (boneUploaded by berichard) [CC BY 2.0] via Wikimedia Commons.

Figure 3.9. Image A modified after image by OpenStax College [CC BY 3.0], via Wikimedia Commons; Image B by Chris Stojanowski, [GSW_closeup] Catalogue No. 77.11.17. Courtesy of Maxwell Museum of Anthropology, University of New Mexico.

Figure 3.10. Image by OpenStax Anatomy and Physiology OpenStax [CC BY 4.0], via Wikimedia Commons.

Figure 3.11. Images by Chris Stojanowski, [epiphyses] Catalogue No. 78.23.12. Courtesy of Maxwell Museum of Anthropology, University of New Mexico.

Figure 3.12. Modified after image by Laboratoire Servier [CC BY-SA 3.0], via Wikimedia Commons.

Figure 3.13. Image A modified after image by Bill Rhodes from Asheville [CC BY 2.0], via Wikimedia Commons; Image B by Robert M. Hunt (Original work of Robert M. Hunt) [public domain], via Wikimedia Commons.

Figure 3.14. Image by Chris Stojanowski, [woven_bone] Catalogue No. 80.7.33. Courtesy of Maxwell Museum of Anthropology, University of New Mexico.

Figure 3.15. Image A modified after image by Laboratoires Servier [CC BY-SA 3.0], via Wikimedia Commons; Image B by

Doc. RNDr. Josef Reischig, CSc. (Archiv autora) [GFDL or CC BY 3.0], via Wikimedia Commons.

Figure 3.16. Images: Flexion/Extension, Abduction/Adduction, Pronation/Supination, all by Connexions (http://cnx.org) [CC BY 3.0], via Wikimedia Commons.

Figure 3.17. Modified after image by OpenStax College [CC BY 3.0], via Wikimedia Commons.

Figure 3.18. Modified after image by OpenStax College [CC BY 3.0], via Wikimedia Commons.

Figure 3.19. Modified after image by OpenStax College [CC BY 3.0], via Wikimedia Commons.

Figure 3.20. Modified after image by Young Lae Moon [CC BY 3.0], via Wikimedia Commons.

Figure 3.21. Image by Nationalmuseet [CC BY-SA 3.0], via Wikimedia Commons.

Figure 3.22. Image by Chris Stojanowski. Texas State Donated Skeletal Collection 2015.057. Courtesy of Forensic Anthropology Center at Texas State University.

Figure 3.23. Modified after image by Nevit Dilmen [CC BY-SA 3.0], via Wikimedia Commons.

Figure 4.1. Modified after image by Wellcome Library, London. Wellcome Images images@wellcome.ac.uk http://wellcomeimages.org Tabula B text Darstellung des Knochenbaues von dem menschlichen Korper Fischer, Johann Martin Published: 1806 [CC BY 4.0], via Wikimedia Commons.

Figure 4.2. Image A by MAKY.OREL [CC BY-SA 4.0], via Wikimedia Commons; Image B, D, F, G, and H by MAKY.OREL [public domain], via Wikimedia Commons; Image C by Bone Clones [CC BY-SA 3.0], via Wikimedia Commons; Image E by Brian C. Goss [public domain], via Wikimedia Commons.

Figure 4.3. Images modified after Sobotta (1909), Figures 37 and 40.

Figure 4.4. Images by Chris Stojanowski. Texas State Donated Skeletal Collection 2016.015. Courtesy of Forensic Anthropology Center at Texas State University.

Figure 4.5. Images by Chris Stojanowksi. Texas State Donated Skeletal Collection 2016.015. Courtesy of Forensic Anthropology Center at Texas State University.

Figure 4.6. Images modified after Sobotta (1909), Figures 37 and 40.

Figure 4.7. Modified after image by MAKY.OREL [CC BY-SA 4.0], via Wikimedia Commons.

Figure 4.8. Image A by Chris Stojanowski. [nasal_closeup] Catalogue No. 79.27.11. Courtesy of Maxwell Museum of Anthropology, University of New Mexico. Image B modified after Sobotta (1909), Figure 101.

Figure 4.9. Image by OpenStax College [CC BY 3.0], via Wikimedia Commons.

Figure 4.10. Images by Chris Stojanowski.

Figure 4.11. Image modified after Image by I, RosarioVanTulpe [GFDL, CC-BY-SA-3.0, or CC BY-SA 2.5; all images under CC BY-SA are licensed under the following terms: https://creativecommons.org/licenses/by-sa/2.5/legalcode], via Wikimedia Commons.

Figure 4.12. Images by Chris Stojanowski.

Figure 4.13. Image by Chris Stojanowski.

Figure 4.14. Images by Chris Stojanowski.

Figure 4.15. Image by Chris Stojanowski.

Figure 4.16. Images by Chris Stojanowski.

Figure 4.17. Modified after images by MAKY.OREL [public domain], via Wikimedia Commons.

Figure 4.18. Image by Brian C. Goss [public domain], from Wikimedia Commons.

Figure 4.19. Image A modified after Sobotta (1909), Figure 131; Image B modified after image by Brian C. Goss [public domain], via Wikimedia Commons.

Figure 4.20. Images by Chris Stojanowski.

Figure 4.21. Images by Chris Stojanowski. Image A, [os coxa] Catalogue No. 78.23.12; Image B, [os coxa] Catalogue No. 79.26.16. Courtesy of Maxwell Museum of Anthropology, University of New Mexico.

Figure 4.22. Modified after images by MAKY.OREL [public domain], via Wikimedia Commons.

Figure 4.23. Image A modified after Sobotta (1909), Figure 154; Image B by Chris Stojanowski.

Figure 4.24. Modified after image by Wellcome Library, London. Wellcome Images images@wellcome.ac.uk http://wellcomeimages.org Tabula B text Darstellung des Knochenbaues von dem menschlichen Korper Fischer, Johann Martin Published: 1806 [CC BY 4.0], via Wikimedia Commons.

Figure 4.25. All images modified after images by MAKY.OREL [public domain], via Wikimedia Commons.

Figure 4.26. Images A and B modified after images by MAKY.OREL [public domain], via Wikimedia Commons; Image C by Chris Stojanowski.

Figure 4.27. Modified after image by Richard Mortel [CC BY 2.0], via Wikimedia Commons.

Figure 4.28. Image by Chris Stojanowski. Texas State Donated Skeletal Collection 2013.059. Courtesy of Forensic Anthropology Center at Texas State University.

Figure 4.29. Image by Chris Stojanowski.

Figure 4.30. Image by Chris Stojanowski, [radius_burned] Catalogue No. 80.18.6. Courtesy of Maxwell Museum of Anthropology, University of New Mexico.

Figure 5.1. Image A [public domain], via Wikimedia Commons; Image B by Forest and Kim Starr [CC BY-SA 3.0], via Wikimedia Commons; Image C by West Yorkshire Archaeology Advisory Service, Philip Holmes, 2019-12-22 10:56:06 [CC BY 2.0], via Wikimedia Commons; Image D. Courtesy of Laura Fulginiti.

Figure 5.2. Image A by The Portable Antiquities Scheme, Kevin Leahy, 2017-07-11 20:18:44 [CC BY 2.0], via Wikimedia Commons; Image B by Chris Stojanowski. [skull_cross-section]

Catalogue No. 77.13.25. Courtesy of Maxwell Museum of Anthropology, University of New Mexico.

Figure 5.3. Image by Paul Gamiche [CC BY-SA 4.0], via Wikimedia Commons.

Figure 5.4. Image by Berkshire Community College Bioscience Image Library [public domain], via Creative Commons.

Figure 5.5. Image A by Brooklyn Museum, CONS.52.52_xrs_view01.jpg [CC BY 3.0], via Creative Commons; Image B by Brooklyn Museum, CONS.1992.73.1_1994_xrs_detail01.jpg [CC BY 3.0], via Creative Commons; Image C by Birmingham Museums Trust, Peter Reavill, 2015-08-06 22:23:49 [CC BY 2.0], via Wikimedia Commons.

Figure 5.6. Image by Andrew Seidel after data from Zimmerman (2013).

Figure 5.7. Image A by North Lincolnshire Museum, Martin Foreman, 2017-05-15 14:04:24 [CC BY 2.0], via Wikimedia Commons; Image B designated as public domain, via Creative Commons; Image C by North Lincolnshire Museum, Martin Foreman, 2014-10-08 11:26:57 [CC BY-SA 4.0], via Wikimedia Commons.

Figure 5.8. Image A by James St. John [CC BY 2.0], via Wikimedia Commons; Image B by Lee Edwin Coursey [CC BY 2.0], via Creative Commons; Image C by Michael Coghlan [CC BY-SA 2.0], via Creative Commons.

Figure 5.9. Images A and B modified after images by James St. John [CC BY 2.0], via Wikimedia Commons; Image C by Chris Stojanowski, Texas State Donated Skeletal Collection 2015.057. Courtesy of Forensic Anthropology Center at Texas State University.

Figure 5.10. All images designated public domain, via Creative Commons.

Figure 5.11. Image A by Didier Descouense [CC BY-SA 4.0], via Wikimedia Commons; Image B designated public domain, via Creative Commons.

Figure 5.12. Image A by Gary Lee Todd, PhD [public domain], via Creative Commons; Image B by Andrew Seidel.

Figure 5.13. Image A by MAKY.OREL [public domain], via Wikimedia Commons. Images B and C by Chris Stojanowski. [subadult_bones] Catalogue No. DOC-78.23.12. Courtesy of Maxwell Museum of Anthropology, University of New Mexico.

Figure 5.14. Images. Courtesy of Laura Fulginiti and Andrew Seidel.

Figure 5.15. Image A modified after picture by Berkshire Community College Bioscience Image Library [public domain], via Creative Commons; Image B modified after Bradfield (2018), Figure 2; Image C modified after Veitschegger et al. (2018), Figure 3.

Figure 5.16. Image modified after Perez et al. (2017), Figure 2 and Figure 5 [CC BY 4.0].

Figure 5.17. Image A modified after image by A. Bolton [public domain], via Creative Commons; Image B. Courtesy of Laura Fulginiti and Andrew Seidel.

Figure 5.18. Image by Matija Križnar [CC BY-SA 4.0], via Wikimedia Commons.

Figure 5.19. Image by Chris Stojanowski. [adipocere] Catalogue No. 88.2.24. Courtesy of Maxwell Museum of Anthropology, University of New Mexico.

Figure 5.20. Image by Chris Stojanowski.

Figure 5.21. Image A by Manfred Werner—Tsui [CC BY-SA 3.0], via Wikimedia Commons; Image B by the Swedish History Museum, Stockholm [CC BY 2.0], via Wikimedia Commons; Image C by Wolfgang Sauber [CC BY-SA 4.0], via Wikimedia Commons; Image D modified after image by Gary Todd [public domain], via Wikimedia Commons; Image E modified after image by the Swedish History Museum, Stockholm [CC BY 2.0], via Wikimedia Commons.

Figure 5.22. Image by the Portable Antiquities Scheme, Dot Boughton, 2008-12-02 12:11:42 [CC BY-SA 4.0], via Wikimedia Commons.

Figure 5.23. Image by Nevit Dilmen [CC BY-SA 3.0], via Wikimedia Commons.

Figure 5.24. Image A by the kirbster [CC BY 2.0], via Creative Commons; Image B by Karen Roe [CC BY 2.0], via Creative Commons; Image C by Lamiot [CC BY-SA 4.0], via Wikimedia Commons; Image D by cogdogblog [CC BY 2.0], via Wikimedia Commons.

Figure 5.25. Image A modified after image by Judgefloro [public domain], via Wikimedia Commons; Image B modified after image by Gary Todd [public domain], via Wikimedia Commons; Image C modified after image by the Swedish History Museum, Stockholm [CC BY 2.0], via Wikimedia Commons.

Figure 5.26. Image A modified after Stagno et al. (2021), Figure 2a; Image B modified after Scheyer (2009), Figure 2; Image C modified after image by Berkshire Community College Bioscience Image Library [public domain], via Wikimedia Commons.

Figure 5.27. Image A by Couraco [CC BY-SA 3.0] via Wikimedia Commons; Image B by Hellerhoff [CC BY-SA 3.0], via Wikimedia Commons.

Figure 6.1. Image by Andrew Seidel.

Figure 6.2. Images by Andrew Seidel.

Figure 6.3. Images by Chris Stojanowski.

Figure 6.4. Modified after image by MAKY.OREL [public domain], via Wikimedia Commons.

Figure 6.5. Images modified after Sobotta (1909), Figures 208–211, via Wikimedia Commons.

Figure 6.6. Image by Andrew Seidel.

Figure 6.7. Image by Andrew Seidel and Chris Stojanowski.

Figure 6.8. Image by Andrew Seidel.

Figure 6.9. Images by Chris Stojanowski.

Figure 6.10. Image by Chris Stojanowski.

Figure 6.11. Images by Chris Stojanowski. Texas State Donated Skeletal Collection 2012.024. Courtesy of Forensic Anthropology Center at Texas State University.

Figure 6.12. Images by Chris Stojanowski. Texas State Donated Skeletal Collection 2015.048. Courtesy of Forensic Anthropology Center at Texas State University.

Figure 6.13. Images by Chris Stojanowski. Texas State Donated Skeletal Collection 2015.048. Courtesy of Forensic Anthropology Center at Texas State University.

Figure 6.14. Images by Chris Stojanowski. Texas State Donated Skeletal Collection 2015.048. Courtesy of Forensic Anthropology Center at Texas State University.

Figure 6.15. Top image modified after image by MAKY.OREL [public domain], via Wikimedia Commons; Bottom image by Chris Stojanowski.

Figure 6.16. Image by Chris Stojanowski. [skull_right] Catalogue No. 96.31.10. Courtesy of Maxwell Museum of Anthropology, University of New Mexico.

Figure 6.17. Image by Chris Stojanowski. Texas State Donated Skeletal Collection 2010.012. Courtesy of Forensic Anthropology Center at Texas State University.

Figure 6.18. Image by Chris Stojanowski. Texas State Donated Skeletal Collection 2013.017. Courtesy of Forensic Anthropology Center at Texas State University.

Figure 6.19. Image by Andrew Seidel.

Figure 7.1. Modified after Image by Indolences via Free Art Licences, via Wikimedia Commons.

Figure 7.2. Image by Andrew Seidel.

Figure 7.3. Modified after image by w: Coronation Dental Group [CC BY 3.0], via Wikimedia Commons.

Figure 7.4. After AlQahtani et al. (2010).

Figure 7.5. After AlQahtani et al. (2010).

Figure 7.6. After AlQahtani et al. (2010).

Figure 7.7. Image by Chris Stojanowski.

Figure 7.8. After AlQahtani et al. (2010).

Figure 7.9. Images by Chris Stojanowski.

Figure 7.10. Image by Chris Stojanowski. [humerus-unfused] Catalogue No. 79.28.21. Courtesy of Maxwell Museum of Anthropology, University of New Mexico.

Figure 7.11. Image by Chris Stojanowski.

Figure 7.12. Image A by Chris Stojanowski. [tibia-unfused] Catalogue No. 79.28.21. Courtesy of Maxwell Museum of Anthropology, University of New Mexico. Image B by Chris Stojanowski.

Figure 7.13. Image A modified after image by Gang65 [CC BY-SA 4.0], via Wikimedia Commons; Image B modified after image by w: Coronation Dental Group [CC BY 3.0], via Wikimedia Commons.

Figure 7.14. Images by Chris Stojanowski. Image C [mandible] Catalogue No. 78.23.12. Courtesy of Maxwell Museum of Anthropology, University of New Mexico.

Figure 8.1. Image by Chris Stojanowski.

Figure 8.2. Image by Chris Stojanowski. [os coxa_pubic-face] Catalogue No. 96.31.10. Courtesy of Maxwell Museum of Anthropology, University of New Mexico.

Figure 8.3. Left image by Chris Stojanowski, based on a 3D scan of Suchey-Brooks cast with permission of France Casting. Right image by Andrew Seidel.

Figure 8.4. Image by Andrew Seidel.

Figure 8.5. Image by Andrew Seidel.

Figure 8.6. Image by Andrew Seidel.

Figure 8.7. Image by Andrew Seidel.

Figure 8.8. Images. Courtesy of Laura Fulginiti and Andrew Seidel.

Figure 8.9. Image by Andrew Seidel.

Figure 8.10. Image by Chris Stojanowski.

Figure 8.11. Images by Chris Stojanowski. Image B, Texas State Donated Skeletal Collection 2016.031. Courtesy of Forensic Anthropology Center at Texas State University.

Figure 8.12. Image by Chris Stojanowski.

Figure 8.13. Image modified after Sobotta (1909), Figure 40 [public domain], via Wikimedia Commons.

Figure 8.14. Images by Chris Stojanowski. Image A, Texas State Donated Skeletal Collection 2014.039; Image B, Texas State Donated Skeletal Collection 2014.008; Image C, Texas State Donated Skeletal Collection 2010.010; Image D, Texas State Donated Skeletal Collection 2012.024. Courtesy of Forensic Anthropology Center at Texas State University.

Figure 8.15. Images by Chris Stojanowski. Image A Texas State Donated Skeletal Collection 2013.053; Image B Texas State Donated Skeletal Collection 2015.055. Courtesy of Forensic Anthropology Center at Texas State University.

Figure 8.16. Image by Andrew Seidel.

Figure 8.17. Image by Andrew Seidel.

Figure 8.18. Images A and B by Chris Stojanowski. Texas State Donated Skeletal Collection 2015.056. Courtesy of Forensic Anthropology Center at Texas State University. Image C redrawn by Chris Stojanowski after data in Buikstra and Ubelaker (1994).

Figure 9.1. Image modified after picture by Dave Auguste [CC BY-SA 4.0], via Wikimedia Commons.

Figure 9.2. Image modified after picture by S25454541 [CC BY-SA 4.0], via Wikimedia Commons.

Figure 9.3. Image by Andrew Seidel.

Figure 9.4. Adapted with the written permission of Joseph Hefner from Hefner (2007).

Figure 9.5. Adapted with the written permission of Joseph Hefner from Hefner (2007).

Figure 9.6. Image created by Andrew Seidel.

Figure 9.7. Images modified after Sobotta (1909), Figures 37, 40, and 41 [public domain], via Wikimedia Commons.

Figure 9.8. Images modified after Sobotta (1909), Figures 37, 40, and 41 [public domain], via Wikimedia Commons.

Figure 9.9. Image by Andrew Seidel.

Figure 9.10. Image created by Andrew Seidel incorporating output from Fordisc 3.1.

Figure 9.11. Image created by Andrew Seidel incorporating output from Fordisc 3.1.

Figure 9.12. Images by Chris Stojanowski. Image A, Texas State Donated Skeletal Collection 2012.024; Image B, Texas State Donated Skeletal Collection 2014.055; Image C, Texas State Donated Skeletal Collection 2016.031. Courtesy of Forensic Anthropology Center at Texas State University.

Figure 9.13. Images by Chris Stojanowski. Image A, Texas State Donated Skeletal Collection 2013.053; Image B Texas State Donated Skeletal Collection 2011.016. Courtesy of Forensic Anthropology Center at Texas State University.

Figure 9.14. Image by Andrew Seidel incorporating output from Fordisc 3.1.

Figure 9.15. Image by Andrew Seidel incorporating output from Fordisc 3.1.

Figure 9.16. Images by Chris Stojanowksi. Texas State Donated Skeletal Collection 2012.024. Courtesy of Forensic Anthropology Center at Texas State University.

Figure 9.17. Image by Andrew Seidel incorporating output from Fordisc 3.1.

Figure 9.18. Image by Andrew Seidel incorporating output from Fordisc 3.1.

Figure 9.19. Image by Andrew Seidel incorporating output from Fordisc 3.1.

Figure 10.1. Central image modified after Barclay (1824), Plate 1 [public domain]; Image A modified after Morris & McMurrich (1907), Figure 1 [public domain]; Image B modified after Braus (1921), Figure 70 [public domain]; Image C modified after Braus (1921), Figure 48 [public domain]; Image D modified after Barclay (1824), Plate 15 [public domain]; Image E modified after Barclay (1824), Plate 23 [public domain]; Image F modified after Barclay (1824), Plate 24 [public domain]; Image G modified after Barclay (1824), Plate 26 [public domain].

Figure 10.2. Image A modified after Dixon (1912), Figure 131 [public domain]; Image B by Andrew Seidel; Images C and D by Chris Stojanowski Texas State Donated Skeletal Collection 2013.017. Courtesy of Forensic Anthropology Center at Texas State University.

Figure 10.3. Images modified after Dixon (1912), Figure 7 (A) and Figure 2 (B) [public domain].

Figure 10.4. Image modified after Dixon (1912), Figure 13 [public domain].

Figure 10.5. Image modified after Dixon (1912), Figure 63 [public domain].

Figure 10.6. Image modified after Dixon (1912), Figure 71 [public domain].

Figure 10.7. Image modified after Dixon (1912), Figure 98 [public domain].

Figure 10.8. Image by Laerd Statistics [CC BY-SA 4.0], via Wikimedia Commons.

Figure 10.9. Image by Chris Stojanowski incorporating output from Fordisc 3.1.

Figure 10.10. Images modified after Braus (1921), Figure 1 [public domain].

Figure 11.1. Image by Chris Stojanowski. Texas State Donated Skeletal Collection 2016.031. Courtesy of Forensic Anthropology Center at Texas State University.

Figure 11.2. Image modified after picture by Hellerhoff [CC BY-SA 3.0], via Wikimedia Commons.

Figure 11.3. Image. Courtesy of Laura Fulginiti and Andrew Seidel.

Figure 11.4. Image by Chris Stojanowski.

Figure 11.5. Image. Courtesy of Laura Fulginiti.

Figure 11.6. Image. Courtesy of Laura Fulginiti and Andrew Seidel.

Figure 11.7. Image by Chris Stojanowski.

Figure 11.8. Image by Chris Stojanowski.

Figure 11.9. Image. Courtesy of Laura Fulginiti and Andrew Seidel.

Figure 11.10. Image by Chris Stojanowski [skull_superior] Catalogue No. 77.11.6. Courtesy of Maxwell Museum of Anthropology, University of New Mexico.

Figure 11.11. Image by Chris Stojanowski. Texas State Donated Skeletal Collection 2008.001. Courtesy of Forensic Anthropology Center at Texas State University.

Figure 11.12. Image by Chris Stojanowski [sternum] well-healed surgery (3)] Catalogue No. 88.2.15. Courtesy of Maxwell Museum of Anthropology, University of New Mexico.

Figure 11.13. Image by Chris Stojanowski. Texas State Donated Skeletal Collection 2014.052. Courtesy of Forensic Anthropology Center at Texas State University.

Figure 11.14. Image. Courtesy of Laura Fulginiti and Andrew Seidel.

Figure 11.15. Image. Courtesy of Laura Fulginiti and Andrew Seidel.

Figure 11.16. Images by Chris Stojanowski: A) [epiphyseal union] Catalogue No. DOC 71 79.28.21. Courtesy of Maxwell Museum of Anthropology, University of New Mexico; B) Texas State Donated Skeletal Collection 2012.024; C) Texas State Donated Skeletal Collection 2013.053; D) Texas State Donated Skeletal Collection 2014.052. Courtesy of Forensic Anthropology Center at Texas State University.

Figure 11.17. Image A modified after image by Hellerhoff [CC BY-SA 3.0], via Wikimedia Commons; Image B modified after image by Rahim Packir Saibo [CC BY 2.0], via Creative Commons.

Figure 11.18. Image A modified after image by MAKY.OREL [public domain], via Wikimedia Commons; Image B modified after image by National Institutes of Health [public domain], via Wikimedia Commons.

Figure 11.19. Image by Andrew Seidel.

Figure 11.20. Image modified after image "Crazy digital dental x-ray, London, UK.jpg" by gruntzooki [CC BY-SA 2.0], via Creative Commons.

Figure 11.21. Images by Bin Im Garten [CC BY-SA 3.0], via Wikimedia Commons.

Figure 11.22. Image A by icethim [CC BY 2.0], via Creative Commons; Image B by gruntzooki [CC BY-SA 2.0], via Creative Commons.

Figure 11.23. Image "Stirnhoehle frontal" by Hellerhoff [CC BY-SA 3.0], via Wikimedia Commons.

Figure 11.24. Image A by Hellerhoff [CC BY-SA 4.0], via Wikimedia Commons; Image B by Hellerhoff [CC BY-SA 3.0], via Wikimedia Commons.

Figure 11.25. Image. Courtesy of University of Glasgow Archives & Special Collections, Forensic Medicine collection, GB 248 GUA/FM/2A/25/249.

Figure 11.26. Image modified after picture by Stefano Ricci Cortili [CC BY 4.0], via Wikimedia Commons.

Figure 11.27. Image. Courtesy of Laura Fulginiti and Andrew Seidel.

Figure 12.1. Image by Andrew Seidel, incorporating image of femur modified after Dixon (1912), Figure 63 [public domain], via Wikimedia Commons.

Figure 12.2. Image by Andrew Seidel.

Figure 12.3. Image modified after picture by OpenStax College [CC BY 4.0], via Wikimedia Commons.

Figure 12.4. Image by National Institutes of Health, Health & Human Services [public domain], via Wikimedia Commons.

Figure 12.5. Image. Courtesy of Laura Fulginiti and Andrew Seidel.

Figure 12.6. Image by Chris Stojanowski.

Figure 12.7. Image A by Chris Stojanowski. [GSW-frontal] Catalogue No. 77.11.17. Courtesy of Maxwell Museum of Anthropology, University of New Mexico; Image B by Chris Stojanowski. [BFT-closeup] Catalogue No. 77.13.4. Courtesy of Maxwell Museum of Anthropology, University of New Mexico; Image C. Courtesy of Laura Fulginiti and Andrew Seidel.

Figure 12.8. Images. Courtesy of Laura Fulginiti and Andrew Seidel.

Figure 12.9. Images by Chris Stojanowski. Image A [GSW-closeup] Catalogue No. 78.20.3. Courtesy of Maxwell Museum of Anthropology, University of New Mexico. Images B and C of France Casting Civil War gunshot wound casts. Image D [healed_bone] Catalogue No. 77.11.16. Courtesy of Maxwell Museum of Anthropology, University of New Mexico.

Figure 12.10. Images A, C, and D. Courtesy of Laura Fulginiti and Andrew Seidel; Image B by Chris Stojanowski. Texas State Donated Skeletal Collection 2012.024. Courtesy of Forensic Anthropology Center at Texas State University.

Figure 12.11. Image by Andrew Seidel, incorporating image of femur modified after Dixon (1912), Figure 63 [public domain], via Wikimedia Commons.

Figure 12.12. Image by Chris Stojanowski [BFT_closeup] Catalogue No. 77.12.5. Courtesy of Maxwell Museum of Anthropology, University of New Mexico.

Figure 12.13. Image A by Andrew Seidel; Image B by Chris Stojanowski [fracture-closeup] Catalogue No. 79.27.11. Courtesy of Maxwell Museum of Anthropology, University of New Mexico; Image C modified after image by MAKY.OREL [public domain], via Wikimedia Commons.

Figure 12.14. Images A and D. Courtesy of Laura Fulginiti and Andrew Seidel. Images B and C modified after images by MAKY.OREL [public domain], via Wikimedia Commons.

Figure 12.15. Images A, B, and D. Courtesy of Laura Fulginiti and Andrew Seidel. Image C by Chris Stojanowski. [bone_fracture] Catalogue No. 78.20.1. Courtesy of Maxwell Museum of Anthropology, University of New Mexico.

Figure 12.16. Image A by Chris Stojanowski. [trauma_closeup] Catalogue No. 77.12.5. Courtesy of Maxwell Museum of Anthropology, University of New Mexico. Images B, C, and D. Courtesy of Laura Fulginiti and Andrew Seidel.

Figure 12.17. Image. Courtesy of Laura Fulginiti and Andrew Seidel.

Figure 12.18. Image. Courtesy of Laura Fulginiti and Andrew Seidel.

Figure 12.19. Images. Courtesy of Laura Fulginiti and Andrew Seidel.

Figure 13.1. Image by Chris Stojanowski [GSW_skull_inferior] Catalogue No. 92.13.3. Courtesy of Maxwell Museum of Anthropology, University of New Mexico.

Figure 13.2 Modified after image by Quadrell derivative work: Indy muaddib (Bullet.svg) [public domain], via Wikimedia Commons.

Figure 13.3. Image A [public domain], via Wikimedia Commons; Image B by Chris Stojanowski, [GSW-closeup] Catalogue No. 77.11.18. Courtesy of Maxwell Museum of Anthropology, University of New Mexico; Image C by Chris Stojanowski, [GSW_closeup_exit] Catalogue No. 77.11.17. Courtesy of Maxwell Museum of Anthropology, University of New Mexico.

Figure 13.4. Image by Chris Stojanowski, [GSW_closeup] Catalogue No. 77.10.8. Courtesy of Maxwell Museum of Anthropology, University of New Mexico.

Figure 13.5. Images by Chris Stojanowski, [GSW_radiating_fractures] Catalogue No. 77.11.17. Courtesy of Maxwell Museum of Anthropology, University of New Mexico.

Figure 13.6. Image. Courtesy of Laura Fulginiti and Andrew Seidel.

Figure 13.7. Image by Chris Stojanowski, [keyhole_GSW] Catalogue No. 96.31.10. Courtesy of Maxwell Museum of Anthropology, University of New Mexico.

Figure 13.8. Image. Courtesy of Laura Fulginiti and Andrew Seidel.

Figure 13.9. Image by Chris Stojanowski, [multiple_GSWs] Catalogue No. 77.13.15. Courtesy of Maxwell Museum of Anthropology, University of New Mexico.

Figure 13.10. Image by Chris Stojanowski.

Figure 13.11. Images A and B. Courtesy of Laura Fulginiti and Andrew Seidel; Image C by National Institutes of Health, Health and Human Services [public domain], via Wikimedia Commons.

Figure 13.12. Images A, B, and C. Courtesy of Laura Fulginiti and Andrew Seidel. Image D by Chris Stojanowski, [GSW_closeup] Catalogue No. 90.8.18. Courtesy of Maxwell Museum of Anthropology, University of New Mexico.

Figure 13.13. Image by Chris Stojanowski. Texas State Donated Skeletal Collection 2012.019. Courtesy of Forensic Anthropology Center at Texas State University.

Figure 13.14. Image by Chris Stojanowski, [GSW_posterior] Catalogue No. 96.31.10. Courtesy of Maxwell Museum of Anthropology, University of New Mexico.

Figure 13.15. Images. Courtesy of Laura Fulginiti and Andrew Seidel.

Figure 13.16. Images. Courtesy of Laura Fulginiti and Andrew Seidel.

Figure 13.17. Images A and B. Courtesy of Laura Fulginiti and Andrew Seidel; Image C by Chris Stojanowksi, [bullet_embedded] Catalogue No. 2003.6.6. Courtesy of Maxwell Museum of Anthropology, University of New Mexico.

Figure 13.18. Modification, original image credit: Bronze Age skull from Jericho, Palestine, 2200–2000. Science Museum, London [CC BY 4.0], via Wellcome Collection.

Figure 14.1. Image by Chris Stojanowski, [BFT_skull] Catalogue No. 88.2.23. Courtesy of Maxwell Museum of Anthropology, University of New Mexico.

Figure 14.2. Images. Courtesy of Laura Fulginiti and Andrew Seidel.

Figure 14.3. Image A modified after image by RosarioVanTulpe [*GFDL, CC-BY-SA-3.0* or *CC BY-SA 2.5-2.0-1.0*], via Wikimedia Commons; Image B modified after image by Internet Archive Book Images [No restrictions], via Wikimedia Commons.

Figure 14.4. Image by National Institutes of Health, Health & Human Services [public domain], via Wikimedia Commons.

Figure 14.5. Image A modified after image by RosarioVanTulpe [GFDL,CC-BY-SA-3.0orCC BY-SA 2.5-2.0-1.0], via Wikimedia Commons; Image B modified after image by Chelepepino (Own work) [GFDL or CC BY 3.0], via Wikimedia Commons.

Figure 14.6. Image A by Chris Stojanowski, [rib_fracture] Catalogue No. 77.13.4. Courtesy of Maxwell Museum of Anthropology, University of New Mexico; Image B modified after image by James Heilman, MD [CC BY-SA 3.0 or GFDL], via Wikimedia Commons.

Figure 14.7. Image by Andrew Seidel, incorporating image of femur modified after Dixon (1912), Figure 63 [public domain], via Wikimedia Commons.

Figure 14.8. Image. Courtesy of Laura Fulginiti and Andrew Seidel.

Figure 14.9. Image. Courtesy of Laura Fulginiti and Andrew Seidel.

Figure 14.10. Image. Courtesy of Laura Fulginiti and Andrew Seidel.

Figure 14.11. Image. Courtesy of Laura Fulginiti and Andrew Seidel.

Figure 14.12. Image by Chris Stojanowski, [SFTs_cranium] Catalogue No. 77.11.11. Courtesy of Maxwell Museum of Anthropology, University of New Mexico.

Figure 14.13. Image A by Chris Stojanowski, [SFT_orbit] Catalogue No. 2690.593.1. Courtesy of Maxwell Museum of Anthropology, University of New Mexico; Image B by Chris Stojanowski, [puncture] Catalogue No. 2005.3.1. Courtesy of Maxwell Museum of Anthropology, University of New Mexico.

Figure 14.14. Image by Chris Stojanowski and Andrew Seidel.

Figure 14.15. Images by Chris Stojanowski.

Figure 14.16. Image A by Chris Stojanowski, [cranial_trauma] Catalogue No. 96.31.7. Courtesy of Maxwell Museum of Anthropology, University of New Mexico; Image B by Chris Stojanowski, [cranial_trauma] Catalogue No. 78.23.12. Courtesy of Maxwell Museum of Anthropology, University of New Mexico.

Figure 14.17. Image A by Chris Stojanowski of a cast by France casting, used with the permission of France Casting; Image B by Chris Stojanowski, [blunt_force_trauma] Catalogue No. 77.11.11. Courtesy of Maxwell Museum of Anthropology, University of New Mexico.

Figure 14.18. Image A by Chris Stojanowski, [BFT] Catalogue No. 77.13.25. Courtesy of Maxwell Museum of Anthropology, University of New Mexico; Image B by Chris Stojanowski, [BFT] Catalogue No. 88.2.23. Courtesy of Maxwell Museum of Anthropology, University of New Mexico.

Figure 14.19. Image A by Chris Stojanowski, [calotte] Catalogue No. 2690.593.1. Courtesy of Maxwell Museum of Anthropology, University of New Mexico; Image B by Chris Stojanowski, [frontal_trauma] Catalogue No. 77.11.11. Courtesy of Maxwell Museum of Anthropology, University of New Mexico.

Figure 14.20. Image A by Chris Stojanowski, [BFT_skull] Catalogue No. 88.2.14. Courtesy of Maxwell Museum of Anthropology, University of New Mexico; Image B by Chris Stojanowski, [triangle_wound] Catalogue No. 77.13.4. Courtesy of Maxwell Museum of Anthropology, University of New Mexico; Image C by Chris Stojanowski, [circular_puncture] Catalogue No. 2005.3.1. Courtesy of Maxwell Museum of Anthropology, University of New Mexico; Image D. Courtesy of Laura Fulginiti and Andrew Seidel.

Figure 14.21. Image A by Chris Stojanowski, [BFT] Catalogue No. 77.12.5. Courtesy of Maxwell Museum of Anthropology, University of New Mexico; Image B by Chris Stojanowski, [BFT] Catalogue No. 88.2.23. Courtesy of Maxwell Museum of Anthropology, University of New Mexico.

Figure 14.22. Image A by Blue tooth7 (Own work) [CC BY-SA 4.0], via Wikimedia Commons. Image B by Darrin Warner [public domain], via Wikimedia Commons.

Figure 14.23. Image by Chris Stojanowski.

Figure 14.24. Image by Chris Stojanowski, [cut_bone]

Catalogue No. 2690.593.1. Courtesy of Maxwell Museum of Anthropology, University of New Mexico.

Figure 14.25. Image by Chris Stojanowski, [cut_bone] Catalogue No. 2690.593.1. Courtesy of Maxwell Museum of Anthropology, University of New Mexico.

Figure 14.26. Image A by Chris Stojanowski, [blade_set] Catalogue No. 2690.593.1. Courtesy of Maxwell Museum of Anthropology, University of New Mexico; Image B modified after image by Ludraman Vector: Ufo karadagli (File:Saw blade.png) [public domain], via Wikimedia Commons; Image C. Courtesy of Laura Fulginiti and Andrew Seidel.

Figure 14.27. Image A by Chris Stojanowski, [cut_bone] Catalogue No. 2690.593.1. Courtesy of Maxwell Museum of Anthropology, University of New Mexico; Image B by Chris Stojanowski.

Figure 14.28. Image by Chris Stojanowski, [kerf] Catalogue No. 78.20.2. Courtesy of Maxwell Museum of Anthropology, University of New Mexico.

Figure 14.29. Image. Courtesy of Laura Fulginiti and Andrew Seidel.

Figure 14.30. Images. Courtesy of Laura Fulginiti and Andrew Seidel.

Figure 14.31. Images. Courtesy of Laura Fulginiti and Andrew Seidel.

Figure 14.32. Image A. Courtesy of Laura Fulginiti and Andrew Seidel. Image B by MAKY.OREL [public domain], via Wikimedia Commons.

Figure 15.1. Images (*1, 2, 3, 4, 5, 6*) by Wellcome Collection [*CC BY 4.0*], via Wikimedia Commons.

Figure 15.2. Image by Ngeokor [*CC BY-SA 4.0*], via Wikimedia Commons.

Figure 15.3. Images by Hbreton19 [*CC BY-SA 3.0*) or *GFDL*], via Wikimedia Commons.

Figure 15.4. Image by Hbreton19 [*CC BY-SA 3.0* or *GFDL*], via Wikimedia Commons.

Figure 15.5. Images A, B, C, D, and E by Chris Stojanowski; Image D, [weathered_bone] Catalogue No. 74.51.2. Courtesy of Maxwell Museum of Anthropology, University of New Mexico; Image F by MOs810 (Own work) [CC BY-SA 4.0], via Wikimedia Commons.

Figure 15.6. Image A by Chris Stojanowski, [weathered_bone] Catalogue No. 74.51.2. Courtesy of Maxwell Museum of Anthropology, University of New Mexico; Image B by Chris Stojanowksi, [weathered_bone] Catalogue No. 74.51.2. Courtesy of Maxwell Museum of Anthropology, University of New Mexico.

Figure 15.7. Image by Chris Stojanowski, [carnivore_marks] Catalogue No. 90.8.18. Courtesy of Maxwell Museum of Anthropology, University of New Mexico.

Figure 15.8. Image A by Alexey Krasavin (originally posted to Flickr as Зубке (Tooth)) [CC BY-SA 2.0], via Wikimedia Commons; Image B by Chris Stojanowski, [rodent_tooth_marks] Catalogue No. 88.2.28. Courtesy of Maxwell Museum of Anthropology, University of New Mexico.

Figure 15.9. Image A by Chris Stojanowski, [bone_damage] Catalogue No. 78.20.1. Courtesy of Maxwell Museum of Anthropology, University of New Mexico.; Image B by Chris Stojanowski, [bone_damage] Catalogue No. 90.8.18. Courtesy of Maxwell Museum of Anthropology, University of New Mexico.

Figure 15.10. Image A by Chris Stojanowski, [burned_bone] Catalogue No. 80.18.25. Courtesy of Maxwell Museum of Anthropology, University of New Mexico; Image B Chris Stojanowski. [burned_skull] Catalogue No. 90.8.2. Courtesy of Maxwell Museum of Anthropology, University of New Mexico.

Figure 15.11. Image by Chris Stojanowski, [burned_frontal] Catalogue No. 80.18.27. Courtesy of Maxwell Museum of Anthropology, University of New Mexico.

Figure 15.12. Image A by Chris Stojanowski, [bone_staining] Catalogue No. 89.14.26. Courtesy of Maxwell Museum of Anthropology, University of New Mexico; Image B by Chris Stojanowski, [bone_damage] Catalogue No. 2005.3.2. Courtesy of Maxwell Museum of Anthropology, University of New Mexico; Image C by Chris Stojanowski.

Figure 15.13. Image by Chris Stojanowski.

Figure 15.14. Image A by Chris Stojanowski; Image B by Chris Stojanowski, [animal_marks] Catalogue No. 2005.3.2. Courtesy of Maxwell Museum of Anthropology, University of New Mexico; Image C modified after image by the Swedish History Museum, Stockholm [CC BY 2.0], via Wikimedia Commons.

Figure 15.15. Images. Courtesy of Laura Fulginiti and Andrew Seidel.

References

Adams BJ. 2003a. Establishing personal identification based on specific patterns of missing, filled, and unrestored teeth. *Journal of Forensic Sciences* 48: 487–496.

Adams BJ. 2003b. The diversity of adult dental patterns in the United States and the implications for personal identification. *Journal of Forensic Sciences* 48: 497–503.

Adams BJ, Crabtree PJ. 2008. *Comparative Skeletal Anatomy: A Photographic Atlas for Medical Examiners, Coroners, Forensic Anthropologists, and Archaeologists.* Totowa, NJ: Humana Press.

Adams BJ, Maves RC. 2002. Radiographic identification using the clavicle of an individual missing from the Vietnam conflict. *Journal of Forensic Sciences* 47: 369–373.

Adams DM, Ralston CE, Sussman RA, Heim K, Bethard JD. 2019. Impact of population-specific dental development on age estimation using dental atlases. *American Journal of Physical Anthropology* 168: 190–199.

Albanese J, Saunders SR. 2006. Is it possible to escape racial typology in forensic identification? In: Schmitt A, Cunha E, Pinheiro J (Eds.), *Forensic Anthropology and Medicine: Complementary Sciences from Recovery to Cause of Death.* Totowa, NJ: Humana Press. p. 281–316.

Algee-Hewitt BFB, Hughes CE, Anderson BE. 2018. Temporal, geographic and identification trends in craniometric estimates of ancestry for persons of Latin American origin. *Forensic Anthropology* 1: 4–17.

AlQahtani SJ, Hector MP, Liversidge HM. 2010. Brief communication. The London atlas of human tooth development and eruption. *American Journal of Physical Anthropology* 142: 481–490.

AlQahtani SJ, Hector MP, Liversidge HM. 2014. Accuracy of dental age estimation charts: Schour and Massler, Ubelaker, and the London Atlas. *American Journal of Physical Anthropology* 154: 70–78.

Anderson BE. 2008. Identifying the dead: methods utilized by the Pima County (Arizona) Office of the Medical Examiner for undocumented border crossers: 2001–2006. *Journal of Forensic Sciences* 53: 8–15.

Angel JL. 1974. Bones can fool people. *FBI Law Enforcement Bulletin* 43: 16–20.

Angyal M, Dérczy K. 1998. Personal identification on the basis of antemortem and postmortem radiographs. *Journal of Forensic Sciences* 43: 1089–1093.

Armelagos GJ, Goodman AH. 1998. Race, racism, and anthropology. In: Goodman AH, Leatherman TL (Eds.), *Building a New Biocultural System: Political-Economic Perspectives on Human Biology.* Ann Arbor: University of Michigan Press. p. 359–377.

Auerbach BM. 2011. Methods for estimating missing human skeletal element osteometric dimensions employed in the revised Fully technique for estimating stature. *American Journal of Physical Anthropology* 145: 67–80.

Aufderheide AC, Rodríguez-Martín C. 1998. *The Cambridge Encyclopedia of Human Paleopathology.* Cambridge: Cambridge University Press.

Austin-Smith D, Maples WR. 1994. The reliability of skull/photograph superimposition in individual identification. *Journal of Forensic Sciences* 39: 446–455.

Ayers HG, Jantz RL, Moore-Jansen PH. 1990. Giles & Elliot race discriminant functions revisited: a test using recent forensic cases. In: Gill GW, Rhine S (Eds.), *Skeletal Attribution of Race: Methods for Forensic Anthropology.* Albuquerque, NM: Maxwell Museum of Anthropology Anthropological Papers No. 4. p. 65–71.

Baht GS, Vi L, Alman BA. 2018. The role of the immune cells in fracture healing. *Current Osteoporosis Reports* 16: 138–145.

Baik SO, Uku JM, Sikirica M. 1991. A case of external beveling with an entrance gunshot wound to the skull made by a small caliber rifle bullet. *American Journal of Forensic Medicine and Pathology* 12: 334–336.

Baker SJ, Gill GW, Kieffer DA. 1990. Race & sex determination from the intercondylar notch of the distal femur. In: Gill GW, Rhine S (Eds.), *Skeletal Attribution of Race: Methods for Forensic Anthropology.* Albuquerque, NM: Maxwell Museum of Anthropology Anthropological Papers No. 4. p. 91–96.

Barbujani G, Colonna V. 2010. Human genome diversity: frequently asked questions. *Trends in Genetics* 26: 285–295.

Barclay J. 1824. *A Series of Engravings Representing the Bones of the Human Skeleton; with the Skeletons of Some of the Lower Animals, and Explanatory References,* Second Edition. Edinburgh, Scotland: Printed for MacLachlan and Stewart.

Barnes E. 1994. *Developmental Defects of the Axial Skeleton in Paleopathology.* Niwot, CO: University Press of Colorado.

Bass WM. 1997. Outdoor decomposition rates in Tennessee. In: Haglund WD, Sorg MH (Eds.), *Forensic Taphonomy. The Postmortem Fate of Human Remains.* Boca Raton, FL: CRC Press. p. 181–186.

Behrensmeyer AK. 1978. Taphonomic and ecological information from bone weathering. *Paleobiology* 4: 150–162.

Berg GE. 2008. Pubic bone age estimation in adult women. *Journal of Forensic Sciences* 53: 569–577.

Berg GE, Kenyhercz MW. 2017. Introducing human mandible identification [(huMANid]: a free, web-based GUI to classify human mandibles. *Journal of Forensic Sciences* 62: 1592–1598.

Berg GE, Ta'Ala SC, Kontanis EJ, Leney SS. 2007. Measuring the intercondylar shelf angle using radiographs: intra- and inter-observer error tests of reliability. *Journal of Forensic Sciences* 52: 1020–1024.

Berryman HE, Symes SA. 1998. Recognizing gunshot and blunt cranial trauma through fracture interpretations. In: Reichs KJ (Ed.), *Forensic Osteology. Advances in the Identification of Human Remains,* Second Edition. Springfield, IL: Charles C. Thomas. p. 333–352.

Berryman HE, Smith OC, Symes SA. 1995. Diameter of cranial gunshot wounds as a function of bullet caliber. *Journal of Forensic Sciences* 40: 751–754.

Berryman HE, Lanfear AK, Shirley NR. 2012. The biomechanics of gunshot trauma to bone: research considerations within the present judicial climate. In: Dirkmaat DC (Ed.), *A Companion to Forensic Anthropology.* London: Blackwell. p. 390–399.

Berryman HE, Shirley NR, Lanfear AK. 2013. Low-velocity trauma. In: Tersigni-Tarrant MA, Shirley NR (Eds.), *Forensic Anthropology: An Introduction.* Boca Raton, FL: CRC Press. p. 271–290.

Berthaume MA, Di Federico E, Bull AMJ. 2019. Fabella prevalence rate increases over 150 years, and rates of other sesamoid bones remain constant: a systematic review. *Journal of Anatomy* 235: 67–79.

Bethard JD, DiGangi EA. 2020. Letter to the editor—moving beyond a lost cause: forensic anthropology and ancestry estimates in the United States. *Journal of Forensic Sciences* 65: 1791–1792.

Bhoopat T. 1995. A case of internal beveling with an exit gunshot wound to the skull. *Forensic Science International* 71: 97–101.

Bidmos MA, Manger PR. 2012. New soft tissue correction factors for stature estimation: results from magnetic resonance imaging. *Forensic Science International* 214: 212e1–212e7.

Birkby WH. 1966. An evaluation of race and sex identification from cranial measurements. *American Journal of Physical Anthropology* 24: 21–27.

Blakey ML. 1999. Scientific racism and the biological concept of race. *Literature and Psychology* 45: 29–43.

Boldsen JL, Milner GR, Konigsberg LW, Wood JW. 2002. Transition analysis: a new method for estimating age from skeletons. In: Hoppa RD, Vaupel J (Eds.), *Palaeodemography: Age Distributions from Skeletal Samples.* Cambridge: Cambridge University Press. p. 73–106.

Brace CL. 1995. Region does not mean "race"—reality versus convention in forensic anthropology. *Journal of Forensic Sciences* 40: 171–175.

Brace CL. 2005. *"Race" Is a Four-Letter Word: The Genesis of the Concept.* Oxford: Oxford University Press.

Bradfield J. 2018. Identifying animal taxa used to manufacture bone tools during the Middle Stone Age at Sibudu, South Africa: results of a CT-rendered histological analysis. *PLoS ONE* 13: e0208319. Available from: https://doi.org/10.1371/journal.pone.0208319.

Braus H. 1921. *Anatomie des Menschen: ein Lehrbuch für Studierende und Ärzte.* Berlin: Julius Springer.

Brogdon BG. 1998. Radiological identification of individual remains. In: Brogdon BG (Ed.), *Forensic Radiology.* Boca Raton, FL: CRC Press. p. 149–187.

Brogdon BG. 2011. Radiological identification of individual remains. In: Thali MJ, Viner MD, Brogdon BG (Eds.), *Brogdon's Forensic Radiology,* Second Edition. Boca Raton, FL: CRC Press. p. 153–176.

Brooks ST, Suchey JM. 1990. Skeletal age determination based on the os pubis: a comparison of the Acsádi-Nemeskéri and Suchey-Brooks methods. *Human Evolution* 5: 227–238.

Buckberry JL, Chamberlain AT. 2002. Age estimation from the auricular surface of the ilium: a revised method. *American Journal of Physical Anthropology* 119: 231–239.

Buikstra JE (Ed.). 2019. *Ortner's Identification of Pathological Conditions in Human Skeletal Remains,* Third Edition. London: Academic Press.

Buikstra JE, Ubelaker DH (Eds.). 1994. *Standards for Data Collection from Human Skeletal Remains.* Fayetteville: Arkansas Archeological Survey.

Byers SN. 2017. *Introduction to Forensic Anthropology,* Fifth Edition. London: Routledge.

Caccia G, Magli F, Tagi VM, Porta DGA, Cummaudo M, Márquez-Grant N, Cattaneo C. 2016. Histological determination of the human origin from dry bone: a cautionary note for subadults. *International Journal of Legal Medicine* 130: 299–307.

Cameriere R, Ferrante L, Mirtella D, Rollo FU, Cingolani M. 2005. Frontal sinuses for identification: quality of classifications, possible error and potential corrections. *Journal of Forensic Sciences* 50: 770–773.

Caplova Z, Obertova Z, Gibelli DM, Mazzarelli D, Fracasso T, Vanezis P, Sforza C, Cattaneo C. 2017. The reliability of facial recognition of deceased persons on photographs. *Journal of Forensic Sciences* 62: 1286–1291.

Caspari R. 2009. 1918: three perspectives on race and human variation. *American Journal of Physical Anthropology* 139: 5–15.

Caspari R. 2010. Deconstructing race: racial thinking, geographic variation, and implications for biological anthropology. In: Larsen CS (Ed.), *A Companion to Biological Anthropology.* Malden, MA: Blackwell. p. 104–123.

Cattaneo C, De Angelis D, Porta D, Grandi M. 2006. Personal identification of cadavers and human remains. In: Schmitt A, Cunha E, Pinheiro J (Eds.), *Forensic Anthropology and Medicine: Complementary Sciences from Recovery to Cause of Death.* Totowa, NJ: Humana Press. p. 359–379.

Cattaneo C, Porta D, Gibelli D, Gamba C. 2009. Histological determination of the human origin of bone fragments. *Journal of Forensic Sciences* 54: 531–533.

Christensen AM. 2004. The impact of *Daubert:* implications for testimony and research in forensic anthropology (and the use of frontal sinuses in personal identification). *Journal of Forensic Sciences* 49: 427–430.

Christensen AM. 2005. Testing the reliability of frontal sinuses in positive identification. *Journal of Forensic Sciences* 50: 18–22.

Christensen AM, Crowder CM. 2009. Evidentiary standards for forensic anthropology. *Journal of Forensic Sciences* 54: 1211–1216.

Christensen AM, Hatch GM. 2016. Quantification of radiologic identification (RADid) and the development of a population

frequency data repository. *Journal of Forensic Radiology and Imaging* 7: 14–16.
Christensen AM, Hatch GM. 2018. Advances in the use of frontal sinuses for human identification. In: Latham K, Bartelink E, Finnegan M (Eds.), *New Perspectives in Forensic Human Skeletal Identification*. London: Academic Press. p. 227–240.
Christensen AM, Smith MA, Thomas RM. 2012. Validation of X-ray fluorescence spectrometry for determining osseous or dental origin of unknown material. *Journal of Forensic Sciences* 57: 47–51.
Christensen AM, Passalacqua NV, Bartelink EJ. 2014. *Forensic Anthropology. Current Methods and Practice.* Oxford: Elsevier.
Christensen AM, Passalacqua NV, Bartelink EJ. 2019. *Forensic Anthropology: Current Methods and Practice,* Second Edition. London: Academic Press.
Clement JG. 2013. Odontology. In: Siegel JA, Saukko PJ (Eds.), *Encyclopedia of Forensic Sciences*. Waltham, MA: Academic Press. p. 106–113.
Coe JI. 1982. External beveling of entrance wounds by handguns. *American Journal of Forensic Medicine and Pathology* 3: 215–219.
Colard T, Delannoy Y, Naji S, Gosset D, Hartnett K, Bécart A. 2015. Specific patterns of canine scavenging in indoor settings. *Journal of Forensic Sciences* 60: 495–500.
Collins JW Jr, David RJ, Handler A, Wall S, Andes S. 2004. Very low birthweight in African American infants: the role of maternal exposure to interpersonal racial discrimination. *American Journal of Public Health* 94: 2132–2138.
Cox K, Tayles NG, Buckley HR. 2006. Forensic identification of "race": the issues in New Zealand. *Current Anthropology* 47: 869–874.
Culbert WL, Law FM. 1927. Identification by comparison of roentgenograms of nasal accessory sinuses and mastoid processes. *Journal of the American Medical Association* 88: 1634–1636.
Cunha E. 2006. Pathology as a factor of personal identity in forensic anthropology. In: Schmitt A, Cunha E, Pinheiro J (Eds.), *Forensic Anthropology and Medicine: Complementary Sciences from Recovery to Cause of Death*. Totowa, NJ: Humana Press. p. 333–358.
Currey JD. 1970. The mechanical properties of bone. *Clinical Orthopedics and Related Research* 73: 210–231.
Daubert v. Merrell Dow Pharmaceuticals, Inc., 509 U.S. 579, 1993.
Del Papa MC, Perez SI. 2007. The influence of artificial cranial vault deformation on the expression of cranial nonmetric traits: its importance in the study of evolutionary relationships. *American Journal of Physical Anthropology* 134: 251–262.
DiCaudo DJ. 2006. Coccidioidomycosis: a review and update. *Journal of the American Academy of Dermatology* 55: 929–942.
Didia BC, Nduka EC, Adele O. 2009. Stature estimation formulae for Nigerians. *Journal of Forensic Sciences* 54: 20–21.
DiGangi EA, Bethard JD. 2021. Uncloaking a lost cause: decolonizing ancestry estimation in the United States. *American Journal of Physical Anthropology* 175: 422–436.
DiGangi EA, Bethard JD, Kimmerle EH, Konigsberg LW. 2009. A new method for estimating age-at-death from the first rib. *American Journal of Physical Anthropology* 138: 164–176.
Dikötter F. 1992. *The Discourse of Race in Modern China*. Hong Kong: Hong Kong University Press.
DiMaio VJM. 2021. *Gunshot Wounds: Practical Aspects of Firearms, Ballistics, and Forensic Techniques,* Third Edition. Boca Raton, FL: CRC Press.
Dixon AF. 1912. *Manual of Human Osteology*. New York, NY: William Wood and Co.
Dixon DS. 1982. Keyhole lesions in gunshot wounds of the skull and direction of fire. *Journal of Forensic Sciences* 27: 555–566.
Dixon DS. 1984a. Exit keyhole lesion and direction of fire in a gunshot wound of the skull. *Journal of Forensic Sciences* 29: 336–339.
Dixon DS. 1984b. Pattern of intersecting fractures and direction of fire. *Journal of Forensic Sciences* 29: 651–654.
d'Oliveira Coelho J, Navega D. 2019. *hefneR: Cranial Nonmetric Traits Ancestry Estimation*. Available from: http://osteomics.com/hefneR.
Dorion RB. 1983. Photographic superimposition. *Journal of Forensic Sciences* 28: 724–734.
Dudzik B, Jantz RL. 2016. Misclassification of Hispanics using Fordisc 3.1: Comparing cranial morphology in Asian and Hispanic populations. *Journal of Forensic Sciences* 61: 1311–1318.
Dunn RR, Spiros MC, Kamnikar KR, Plemons AM, Hefner JT. 2020. Ancestry estimation in forensic anthropology: a review. *WIREs Forensic Science* 2: e1369. Available from: https://doi.org/10.1002/wfs2.1369.
Earnshaw VA, Rosenthal L, Lewis JB, Stasko EC, Tobin JN, Lewis TT, Reid AE, Ickovics JR. 2013. Maternal experiences with everyday discrimination and infant birth weight: a test of mediators and moderators among young, urban women of color. *Annals of Behavioral Medicine* 45: 13–23.
Edgar HJH. 2005. Prediction of race using characteristics of dental morphology. *Journal of Forensic Sciences* 50: 269–273.
Edgar HJH. 2009. Biohistorical approaches to "race" in the United States: biological distances among African Americans, European Americans, and their ancestors. *American Journal of Physical Anthropology* 139: 58–67.
Edgar HJH. 2013. Estimation of ancestry using dental morphological characteristics. *Journal of Forensic Sciences* 58: S3–S8.
Edgar HJH. 2014. Dental morphological estimation of ancestry in forensic contexts. In: Berg G, Ta'ala SC (Eds.), *Biological Affinity in Forensic Identification of Human Skeletal Remains: Beyond Black and White*. Boca Raton, FL: CRC Press. p. 191–207.
Elhaik E. 2012. Empirical distributions of F_{ST} from large-scale human polymorphism data. *PLoS ONE* 7(11): e49837. doi: 10.1371/journal.pone.0049837.
Fazekas GI, Kósa F. 1978. *Forensic Fetal Osteology*. Budapest: Akadémiai Kiadó.
Federal Interagency Working Group for Research on Race and Ethnicity. *Interim Report to the Office of Management and Budget: Review of Standards for Maintaining, Collecting, and Presenting Federal Data on Race and Ethnicity* (March 3, 2017). Available from: https://www.whitehouse.gov/wp-content/uploads/legacy_drupal_files/briefing-room/

presidential-actions/related-omb-material/r_e_iwg_interim_report_022417.pdf

Fischer JF, Nickell J. 1986. “Keyhole” skull wounds: the problem of bullet-caliber determination. *Identification News* December: 8–10.

Fisher TD, Gill GW. 1990. Application of the Giles & Elliot discriminant function formulae to a cranial sample of Northwestern Plains Indians. In: Gill GW, Rhine S (Eds.), *Skeletal Attribution of Race: Methods for Forensic Anthropology.* Albuquerque, NM: Maxwell Museum of Anthropology Anthropological Papers No. 4. p. 59–63.

Forbes SL, Perrault KA, Comstock JL. 2017. Microscopic post-mortem changes: the chemistry of decomposition. In: Schotsmans EMJ, Márquez-Grant N, Forbes SL (Eds.), *Taphonomy of Human Remains: Forensic Analysis of the Dead and the Depositional Environment.* New York, NY: John Wiley & Sons. p. 26–38.

Forrest AS, Wu HY. 2010. Endodontic imaging as an aid to forensic personal identification. *Australian Endodontic Journal* 36: 87–94.

France DL. 2009. *Human and Non-Human Bone Identification: A Color Atlas.* Boca Raton, FL: CRC Press.

France DL. 2011. *Human and Non-Human Bone Identification: A Concise Field Guide.* Boca Raton, FL: CRC Press.

Fulginiti LC, Seidel AC, Bolhofner KL. 2021. Fusion and fracture: forensic implications of the hyoid bone. *Proceedings of the 73rd Annual Scientific Meeting of the American Academy of Forensic Sciences:* 33.

Fully G. 1956. Une nouvelle méthode de détermination de la taille. *Annales de Medicine Legale* 36: 266–273.

Fully G, Pineau H. 1960. Détermination de la stature au moyen du squelette. *Annales de Medicine Legale* 40: 145–154.

Galloway A. 1988. Estimating actual height in the older individual. *Journal of Forensic Sciences* 33: 126–136.

Galloway A. 1997. The process of decomposition: a model from the Arizona-Sonoran Desert. In: Haglund WD, Sorg MH (Eds.), *Forensic Taphonomy: The Postmortem Fate of Human Remains.* Boca Raton, FL: CRC Press. p. 139–150.

Galloway A. 1998. Estimating actual height in the older individual. *Journal of Forensic Sciences* 33: 126–136.

Galloway A, Zephro L, Wedel VL. 2014a. Classification of fractures. In: Wedel VL and Galloway A (Eds.), *Broken Bones: Anthropological Analysis of Blunt Force Trauma.* Springfield, IL: Charles C. Thomas. p. 59–72.

Galloway A, Zephro L, Wedel VL. 2014b. Diagnostic criteria for the determination of timing and fracture mechanism. In: Wedel VL, Galloway A (Eds.), *Broken Bones: Anthropological Analysis of Blunt Force Trauma.* Springfield, IL: Charles C. Thomas. p. 47–58.

Garvin HM, Passalacqua NV. 2012. Current practices by forensic anthropologists in adult skeletal age estimation. *Journal of Forensic Sciences* 57: 427–433.

Gerasimov MM. 1971. *The Face Finder.* Philadelphia, PA: JB Lippincott.

Gilbert BM, McKern TW. 1973. A method for aging the female *Os pubis. American Journal of Physical Anthropology* 38: 31–38.

Gilbert R, Gill GW. 1990. A metric technique for identifying American Indian femora. In: Gill GW, Rhine S (Eds.), *Skeletal Attribution of Race: Methods for Forensic Anthropology.* Albuquerque, NM: Maxwell Museum of Anthropology Anthropological Papers No. 4. p. 97–99.

Giles E. 1991. Corrections for age in estimating older adults' stature from long bones. *Journal of Forensic Sciences* 36: 898–901.

Giles E, Elliot O. 1962. Race identification from cranial measurements. *Journal of Forensic Sciences* 7: 147–157.

Giles E, Hutchinson DL. 1991. Stature- and age-related bias in self-reported stature. *Journal of Forensic Sciences* 36: 765–780.

Giles E, Klepinger LL. 1988. Confidence intervals for estimates based on linear regression in forensic anthropology. *Journal of Forensic Sciences* 33: 1218–1222.

Gravlee CC. 2009. How race becomes biology: embodiment of social inequality. *American Journal of Physical Anthropology* 139: 47–57.

Grivas CR, Komar DA. 2008. *Kumho, Daubert,* and the nature of scientific inquiry: implications for forensic anthropology. *Journal of Forensic Sciences* 53: 771–776.

Hanihara T, Ishida H, Dodo Y. 1998. Os zygomaticum bipartitum: frequency distribution in major human populations. *Journal of Anatomy* 192: 539–555.

Harris AM, Wood RE, Nortjé CJ, Thomas CJ. 1987. Gender and ethnic differences of the radiographic image of the frontal region. *The Journal of Forensic Odonto-Stomatology* 5: 51–57.

Harris EF. 2008. Statistical applications in dental anthropology. In: Irish JD, Nelson GC (Eds.), *Technique and Application in Dental Anthropology.* Cambridge: Cambridge University Press. p. 35–67.

Hart GO. 2005. Fracture pattern interpretation in the skull: differentiating blunt force from ballistics trauma using concentric fractures. *Journal of Forensic Sciences* 50: 1276–1281.

Hartnett KM. 2007. *A Re-evaluation and Revision of Pubic Symphysis and Fourth Rib Aging Techniques.* PhD dissertation, School of Human Evolution and Social Change, Arizona State University, Tempe.

Hartnett KM. 2010a. Analysis of age-at-death estimation using data from a new, modern autopsy sample—part I: pubic bone. *Journal of Forensic Sciences* 55: 1145–1151.

Hartnett KM. 2010b. Analysis of age-at-death estimation using data from a new, modern autopsy sample—part II: sternal end of the fourth rib. *Journal of Forensic Sciences* 55: 1152–1156.

Hatch GM, Dedouit F, Christensen AM, Thali MJ, Ruder TD. 2014. RADid: a pictorial review of radiologic identification using postmortem CT. *Journal of Forensic Radiology and Imaging* 2: 52–59.

Hefner JT. 2007. *The Statistical Determination of Ancestry Using Nonmetric Traits.* PhD Dissertation, University of Florida.

Hefner JT. 2009. Cranial nonmetric variation and estimating ancestry. *Journal of Forensic Sciences* 54: 985–995.

Hefner JT, Linde KC. 2018. *Atlas of Human Cranial Macromorphoscopic Traits.* Cambridge, MA: Elsevier.

Hefner JT, Ousley SD. 2014. Statistical classification methods for estimating ancestry using morphoscopic traits. *Journal of Forensic Sciences* 59: 883–890.

Hefner JT, Spradley MK, Anderson B. 2014. Ancestry assessment using random forest modeling. *Journal of Forensic Sciences* 59: 583–589.

Herrmann NP, Plemons A, Harris EF. 2016. Estimating ancestry of fragmentary remains via multiple classifier systems: a study of the Mississippi State Asylum skeletal assemblage. In: Pilloud MA, Hefner JT (Eds.), *Biological Distance Analysis: Forensic and Bioarchaeological Perspectives.* London: Elsevier. p. 285–299.

Hillier ML, Bell LS. 2007. Differentiating human bone from animal bone: a review of histological methods. *Journal of Forensic Sciences* 52: 249–263.

Hochman A. 2013. Racial discrimination: how not to do it. *Studies in History and Philosophy of Biological and Biomedical Sciences* 44: 278–286.

Hoffman JM. 1984. Identification of nonskeletonized bear paws and human feet. In: Rathbun TA, Buikstra JE (Eds.), *Human Identification: Case Studies in Forensic Anthropology.* Springfield, IL: Charles C. Thomas. p. 96–106.

Hogge JP, Messmer JM, Doan QN. 1994. Radiographic identification of unknown human remains and interpreter experience level. *Journal of Forensic Sciences* 39: 373–377.

Holcomb SMC, Konigsberg LW. 1995. Statistical study of sexual dimorphism in the human fetal sciatic notch. *American Journal of Physical Anthropology* 97: 113–125.

Holliday TW, Falsetti AB. 1999. A new method for discriminating African-American from European-American skeletons using postcranial osteometrics reflective of body shape. *Journal of Forensic Sciences* 44: 926–930.

Hubbard AR. 2017. Teaching race (bioculturally) matters: a visual approach for college biology courses. *The American Biology Teacher* 79: 516–524.

Hughes CE, Tise ML, Trammel LH, Anderson BE. 2013. Cranial morphological variation among contemporary Mexicans: regional trends, ancestral affinities, and genetic comparisons. *American Journal of Physical Anthropology* 151: 506–517.

Hurst CV, Soler A, Fenton TW. 2013. Personal identification in forensic anthropology. In: Siegel JA, Saukko PJ (Eds.), *Encyclopedia of Forensic Sciences.* Waltham, MA: Academic Press. p. 68–75.

Igarashi Y, Uesu K, Wakebe T, Kanazawa F. 2005. New method for estimation of adult skeletal age at death from the morphology of the auricular surface of the ilium. *American Journal of Physical Anthropology* 128: 324–339.

Iliescu FM, Chaplin G, Rai N, Jacobs GS, Mallick CB, Mishra A, Thangaraj K, Jablonski NG. 2018. The influences of genes, the environment, and social factors in the evolution of skin color diversity in India. *American Journal of Human Biology* 30: e23170.

Irish JD. 2015. Dental nonmetric variation around the world: using key traits in populations to estimate ancestry in individuals. In: Berg GE, Ta'Ala SC (Eds.), *Biological Affinity in Forensic Identification of Human Skeletal Remains.* Boca Raton, FL: CRC Press. p. 165–190.

İşcan MY, Loth SR, Wright RK. 1984a. Metamorphosis at the sternal rib end: a new method to estimate age at death in white males. *American Journal of Physical Anthropology* 65: 147–156.

İşcan MY, Loth SR, Wright RK. 1984b. Age estimation from the rib by phase analysis: white males. *Journal of Forensic Sciences* 29: 1094–1104.

İşcan MY, Loth SR, Wright RK. 1985. Age estimation from the rib by phase analysis: white females. *Journal of Forensic Sciences* 30: 853–863.

İşcan MY, Loth SR, Wright RK. 1987. Racial variation in the sternal extremity of the rib and its effect on age determination. *Journal of Forensic Sciences* 32: 452–466.

Jablonski NG. 2012. *Living Color: The Biological and Social Meaning of Skin Color.* Berkeley: University of California Press.

Jablonski NG, Chaplin G. 2010. Human skin pigmentation as an adaptation to UV radiation. *Proceedings of the National Academy of Sciences of the United States of America* 107: 8962–8968.

Jantz LM, Jantz RL. 1999. Secular change in long bone length and proportion in the United States, 1800–1970. *American Journal of Physical Anthropology* 110: 57–67.

Jantz RL, Ousley SD. 2005. *FORDISC 3.0: Personal Computer Forensic Discriminant Functions.* Knoxville: University of Tennessee.

Jantz RL, Hunt DR, Meadows L. 1994. Maximum length of the tibia. How did Trotter measure it? *American Journal of Physical Anthropology* 93: 525–528.

Jantz RL, Hunt DR, Meadows L. 1995. The measure and mismeasure of the tibia: implications for stature estimation. *Journal of Forensic Sciences* 40: 758–761.

Kennedy KAR. 1995. But professor, why teach race identification if races don't exist? *Journal of Forensic Sciences* 40: 797–800.

Kenyhercz MW, Klales AR, Rainwater CW, Fredette SM. 2017. The optimized summed scored attributes method for the classification of U.S. Blacks and whites: a validation study. *Journal of Forensic Sciences* 62: 174–180.

Kirk NJ, Wood RE, Goldstein M. 2002. Skeletal identification using the frontal sinus region: a retrospective study of 39 cases. *Journal of Forensic Sciences* 47: 318–323.

Klales AR, Ousley SD, Vollner JM. 2012. A revised method of sexing the human innominate using Phenice's nonmetric traits and statistical methods. *American Journal of Physical Anthropology* 149: 104–114.

Klecka, WR. 1980. *Discriminant Analysis.* Newbury Park, CA: SAGE Publications, Inc.

Kobayashi M, Togo M. 1993. Twice daily measurements of stature and body weight in two children and one adult. *American Journal of Human Biology* 5: 193–201.

Komar DA, Buikstra JE. 2008. *Forensic Anthropology: Contemporary Theory and Practice.* Oxford: Oxford University Press.

Koot MG, Sauer NJ, Fenton TW. 2005. Radiographic human identification using bones of the hand: a validation study. *Journal of Forensic Sciences* 50: 263–268.

Krogman WM, İşcan MY. 1986. *The Human Skeleton in Forensic Medicine,* Second Edition. Springfield, IL: Charles C. Thomas.

Kuzawa CW, Sweet E. 2009. Epigenetics and the embodiment of race: developmental origins of US racial disparities in cardiovascular health. *American Journal of Human Biology* 21: 2–15.

Langley NR, Meadows Jantz L, Ousley SD, Jantz RL, Milner G. 2016. *Data Collection Procedures for Forensic Skeletal Material 2.0.* Knoxville: University of Tennessee. Available from: http://fac.utk.edu/wp-content/uploads/2016/03/DCP20_webversion.pdf.

Lease LR, Sciulli PW. 2005. Brief communication: discrimination between European-American and African-American children based on deciduous dental metrics and morphology. *American Journal of Physical Anthropology* 126: 56–60.

Le Fort R. 1901. Etude experimentale sur les fractures de la machoire superieure. *Revue de Chirurgie Paris* 23: 208–227, 360–379, 479–507.

Lesciotto KM. 2015. The impact of *Daubert* on the admissibility of forensic anthropology expert testimony. *Journal of Forensic Sciences* 60: 549–555.

Lewis JM, Senn DR. 2010. Dental age estimation utilizing third molar development: a review of principles, methods, and population studies in the United States. *Forensic Science International* 201: 79–83.

Li JZ, Absher DM, Tang H, Southwick AM, Casto AM, Ramachandran S, Cann HM, Barsh GS, Feldman M, Cavalli-Sforza LL, Myers RM. 2008. Worldwide human relationships inferred from genome-wide patterns of variation. *Science* 319: 1100–1104.

Lipton BE, Murmann DC, Pavlik EJ. 2013. History of forensic odontology. In: Senn DR, Weems RA (Eds.), *Manual of Forensic Odontology,* Fifth Edition. Boca Raton, FL: CRC Press. p. 1–39.

Loth SR, Henneberg M. 2001. Sexually dimorphic mandibular morphology in the first few years of life. *American Journal of Physical Anthropology* 115: 179–186.

Lovejoy CO, Meindl RS, Pryzbeck TR, Mensforth RP. 1985. Chronological metamorphosis of the auricular surface of the ilium: a new method for the determination of adult skeletal age at death. *American Journal of Physical Anthropology* 68: 15–28.

Lowenstein JM, Reuther JD, Hood DG, Scheuenstuhl G, Gerlach SC, Ubelaker DH. 2006. Identification of animal species by protein radioimmunoassay of bone fragments and bloodstained stone tools. *Forensic Science International* 159: 182–188.

Lukachko A, Hatzenbuehler ML, Keyes KM. 2014. Structural racism and myocardial infarction in the United States. *Social Science & Medicine* 103: 42–50.

Madea B, Staak M. 1988. Determination of the sequence of gunshot wounds of the skull. *Journal of the Forensic Science Society* 28: 321–328.

Mahakkanukrauh P, Khanpetch P, Prasitwattanseree S, Vichairit K, Case DT. 2011. Stature estimation from long bone lengths in a Thai population. *Forensic Science International* 210: 279e1–279e7.

Maier C, Craig A, Adams DM. 2021. Language use in ancestry research and estimation. *Journal of Forensic Sciences* 66: 11–24.

Manica A, Amos W, Balloux F, Hanihara T. 2007. The effect of ancient population bottlenecks on human phenotypic variation. *Nature* 448: 346–348.

Martiniaková M, Grosskopf B, Omelka R. Vondráková M, Bauerová M. 2006. Differences among species in compact bone tissue microstructure of mammalian skeleton: use of a discriminant function analysis for species identification. *Journal of Forensic Sciences* 51: 1235–1239.

Mays S. 2021. *The Archaeology of Human Bones,* Third Edition. London: Routledge.

McKern TW, Stewart TD. 1957. Skeletal age changes in young American males. In *Headquarters Quartermaster Research and Development and Command Technical Report RP-45.* Natick: MA.

Meadows L, Jantz RL. 1995. Allometric secular change in the long bones from the 1800s to the present. *Journal of Forensic Sciences* 40: 762–767.

Meeusen RA, Christensen AM, Hefner JT. 2015. The use of femoral neck axis length to estimate sex and ancestry. *Journal of Forensic Sciences* 60: 1300–1304.

Megyesi MS, Nawrocki SP, Haskell NH. 2005. Using accumulated degree-days to estimate the postmortem interval from decomposed human remains. *Journal of Forensic Sciences* 50: 618–626.

Meindl RS, Lovejoy CO. 1985. Ectocranial suture closure: a revised method for the determination of skeletal age at death based on the lateral-anterior sutures. *American Journal of Physical Anthropology* 68: 57–66.

Milligan CF, Finlayson JE, Cheverko CM, Zarenko KM. 2018. Advances in the use of craniofacial superimposition for human identification. In: Latham K, Bartelink E, Finnegan M (Eds.), *New Perspectives in Forensic Human Skeletal Identification.* London: Academic Press. p. 241–250.

Mittler DM, Sheridan SG. 1992. Sex determination in subadults using auricular surface morphology: a forensic science perspective. *Journal of Forensic Sciences* 37: 1068–1075.

Moore-Jansen PH, Jantz RL. 1986. *A Computerized Skeletal Data Bank for Forensic Anthropology.* Knoxville: University of Tennessee.

Moorrees CFA, Fanning EA, Hunt EE Jr. 1963a. Formation and resorption of three deciduous teeth in children. *American Journal of Physical Anthropology* 21: 205–213.

Moorrees CFA, Fanning EA, Hunt EE Jr. 1963b. Age formation stages for ten permanent teeth. *Journal of Dental Research* 42: 1490–1502.

Morris H, McMurrich JP (Eds.) 1907. *Morris's Human Anatomy: a Complete and Systematic Treatise by English and American Authors,* Fourth Edition. Philadelphia, PA: P. Blakiston's Son & Co.

Mulhern DM, Ubelaker DH. 2001. Differences in osteon banding between human and nonhuman bone. *Journal of Forensic Sciences* 46: 220–222.

Mulhern DM, Ubelaker DH. 2012. Differentiating human from nonhuman bone microstructure. In: Crowder C, Stout S (Eds.), *Bone Histology. An Anthropological Perspective.* Boca Raton, FL: CRC Press. p. 109–134.

Murray EA, Anderson BE, Clark SC, Hanzlick RL. 2018. The history and use of the National Missing and Unidentified Persons System (NamUs) in the identification of unknown persons. In: Latham K, Bartelink E, Finnegan M (Eds.), *New Perspectives in Forensic Human Skeletal Identification.* London: Academic Press. p. 115–126.

Nawrocki SP. 1998. Regression formulae for the estimation of age from cranial suture closure. In: Reichs KJ (Ed.), *Forensic*

Osteology: Advances in the Identification of Human Remains, Second Edition. Springfield, IL: Charles C. Thomas. p. 276–292.

Office of Management and Budget. *Revisions to the Standards for the Classification of Federal Data on Race and Ethnicity,* 62 Fed. Reg. 58782 (October 30, 1997).

Osborne DL, Simmons TL, Nawrocki SP. 2004. Reconsidering the auricular surface as an indicator of age at death. *Journal of Forensic Sciences* 49: 905–911.

Osborne-Gustavson AE, McMahon T, Josserand M, Spamer BJ. 2018. The utilization of databases for the identification of human remains. In: Latham K, Bartelink E, Finnegan M (Eds.), *New Perspectives in Forensic Human Skeletal Identification.* London: Academic Press. p. 129–139.

Ousley S. 1995. Should we estimate biological or forensic stature? *Journal of Forensic Sciences* 40: 768–773.

Ousley SD, Jantz RL. 1998. The Forensic Data Bank: documenting skeletal trends in the United States. In: Reichs KJ (Ed.), *Forensic Osteology: Advances in the Identification of Human Remains,* Second Edition. Springfield, IL: Charles C. Thomas. p. 441–458.

Ousley SD, Jantz RL. 2012. Fordisc 3 and statistical methods for estimating sex and ancestry. In: Dirkmaat DC (Ed.), *A Companion to Forensic Anthropology.* London: Blackwell. p. 311–329.

Ousley SD, Billeck WT, Hollinger RE. 2005. Federal repatriation legislation and the role of physical anthropology in repatriation. *Yearbook of Physical Anthropology* 48: 2–32.

Ousley S, Jantz R, Fried D. 2009. Understanding race and human variation: why forensic anthropologists are good at identifying race. *American Journal of Physical Anthropology* 139: 68–76.

Owsley DW, Mann RW. 1992. Positive personal identity of skeletonized remains using abdominal and pelvic radiographs. *Journal of Forensic Sciences* 37: 332–336.

Owsley DW, Mires AM, Keith MS. 1985. Case involving differentiation of deer and human bone fragments. *Journal of Forensic Sciences* 30: 572–578.

Parra EJ, Marcini A, Akey J, Martinson J, Batzer MA, Cooper R, Forrester T, Allison DB, Deka R, Ferrell RE, Shriver MD. 1998. Estimating African American admixture proportions by use of population-specific alleles. *American Journal of Human Genetics* 63: 1839–1851.

Parsons HR. 2017. *The Accuracy of the Biological Profile in Casework: An Analysis of Forensic Anthropology Reports in Three Medical Examiners' Offices.* PhD Dissertation, University of Tennessee, Knoxville.

Paschall A, Ross AH. 2017. Bone mineral density and wounding capacity of handguns: implications for estimation of caliber. *International Journal of Legal Medicine* 131: 161–166.

Pérez MJ, Barquez RM, Díaz MM. 2017. Morphology of the limbs in the semi-fossorial desert rodent species of *Tympanoctomys* (Octodontidae, Rodentia). *ZooKeys* 710: 77–96.

Perini TA, Lameira de Oliveira G, dos Santos Ornellas J, Palha de Oliveira F. 2005. Technical error of measurement in anthropometry. *Revista Brasileira de Medicina do Esporte* 11: 86–90.

Peterson BL. 1991. External beveling of cranial gunshot entrance wounds. *Journal of Forensic Sciences* 36: 1592–1595.

Phenice T. 1969. A newly developed visual method of sexing in the os pubis. *American Journal of Physical Anthropology* 30: 297–301.

Pilloud MA, Hefner JT, Hanihara T, Hayashi A. 2014. The use of tooth crown measurements in the assessment of ancestry. *Journal of Forensic Sciences* 59: 1493–1501.

Pilloud MA, Maier C, Scott GR, Hefner JT. 2018. Advances in cranial macromorphoscopic trait and dental morphology analysis for ancestry estimation. In: Latham K, Bartelink E, Finnegan M (Eds.), *New Perspectives in Forensic Human Skeletal Identification.* London: Academic Press. p. 23–34.

Pilloud MA, Adams DM, Hefner JT. 2019. Observer error and its impact on ancestry estimation using dental morphology. *International Journal of Legal Medicine* 133: 949–962.

Pollanen MS, Chiasson DA. 1996. Fracture of the hyoid in strangulation. Comparison of fractured and unfractured hyoids from victims of strangulation. *Journal of Forensic Sciences* 41: 110–113.

Pratt BM, Hixson L, Jones NA. n.d. Infographic: measuring race and ethnicity across the decades, 1790–2010. United States Census Bureau. Available from https://www.census.gov/data-tools/demo/race/MREAD_1790_2010.html.

Quatrehomme G, Alunni V. 2013. Bone trauma. In: Siegel JA, Saukko PJ (Eds.), *Encyclopedia of Forensic Sciences.* Waltham, MA: Academic Press. p. 89–96.

Quillen EE, Norton HL, Parra EJ, Lona-Durazo F, Ang KC, Iliescu FM, Pearson LN, Shriver MD, Lasisi T, Gokcumen O, Starr I, Lin Y-L, Martin AR, Jablonski NG. 2019. Shades of complexity: new perspectives on the evolution and genetic architecture of human skin. *American Journal of Physical Anthropology* 168 (S67): 4–26.

Raxter MH, Auerbach BM, Ruff CB. 2006. Revision of the Fully technique for estimating statures. *American Journal of Physical Anthropology* 130: 374–384.

Reconstruct body to solve murders. *New York Times,* 25 September 1916. p. 1.

Reichs KJ. 1998. Postmortem dismemberment: recovery, analysis and interpretation. In: Reichs KJ (Ed.). *Forensic Osteology. Advances in the Identification of Human Remains,* Second Edition. Springfield, IL: Charles C. Thomas. p. 353–388.

Relethford JH. 2004. Boas and beyond: migration and craniometric variation. *American Journal of Human Biology* 16: 379–386.

Relethford JH. 2009. Race and global patterns of phenotypic variation. *American Journal of Physical Anthropology* 139: 16–22.

Relethford JH. 2010. The study of human population genetics. In: Larsen CS (Ed.), *A Companion to Biological Anthropology.* Malden, MA: Blackwell Publishing. p. 74–87.

Rhine S. 1990. Non-metric skull racing. In: Gill GW, Rhine S (Eds.), *Skeletal Attribution of Race: Methods for Forensic Anthropology.* Albuquerque, NM: Maxwell Museum of Anthropology Anthropological Papers No. 4. p. 9–20.

Rhine S, Sperry K. 1991. Radiographic identification by mastoid sinus and arterial pattern. *Journal of Forensic Sciences* 36: 272–279.

Riddick L, Brogdon BG, Lasswell-Hoff J, Delmas B. 1983. Radiographic identification of charred human remains through use of the dorsal defect of the patella. *Journal of Forensic Sciences* 28: 263–267.

Rodriguez WC, Bass WM. 1983. Insect activity and its relationship to decay rates of human cadavers in East Tennessee. *Journal of Forensic Sciences* 28: 423–432.

Rodriguez WC, Bass WM. 1985. Decomposition of buried bodies and methods that may aid in their location. *Journal of Forensic Sciences* 30: 836–852.

Rogers TL, Allard TT. 2004. Expert testimony and positive identification of human remains through cranial suture patterns. *Journal of Forensic Sciences* 49: 203–207.

Rogers T, Saunders S. 1994. Accuracy of sex determination using morphological traits of the human pelvis. *Journal of Forensic Sciences* 39: 1047–1056.

Ross AH. 1996. Caliber estimation from cranial entrance defect measurements. *Journal of Forensic Sciences* 41: 629–633.

Ross AH, Cunningham SL. 2011. Time-since-death and bone weathering in a tropical environment. *Forensic Science International* 204: 126–133.

Ross AH, Pilloud M. 2021. The need to incorporate human variation and evolutionary theory in forensic anthropology: a call for reform. *American Journal of Physical Anthropology* 176: 672–683.

Rougé D, Telmon N, Arrue P, Larrouy G, Arbus L. 1993. Radiographic identification of human remains through deformities and anomalies of post-cranial bones: a report of two cases. *Journal of Forensic Sciences* 38: 997–1007.

Saks MJ, Koehler JJ. 2005. The coming paradigm shift in forensic identification science. *Science* 309: 892–895.

Sauer NJ. 1992. Forensic anthropology and the concept of race: if races don't exist, why are forensic anthropologists so good at identifying them? *Social Science & Medicine* 34: 107–111.

Sauer NJ, Brantley RE, Barondess DA. 1988. The effects of aging on the comparability of antemortem and postmortem radiographs. *Journal of Forensic Sciences* 33: 1223–1230.

Sauer NJ, Wankmiller JC, Hefner JT. 2016. The assessment of ancestry and the concept of race. In: Blau S, Ubelaker DH (Eds.), *Handbook of Forensic Anthropology and Archaeology,* Second Edition. New York, NY: Routledge. p. 243–260.

Schaefer M, Black S, Scheuer L. 2009. *Juvenile Osteology: A Laboratory and Field Manual.* San Diego: Academic Press.

Scheuer L, Black S. 2000. *Developmental Juvenile Osteology.* San Diego, CA: Academic Press.

Scheyer TM. 2009. Conserved bone microstructure in the shells of long-necked and short-necked chelid turtles (Testudinata, Pleurodira). *Fossil Record* 12: 47–57.

Schour I, Massler M. 1941a. The development of the human dentition. *Journal of the American Dental Association* 28: 1153–1160.

Schour I, Massler M. 1941b. *Development of Human Dentition Chart,* Second Edition. Chicago: American Dental Association.

Schutkowski H. 1993. Sex determination of infant and juvenile skeletons I: Morphognostic features. *American Journal of Physical Anthropology* 90: 199–205.

Scientific Working Group for Forensic Anthropology (SWGANTH). 2013. Ancestry assessment, June 12, 2013. Available from: https://www.nist.gov/system/files/documents/2018/03/13/swganth_ancestry_assessment.pdf.

Scott GR, Pilloud MA, Navega D, d'Oliveira Coelho J, Cunha E, Irish JD. 2018. rASUDAS: a new web-based application for estimating ancestry from tooth morphology. *Forensic Anthropology* 1: 18–31.

Shennan S. 1997. *Quantifying Archaeology,* Second Edition. Iowa City: University of Iowa Press.

Simmons T. 2017. Post-mortem interval estimation: an overview of techniques. In: Schotsmans EMJ, Márquez-Grant N, Forbes SL (Eds.), *Taphonomy of Human Remains: Forensic Analysis of the Dead and the Depositional Environment.* New York, NY: John Wiley & Sons. p. 134–142.

Sims ME. 2007. Comparison of black bear paws to human hands and feet. Identification Guides for Wildlife Law Enforcement No. 11. USFWS, National Fish and Wildlife Forensics Laboratory, Ashland, OR.

Smay D, Armelagos G. 2000. Galileo wept: a critical assessment of the use of race in forensic anthropology. *Transforming Anthropology* 9: 19–29.

Smith BH. 1991. Standards of human tooth formation and dental age assessment. In: Kelley MA, Larsen CS (Eds.), *Advances in Dental Anthropology.* New York, NY: Wiley-Liss. p. 143–168.

Smith OC, Berryman HE, Lahren CH. 1987. Cranial fracture patterns and estimate of direction from low velocity gunshot wounds. *Journal of Forensic Sciences* 32: 1416–1421.

Smith OC, Berryman HE, Symes SA, Francisco JT, Hnilica V. 1993. Atypical gunshot exit defects to the cranial vault. *Journal of Forensic Sciences* 38: 339–343.

Smith-Bindman R, Miglioretti DL, Larson EB. 2008. Rising use of diagnostic medical imaging in a large integrated health system. *Health Affairs* 27: 1491–1502.

Snow CC, Gatliff BP, McWilliams KR. 1970. Reconstruction of facial features from the skull: an evaluation of its usefulness in forensic anthropology. *American Journal of Physical Anthropology* 33: 221–228.

Sobotta J. 1909. *Atlas and Text-Book of Human Anatomy.* Philadelphia, PA: W.B. Saunders Co.

Spiros MC, Hefner JT. 2020. Ancestry estimation using cranial and postcranial macromorphoscopic traits. *Journal of Forensic Sciences* 65: 921–929.

Spradley MK. 2014. Toward estimating geographic origin of migrant remains along the United States–Mexico border. *Annals of Anthropological Practice* 38: 101–110.

Spradley MK, Jantz RL. 2011. Sex estimation in forensic anthropology: skull versus postcranial elements. *Journal of Forensic Sciences* 56: 289–296.

Stagno V, Mailhiot S, Capuani S, Galotta G, Telkki V-V. 2021. Testing 1D and 2D single-sided NMR on Roman age waterlogged woods. *Journal of Cultural Heritage* 50: 95–105.

Steadman DW, Adams BJ, Konigsberg LW. 2006. Statistical basis for positive identification in forensic anthropology. *American Journal of Physical Anthropology* 131: 15–26.

Steele DG. 1970. Estimation of stature from fragments of long limb bones. In: Stewart TD (Ed.), *Personal Identification in Mass Disasters.* Washington, DC: Smithsonian Institution. p. 85–97.

Stephan CN. 2013. Facial approximation. In: Siegel JA, Saukko PJ (Eds.), *Encyclopedia of Forensic Sciences.* Waltham, MA: Academic Press. p. 60–67.

Stephan CN. 2014. The application of the central limit theorem and the law of large numbers to facial soft tissue depths: T-table robustness and trends since 2008. *Journal of Forensic Sciences* 59: 454–462.

Stephan CN. 2017. Estimating the skull-to-camera distance from facial photographs for craniofacial superimposition. *Journal of Forensic Sciences* 62: 850–860.

Stephan CN, Cicolini J. 2008. Measuring the accuracy of facial approximations: a comparative study of resemblance rating and face array methods. *Journal of Forensic Sciences* 53: 58–64.

Stephan CN, Henneberg M. 2001. Building faces from dry skulls: are they recognized above chance rates? *Journal of Forensic Sciences* 46: 432–440.

Stephan CN, Simpson EK. 2008. Facial soft tissue depths in craniofacial identification (Part I): an analytical review of the published adult data. *Journal of Forensic Sciences* 53: 1257–1272.

Stephan CN, Winburn AP, Christensen AF, Tyrell AJ. 2011. Skeletal identification by radiographic comparison: blind tests of a morphoscopic method using antemortem chest radiographs. *Journal of Forensic Sciences* 56: 320–332.

Stephan CN, Simpson EK, Byrd JE. 2013. Facial soft tissue depth statistics and enhanced point estimators for craniofacial identification: the debut of the shorth and the 75-shormax. *Journal of Forensic Sciences* 58: 1439–1457.

Stewart TD. 1959. Bear paw remains closely resemble human bones. *FBI Law Enforcement Bulletin* 28: 18–22.

Stewart TD. 1979. *Essentials of Forensic Anthropology, Especially as Developed in the United States.* Springfield, IL: Charles C. Thomas.

St. Hoyme LE, İşcan MY. 1989. Determination of sex and race: accuracy and assumptions. In: İşcan MY, Kennedy KAR (Eds.), *Reconstruction of Life from the Skeleton.* New York, NY: Alan R. Liss, Inc. p. 53–94.

Stock MK, Rubin KM. 2020. Race and the role of sociocultural context in forensic anthropological ancestry assessment. In: Garvin HM, Langley NR (Eds.), *Case Studies in Forensic Anthropology: Bonified Skeletons.* Boca Raton, FL: CRC Press. p. 39–50.

Streetman E, Fenton TW. 2018. Comparative medical radiography: practice and validation. In: Latham K, Bartelink E, Finnegan M (Eds.), *New Perspectives in Forensic Human Skeletal Identification.* London: Academic Press. p. 251–264.

Stull KE, L'Abbé EN, Ousley SD. 2014. Using multivariate adaptive regression splines to estimate subadult age from diaphyseal dimensions. *American Journal of Physical Anthropology* 154: 376–386.

Stull KE, Cirillo LE, Cole SJ, Hulse CN. 2020. Subadult sex estimation and KidStats. In: Klales A (Ed.), *Sex Estimation of the Human Skeleton: History, Methods, and Emerging Techniques.* London: Academic Press. p. 219–242.

Stull KE, Bartelink EJ, Klales AR, Berg GE, Kenyhercz MW, L'Abbé EN, Go MC, McCormick K, Mariscal C. 2021. Commentary on: Bethard JD, DiGangi EA. Letter to the editor—moving beyond a lost cause: forensic anthropology and ancestry estimates in the United States. *Journal of Forensic Sciences* 66: 417–420.

Sudimack JR, Lewis BJ, Rich J, Dean DE, Fardal PM. 2002. Identification of decomposed human remains from radiographic comparisons of an unusual foot deformity. *Journal of Forensic Sciences* 47: 218–220.

Symes SA, L'Abbé EN, Chapman EN, Wolff I, Dirkmaat DC. 2012. Interpreting traumatic injury to bone in medicolegal investigations. In: Dirkmaat DC (Ed.), *A Companion to Forensic Anthropology.* London: Blackwell. p. 340–389.

Symes SA, L'Abbé EN, Stull KE, Lacroix M, Pokines JT. 2014. Taphonomy and the timing of bone fractures in trauma analysis. In: Pokines JT, Symes SA (Eds.), *Manual of Forensic Taphonomy.* Boca Raton, FL: CRC Press. p. 341–365.

Telles E, Paschel T. 2014. Who is Black, white, or mixed race? How skin color, status, and nation shape racial classification in Latin America. *American Journal of Sociology* 120: 864–907.

Templeton AR. 2013. Biological races in humans. *Studies in History and Philosophy of Biological and Biomedical Sciences* 44: 262–271.

Thomas RM, Parks CL, Richard AH. 2017. Accuracy rates of ancestry estimation by forensic anthropologists using identified forensic cases. *Journal of Forensic Sciences* 62: 971–974.

Todd TW. 1920. Age changes in the pubic bone: I. The male white pubis. *American Journal of Physical Anthropology* 3: 285–334.

Todd TW. 1921. Age changes in the pubic bone. *American Journal of Physical Anthropology* 4: 1–70.

Trotter M. 1970. Estimation of stature from intact long limb bones. In: Stewart TD (Ed.), *Personal Identification in Mass Disasters.* Washington, DC: Smithsonian Institution. p. 71–83.

Trotter M, Gleser G. 1951. The effect of aging on stature. *American Journal of Physical Anthropology* 9: 311–324.

Trotter M, Gleser GC. 1952. Estimation of stature from long bones of American Whites and Negroes. *American Journal of Physical Anthropology* 10: 463–514.

Trotter M, Gleser GC. 1958. A re-evaluation of estimation of stature based on measurements of stature taken during life and of long bones after death. *American Journal of Physical Anthropology* 16: 79–123.

Ubelaker DH. 1978. *Human Skeletal Remains: Excavation, Analysis, Interpretation.* Washington, DC: Taraxacum.

Ubelaker DH. 1984. Positive identification from the radiographic comparison of frontal sinus patterns. In: Rathbun TA, Buikstra JE (Eds.), *Human Identification: Case Studies in Forensic Anthropology.* Springfield, IL: Charles C. Thomas. p. 399–411.

Ubelaker DH. 1987. Estimating age at death from immature human skeletons: an overview. *Journal of Forensic Sciences* 32: 1254–1263.

Ubelaker DH. 1991a. Perimortem and postmortem modification of human bone. Lessons from forensic anthropology. *Anthropologie* 29: 171–174.

Ubelaker DH. 1991b. Hyoid fracture and strangulation. *Journal of Forensic Sciences* 37: 1216–1222.

Ubelaker DH. 1998. The evolving role of the microscope in forensic anthropology. In: Reichs KJ (Ed.), *Forensic Osteology, Advances in the Identification of Human Remains,* Second Edition. Springfield, IL: Charles C Thomas. p. 514–532.

Ubelaker DH. 2014. Radiocarbon analysis of human remains: a review of forensic applications. *Journal of Forensic Sciences* 59: 1466–1472.

Ubelaker DH, Adams BJ. 1995. Differentiation of perimortem and postmortem trauma using taphonomic indicators. *Journal of Forensic Sciences* 40: 509–512.

Ubelaker DH, Jacobs CH. 1995. Identification of orthopedic device manufacturer. *Journal of Forensic Sciences* 40: 168–170.

Ubelaker DH, Bubniak E, O'Donnell G. 1992. Computer-assisted photographic superimposition. *Journal of Forensic Sciences* 37: 750–762.

Ubelaker DH, Ward DC, Braz VS, Stewart J. 2002. The use of SEM/EDS analysis to distinguish dental and osseous from other materials. *Journal of Forensic Sciences* 47: 940–943.

Ubelaker DH, Lowenstein JM, Hood DG. 2004. Use of solid-phase double-antibody radioimmunoassay to identify species from small skeletal fragments. *Journal of Forensic Sciences* 49: 924–929.

Urbanová P, Novotný V. 2005. Distinguishing between human and non-human bones: histometric method for forensic anthropology. *Anthropologie* 43: 77–85.

Vass AA. 2001. Beyond the grave—understanding human decomposition. *Microbiology Today* 28: 190–192.

Vass AA. 2011. The elusive universal post-mortem interval formula. *Forensic Science International* 204: 34–40.

Vass AA, Bass WM, Wolt JD, Foss JE, Ammons JT. 1992. Time since death determinations of human cadavers using soil solution. *Journal of Forensic Sciences* 37: 1236–1253.

Veitschegger K, Kolb C, Amson E, Scheyer TM, Sánchez-Villagra MR. 2018. Palaeohistology and life history evolution in cave bears, *Ursus spelaeus* sensu lato. *PLoS ONE* 13: e0206791. Available from: https://doi.org/10.1371/journal.pone.0206791.

Vidoli GM, Steadman DW, Devlin JB, Meadows Jantz L. 2017. History and development of the first Anthropology Research Facility, Knoxville, Tennessee. In: Schotsmans EMJ, Márquez-Grant N, Forbes SL (Eds.), *Taphonomy of Human Remains: Forensic Analysis of the Dead and the Depositional Environment.* New York, NY: John Wiley & Sons. p. 463–475.

Viner M. 2018. Overview of advances in forensic radiological methods of human identification. In: Latham K, Bartelink E, Finnegan M (Eds.), *New Perspectives in Forensic Human Skeletal Identification.* London: Academic Press. p. 217–226.

Vlak D, Roksandic M, Schillaci MA. 2008. Greater sciatic notch as a sex indicator in juveniles. *American Journal of Physical Anthropology* 137: 309–315.

Walker PL. 1995. Problems of preservation and sexism in sexing: some lessons from historical collections for palaeodemographers. In: Saunders SR, Herring A (Eds.), *Grave Reflections: Portraying the Past through Cemetery Studies.* Toronto: Canadian Scholars' Press. p. 31–47.

Walker PL. 2008. Sexing skulls using discriminant function analysis of visually assessed traits. *American Journal of Physical Anthropology* 136: 39–50.

Walsh-Haney H, Boys S. 2015. Creating the biological profile: the question of race and ancestry. In: Crossland Z, Joyce RA (Eds.), *Disturbing Bodies: Perspectives on Forensic Anthropology.* Santa Fe, NM: SAR Press. p. 121–135.

Weaver DS. 1980. Sex differences in the ilia of a known sex and age sample of fetal and infant skeletons. *American Journal of Physical Anthropology* 52: 191–195.

Weir BS, Cardon LR, Anderson AD, Nielsen DM, Hill WG. 2005. Measures of human population structure show heterogeneity among genomic regions. *Genome Research* 15: 1468–1476.

White TD, Black MT, Folkens PA. 2012. *Human Osteology,* Third Edition. Burlington, MA: Academic Press.

Willey P. 2016. Stature estimation. In: Blau S, Ubelaker DH (Eds.), *Handbook of Forensic Anthropology and Archaeology,* Second Edition. London: Routledge. p. 308–321.

Willey P, Falsetti T. 1991. Inaccuracy in height information on driver's licenses. *Journal of Forensic Sciences* 36: 813–819.

Wilson RJ, Bethard JD, DiGangi EA. 2011. The use of orthopedic surgical devices for forensic identification. *Journal of Forensic Sciences* 56: 460–469.

Wilson RJ, Herrmann NP, Jantz LM. 2010. Evaluation of stature estimation from the Database for Forensic Anthropology. *Journal of Forensic Sciences* 55: 684–689.

Workshop of European Anthropologists. 1980. Recommendations for age and sex diagnoses of skeletons. *Journal of Human Evolution* 9: 517–549.

Yoder C, Ubelaker DH, Powell JF. 2001. Examination of variation in sternal rib end morphology relevant to age assessment. *Journal of Forensic Sciences* 46: 223–227.

Yoshino M, Miyasaka S, Sato H, Tsuzuki Y, Seta S. 1989. Classification system of frontal sinus patterns. *Canadian Society of Forensic Science Journal* 22: 135–146.

Zephro L, Galloway A. 2014. The biomechanics of fracture production. In: Wedel VL, Galloway A (Eds.), *Broken Bones: Anthropological Analysis of Blunt Force Trauma.* Springfield, IL: Charles C. Thomas. p. 33–45.

Zimmerman HA. 2013. *Preliminary Validation of Handheld X-Ray Fluorescence (HHXRF) Spectrometry: Distinguishing Osseous and Dental Tissue from Non-Bone Material of Similar Chemical Composition.* MA Thesis, University of Central Florida, Orlando.

Christopher M. Stojanowski is professor of anthropology in the School of Human Evolution and Social Change at Arizona State University. He is coeditor of *Studies in Forensic Biohistory* and the author of *The Bioarchaeology of Ethnogenesis in the Colonial Southeast.*

Andrew C. Seidel received his certification from the American Board of Forensic Anthropology in 2022. He currently serves as the forensic anthropologist for the state of Washington.